Microorganisms for Sustainability

Volume 65

Series Editor

Naveen Kumar Arora, Environmental Microbiology, School for Environmental Science, Babasaheb Bhimrao Ambedkar University, Lucknow, India

Microorganisms have been in existence since the origin of life on earth and can survive the most extreme habitats or conditions on earth. Microorganisms are involved in regulating biogeochemical cycles, maintaining plant and animal health, and sustaining the global food chain. Moreover, they play crucial roles in addressing the challenges of climate change and achieving the targets of Sustainable Development Goals (SDGs).

This multidisciplinary book series captures the role of microbes towards building a sustainable world, while encompassing cutting-edge technologies and current needs across various fields such as agriculture sustainability, bioremediation, restoration of degraded habitats and wastelands, and food security. Additionally, this series explores microbial applications in industries, and building their utilization in clean and green energy solutions. Furthermore, themes like microbial secondary metabolites, extremophilic microbes and modern omics, including next generation sequencing and metagenomics, are also covered in this series.

With contributions from researchers across the globe, this series addresses the important call of 'One Planet-One Health-One Future'. It comprises a collection of diverse volumes that provides insights for scientists, young researchers, educators and decision-makers in the government, private sector, and non-governmental organizations, empowering their efforts to achieve the global goals.

The series invites, evaluates, and accepts book proposals to ensure a diverse, inclusive, and evolving program. The final decision regarding acceptance rests with the series editor.

Peer-review

This book series follows a stratified review process. Proposals for individual volumes are reviewed by the series editor and then the editorial board members. On a case-to-case basis, external reviewers are also invited for further evaluation of the book proposal. Review of the chapters is the responsibility of the volume editor(s). A manuscript submission platform has been recently made available to the authors, volume editors and the series editor.

Samina Mehnaz • Izzah Shahid
Editors

Frontiers in Rhizoengineering

Genomics, Metabolomics and Beyond

Springer

Editors
Samina Mehnaz
Kauser Abdulla Malik School
of Life Sciences
Forman Christian College
(A Chartered University)
Lahore, Pakistan

Izzah Shahid
Institute of Multidisciplinary Research
in Applied Biology
Public University of Navarra
Pamplona, Navarra, Spain

ISSN 2512-1901 ISSN 2512-1898 (electronic)
Microorganisms for Sustainability
ISBN 978-981-92-3664-0 ISBN 978-981-92-3665-7 (eBook)
https://doi.org/10.1007/978-981-92-3665-7

This Springer imprint is published by the registered company Springer Nature Singapore Pte Ltd.
The registered company address is: 152 Beach Road, #21-01/04 Gateway East, Singapore 189721, Singapore

Preface

The rhizosphere, a narrow zone of soil directly influenced by root secretions, is one of the most complex and dynamic ecosystems on Earth. It is a theatre of intense biological interactions, where a whisper of chemical signals can determine the fate of a plant, dictating its growth, health, and resilience. For decades, our understanding of this hidden world was limited to what we could culture in a petri dish, a mere fraction of its true diversity. The advent of third-generation sequencing platforms and the proliferation of OMICS technologies have shattered these barriers, unveiling a universe of unculturable microbes, intricate metabolic networks, and sophisticated genetic dialogues that govern plant-microbe relationships. This wealth of data has fueled the ambitious field of rhizoengineering, promising a future in which the rhizomicrobiome can be intentionally designed and manipulated to create self-sustaining, resilient agroecosystems.

However, as we stand at this exciting juncture, a critical paradox confronts us. The scientific literature is replete with groundbreaking insights from genomics, metagenomics, transcriptomics, and metabolomics, detailing the mechanisms of plant growth-promoting rhizobacteria (PGPR), inter-kingdom signaling, and genetic determinants of successful microbial consortia. However, the journey from promising laboratory findings to a reliable, commercially successful bioinoculum remains fraught with unforeseen obstacles. Most commercially available biofertilizers and biostimulants are single-strain inocula developed using conventional isolation and screening approaches. The promise of a new generation of "climate-smart" bioinoculants, designed using sophisticated rhizoengineering principles, remains largely unfulfilled.

This book, *Emerging Frontiers in Rhizoengineering: Genomics, Metabolomics, and Beyond*, was conceived to address this critical gap. It is not merely another compilation of the successes of OMICS technologies; rather, it is a truthful and comprehensive exploration of the challenges, bottlenecks, and practical limitations that hinder the translation of this vast knowledge into tangible agricultural solutions. Our primary focus is to dissect the reasons behind the stalled pipeline from the bench to the field and propose a rationalized path forward. We have brought together leading experts to critically examine the foundations upon which the next generation of microbial inoculants must be built.

The book is structured to guide readers through this complex landscape. The initial chapters lay the groundwork by establishing the current state of the art, delving into the genetics of host plants as a primary determinant of the rhizomicrobiome, and exploring the nuances of microbe-dependent heterosis. We then confront the practical realities with dedicated sections on the limitations and bottlenecks of current rhizoengineering techniques, including a comparative analysis of why contemporary approaches have yet to yield commercially viable products. A significant portion of the work is devoted to the pressing need for climate-smart solutions, featuring in-depth discussions on the effects of drought on plant physiology, from leaf anatomy to chlorophyll dynamics, and the development of innovative, biodegradable inorganic biostimulants that can work in concert with microbial inocula.

The latter part of the book serves as a strategic road map. It offers a critical reevaluation of traditional approaches, distinguishing between what remains useful and what must be superseded. We propose a rationalization of contemporary rhizoengineering targets, moving beyond purely descriptive studies toward hypothesis-driven engineering with defined and quantifiable outcomes. Our goal is to provide a balanced perspective that celebrates the immense power of genomics and metabolomics while advocating for a more pragmatic, systems-level approach that integrates plant genetics, environmental context, and robust formulation technologies.

This volume is the result of a collaborative endeavor, bringing together the insights of researchers from across the globe, with a significant and invaluable contribution from the Indian subcontinent, where the challenge of sustainable agriculture is both urgent and deeply felt. We extend our deepest gratitude to all the contributing authors, whose scholarly articles and critical perspectives form the backbone of this work. Their willingness to share not only their successes but also their insights into the challenges they faced make this book unique. We are also profoundly thankful to our colleagues who generously volunteered their time to review the manuscripts, ensuring the highest quality and rigor of the content. Finally, we acknowledge the Editorial Team at Springer for their exceptional coordination and guidance throughout the publication process.

We sincerely hope that this book serves as an essential guide for researchers, academicians, and industry professionals who are either deeply engaged in or are just embarking on projects related to rhizoengineering, microbiome manipulation, and sustainable agriculture. By honestly addressing the hurdles and charting a course for rational innovation, we aim to bridge the gap between the frontier of science and the reality of the field, paving the way for the development of successful next-generation bioinocula that can truly contribute to a resilient and food-secure future.

Lahore, Pakistan Samina Mehnaz
Pamplona, Navarra, Spain Izzah Shahid

Contents

Editors and Contributors

About the Editors

Samina Mehnaz is a professor at the KAM School of Life Sciences, Forman Christian College University (FCCU), where she has been serving since September 2012. A microbiologist since 1991, her research focuses on bioformulations (biofertilizers and biopesticides), microbial metabolites, bioplastics, and plastic biodegradation. She has been the recipient of several prestigious international fellowships, including the Georg Foster Research Fellowship from the Alexander von Humboldt (AvH) Foundation (Germany), National Science and Engineering Research Council (NSERC) Fellowship (Canada), International Atomic Energy Agency (IAEA) Fellowship (Vienna), and a Postdoctoral Fellowship in the USA. Dr. Mehnaz has an extensive research portfolio, with 81 peer-reviewed publications and 13 book chapters, and her work has received more than 7000 citations. She has supervised numerous BS, MS, and PhD students and actively contributed to the scientific community through editorial and review work. She is the editor of two books published by Springer, currently serves on the editorial boards of three international journals, and has edited more than 100 manuscripts. In addition, she has reviewed approximately 400 manuscripts for 80 scientific journals and has presented her research at over 70 national and international conferences. She served as a Humboldt Ambassador Scientist for Pakistan (2016–2021), representing the Alexander von Humboldt Foundation, Germany. In October 2022, she was appointed as the vice president of the Pakistan Regional Branch of the Asian Federation of Biotechnology (AFOB). Dr. Mehnaz has also been recognized among the world's top 2% most highly cited researchers by Stanford-Elsevier for four consecutive years since 2020.

Izzah Shahid is a Marie Skłodowska-Curie Postdoctoral Fellow at the Institute of Multidisciplinary Research in Applied Biology, Public University of Navarra (UPNA), Pamplona, Spain. She holds a PhD in Biotechnology from Forman Christian College (A Chartered University), Pakistan. Her research focuses on rhizosphere microbiology, microbial metabolites, rhizoengineering, and plant–microbiome interactions for sustainable agriculture. Dr. Shahid has experience in microbial genomics, metabolomics, and the development of microbial biofertilizers and biocontrol strategies for climate-resilient agricultural practices. She has published 20+ peer-reviewed research articles, reviews, and book chapters, with over 1000 citations and an h index of 12. Her work includes highly cited studies on plant growth-promoting rhizobacteria and microbial secondary metabolites. Dr. Shahid has served as a topic and guest editor for several international journals and is actively involved in international research networks, such as COST Actions on plant microbiomes and synthetic microbial communities, and contributes to advancing microbiome-based solutions for sustainable crop production.

Contributors

Debora D. M. Ambarita Faculty of Agriculture, Universitas Bangka Belitung, West Java, Indonesia

Elena Andriūnaitė Institute of Horticulture, Lithuanian Research Centre for Agriculture and Forestry, Kaunas, Lithuania

Rhea Aqueel Kauser Abdulla Malik School of Life Sciences, Forman Christian College (A Chartered University), Lahore, Pakistan

Department of Biological Sciences, Royal Holloway University of London, Egham, Surrey, United Kingdom

Naveen Kumar Arora Department of Environmental Science, Babasaheb Bhimrao Ambedkar University, Lucknow, Uttar Pradesh, India

Danas Baniulis Institute of Horticulture, Lithuanian Research Centre for Agriculture and Forestry, Kaunas, Lithuania

Salma Batool Department of Basic and Applied Chemistry, Faculty of Science & Technology, University of Central Punjab, Lahore, Pakistan

Ankita Bhattacharya Department of Environmental Science, Babasaheb Bhimrao Ambedkar University, Lucknow, Uttar Pradesh, India

Odeta Buzaitė Department of Biochemistry, Vytautas Magnus University, Kaunas, Lithuania

Subhradeep Chakraborty Soil and Environment Microbiology Laboratory, Department of Microbiology, Assam University, Silchar, Assam, India

Julia Chmiel Faculty of Chemistry, Silesian University of Technology, Gliwice, Poland

Ankita Das Soil and Environment Microbiology Laboratory, Department of Microbiology, Assam University, Silchar, Assam, India

Nandita Das Soil and Environment Microbiology Laboratory, Department of Microbiology, Assam University, Silchar, Assam, India

Farah Deeba Department of Biochemistry & Biotechnology, The Women University, Multan, Pakistan

Tehreem Ghafoor Institute of Botany, Bahauddin Zakariya University, Multan, Pakistan

Muhammad Fahad Hakim Institute of Horticulture, Lithuanian Research Centre for Agriculture and Forestry, Kaunas, Lithuania

Anna Hawkins A & L Biological Agroecology Services Inc, London, ON, Canada

Department of Biology, University of Western Ontario, London, ON, Canada

Diyan Herdiyantoro Faculty of Agriculture, Department of Soil Sciences and Land Resources, Universitas Padjadjaran, West Java, Indonesia

Fairus Hisanah Hibatullah Faculty of Agriculture, Department of Soil Sciences and Land Resources, Universitas Padjadjaran, West Java, Indonesia

Jesse Hoage A & L Biological Agroecology Services Inc, London, ON, Canada

Hayden Hornick-Martyk A & L Biological Agroecology Services Inc, London, ON, Canada

Department of Biology, University of Western Ontario, London, ON, Canada

Irwandhi Faculty of Agriculture, Department of Soil Sciences and Land Resources, Universitas Padjadjaran, West Java, Indonesia

Iqra Jabeen Department of Biochemistry & Biotechnology, The Women University, Multan, Pakistan

Hafsa Jekeen Department of Basic & Applied Chemistry, Faculty of Science & Technology, University of Central Punjab, Lahore, Pakistan

Ryan Joyce A & L Biological Agroecology Services Inc, London, ON, Canada

Department of Biology, University of Western Ontario, London, ON, Canada

Nadia N. Kamaluddin Faculty of Agriculture, Department of Soil Sciences and Land Resources, Universitas Padjadjaran, West Java, Indonesia

Fiqriah Hanum Khumairah Research Center of Applied Microbiology, National Research and Innovation Agency, West Java, Indonesia

Nikodem Kuźnik Faculty of Chemistry, Silesian University of Technology, Gliwice, Poland

Saima Majeed Department of Biotechnology, Faculty of Chemical and Biological Sciences, The Islamia University of Bahawalpur, Bahawalpur, Pakistan

Kauser Abdulla Malik Kauser Abdulla Malik School of Life Sciences, Forman Christian College (A Chartered University), Lahore, Pakistan

Roomina Mazhar Department of Botany, Shaheed Benazir Bhutto Women University, Peshawar, Pakistan

Samina Mehnaz Kauser Abdulla Malik School of Life Sciences, Forman Christian College (A Chartered University), Lahore, Pakistan

Umarah Mubeen A & L Biological Agroecology Services Inc, London, ON, Canada

Department of Biology, University of Western Ontario, London, ON, Canada

Anne Nurbaity Faculty of Agriculture, Department of Soil Sciences and Land Resources, Universitas Padjadjaran, West Java, Indonesia

Piyush Pandey Soil and Environment Microbiology Laboratory, Department of Microbiology, Assam University, Silchar, Assam, India

Malik Abdul Rahim Department of Basic & Applied Chemistry, Faculty of Science & Technology, University of Central Punjab, Lahore, Pakistan

Soledad Saldias A & L Biological Agroecology Services Inc, London, ON, Canada

Izzah Shahid Institute of Multidisciplinary Research in Applied Biology, Public University of Navarra, Pamplona, Spain

Kausar Hussain Shah Institute of Botany, Bahauddin Zakariya University, Multan, Pakistan

Ayesha Shehzadi Department of Biotechnology, Faculty of Chemical and Biological Sciences, The Islamia University of Bahawalpur, Bahawalpur, Pakistan

Tualar Simarmata Faculty of Agriculture, Department of Soil Sciences and Land Resources, Universitas Padjadjaran, West Java, Indonesia

Juli Sinaga Faculty of Agriculture, Department of Soil Sciences and Land Resources, Universitas Padjadjaran, West Java, Indonesia

Silpi Sorongpong Soil and Environment Microbiology Laboratory, Department of Microbiology, Assam University, Silchar, Assam, India

Inga Tamošiūnė Institute of Horticulture, Lithuanian Research Centre for Agriculture and Forestry, Kaunas, Lithuania

Diyah Sri Utami Faculty of Agriculture, Department of Soil Sciences and Land Resources, Universitas Padjadjaran, West Java, Indonesia

Priyanka Verma Department of Environmental Science Babasaheb Bhimrao Ambedkar University, Lucknow, Uttar Pradesh, India

Raheela Waheed Department of Biosciences and Technology, Emerson University, Multan, Pakistan

Nimalka Weerasuriya A & L Biological Agroecology Services Inc, London, ON, Canada

Krystian Wolski Faculty of Chemistry, Silesian University of Technology, Gliwice, Poland

Tahira Younis Department of Biochemistry & Biotechnology, The Women University, Multan, Pakistan

Abbreviations

16S rRNA	16S Ribosomal ribonucleic acid
2,4-DAPG	2,4-Diacetylphloroglucinol
ALA	5-Aminolevulinic acid
ABA	Abscisic acid
ACC	1-Aminocyclopropane-1-carboxylate
AHLs	N-acyl-homoserine lactones
AI	Artificial intelligence
AOS	Active oxygen species
APX	Ascorbate peroxidase
BGCs	Biosynthetic gene clusters
BIS	Bureau of Indian Standards
BLAST	Basic local alignment search tool
BNF	Biological nitrogen fixation
BSMs	Beneficial soil microorganisms
BWA-MEM	Burrows-Wheeler aligner-maximal exact-matches
CAMTAs	Calmodulin-binding transcription activators
cDNA	Complementary deoxyribonucleic acid
CEC	Cation exchange capacity
CF	Chlorophyll fluorescence
CFU	Colony-forming units
CK	Cytokinin
CRISPR	Clustered regularly interspaced short palindromic repeats
CNGCs	Cyclic nucleotide-gated channels
CO_2	Carbon dioxide
CSA	Climate-smart agriculture
COGs	Clusters of orthologous genes
CUE	Carbon use efficiency
DEGs	Differentially expressed genes
DMDS	Dimethyl disulfide
DOPE-FISH	Double-labeled oligonucleotide probes–Fluorescent in situ hybridization

DREB	Dehydration responsive element binding proteins
DT-SynCom	Drought Tolerant Synthetic Microbial Community
EBIC	European Biostimulants Industry Council
EC	Electrical conductivity
ECCO	European Culture Collections' Organization
ePGPR	Extracellular plant growth-promoting rhizobacteria
EPS	Exopolysaccharide
EST	Expressed sequence tags
ET	Ethylene
FA	Fulvic acid
FAO	Food and Agriculture Organization
FC	Field capacity
GABA	γ-Aminobutyric acid
GALDA	Global assessment of land degradation and improvement
GC-MS	Gas chromatography coupled with mass spectrometry
GEM	Genome-edited microorganism
GEMs	Genome-scale metabolic models
GHG	Greenhouse gas
GMMs	Genetically modified microorganisms
GO	Gene ontology
GS	Genomic selection
GWAS	Genome-wide association study
H_2O_2	Hydrogen peroxide
HA	Humic acids
HGT	Horizontal gene transfer
HKT	High-affinity potassium transporter
HMME	Host-mediated microbiome engineering
HPLC	High-performance liquid chromatography
HS	Humic substances
HTS	High-throughput third-generation sequencing
HT-SIP	High-throughput stable isotope probing
IAA	Indole-3-acetic acid
IBS	Inorganic biostimulants
ICA	Indole-3-carboxylic acid
ILA	Indole-3-lactic acid
IoT	Internet of Things
IPCC	Intergovernmental panel on climate change
IPyA	Indole-3-pyruvic acid
iPGPR	Intracellular plant growth-promoting rhizobacteria
ISR	Induced systemic resistance
JA	Jasmonic acid
KEGG	Kyoto encyclopedia of genes and genomes
KSB	Potassium solubilizing bacteria
LC-MS/MS	Liquid chromatography coupled with tandem mass spectrometry
LEA	Late embryogenesis abundant proteins

LOC	Labile organic carbon
MAGs	Metagenome-assembled genomes
MAOM	Mineral-associated organic matter
MAPK	Mitogen-activated protein kinase
MAS	Marker-assisted selection
MDA	Malondialdehyde
ML	Machine learning
MSiNPs	Mesoporous silica nanoparticles
NatComs	Natural community
NbS	Nature-based solutions
NCC	Nano-calcium carbonate
NGS	Next-generation sequencing
NMR	Nuclear magnetic resonance
NPK	Nitrogen, phosphorus, and potassium
NPs	Nanoparticles
NRPS	Non-ribosomal peptide synthetases
NUE	Nitrogen use efficiency
OA	Osmolyte accumulation
OECD	Organization for economic co-operation and development
OMICs	Collective term for genomics, transcriptomics, proteomics, metabolomics
PacBio	Pacific biosciences
PGPB	Plant growth-promoting bacteria
PGPM	Plant growth-promoting microbes
PGPR	Plant growth-promoting rhizobacteria
PH	Protein hydrolysates
PHA	Process hazards analysis
POD	Peroxidase
POM	Particulate organic matter
POPs	Persistent organic pollutants
PQQ	Pyrroloquinoline quinone
PRPs	Pathogenesis-related proteins
PSB	Phosphate solubilizing bacteria
PSM	Phosphate-solubilizing microorganisms
QC	Quality control
QIIME	Quantitative insights into microbial ecology
qPCR	Quantitative polymerase chain reaction
QS	Quorum sensing
qSIP	Quantitative stable isotope probing
QTLs	Quantitative trait loci
QQ	Quorum quenching
RBD	Randomized block design
RIDER	Rhizobacteria-induced drought endurance and resilience
ROC	Recalcitrant organic carbon
ROS	Reactive oxygen species

RuBisCO	Ribulose-1,5-bisphosphate carboxylase/oxygenase
RWC	Relative water content
SA	Salicylic acid
SAR	Systemic acquired resistance
SC-SIP	Single-cell stable isotope probing
SGDs	Sustainable development goals
SiNPs	Silica nanoparticles
SMs	Secondary metabolites
SNNMs	Siliceous natural nanomaterials
SNPs	Single-nucleotide polymorphisms
SOBARI	System of organic-based aerobic rice intensification
SOC	Soil organic carbon
SOD	Superoxide dismutase
SOM	Soil organic matter
SPAD	Soil plant analysis development
SynCom	Synthetic community
TEs	Transposable elements
UD	Uridine diphosphate
UNDE	United nations department of economic and social affairs
US EPA	United States environmental protection agency
VOCs	Volatile organic compounds
WFCC	World federation for culture collections
WUE	Water use efficiency
XCMS	Software platform for LC-MS/MS data analysis
ZnNPs	Zinc oxide nanoparticles

Part I
Limitations and Bottlenecks in Rhizoengineering Approaches

Chapter 1
Engineering the Rhizosphere: Microbial Pathways for Soil Health, Resilient Crops, and Climate-Smart Agriculture

Umarah Mubeen, Nimalka Weerasuriya, Jesse Hoage, Anna Hawkins, Hayden Hornick-Martyk, Ryan Joyce, and Soledad Saldias

Abstract Soil health is foundational to sustainable agriculture and is shaped by complex interactions between soil microbes, plants, and environmental factors. Microbial communities regulate nutrient cycling, structure formation, and organic matter stabilization, all of which are essential for productive and resilient ecosystems. This chapter explores the pivotal roles of soil microorganisms in enhancing soil health through nutrient recycling, drought resilience, carbon sequestration, and pollutant degradation. Microbial interactions, strongly influenced by root exudates and environmental conditions, shape soil structure and functionality. Recent developments in microbial biostimulants and metabolite-driven changes in soil microbial communities are promising tools for improving crop productivity and reducing reliance on chemical inputs. Harnessing microbial processes provides a sustainable path for agricultural resilience and environmental protection. Integrating native plant growth-promoting microbes, mycorrhizal fungi, and microbial metabolites into agricultural systems can simultaneously improve yields, reduce greenhouse gas emissions, and remediate chemical pollutants. Strategic manipulation of the soil microbiome is an emerging frontier in sustainable land management.

Keywords Soil microbiome · Plant growth-promoting microbes (PGPM) · Carbon sequestration · Microbial biostimulants · Sustainable agriculture · Soil health management

U. Mubeen (✉) · A. Hawkins · H. Hornick-Martyk · R. Joyce
A & L Biological Agroecology Services Inc, London, ON, Canada

Department of Biology, University of Western Ontario, London, ON, Canada
e-mail: umubeen@albiological.com; uusman3@uwo.ca

N. Weerasuriya · J. Hoage · S. Saldias
A & L Biological Agroecology Services Inc, London, ON, Canada

S. Mehnaz, I. Shahid (eds.), *Frontiers in Rhizoengineering*, Microorganisms for Sustainability 65, https://doi.org/10.1007/978-981-92-3665-7_1

1.1 Soil Health

Soil health is a multifaceted concept encompassing the physical, chemical, and biological properties of soil, as well as its capacity to support plant growth, maintain ecosystem functions, and sustain agricultural productivity. Soil health is defined as the capacity of soil to function as a vital living system that sustains plant, animal, and human health. It is characterized by dynamic and balanced interactions among physical, chemical, and biological properties that support diverse and resilient ecosystems, maintain nutrient cycling, promote plant growth, and enhance environmental sustainability (Doran and Zeiss 2000; Lehmann and Kleber 2015). In recent years, the One Health concept has evolved, advocating that the health of humans, animals, and the environment are all connected (Lehmann et al. 2020), defining soil health as the continued capacity of soil to function as a vital living ecosystem that sustains plants, animals, and humans, as well as maintaining the socio-ecological functions of the ecosystem in which it is located (Lehmann et al. 2020).

1.2 Role of Microbes in Soil Health

1.2.1 Nutrient Recycling and Stabilization

Soil microorganisms, including bacteria and fungi, play central roles in structuring soils and stabilizing nutrients across scales, from microsites on individual soil particles to whole-ecosystem interactions with plants. Direct mechanisms include efficient nutrient uptake, biofilm and mycelial formation, and high microbial carbon use efficiency (CUE), all of which enhance aggregate stability and nutrient retention (Domeignoz-Horta et al. 2020). Microbes indirectly interact with soil minerals and organic inputs, reinforcing nutrient stabilization and soil structure. This positive feedback loop underpins improved plant productivity and litter deposition (i.e., root growth and residue input), which is particularly evident in regenerative farming systems. Conversely, conventional intensive cropping, with disturbance and nutrient depletion, diminishes microbial diversity and abundance, impairing these functional processes (Domeignoz-Horta et al. 2020).

Beyond immediate nutrient cycling, microbes determine the persistence of carbon in soils by regulating the transformation of soil organic matter (SOM). The rapid mineralization of organic inputs, their incorporation into particulate organic matter (POM), or their stabilization as mineral-associated organic matter (MAOM) largely depend on microbial processing (Peng et al. 2025; Whalen et al. 2024). Thus, microbial activity is a critical link between short-term nutrient availability and long-term soil fertility.

1.2.2 Soil Organic Matter (SOM) Pools

SOM comprises two major fractions:

- **Particulate Organic Matter (POM):** Coarse residues (e.g., plant debris) with rapid turnover (<10 years), which release nutrients quickly and contribute to soil structure through fungal hyphae and microbial exudates.
- **Mineral-Associated Organic Matter (MAOM):** Fine organic compounds bound to minerals, characterized by slower turnover and long-term nutrient storage (Cotrufo and Lavallee 2022).

Field trials have shown that residue retention and reduced tillage increase POM accumulation, which in turn promotes MAOM formation. Microbial necromass, comprising dead cells and secreted by-products, binds to mineral surfaces and is a key precursor of MAOM (Angst et al. 2024; Zhao et al. 2025). Highly decomposed microbial inputs accelerate MAOM accrual, making microbial activity essential for balancing POM inputs, necromass turnover, and mineral sorption capacity (Cotrufo and Lavallee 2022).

1.2.3 Labile Versus Recalcitrant Carbon: Microbial Community Dynamics

Microbial processes determine the quantity and quality of carbon retained in soils. Labile organic carbon (LOC), such as root exudates and fresh litter, fuels copiotrophic microbes that grow rapidly but quickly deplete resources. Upon senescence, their necromass, rich in residues such as N-acetylmuramic acid and N-acetylglucosamine, contributes to recalcitrant organic carbon (ROC), which persists for centuries to millennia (Angst et al. 2024).

ROC selectively favors oligotrophic microbes that recycle necromass into biomass and secrete extracellular enzymes to mine stable organic matter and minerals, driving slower but more stable nutrient cycling (Kästner et al. 2021; Ni et al. 2021; Yin et al. 2024).

1.2.4 Microbial Carbon Use Efficiency (CUE)

The ratio of carbon allocated to growth versus respiration determines whether soil carbon is sequestered or lost as carbon dioxide (CO_2). High CUE supports microbial biomass accumulation and greater soil organic carbon (SOC) storage, and global models have identified CUE as a dominant regulator of SOC, exceeding the effects of input rates or decomposition alone (Tao et al. 2023). Environmental factors, such as nutrient availability, temperature, moisture, pH, community composition, and substrate quality, modulate CUE at local and global scales (Yu et al. 2025). High CUE therefore not only promotes carbon storage but also fosters structured, biologically active soils, reinforcing positive plant-microbe feedbacks that underpin sustainable agriculture.

Having established the fundamental roles of microbes in soil health, including nutrient recycling, soil organic matter stabilization, carbon sequestration, and microbial carbon use efficiency, it becomes clear that soil microorganisms not only regulate soil fertility but also strongly influence how plants respond to environmental stressors. Drought is one of the most pressing challenges in modern agriculture and has profound effects on plant growth, soil structure, and microbial community dynamics. Plant growth-promoting microbes (PGPM), including bacteria and fungi, occupy a pivotal position at thc interface between soil health and plant stress resilience. Building upon their established roles in nutrient cycling and soil carbon stabilization, PGPM extend their benefits to plants by enhancing drought tolerance through hormonal regulation, improved root growth, osmotic balance, and antioxidant activity (Kumar et al. 2019; Zhang et al. 2019).

1.3 Plant Growth-Promoting Microbes (PGPM) and Stress Resilience

Building on the roles of soil microbes in nutrient recycling, SOM stabilization (POM → MAOM), and high microbial CUE outlined above, plant growth-promoting microbes (PGPM) channel these processes into the physiological resilience of the host plant, modulating hormone signaling, root system architecture, osmotic balance, and antioxidant defenses that buffer crops against water limitation and other abiotic stresses. For broad syntheses of PGPM-mediated stress tolerance, see the reviews by Lata and Gond (2019) and Omomowo and Babalola (2019).

Water deficiency, salinity, and high temperatures are among the most critical abiotic stresses that disrupt plant physiology, often leading to wilting, reduced nutrient uptake, oxidative stress, and, ultimately, yield loss (Nemeskéri and Helyes 2019; Schneider et al. 2019; Sofi et al. 2018). Soil-dwelling PGPM, including bacteria and fungi, play a central role in mitigating these stressors on plants. PGPM mitigate oxidative and osmotic stress in plants and promotes recovery from water scarcity by producing phytohormones, such as indole acetic acid (IAA), gibberellic acid (GA), salicylic acid (SA), and abscisic acid (ABA), solubilizing phosphate, releasing antioxidants, and activating enzymes such as ACC deaminase (Castro-Camba et al. 2022; Khan et al. 2021).

1.3.1 Soil Resilience to Drought

Under drought conditions, abscisic acid (ABA) accumulation triggers stomatal closure via aquaporin regulation (Acharya and Assmann 2009) and alters reproductive development (Riboni et al. 2016). In parallel, ABA-independent pathways involving salicylic acid (SA), jasmonic acid (JA), ethylene, and auxins coordinate stress responses (Mathur and Roy 2021). JA decreases transpiration through stomatal regulation (De Ollas et al. 2018), while stress-induced ethylene can accelerate senescence and restrict growth, but also enhances drought tolerance when balanced with cytokinin and JA (Egea et al. 2018; Kazan 2015). Many PGPR produce IAA, which stimulates ACC deaminase activity to suppress stress ethylene, thereby sustaining root development and enabling faster recovery under water limitation (Glick 2012; Mayak et al. 2004).

Beyond hormonal signaling, PGPM mediate structural and metabolic adaptations that enhance drought tolerance. Rhizosheath formation, a soil layer adhering to roots, improves water retention and protects both roots and microbes under desiccation (Zhang et al. 2020). Biofilms and exopolysaccharides (EPS) secreted by bacteria retain water in root-associated microenvironments (Chang et al. 2007; Timmusk and Nevo 2011). Compatible solutes, such as trehalose, glycine betaine, and proline, produced by microbes or stimulated in plants, stabilize membranes, balance osmotic pressure, and scavenge reactive oxygen species (ROS) (Kour et al. 2020; Yang et al. 2013). These physiological adjustments protect the photosynthetic machinery, reduce electrolyte leakage, and sustain chlorophyll content under stress.

Drought resilience in agroecosystems emerges from plant × soil physicochemistry × microbiome × climate interactions. Because water availability is a dominant evolutionary and agronomic constraint, breeding/transgenics/chemistry alone rarely suffices where drought frequency and intensity are rising (global yield limits). Targeting the soil microbiome complements these strategies by leveraging the same microbial processes that sustain soil function (CUE and SOM formation) to stabilize plant performance under water deficit (Kijne 2006; Sofi et al. 2018). Across diverse systems, PGPM (bacteria and fungi) re-tune phytohormone networks (ABA, SA, JA, ethylene, auxin, and cytokinin), regulate stomatal conductance and aquaporins, and enhance antioxidant capacity, thereby reducing ROS damage and maintaining photosynthesis and nutrient uptake during drought. ABA levels generally increase under stress and control stomatal closure via aquaporin regulation, and ABA-independent routes (e.g., SA/JA/ethylene) also contribute. ACC-deaminase-producing PGPR lower stress ethylene, supporting root growth and faster recovery from transient water deficits (Ahmad et al 2022; Mahmood et al. 2019; Murali et al. 2021).

PGPM also enhance drought tolerance through structural and metabolic adjustments, which include:

- Rhizosheath formation protects the roots and microbes from desiccation (Zhang et al. 2019).

- Biofilms and polysaccharides retain water in microenvironments (Chang et al. 2007; Timmusk and Nevo 2011).
- Rhizobia stimulate lateral root formation and biomass accumulation (Sofi et al. 2017).
- Compatible solutes (trehalose, glycine, and betaine) improve osmotic balance and membrane stability (Kour et al. 2020).
- Proline accumulation scavenges ROS and stabilizes proteins (Yang et al. 2013).

1.3.2 PGPM and Abiotic Stresses

While drought has received much attention, PGPM also confer resilience against other abiotic stresses, particularly salinity, heat, and heavy metal contamination. Under salinity stress, high Na^+ and Cl^- disrupt nutrient uptake and water potential; however, halotolerant microbes such as *Bacillus*, *Pseudomonas*, and *Halomonas* spp. enhance K^+/Na^+ ratios, stimulate osmolyte accumulation, and maintain photosynthetic activity (Verma et al. 2014). Heat stress, which destabilizes proteins and membranes, is mitigated by the microbial induction of heat shock proteins, antioxidants, and secondary metabolites. *Trichoderma* spp. and *Bacillus* spp. produce volatile organic compounds (VOCs) and metabolites that enhance thermotolerance, whereas *Piriformospora indica* improves membrane stability and photosynthesis at high temperatures (Ji et al. 2024; Senizza et al. 2023).

1.3.3 PGPM and Heavy Metal Stress

Heavy metal stress poses another serious challenge, impairing plant metabolism and soil microbial communities. PGPM alleviate toxicity via biosorption and bioaccumulation, as demonstrated in *Pseudomonas aeruginosa* and *Bacillus subtilis*, which immobilize toxic ions (Wang et al. 2022). Exopolysaccharide production further protects plants by binding metals in the rhizosphere (Zhang et al. 2025), whereas arbuscular mycorrhizal fungi (AMF), such as *Glomus* spp., sequester heavy metals in their hyphae, reducing their translocation to aerial tissues (Ma et al. 2022).

1.3.4 PGPM in Biotic Stress Management

In addition to abiotic stress mitigation, PGPM play vital roles in plant defense against pathogens, insects, and nematodes. Many bacteria, such as *Pseudomonas fluorescens* and *Bacillus subtilis*, suppress pathogens through the production of antibiotics, siderophores, and lytic enzymes while also inducing systemic resistance (ISR) in host plants via jasmonate and ethylene pathways (Beneduzi et al. 2012;

Mehmood et al. 2023). Fungi, such as *Trichoderma* spp., combat soil-borne pathogens through mycoparasitism, antibiosis, and ISR induction (Chen et al. 2025; Tyśkiewicz et al. 2022).

PGPM are also effective against insect pests and plant-parasitic nematodes. *Bacillus thuringiensis* produces Cry toxins that disrupt the gut lining of insect larvae (Palma et al. 2014), whereas nematophagous fungi such as *Paecilomyces lilacinus* and *Pochonia chlamydosporia* parasitize nematode eggs and release proteases, thereby reducing nematode populations in the soil (Meel and Saharan 2025). These examples demonstrate that PGPM function as natural biocontrol agents, reducing dependence on chemical pesticides while simultaneously supporting plant growth.

Taken together, PGPM act as multifunctional allies in crop resilience, enhancing drought tolerance and buffering against salinity, heat, heavy metals, pathogens, insects, and nematodes. These mechanisms range from hormonal modulation and osmotic adjustment to detoxification, ISR induction, and symbiotic nutrient exchange. Representative microbes, such as *Azospirillum*, *Rhizobium*, *Bacillus*, *Pseudomonas*, *Rhizophagus*, *Trichoderma*, and *Piriformospora*, illustrate the diversity of microbial strategies deployed under stress conditions. A concise summary of these microbial groups, their target stresses, and their mechanisms of action is presented in Table 1.1.

Table 1.1 Plant growth-promoting microbes (PGPM) mitigating abiotic and biotic stresses

Microbe (representative species)	Kingdom/class	Stress target(s)	Mechanism(s) of stress alleviation	References
Acinetobacter calcoaceticus	Bacteria/ Gammaproteobacteria	Drought, salinity	Enhances osmotic balance, water retention	Kour et al. (2020)
Arthrobacter sp.	Bacteria/ Actinomycetia	Drought, heavy metals	ROS scavenging, sporulation, metal detoxification	Hone et al. (2021)
Azospirillum brasilense	Bacteria/ Alphaproteobacteria	Drought, salinity	Root proliferation, nodulation, osmolyte accumulation	German et al. (2000), Steiner et al. (2020)
Bacillus spp. (*B. subtilis*, *B. amyloliquefaciens*, *B. thuringiensis*)	Bacteria/Bacilli	Drought, salinity, heat, pathogens, insects	Biofilms and EPS, IAA, ACC deaminase, VOCs, cry toxins, ISR induction	Palma et al. (2014), Prudent et al. (2015)
Bradyrhizobium spp. (*B. japonicum*, *B. Elkani*)	Bacteria/ Alphaproteobacteria	Drought, salinity	Symbiotic N-fixation, nodulation, improved root biomass	Tittabutr et al. (2013)

(continued)

Table 1.1 (continued)

Microbe (representative species)	Kingdom/class	Stress target(s)	Mechanism(s) of stress alleviation	References
Rhizobium spp. (*R. etli, R. gallicum, R. tropici*)	Bacteria/ Alphaproteobacteria	Drought, salinity	Nodulation, osmotic protection, higher WUE, reduced flower abortion	del-Canto et al. (2023)
Pseudomonas spp. (*P. fluorescens, P. putida, P. aeruginosa*)	Bacteria/ Gammaproteobacteria	Heavy metals	Antibiotics, siderophores, IAA, antioxidant enzymes, synergism with AMF	Wang et al. (2022)
Variovorax paradoxus	Bacteria/ Betaproteobacteria	Drought	Auxin modulation, improved root system architecture	Belimov et al. (2009)
Halomonas spp.	Bacteria/ Gammaproteobacteria	Salinity	Ion homeostasis (K^+/Na^+), osmolyte production	Verma et al. (2014)
Diversispora versiformis	Fungi/ Glomeromycetes (AMF)	Drought	Improves root growth, nutrient uptake, drought resilience	Zhang et al. (2019)
Rhizophagus spp. (*R. intraradices, R. irregularis, R. mosseae*)	Fungi/ Glomeromycetes (AMF)	Drought, heavy metals	Hyphal water/ nutrient exchange, metal sequestration	Azizi et al. (2022), Ma et al. (2022)
Penicillium minioluteum	Fungi/Eurotiomycetes	Salinity, ROS stress	Osmotic balance, ROS scavenging, metabolite production	González-Teuber et al. (2018), Kour et al. (2020)
Piriformospora indica	Fungi/Agaricomycetes	Heat, pathogens	Enhances photosynthesis, membrane stability, induces systemic tolerance	Gill et al. (2016)
Trichoderma spp.	Fungi/ Sordariomycetes	Heat, pathogens	VOCs, ISR induction, mycoparasitism	Chen et al. (2025), Poveda (2020)
Paecilomyces lilacinus, Pochonia chlamydosporia	Fungi/ Sordariomycetes	Nematodes	Egg parasitism, protease production	Meel and Saharan (2025)

1.4 Soil Health and GHG Emissions

Soil health and greenhouse gas (GHG) emissions are inherently linked through the biological activity of soil microorganisms, which govern the cycling of carbon, nitrogen, and other essential elements in soil. Healthy, microbially active soils promote carbon sequestration and efficient nutrient turnover, whereas degraded soils with low microbial diversity tend to emit more CO_2, CH_4, and N_2O (Zewdie et al. 2021). These gases, carbon dioxide (CO_2), methane (CH_4), and nitrous oxide (N_2O), collectively drive global warming by trapping heat in the atmosphere through the *greenhouse effect* (Intergovernmental Panel on Climate Change 2023). *According to the Intergovernmental Panel on Climate Change (IPCC), global* GHG emissions have increased sharply in recent decades, increasing by 12% between 2010 and 2019 and by 54% since 1990. Agricultural activities, including *fertilizer application, residue management, and soil disturbance,* are responsible for approximately 13–21% of the total emissions (IPCC 2023).

Intensive nitrogen fertilization enhances soil N_2O fluxes due to accelerated microbial nitrification and denitrification processes (Rasconi et al. 2014). However, mounting evidence indicates that optimizing soil microbial activity through the use of *native plant growth-promoting microbes (PGPM)* and *biofertilizers* can significantly reduce emissions while sustaining or even improving crop yield (Geng et al. 2021; Sun et al. 2020). This microbe-based approach integrates soil health restoration with the global emission reduction targets outlined in the *Paris Agreement* (United Nations 2015), offering a sustainable pathway toward climate-resilient agriculture.

1.4.1 Nitrogen Fixation and Microbial Regulation of N_2O Emissions

Nitrogen (N) is an essential macronutrient that supports plant growth; however, its management in agriculture has major implications for both productivity and greenhouse gas (GHG) emissions. Excessive use of synthetic nitrogen fertilizers stimulates microbial nitrification and denitrification, leading to elevated emissions of nitrous oxide (N_2O), a gas with nearly 300 times the global warming potential of CO_2 (IPCC 2023). These processes are primarily microbial, involving nitrifiers and denitrifiers that convert ammonium (NH_4^+) to nitrate (NO_3^-) and gaseous N_2O/N_2 through sequential redox reactions (Roussel et al. 2014).

Biological nitrogen fixation (BNF) provides a natural alternative to chemical fertilizers by converting atmospheric nitrogen (N_2) into bioavailable ammonium

(NH_4^+) through the activity of diazotrophic bacteria, such as *Rhizobium*, *Azospirillum*, and *Bradyrhizobium*. This process not only supplies plant-available nitrogen but also minimizes N_2O emissions by optimizing soil nitrogen cycling. However, the balance between nitrification and denitrification strongly depends on the soil microbial community structure, pH, and carbon availability (Houlton et al. 2019).

Recent studies have highlighted that supplementing mineral fertilizers with beneficial microbes can improve nitrogen use efficiency (NUE) while reducing GHG emissions. For example, the coapplication of *Bacillus subtilis* with a 50% urea reduction improved NUE by 11.2% and yield by 5%, while cutting N losses by 54% compared to the full urea treatment (Sun et al. 2020). Similarly, bioorganic fertilizers containing *Trichoderma guizhouense* decreased N_2O emissions by 18.7% relative to NPK fertilizers and simultaneously increased cucumber yield (Geng et al. 2021). *Bacillus amyloliquefaciens* inoculation in oilseed rape reduced N_2O emissions by 42.6% through the stimulation of N_2O-reducing bacterial communities (Wu et al. 2017). These examples demonstrate how optimizing soil microbiomes through microbial inoculants can enhance productivity and mitigate emissions, which is an essential strategy for climate-resilient agriculture.

Carbon fixation is the process of converting inorganic carbon (CO_2) into organic carbon compounds that are used to store energy. During this process, some CO_2 is returned to the atmosphere through plant respiration; however, most CO_2 is converted to organic compounds that reside in plants (Lawson et al. 2022). During and after the growing season, plants begin to die, and crop residues begin to integrate with the soil. As residues decompose, they release CO_2 into the atmosphere; however, much of the carbon can be retained in the soil as soil organic carbon (SOC). SOC plays an important role in ecosystem services and supplies nutrients to the microbial communities in soils, which can help foster long-term soil health (Lal 2014). Research also suggests that a healthy microbiome leads to an increase in CO_2 sequestration in soil, as *cbbL* (ribulose-1,5-biphosphate carboxylase/oxygenase gene) abundance is positively correlated with SOC and microbial assimilation of carbon. This suggests that many soil bacteria fix carbon; however, the abundance and diversity of carbon-fixing microbes are likely variable with soil type and characteristics (Lynn et al. 2017). *Azospirillum lipoferum* is a common microbe known to carry the *cbbL* gene and plays an important role in SOC formation, along with other microbes. Microbial assimilation has been shown to account for up to 0.59% of SOC (Yuan et al. 2012). Additionally, other common bacteria, such as *Pseudomonas fluorescens,* have increased plant health indicators while increasing the rate of photosynthesis and carbon fixation in the host plant. Root and shoot lengths, as well as root and shoot dry masses, significantly increased when *P. fluorescens* was applied to *Sedum alfredii. P. fluorescens* also significantly increased chlorophyll content, which promoted photosynthesis and increased CO_2 assimilation in *S. alfredii* by increasing the expression of specific genes (Wu et al. 2020). Microbes play an important role in soil carbon sequestration and, in some cases, stimulate plant growth.

1.4.2 Carbon Fixation and Microbial Contribution to Soil Carbon Sequestration

The conversion of atmospheric CO_2 into organic compounds is fundamental to both plant productivity and long-term soil carbon storage. Soil microbes play a central role in stabilizing carbon inputs from plant residues in soil organic carbon (SOC) pools (Lal 2014). Microbial carbon use efficiency (CUE) determines whether assimilated carbon becomes biomass or is respired as CO_2 (Tao et al. 2023). Thus, high microbial CUE and efficient decomposition dynamics are essential for enhancing soil carbon sequestration.

Several autotrophic soil bacteria and archaea possess the *cbbL* gene, which encodes ribulose-1,5-bisphosphate carboxylase/oxygenase (RuBisCO) and enables biological CO_2 fixation in soils. The abundance of *cbbL*-carrying microbes positively correlates with SOC content and microbial biomass (Lynn et al. 2017). For instance, *Azospirillum lipoferum* and *Nitrobacter winogradskyi* actively contribute to carbon assimilation and enhance soil carbon stocks in crop systems (Yuan et al. 2012). Moreover, plant-associated bacteria, such as *Pseudomonas fluorescens*, can stimulate plant photosynthesis by upregulating chlorophyll synthesis and CO_2 assimilation genes, leading to improved root and shoot biomass (Wu et al. 2020). Microbial-driven carbon stabilization, particularly through necromass formation and mineral-associated organic matter (MAOM), is increasingly recognized as a critical component of long-term soil carbon storage (Cotrufo and Lavallee 2022). Therefore, managing soils to foster high microbial diversity and balanced C:N stoichiometry can enhance both plant productivity and climate change mitigation.

1.5 Microbial Degradation of Soil Pollutants and Heavy Metal Immobilization

While synthetic inputs, such as fertilizers and pesticides, improve crop productivity, their overuse degrades soil ecosystems and threatens human health. Persistent organic pollutants (POPs) and heavy metals accumulate in soils, disrupt microbial diversity, and impair nutrient cycling (Alengebawy et al. 2021). Pesticides such as chlorpyrifos, maneb, and trifluralin persist for years, altering microbial community composition and soil enzyme activity (Riedo et al. 2022). Pesticide residues in agricultural soils can remain for up to two decades after conversion to organic management (Riedo et al. 2022).

Microbial bioremediation provides a sustainable solution by leveraging the metabolic versatility of soil microbes to degrade or immobilize contaminants. Certain bacteria, including *Pseudomonas*, *Bacillus*, and *Arthrobacter* spp., can metabolize pesticide molecules through dealkylation, dehalogenation, and nitro group reduction, transforming them into less toxic intermediates (Coleman et al. 2020). For example, a microbial consortium of *Bacillus* and *Pseudomonas* species degraded

72% of trifluralin within 100 days, reducing its half-life by 600 days (Lara-Moreno et al. 2022). Similarly, *Bacillus thuringiensis* completely degraded chlorpyrifos through denitrification and oxidation pathways (Ambreen and Yasmin 2021).

In addition to pesticide degradation, soil microbes facilitate heavy metal detoxification through biosorption, bioaccumulation, and precipitation. *Pseudomonas aeruginosa* and *Bacillus subtilis* immobilize toxic ions, such as cadmium and lead, whereas arbuscular mycorrhizal fungi (AMF), such as *Glomus* spp., sequester metals in their hyphae, reducing plant uptake (Putra et al. 2022). These microbial processes restore soil functionality by improving soil enzyme activity, nutrient cycling, and overall soil biodiversity (Haider et al. 2021). Consequently, integrating microbial bioremediation into agricultural management enhances both soil health and environmental safety, advancing a circular, low-input model of sustainable agriculture.

1.6 Soil Microbiome Shaped by Plant–Microbe–Environment Interactions

The soil microbiome underpins terrestrial ecosystem functions and crop performance. Its composition and activity emerge from feedback among plants (especially via root exudates), edaphic factors (pH, texture, nutrients, and moisture), climate, and farm management. Microbial diversity generally supports resilience by distributing functions such as decomposition, nutrient cycling, pathogen suppression, and stress buffering across different taxa (Jiao et al. 2019; Kandasamy et al. 2021).

1.6.1 The Rhizosphere and Root Exudates

Plants actively "garden" their microbiome through root exudates and low- and high-molecular compounds (sugars, amino and organic acids, phenolics, peptides, and mucilage) that feed, select, and signal microbes within millimeters of the root surface. Exudate quantity and chemistry vary by species, stage, and stress, shaping microbial recruitment and activity (Becker and Holz 2021; Canarini et al. 2019; Kuzyakov and Razavi 2019; Sasse et al. 2018). These hotspots of carbon and signaling drive rapid biogeochemical fluxes and community turnover compared to bulk soil (Becker and Holz 2021; Kuzyakov and Razavi 2019). Exudates can selectively enrich taxa with plant-beneficial traits (e.g., P solubilization, hormone modulation, ACC deaminase, and siderophores). They also reprogram microbial gene expression and enzyme production, thereby altering nutrient acquisition and defense signaling (Jamil et al. 2022; Venturi and Keel 2016).

1.6.2 Symbiosis and Beneficial Guilds

Endophytes colonize internal plant tissues and often synthesize phytohormones (IAA, cytokinins, and GAs) and antimicrobials (e.g., lipopeptides), thereby promoting growth and disease suppression in plants. These effects are strain and host specific, emphasizing mechanisms (hormone modulation, ISR priming) over broad generalities (Kamran et al. 2022; Pieterse et al. 2014). *Arbuscular mycorrhizal fungi (AMF)* extend the effective root system via hyphal networks, improving P (and other nutrient) uptake, water relations, and stress tolerance. Multiple meta-analyses have shown that AMF can increase grain yields and improve performance under water deficit, particularly in rainfed systems; however, the effects vary by crop, AMF species, and environment (Jayne and Quigley 2014; Wu et al. 2022; Zhang et al. 2019).

Diazotrophs (e.g., *Azospirillum*) supplement N via associative fixation, whereas phosphate-solubilizing microorganisms (PSM) release organic acids and phosphatases that mobilize inorganic and organic P. Classic and modern reviews remain the best evidence base for agronomic claims (Rodríguez and Fraga 1999; Sharma et al. 2013).

1.6.3 Disease Suppression and Immune Priming

Beneficial bacteria and fungi suppress pathogens via niche competition, antibiosis (e.g., *Bacillus* lipopeptides such as surfactin/iturin/fengycin), mycoparasitism (e.g., *Trichoderma*), and induced systemic resistance (ISR), which sensitizes plant defenses without growth penalties (Ongena and Jacques 2008; Pieterse et al. 2014; Raaijmakers et al. 2010). Recent work with *Bacillus velezensis* has shown reduced *Fusarium* growth and mycotoxins (DON/NIV) and promising field performance, although efficacy is strain-, crop-, and context-dependent (Yeo et al. 2024; Zanon et al. 2024).

1.7 Metabolite Signaling-Based Maneuvering of the Soil Microbial Communities

1.7.1 Microbial Biostimulants Applications to Enhance Soil Health

Microbial and metabolite-based interventions can be used to reprogram rhizosphere functions, such as nutrient mobilization, stress buffering, disease suppression, and organic matter stabilization, without increasing synthetic inputs. Framed as metabolite steering, these strategies leverage substrates and signals to enrich functional guilds (e.g., P-mobilizers, siderophore producers, ACC-deaminase/IAA modulators), thereby improving soil health indicators, such as aggregate stability, nutrient-use efficiency (NUE), microbial carbon use efficiency (CUE), and long-term SOC formation (Colla et al. 2014; du Jardin 2015; Rouphael and Colla 2020; Sharma et al. 2013; Venturi and Keel 2016).

A plant biostimulant is a substance or microorganism applied to seeds, plants, or soils that stimulates natural processes to enhance NUE, abiotic stress tolerance, and/or quality, independent of nutrient content. This definition, now widely adopted in research and policy, also covers commercial products containing mixtures of these components (du Jardin 2015; Rouphael and Colla 2020). Based on this mechanism, biostimulants can be divided into two broad categories:

A. Microbial biostimulants (e.g., *Rhizobium*, *Azospirillum*, AMF, *Trichoderma*, and *Bacillus*) act through associative/symbiotic N fixation, organic acid/phosphatase-mediated P scavenging, phytohormone modulation (IAA, cytokinins, GAs), ACC-deaminase activity, siderophore-mediated Fe capture, volatile organic compounds (VOCs), induced systemic resistance (ISR), and competitive exclusion (Rouphael and Colla 2020; Venturi and Keel 2016).
B. Nonmicrobial biostimulants include protein hydrolysates, humic substances, seaweed extracts, and defined metabolites (amino acids, organic/phenolic acids, and peptides) that modulate both plant physiology and microbial behavior (Colla et al. 2017).

Microbial biostimulants are gaining popularity in agriculture because of their ability to enhance plant growth, improve nutrient uptake, and boost crop resilience. For example, rhizobia are nitrogen-fixing bacteria that form symbiotic relationships with leguminous plants, such as soybeans and peas, converting atmospheric nitrogen into a form that plants can utilize. Rhizobial biostimulants promote nitrogen fixation, leading to improved plant growth and soil fertility (Graham and Vance 2003). Similarly, *Azospirillum* stimulates root development, enhances nutrient uptake, and produces growth-promoting hormones that benefit plant growth. *Azospirillum* biostimulants have been used to increase crop productivity and improve soil health in various agricultural systems (Bashan and De-Bashan 2010). In contrast, mycorrhizal fungi form mutualistic associations with plant roots, enhancing nutrient uptake, water absorption, and overall plant health. Mycorrhizal biostimulants have been shown to improve crop yields, increase stress tolerance, and reduce the need for chemical fertilizers (Smith and Read 2008). *Trichoderma* is known for its ability to suppress plant pathogens, promote root growth, and enhance plant defense mechanisms. *Trichoderma* biostimulants are used to protect crops from diseases, improve soil structure, and stimulate plant growth in sustainable agricultural practices (Hermosa et al. 2012).

In addition to the use of microbes, the use of cell-free supernatants as biostimulants in agriculture is an emerging approach that harnesses the beneficial effects of microbial metabolites and signaling molecules to enhance plant growth, improve nutrient uptake, and boost stress tolerance. Cell-free supernatants are metabolite fractions obtained after the removal of microbial cells from a culture and contain a diverse array of bioactive compounds produced by microbes during growth (Colla et al. 2014; Sharma et al. 2019). For instance, cell-free supernatants from cultures of *Pseudomonas fluorescens*, a common plant growth-promoting rhizobacterium, have been shown to contain various metabolites that can enhance plant growth and induce systemic resistance against pathogens. These supernatants may contain compounds such as siderophores, indole-3-acetic acid (IAA), and antimicrobial peptides that promote root development, improve nutrient uptake, and protect plants from diseases (Sharma et al. 2013).

Similarly, *Trichoderma* species are well-known biocontrol agents that produce cell wall-degrading enzymes and antifungal compounds. Cell-free supernatants from *Trichoderma harzianum* cultures have been used as biostimulants to enhance plant growth, suppress soil-borne pathogens, and improve nutrient availability in the rhizosphere. These supernatants may contain enzymes such as chitinases and glucanases, as well as secondary metabolites that stimulate plant defense responses (Harman et al. 2004). Another example is *Azotobacter vinelandii*, a nitrogen-fixing bacterium that produces plant growth-promoting metabolites, such as vitamins, amino acids, and phytohormones. Cell-free supernatants from *Azotobacter* cultures have been applied as biostimulants to enhance nitrogen availability, promote root development, and improve plant growth in nutrient-deficient soils. These supernatants can also stimulate the growth of beneficial microorganisms in the rhizosphere, leading to a healthier soil microbiome (Bashan et al. 2013). Moreover, *Bacillus* is a beneficial soil bacterium that produces a range of bioactive compounds with plant growth-promoting and biocontrol properties. The cell-free supernatants of *Bacillus* contain antimicrobial peptides, enzymes, and volatile organic compounds that can enhance plant growth, induce systemic resistance, and improve nutrient uptake efficiency. These supernatants have been used as biostimulants to promote crop health and productivity in various agricultural systems (Valenzuela Ruiz et al. 2024).

Given the role of healthy soil microbial communities in determining various soil functions and setting the foundation for sustainable agriculture, researchers have explored the potential of manipulating microbial communities using metabolite applications (Olanrewaju et al. 2024). Metabolites can serve as carbon or energy sources for soil microbes, influencing the abundance of microbial populations that utilize these metabolites for growth and metabolism. For example, the addition of simple sugars, such as glucose, can stimulate the proliferation of sugar-consuming bacteria and fungi, leading to shifts in the microbial community structure. Furthermore, they mediate interactions between different microbial species in the soil. Some metabolites act as signaling molecules that trigger communication and cooperation among microbes, whereas others may have antimicrobial properties that influence competitive dynamics within microbial populations (Hibbing et al. 2010; de Wit 2015). Metabolite application also alters the functional diversity of soil microbial communities by selecting specific metabolic pathways or enzymatic activities (Philippot et al. 2013). Finally, certain metabolites can enhance plant-microbe interactions by promoting the growth of beneficial rhizosphere microbes that support plant growth and health. Plant-derived metabolites, such as root exudates, attract specific microbial populations to the rhizosphere, shaping the microbial community composition around plant roots (Bais et al. 2006).

The application of beneficial metabolites to plants not only acts as a positive growth signal for plants but also establishes rhizosphere microbial communities, leading to shifts in microbial diversity, functionality, and ecosystem processes. By providing microbial communities with targeted metabolites, researchers can selectively promote the growth of beneficial microbes, suppress pathogens, and enhance nutrient cycling in the soil. For instance, the impact of glucose addition on soil microbial communities and carbon cycling processes was evaluated by Kaiser et al. (2015). The addition of glucose can stimulate the growth of sugar-consuming microbes, leading to changes in microbial community composition and metabolic activities in the soil. Similarly, the addition of amino acids changes the microbial community because amino acids can serve as nitrogen sources for microbes, influencing the abundance of nitrogen-cycling bacteria (Ramirez et al. 2012), while they serve as positive growth signals for plants (Liu et al. 2021), leading to high crop yields (Hassan et al. 2020). Another study examined how organic acid addition changes soil microbial diversity and community structure along an altitudinal gradient (Siles and Margesin 2016), where, in addition to shifting microbial communities, they changed the soil pH, affecting the chemical and physical properties of the soil.

1.8 Conclusion

Soil health emerges from the tight coupling of microbial diversity, plant inputs, and environmental context. In this chapter, we showed how soil microorganisms govern nutrient recycling, stabilize soil organic matter via necromass and MAOM formation, and modulate plant resilience through hormone signaling, osmotic balance, and immune priming. These processes regulate greenhouse gas fluxes and drive the bioremediation of legacy pollutants, placing the soil microbiome at the center of climate-smart, low-input agriculture. Pragmatically, two levers recur: (i) fostering conditions that raise microbial carbon-use efficiency (residue retention, reduced disturbance, and balanced nutrition) and (ii) deploying beneficial guilds, via native consortia, AMF, *Trichoderma*, and other PGPM, to convert short-term fertility into durable soil function. The next step is the deliberate engineering of rhizosphere function by pairing live inoculants with metabolite steering (defined substrates and signals or cell-free supernatants) to enrich target associations, P-mobilizers, siderophore producers, and ACC deaminase/IAA modulators while safeguarding ecological balance. Real progress will hinge on multi-site, multi-year trials that report crop × soil × dose × formulation; QA/QC for composition-defined products; and decision tools that integrate soil diagnostics (pH, Olsen-P, SOC fractions, and enzyme activities) with management. When developed alongside farmers, microbiome-informed, nature-based approaches raise productivity, lower input demand, and fortify resilient farming systems, positioning soil as a long-term asset for food security and climate action.

References

Acharya BR, Assmann SM (2009) Hormone interactions in stomatal function. Plant Mol Biol 69:451–462

Ahmad HM, Fiaz S, Hafeez S, Zahra S, Shah AN, Gul B, Aziz O, Mahmood-Ur-Rahman, Fakhar A, Rafique M, Chen Y, Yang SH, Wang X (2022) Plant growth-promoting rhizobacteria eliminate the effect of drought stress in plants: a review. Front Plant Sci 13:875774. https://doi.org/10.3389/fpls.2022.875774

Alengebawy A, Abdelkhalek ST, Qureshi SR, Wang MQ (2021) Heavy metals and pesticides toxicity in agricultural soil and plants: ecological risks and human health implications. Toxics 9:42. https://doi.org/10.3390/toxics9030042

Ambreen S, Yasmin A (2021) Novel degradation pathways for Chlorpyrifos and 3,5,6-Trichloro-2-pyridinol degradation by bacterial strain *Bacillus thuringiensis* MB497 isolated from agricultural fields of Mianwali, Pakistan. Pestic Biochem Physiol 172:104750. https://doi.org/10.1016/j.pestbp.2020.104750

Angst G, Angst Š, Frouz J, Jabinski S, Jílková V, Kukla J, Li M, Meador TB, Angel R (2024) Stabilized microbial necromass in soil is more strongly coupled with microbial diversity than the bioavailability of plant inputs. Soil Biol Biochem 190:109323. https://doi.org/10.1016/j.soilbio.2024.109323

Azizi S, Tabari M, Abad ARFN, Ammer C, Guidi L, Bader MKF (2022) Soil inoculation with beneficial microbes buffers negative drought effects on biomass, nutrients, and water relations of common myrtle. Front Plant Sci 13. https://doi.org/10.3389/fpls.2022.892826

Bais HP, Weir TL, Perry LG, Gilroy S, Vivanco JM (2006) The role of root exudates in rhizosphere interactions with plants and other organisms. Annu Rev Plant Biol 57:233–266

Bashan Y, De-Bashan LE (2010) How the plant growth-promoting bacterium *Azospirillum* promotes plant growth: a critical assessment. Adv Agron 108:77–136

Bashan Y, Kamnev AA, de-Bashan LE (2013) Tricalcium phosphate is inappropriate as a universal selection factor for isolating and testing phosphate-solubilizing bacteria that enhance plant growth: a proposal for an alternative procedure. Biol Fertil Soils 49:465–479

Becker JN, Holz M (2021) Hot or not? Connecting rhizosphere hotspots to total soil respiration. Plant Soil 464:489–499. https://doi.org/10.1007/s11104-021-04963-4

Belimov AA, Dodd IC, Hontzeas N, Theobald JC, Safronova VI, Davies WJ (2009) Rhizosphere bacteria containing 1-aminocyclopropane-1-carboxylate deaminase increase yield of plants grown in drying soil via both local and systemic hormone signalling. New Phytol 181:413–423

Beneduzi A, Ambrosini A, Passaglia LMP (2012) Plant growth-promoting rhizobacteria (PGPR): their potential as antagonists and biocontrol agents. Genet Mol Biol 35:1044–1051. https://doi.org/10.1590/s1415-47572012000600020

Canarini A, Kaiser C, Merchant A, Richter A, Wanek W (2019) Root exudation of primary metabolites: mechanisms and their roles in plant responses to environmental stimuli. Front Plant Sci 10:157. https://doi.org/10.3389/fpls.2019.00157

Castro-Camba R, Sánchez C, Vidal N, Vielba JM (2022) Interactions of gibberellins with phytohormones and their role in stress responses. Horticulturae 8:241. https://doi.org/10.3390/horticulturae8030241

Chang WS, Van De Mortel M, Nielsen L, Nino de Guzman G, Li X, Halverson LJ (2007) Alginate production by *Pseudomonas putida* creates a hydrated microenvironment and contributes to biofilm architecture and stress tolerance under water-limiting conditions. J Bacteriol 189. https://doi.org/10.1128/jb.00727-07

Chen X, Lu Y, Liu X, Gu Y, Li F (2025) Trichoderma: dual roles in biocontrol and plant growth promotion. Microorganisms 13:1840. https://doi.org/10.3390/microorganisms13081840

Coleman NV, Rich DJ, Tang FHM, Vervoort RW, Maggi F (2020) Biodegradation and abiotic degradation of trifluralin: a commonly used herbicide with a poorly understood environmental fate. Environ Sci Technol 54:10399–10410. https://doi.org/10.1021/acs.est.0c02070

Colla G, Rouphael Y, Canaguier R, Svecova E, Cardarelli M (2014) Biostimulant action of a plant-derived protein hydrolysate produced through enzymatic hydrolysis. Front Plant Sci 5:110917

Colla G, Hoagland L, Ruzzi M, Cardarelli M, Bonini P, Canaguier R, Rouphael Y (2017) Biostimulant action of protein hydrolysates: unraveling their effects on plant physiology and microbiome. Front Plant Sci 8:2202. https://doi.org/10.3389/fpls.2017.02202

Cotrufo MF, Lavallee JM (2022) Soil organic matter formation, persistence, and functioning: a synthesis of current understanding to inform its conservation and regeneration. In: Advances in agronomy. Elsevier, pp 1–66. https://doi.org/10.1016/bs.agron.2021.11.002

De Ollas C, Arbona V, Gómez-Cadenas A, Dodd IC (2018) Attenuated accumulation of jasmonates modifies stomatal responses to water deficit. J Exp Bot 69:2103–2116

de Wit PJ (2015) Plant pathogenic fungi and oomycetes. In: Principles of plant-microbe interactions: microbes for sustainable agriculture. Springer, pp 79–90

del-Canto A, Sanz-Saez Á, Sillero-Martínez A, Mintegi E, Lacuesta M (2023) Selected indigenous drought tolerant rhizobium strains as promising biostimulants for common bean in Northern Spain. Front Plant Sci 14. https://doi.org/10.3389/fpls.2023.1046397

Domeignoz-Horta LA, Pold G, Liu XJA, Frey SD, Melillo JM, DeAngelis KM (2020) Microbial diversity drives carbon use efficiency in a model soil. Nat Commun 11:3684. https://doi.org/10.1038/s41467-020-17502-z

Doran JW, Zeiss MR (2000) Soil health and sustainability: managing the biotic component of soil quality. Appl Soil Ecol 15:3–11

du Jardin P (2015) Plant biostimulants: definition, concept, main categories and regulation. Sci Hortic 196:3–14. https://doi.org/10.1016/j.scienta.2015.09.021

Egea I, Albaladejo I, Meco V, Morales B, Sevilla A, Bolarin MC, Flores FB (2018) The drought-tolerant Solanum pennellii regulates leaf water loss and induces genes involved in amino acid and ethylene/jasmonate metabolism under dehydration. Sci Rep 8:2791. https://doi.org/10.1038/s41598-018-21193-2

Geng Y, Yuan Y, Miao Y, Zhi J, Huang M, Zhang Y, Wang H, Shen Q, Zou J, Li S (2021) Decreased nitrous oxide emissions associated with functional microbial genes under bio-organic fertilizer application in vegetable fields. Pedosphere 31:279–288. https://doi.org/10.1016/S1002-0160(20)60075-3

German MA, Burdman S, Okon Y, Kigel J (2000) Effects of *Azospirillum brasilense* on root morphology of common bean (*Phaseolus vulgaris* L.) under different water regimes. Biol Fertil Soils 32:259–264. https://doi.org/10.1007/S003740000245

Gill SS, Gill R, Trivedi DK, Anjum NA, Sharma KK, Ansari MW, Ansari AA, Johri AK, Prasad R, Pereira E (2016) *Piriformospora indica*: potential and significance in plant stress tolerance. Front Microbiol 7:332. https://doi.org/10.3389/fmicb.2016.00332

Glick BR (2012) Plant growth-promoting bacteria: mechanisms and applications. Scientifica 2012:1

González-Teuber M, Urzúa A, Plaza P, Bascuñán-Godoy L (2018) Effects of root endophytic fungi on response of *Chenopodium quinoa* to drought stress. Plant Ecol 219:231–240

Graham PH, Vance CP (2003) Legumes: importance and constraints to greater use. Plant Physiol 131:872–877

Haider FU, Liqun C, Coulter JA, Cheema SA, Wu J, Zhang R, Wenjun M, Farooq M (2021) Cadmium toxicity in plants: impacts and remediation strategies. Ecotoxicol Environ Saf 211:111887. https://doi.org/10.1016/j.ecoenv.2020.111887

Harman GE, Howell CR, Viterbo A, Chet I, Lorito M (2004) *Trichoderma* species – opportunistic, avirulent plant symbionts. Nat Rev Microbiol 2:43–56

Hassan MU, Islam MM, Wang R, Guo J, Luo H, Chen F, Li X (2020) Glutamine application promotes nitrogen and biomass accumulation in the shoot of seedlings of the maize hybrid ZD958. Planta 251:66. https://doi.org/10.1007/s00425-020-03363-9

Hermosa R, Viterbo A, Chet I, Monte E (2012) Plant-beneficial effects of *Trichoderma* and of its genes. Microbiology 158:17–25

Hibbing ME, Fuqua C, Parsek MR, Peterson SB (2010) Bacterial competition: surviving and thriving in the microbial jungle. Nat Rev Microbiol 8:15–25

Hone H, Mann R, Yang G, Kaur J, Tannenbaum I, Li T, Spangenberg G, Sawbridge T (2021) Profiling, isolation and characterisation of beneficial microbes from the seed microbiomes of drought tolerant wheat. Sci Rep 11. https://doi.org/10.1038/s41598-021-91351-8

Houlton BZ, Almaraz M, Aneja V, Austin AT, Bai E, Cassman KG, Compton JE, Davidson EA, Erisman JW, Galloway JN, Gu B, Yao G, Martinelli LA, Scow K, Schlesinger WH, Tomich TP, Wang C, Zhang X (2019) A world of cobenefits: solving the global nitrogen challenge. Earth's Future 7:865–872. https://doi.org/10.1029/2019EF001222

Intergovernmental Panel on Climate Change (IPCC) (2023) Climate change 2022 – mitigation of climate change: working group III contribution to the sixth assessment report of the intergovernmental panel on climate change. Cambridge University Press, Cambridge. https://doi.org/10.1017/9781009157926

Jamil F, Mukhtar H, Fouillaud M, Dufossé L (2022) Rhizosphere signaling: insights into plant–rhizomicrobiome interactions for sustainable agronomy. Microorganisms 10:899. https://doi.org/10.3390/microorganisms10050899

Jayne B, Quigley M (2014) Influence of arbuscular mycorrhiza on growth and reproductive response of plants under water deficit: a meta-analysis. Mycorrhiza 24:109–119. https://doi.org/10.1007/s00572-013-0515-x

Ji H, Zhang M, Huang C, Lin W, Lu Y, Wang P, Dong B, He B, Wu B, Guo L (2024) The impact of *Piriformospora indica* on plant heat and drought tolerance. Front Plant Sci 15:1479561. https://doi.org/10.3389/fpls.2024.1479561

Jiao S, Chen W, Wei G (2019) Resilience and assemblage of soil microbiome in response to chemical contamination combined with plant growth. Appl Environ Microbiol 85:e02523–e02518. https://doi.org/10.1128/AEM.02523-18

Kaiser C, Kilburn MR, Clode PL, Fuchslueger L, Koranda M, Cliff JB, Solaiman ZM, Murphy DV (2015) Exploring the transfer of recent plant photosynthates to soil microbes: Mycorrhizal pathway vs direct root exudation. New Phytol 205:1537–1551

Kamran M, Imran QM, Ahmed MB, Falak N, Khatoon A, Yun BW (2022) Endophyte-mediated stress tolerance in plants: a sustainable strategy to enhance resilience and assist crop improvement. Cells 11:3292. https://doi.org/10.3390/cells11203292

Kandasamy S, Weerasuriya N, White JF, Patterson G, Lazarovits G (2021) Soil's physical and nutritional balance is essential for establishing a healthy microbiome. In: White J, Kumar A, Droby S (eds) Microbiome stimulants for crops. Woodhead Publishing, pp 381–404. https://doi.org/10.1016/B978-0-12-822122-8.00004-2

Kästner M, Miltner A, Thiele-Bruhn S, Liang C (2021) Microbial necromass in soils—linking microbes to soil processes and carbon turnover. Front Environ Sci 9. https://doi.org/10.3389/fenvs.2021.756378

Kazan K (2015) Diverse roles of jasmonates and ethylene in abiotic stress tolerance. Trends Plant Sci 20:219–229

Khan MA, Hamayun M, Asaf S, Khan M, Yun BW, Kang SM, Lee IJ (2021) Rhizospheric bacillus spp. rescues plant growth under salinity stress via regulating gene expression, endogenous hormones, and antioxidant system of *Oryza sativa* L. Front Plant Sci 12. https://doi.org/10.3389/fpls.2021.665590

Kijne JW (2006) Abiotic stress and water scarcity: identifying and resolving conflicts from plant level to global level. Field Crop Res 97:3–18

Kour D, Rana KL, Yadav AN, Sheikh I, Kumar V, Dhaliwal HS, Saxena AK (2020) Amelioration of drought stress in foxtail millet (*Setaria italica* L.) by P-solubilizing drought-tolerant microbes with multifarious plant growth promoting attributes. Environ Sustain 3:23–34. https://doi.org/10.1007/s42398-020-00094-1

Kumar G, Teli B, Mukherjee A, Bajpai R, Sarma BK (2019) Secondary metabolites from cyanobacteria: a potential source for plant growth promotion and disease management. In: Singh HB, Keswani C, Reddy MS, Sansinenea E, García-Estrada C (eds) Secondary metabolites of plant growth promoting rhizomicroorganisms: discovery and applications. Springer, pp 239–252. https://doi.org/10.1007/978-981-13-5862-3_12

Kuzyakov Y, Razavi BS (2019) Rhizosphere size and shape: temporal dynamics and spatial stationarity. Soil Biol Biochem 135:343–360. https://doi.org/10.1016/j.soilbio.2019.05.011

Lal R (2014) Soil conservation and ecosystem services. Int Soil Water Conserv Res 2:36–47. https://doi.org/10.1016/S2095-6339(15)30021-6

Lara-Moreno A, Morillo E, Merchán F, Madrid F, Villaverde J (2022) Bioremediation of a trifluralin contaminated soil using bioaugmentation with novel isolated bacterial strains and cyclodextrin. Sci Total Environ 840:156695. https://doi.org/10.1016/j.scitotenv.2022.156695

Lata R, Gond SK (2019) Plant growth-promoting microbes for abiotic stress tolerance in plants. In: Kumar A, Singh AK, Choudhary KK (eds) Role of plant growth promoting microorganisms in sustainable agriculture and nanotechnology. Woodhead Publishing/Elsevier, pp 89–105

Lawson T, Emmerson R, Battle M, Pullin J, Wall S, Hofmann TA (2022) Carbon fixation. In: Ruban A, Foyer CH, Murchie EH (eds) Photosynthesis in action. Academic, pp 31–58. https://doi.org/10.1016/B978-0-12-823781-6.00008-3

Lehmann J, Kleber M (2015) The contentious nature of soil organic matter. Nature 528:60–68

Lehmann J, Bossio DA, Kögel-Knabner I, Rillig MC (2020) The concept and future prospects of soil health. Nat Rev Earth Environ 1:544–553. https://doi.org/10.1038/s43017-020-0080-8

Liu Y, Duan X, Zhao X, Ding W, Wang Y, Xiong Y (2021) Diverse nitrogen signals activate convergent ROP2-TOR signaling in *Arabidopsis*. Dev Cell 56:1283–1295. https://doi.org/10.1016/j.devcel.2021.03.022

Lynn TM, Ge T, Yuan H, Wei X, Wu X, Xiao K, Kumaresan D, Yu SS, Wu J, Whiteley AS (2017) Soil carbon-fixation rates and associated bacterial diversity and abundance in three natural ecosystems. Microb Ecol 73:645–657. https://doi.org/10.1007/s00248-016-0890-x

Ma Y, Ankit, Tiwari J, Bauddh K (2022) Plant-mycorrhizal fungi interactions in phytoremediation of geogenic contaminated soils. Front Microbiol 13. https://doi.org/10.3389/fmicb.2022.843415

Mahmood T, Khalid S, Abdullah M, Ahmed Z, Shah MKN, Ghafoor A, Du X (2019) Insights into drought stress signaling in plants and the molecular genetic basis of cotton drought tolerance. Cells 9:105. https://doi.org/10.3390/cells9010105

Mathur P, Roy S (2021) Insights into the plant responses to drought and decoding the potential of root associated microbiome for inducing drought tolerance. Physiol Plant 172:1016–1029. https://doi.org/10.1111/ppl.13338

Mayak S, Tirosh T, Glick BR (2004) Plant growth-promoting bacteria that confer resistance to water stress in tomatoes and peppers. Plant Sci 166:525–530

Meel S, Saharan BS (2025) Microbial warfare against nematodes: a review of nematicidal compounds for horticulture, environment, and biotechnology. Microbe 9:100557. https://doi.org/10.1016/j.microb.2025.100557

Mehmood N, Saeed M, Zafarullah S, Hyder S, Rizvi ZF, Gondal AS, Jamil N, Iqbal R, Ali B, Ercisli S, Kupe M (2023) Multifaceted impacts of plant-beneficial *Pseudomonas* spp. in managing various plant diseases and crop yield improvement. ACS Omega 8:22296–22315. https://doi.org/10.1021/acsomega.3c00870

Murali M, Gowtham HG, Singh SB, Shilpa N, Aiyaz M, Niranjana SR, Amruthesh KN (2021) Bioprospecting of ACC deaminase producing rhizobacteria towards sustainable agriculture: a special emphasis on abiotic stress in plants. Appl Soil Ecol 168:104142. https://doi.org/10.1016/j.apsoil.2021.104142

Nemeskéri E, Helyes L (2019) Physiological responses of selected vegetable crop species to water stress. Agronomy 9:447. https://doi.org/10.3390/agronomy9080447

Ni H, Jing X, Xiao X, Zhang N, Wang X, Sui Y, Sun B, Liang Y (2021) Microbial metabolism and necromass mediated fertilization effect on soil organic carbon after long-term community incubation in different climates. ISME J 15:2561–2573. https://doi.org/10.1038/s41396-021-00950-w

Olanrewaju OS, Glick BR, Babalola OO (2024) Metabolomics-guided utilization of beneficial microbes for climate-resilient crops. Curr Opin Chem Biol 79:102427. https://doi.org/10.1016/j.cbpa.2024.102427

Omomowo OI, Babalola OO (2019) Bacterial and fungal endophytes: tiny giants with immense beneficial potential for plant growth and sustainable agricultural productivity. Microorganisms 7:481

Ongena M, Jacques P (2008) Bacillus lipopeptides: versatile weapons for plant disease biocontrol. Trends Microbiol 16:115–125. https://doi.org/10.1016/j.tim.2007.12.009

Palma L, Muñoz D, Berry C, Murillo J, Caballero P (2014) *Bacillus thuringiensis* toxins: an overview of their biocidal activity. Toxins 6:3296–3325. https://doi.org/10.3390/toxins6123296

Peng Y, Biswas A, Adamowski JF, Zhang X, Li Y, Li L, Peng Y, Cao J (2025) Sources and stability of particulate organic matter (POM) and mineral-associated organic matter (MAOM) on the loess plateau: implications for soil carbon management. Ecol Indic 179:114162. https://doi.org/10.1016/j.ecolind.2025.114162

Philippot L, Raaijmakers JM, Lemanceau P, Van Der Putten WH (2013) Going back to the roots: the microbial ecology of the rhizosphere. Nat Rev Microbiol 11:789–799

Pieterse CMJ, Zamioudis C, Berendsen RL, Weller DM, Wees SCMV, Bakker PAHM (2014) Induced systemic resistance by beneficial microbes. Annu Rev Phytopathol 52:347–375. https://doi.org/10.1146/annurev-phyto-082712-102340

Poveda J (2020) *Trichoderma parareesei* favors the tolerance of rapeseed (*Brassica napus* L.) to salinity and drought due to a chorismate mutase. Agronomy 10:118

Prudent M, Salon C, Souleimanov A, Emery RJN, Smith DL (2015) Soybean is less impacted by water stress using *Bradyrhizobium japonicum* and thuricin-17 from *Bacillus thuringiensis*. Agron Sustain Dev 35:749–757

Putra B, Warly L, Evitayani E, Utama BP (2022) The role of arbuscular mycorrhizal fungi in phytoremediation of heavy metals and their effect on the growth of *Pennisetum purpureum* cv. Mott on gold mine tailings in Muara Bungo, Jambi. Biodiversitas 23. https://doi.org/10.13057/biodiv/d230151

Raaijmakers JM, De Bruijn I, Nybroe O, Ongena M (2010) Natural functions of lipopeptides from *Bacillus* and *Pseudomonas*: more than surfactants and antibiotics. FEMS Microbiol Rev 34:1037–1062. https://doi.org/10.1111/j.1574-6976.2010.00221.x

Ramirez KS, Craine JM, Fierer N (2012) Consistent effects of nitrogen amendments on soil microbial communities and processes across biomes. Glob Change Biol 18:1918–1927

Rasconi S, Grami B, Niquil N, Jobard M, Sime-Ngando T (2014) *Parasitic chytrids* sustain zooplankton growth during inedible algal bloom. Front Microbiol 5. https://doi.org/10.3389/fmicb.2014.00229

Riboni M, Robustelli Test A, Galbiati M, Tonelli C, Conti L (2016) ABA-dependent control of GIGANTEA signalling enables drought escape via up-regulation of flowering locust in *Arabidopsis thaliana*. J Exp Bot 67:6309–6322

Riedo J, Herzog C, Banerjee S, Fenner K, Walder F, van der Heijden MGA, Bucheli TD (2022) Concerted evaluation of pesticides in soils of extensive grassland sites and organic and conventional vegetable fields facilitates the identification of major input processes. Environ Sci Technol 56:13686–13695. https://doi.org/10.1021/acs.est.2c02413

Rodríguez H, Fraga R (1999) Phosphate solubilizing bacteria and their role in plant growth promotion. Biotechnol Adv 17:319–339. https://doi.org/10.1016/S0734-9750(99)00014-2

Rouphael Y, Colla G (2020) Editorial: biostimulants in agriculture. Front Plant Sci 11. https://doi.org/10.3389/fpls.2020.00040

Roussel JM, Perrier C, Erkinaro J, Niemelä E, Cunjak RA, Huteau D, Riera P (2014) Stable isotope analyses on archived fish scales reveal the long-term effect of nitrogen loads on carbon cycling in rivers. Glob Change Biol 20:523–530. https://doi.org/10.1111/gcb.12293

Sasse J, Martinoia E, Northen T (2018) Feed your friends: do plant exudates shape the root microbiome? Trends Plant Sci 23:25–41. https://doi.org/10.1016/j.tplants.2017.09.003

Schneider JR, Caverzan A, Chavarria G (2019) Water deficit stress, ROS involvement, and plant performance. Arch Agron Soil Sci 65:1160–1181. https://doi.org/10.1080/03650340.2018.1556789

Senizza B, Araniti F, Lewin S, Wende S, Kolb S, Lucini L (2023) *Trichoderma* spp.-mediated mitigation of heat, drought, and their combination on the *Arabidopsis thaliana* holobiont: a metabolomics and metabarcoding approach. Front Plant Sci 14. https://doi.org/10.3389/fpls.2023.1190304

Sharma SB, Sayyed RZ, Trivedi MH, Gobi TA (2013) Phosphate solubilizing microbes: sustainable approach for managing phosphorus deficiency in agricultural soils. Springerplus 2:1–14

Sharma A, Shahzad B, Kumar V, Kohli SK, Sidhu GPS, Bali AS, Handa N, Kapoor D, Bhardwaj R, Zheng B (2019) Phytohormones regulate accumulation of osmolytes under abiotic stress. Biomolecules 9:285

Siles JA, Margesin R (2016) Abundance and diversity of bacterial, archaeal, and fungal communities along an altitudinal gradient in alpine forest soils: what are the driving factors? Microb Ecol 72:207–220

Smith S, Read D (2008) Mycorrhizal symbiosis. Academic, New York

Sofi PA, Saba I, Amin Z (2017) Root architecture and rhizobial inoculation in relation to drought stress response in common bean (*Phaseolus vulgaris* L.). J Appl Nat Sci 9:502–507

Sofi PA, Baba ZA, Hamid B, Meena RS (2018) Harnessing soil rhizobacteria for improving drought resilience in legumes. In: Meena RS, Das A, Yadav GS, Lal R (eds) Legumes for soil health and sustainable management. Springer, Singapore, pp 235–275. https://doi.org/10.1007/978-981-13-0253-4_8

Steiner F, da Silva Oliveira CE, Zoz T, Zuffo AM, de Freitas RS (2020) Co-inoculation of common bean with *Rhizobium* and *Azospirillum* enhance the drought tolerance. Russ J Plant Physiol 67:923–932. https://doi.org/10.1134/S1021443720050167

Sun B, Gu L, Bao L, Zhang S, Wei Y, Bai Z, Zhuang G, Zhuang X (2020) Application of biofertilizer containing *Bacillus subtilis* reduced the nitrogen loss in agricultural soil. Soil Biol Biochem 148:107911. https://doi.org/10.1016/j.soilbio.2020.107911

Tao F, Huang Y, Hungate BA, Manzoni S, Frey SD, Schmidt MWI, Reichstein M, Carvalhais N, Ciais P, Jiang L, Lehmann J, Wang YP, Houlton BZ, Ahrens B, Mishra U, Hugelius G, Hocking TD, Lu X, Shi Z, Luo Y, Tao F et al (2023) Microbial carbon use efficiency promotes global soil carbon storage. Nature 618:981–985. https://doi.org/10.1038/s41586-023-06042-3

Timmusk S, Nevo E (2011) Plant root associated biofilms: perspectives for natural product mining. In: Maheshwari DK (ed) Bacteria in agrobiology: plant nutrient management. Springer, Berlin, pp 285–300

Tittabutr P, Piromyou P, Longtonglang A, Noisa-Ngiam R, Boonkerd N, Teaumroong N (2013) Alleviation of the effect of environmental stresses using co-inoculation of mungbean by *Bradyrhizobium* and rhizobacteria containing stress-induced ACC deaminase enzyme. Soil Sci Plant Nutr 59:559–571

Tyśkiewicz R, Nowak A, Ozimek E, Jaroszuk-Ściseł J (2022) *Trichoderma*: the current status of its application in agriculture for the biocontrol of fungal phytopathogens and stimulation of plant growth. Int J Mol Sci 23:2329. https://doi.org/10.3390/ijms23042329

United Nations (2015) The Paris Agreement. United Nations. https://www.un.org/en/climatechange/paris-agreement. Accessed 23 Apr 2026

Valenzuela Ruiz V, Gándara-Ledezma A, Villarreal-Delgado MF, Villa-Rodríguez ED, Parra-Cota FI, Santoyo G, Gómez-Godínez LJ, Cira Chávez LA, de los Santos-Villalobos S (2024) Regulation, biosynthesis, and extraction of *Bacillus*-derived lipopeptides and its implications in biological control of phytopathogens. Stress 4:107–132

Venturi V, Keel C (2016) Signaling in the rhizosphere. Trends Plant Sci 21:187–198. https://doi.org/10.1016/j.tplants.2016.01.005

Verma JP, Yadav J, Tiwari KN, Jaiswal DK (2014) Evaluation of plant growth promoting activities of microbial strains and their effect on growth and yield of chickpea (*Cicer arietinum* L.) in India. Soil Biol Biochem 70:33–37. https://doi.org/10.1016/j.soilbio.2013.12.001

Wang Y, Narayanan M, Shi X, Chen X, Li Z, Natarajan D, Ma Y (2022) Plant growth-promoting bacteria in metal-contaminated soil: current perspectives on remediation mechanisms. Front Microbiol 13:966226. https://doi.org/10.3389/fmicb.2022.966226

Whalen ED, Grandy AS, Geyer KM, Morrison EW, Frey SD (2024) Microbial trait multifunctionality drives soil organic matter formation potential. Nat Commun 15:10209. https://doi.org/10.1038/s41467-024-53947-2

Wu S, Zhuang G, Bai Z, Cen Y, Xu S, Sun H, Han XG, Zhuang X (2017) Mitigation of nitrous oxide emissions from acidic soils by *Bacillus amyloliquefaciens*, a plant growth-promoting bacterium. Glob Change Biol 24. https://doi.org/10.1111/gcb.14025

Wu Y, Ma L, Liu Q, Sikder MM, Vestergård M, Zhou K, Wang Q, Yang X, Feng Y (2020) *Pseudomonas fluorescens* promote photosynthesis, carbon fixation and cadmium phytoremediation of hyperaccumulator *Sedum alfredii*. Sci Total Environ 726:138554. https://doi.org/10.1016/j.scitotenv.2020.138554

Wu S, Shi Z, Chen X, Gao J, Wang X (2022) Arbuscular mycorrhizal fungi increase crop yields by improving biomass under rainfed condition: a meta-analysis. PeerJ 10:e12861. https://doi.org/10.7717/peerj.12861

Yang P, Zhang P, Li B, Hu T (2013) Effect of nodules on dehydration response in alfalfa (*Medicago sativa* L.). Environ Exp Bot 86:29–34. https://doi.org/10.1016/j.envexpbot.2011.05.012

Yeo YJ, Park AR, Vuong BS, Kim JC (2024) Biocontrol of *Fusarium* head blight in rice using *Bacillus velezensis* JCK-7158. Front Microbiol 15:1358689. https://doi.org/10.3389/fmicb.2024.1358689

Yin Q, He K, Collins G, De Vrieze J, Wu G (2024) Microbial strategies driving low concentration substrate degradation for sustainable remediation solutions. NPJ Clean Water 7:52. https://doi.org/10.1038/s41545-024-00348-z

Yu W, Sheng L, Wang X, Tang X, Yuan J, Luo W (2025) Soil microbial carbon use efficiency in natural terrestrial ecosystems. Biology 14:348. https://doi.org/10.3390/biology14040348

Yuan H, Ge T, Chen C, O'Donnell AG, Wu J (2012) Significant role for microbial autotrophy in the sequestration of soil carbon. Appl Environ Microbiol 78:2328–2336. https://doi.org/10.1128/AEM.06881-11

Zanon MSA, Cavaglieri LR, Palazzini JM, Chulze SN, Chiotta ML (2024) *Bacillus velezensis* RC218 and emerging biocontrol agents against *Fusarium graminearum* and *Fusarium poae* in barley: in vitro, greenhouse and field conditions. Int J Food Microbiol 413:110580. https://doi.org/10.1016/j.ijfoodmicro.2024.110580

Zewdie B, Tack AJM, Ayalew B, Adugna G, Nemomissa S, Hylander K (2021) Temporal dynamics and biocontrol potential of a hyperparasite on coffee leaf rust across a landscape in Arabica coffee's native range. Agric Ecosyst Environ 311:107297. https://doi.org/10.1016/j.agee.2021.107297

Zhang Z, Zhang J, Xu G, Zhou L, Li Y (2019) Arbuscular mycorrhizal fungi improve the growth and drought tolerance of *Zenia insignis* seedlings under drought stress. New For 50:593–604

Zhang H, Zhang J, Tang S, Deng Z, Makar RS, Yao L, Han H (2025) Exopolysaccharide-producing strains alter heavy metal fates and bacterial communities in soil aggregates to reduce metal uptake by pakchoi. Front Microbiol 16. https://doi.org/10.3389/fmicb.2025.1595142

Zhang Y, Du H, Xu F, Ding Y, Gui Y, Zhang J, Xu W (2020) Root-bacteria associations boost rhizosheath formation in moderately dry soil through ethylene responses. Plant Physiol 183(2):780–792. https://doi.org/10.1104/pp.19.01020

Zhao Q, Bell S, Kukkadapu R, Richardson J, Cliff J, Bowden M, Leichty S, Hofmockel KS (2025) Accumulation of soil microbial necromass controlled by microbe–mineral interactions. Environ Sci Technol 59:17558–17570. https://doi.org/10.1021/acs.est.5c01482

Chapter 2
Rhizomicrobiome Engineering: A New Tool for Soil and Plant Health

Rhea Aqueel, Samina Mehnaz, and Kauser Abdulla Malik

Abstract Sustainable agricultural practices utilize plant-beneficial microbiota that reside in the rhizospheric part of the soil. These bacteria are known for their plant growth-promoting (PGP) and disease-suppressive properties. Although much remains unknown about the resident microbial communities of the soil rhizosphere, next-generation sequencing (NGS) technologies and metagenomics have made it possible to decipher the diversity patterns and metabolic functions of these microbes. Unraveling these mechanisms will enable the formulation of better agricultural solutions to promote plant health. They can be used in conjunction with synthetically developed microbial formulations, nanotechnology, and other soil-agricultural practices to engineer the rhizobiome to enhance crop productivity with minimal chemical fertilizer usage.

Keywords Rhizosphere · Sustainable agriculture · Biofertilizers · Rhizobiome · Metagenomics · Soil engineering

2.1 Introduction

Soil is vital to the environment as it facilitates the anchoring of plant roots and provides essential elements, such as oxygen, nutrients, and water. Furthermore, soil functions as a "living system" that hosts microorganisms that interact with plant

R. Aqueel
Kauser Abdulla Malik School of Life Sciences, Forman Christian College
(A Chartered University), Lahore, Pakistan

Department of Biological Sciences, Royal Holloway University of London,
Egham, Surrey, United Kingdom

S. Mehnaz · K. A. Malik (✉)
Kauser Abdulla Malik School of Life Sciences, Forman Christian College
(A Chartered University), Lahore, Pakistan
e-mail: kausermalik@fccollege.edu.pk

S. Mehnaz, I. Shahid (eds.), *Frontiers in Rhizoengineering*, Microorganisms for Sustainability 65, https://doi.org/10.1007/978-981-92-3665-7_2

roots. The diverse array of genomes present in the rhizosphere was previously collectively referred to as the "rhizosphere microbiome" (Ali 2017). The rhizosphere is a comprehensive collection of microorganisms that provides the genetic makeup and cellular machinery required for replication, which is commonly known as the "rhizomicrobiome/rhizobiome" (Olanrewaju et al. 2019).

There are 1×10^9 bacterial cells and 1×10^6 distinct bacterial taxa per gram of soil (Bulgarelli et al. 2013). Microbes occur in the plant's surroundings as plant growth-promoting bacteria (PGPB), which are beneficial to the plant, as plant pathogens that harm the plant, or in a neutral relationship with no effect at all (Aqueel et al. 2024a, b). Numerous studies have shown that the microbiota residing in the rhizosphere are distinct from those present in the endosphere or bulk soil (Aqueel et al. 2023; Bakker et al. 2015; Elvia et al. 2021; Fernández-González et al. 2019; Gottel et al. 2011; Ha et al. 2021).

Plant health is greatly affected by the microbes that reside in the surrounding environment. Therefore, it is important to study these microorganisms extensively. Next-generation sequencing techniques have made it possible to decipher the composition, dynamics, and functions of these microbial communities (Richter-Heitmann et al. 2016). Moreover, in the current world of regenerative agriculture, a diverse crop rotation system is important (Giller et al. 2021). This diversity helps the rhizobiome function better, leading to improved crop productivity.

At the dawn of new studies, an interesting question arises: Is the rhizosphere limited to plants in the soil? Recent studies suggest that the rhizosphere extends beyond the soil and is present in the surroundings of the aerial roots (Nazari 2024). This is also supported by the fact that a portion of the rhizospheric microbiome is completely dependent on host genetics, regardless of soil conditions (Xun et al. 2024). In this chapter, we delve deep into the structure of the rhizospheric microbiome and how the interplay of dynamics can be engineered to improve crop health.

2.2 Structure and Dynamics of Rhizosphere Microbiome

The rhizosphere is the most important part of the soil microbiome within 3–5 mm of soil surrounding the roots (Gregory 2006). This unique and dynamic zone is where intricate interactions between plant roots and microorganisms occur. Within this microscopic realm, a diverse community of bacteria, fungi, and other microbes thrives, forming a complex network that supports plant health and growth. The rhizosphere plays a vital role in sustaining the plant ecosystem, as it not only facilitates nutrient uptake and enhances soil structure but is also profoundly influenced by root activities. This relationship is fundamental to the overall well-being of the plant and its surrounding environment (Gupta et al. 2024).

The rhizosheath can be referred to as the "inner rhizosphere soil" as it provides a bridge between the plant root and the soil. It reduces the air gaps between the roots and soil, improving water transport at the root-soil interface. It also fosters a beneficial environment for plant growth-promoting rhizobacteria, which enhances the absorption of water and nutrients from the soil (Li et al. 2024).

The rhizoplane is the outermost layer of a plant root that directly interacts with the soil environment (Choudhary et al. 2022). This delicate surface acts as a crucial interface where essential processes, such as nutrient absorption and water uptake, occur. It is densely populated with microorganisms, including bacteria and fungi, that form a dynamic ecosystem that supports plant health. The rhizoplane not only facilitates the uptake of vital minerals from the soil but also plays a significant role in the plant defense mechanisms against pathogens (Riseh et al. 2025). Overall, it serves as a vibrant zone of activity that profoundly influences the growth, development, and overall vitality of the plant.

Pathogen suppression is correlated with the changes that occur in the rhizosphere due to root exudate secretion, phytohormone production, nitrogen fixation, and phosphate and zinc solubilization (Backer et al. 2018; Mitter et al. 2021; Ortiz and Sansinenea 2022) (Fig. 2.1). However, if the microbial community in the rhizosphere is unbalanced, it can cause the colonization of soil-borne pathogens (Berlanas et al. 2019). Microbial diversity determines the health of rhizosphere microbiomes. The term "Metaorganism" is important in modern agriculture as it treats the plant and its microbiome as one functional biological unit. It is important to understand that plant health and productivity depend on microbial colonizers. Proteobacteria, Firmicutes, and Bacteroidetes are consistently dominant bacterial groups in the rhizosphere. In contrast, Ascomycota and Basidiomycota are the most prevalent fungal colonizers in this environment (Trivedi et al. 2020).

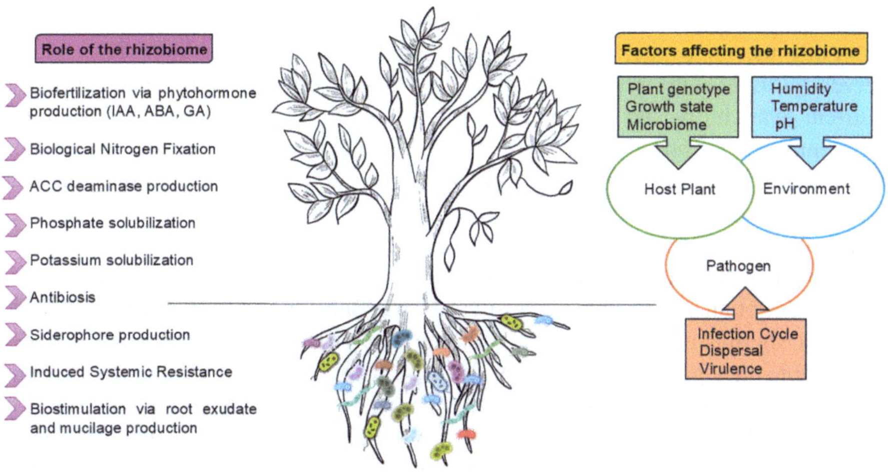

Fig. 2.1 Role of the rhizobiome. The rhizobiome plays a crucial role in the health and functionality of the plant root systems. It consists of a diverse community of microorganisms, including bacteria, fungi, and other microbes that inhabit the soil around plant roots. These organisms interact with plants, facilitating nutrient uptake, enhancing soil structure, and providing protection against pathogens. By fostering a symbiotic relationship, the rhizobiome significantly contributes to plant growth, resilience, and the overall stability of the ecosystem

2.3 Factors Affecting Rhizosphere Microbiome

There are multiple factors, including biotic and abiotic stressors, that affect rhizospheric microbial diversity, which have been discussed in detail in this section.

2.3.1 Biotic Factors

The "rhizosphere effect" arises from the secondary metabolites synthesized by plant roots (Hein et al. 2008), which profoundly influence the unique microbiota present in the rhizosphere, thereby distinguishing it from the surrounding bulk soil. Various biotic factors, such as specific plant species (Barbosa Lima et al. 2015), plant genotypes (Marques et al. 2014), and distinct root phenotypes (Cotton et al. 2019), play crucial roles in shaping the composition and diversity of the rhizosphere microbiome, leading to complex interactions that can enhance plant health and soil vitality. Genotype-specific variations play a significant role in shaping root architecture, influencing exudation patterns, and determining compatibility with various microorganisms. These variations can enhance the adaptability and efficiency of root systems in different environments, ultimately affecting plant health and growth negatively. The developmental stages of plants (Zhang et al. 2018) play a crucial role in influencing the composition and functionality of the rhizobiome, a community of microorganisms residing in the rhizosphere. As plants grow, several factors, including root aging and propagation methods, actively contribute to the dynamic shaping of the microbiome. During this process, plant roots release root cap border cells into the rhizosphere, a phenomenon known as rhizodeposition. This release serves multiple purposes: it enhances the rhizospheric effect by increasing soil nutrient availability and improving soil structure while also creating a favorable environment that promotes the growth of beneficial microorganisms in the soil. Thus, the interaction between plants and their associated microbial communities is critically important for nutrient cycling, disease suppression, and overall plant health. The specificity of microorganisms recruited during this stage can further affect plant growth, resilience, and productivity. Bacterial, fungal, and viral diseases (Berendsen et al. 2018) and insect infestations (Benitez et al. 2017) are biotic factors that significantly influence rhizobiome diversity. The rhizobiome of disease-susceptible plant genotypes demonstrates lower diversity within their core microbial communities than resistant genotypes, where higher diversity confers stability (Aqueel et al. 2024a, b). Soil microbial community composition is also influenced by microbial dynamics, such as competition, cooperation, quorum sensing, and horizontal gene transfer (Fig. 2.1).

2.3.2 *Abiotic Factors*

Several abiotic factors affect rhizospheric microbial communities (Fig. 2.1). Soil properties, such as clay content, porosity, and compaction, significantly influence water retention, aeration, and microbial habitats. Sandy soils are known to harbor different microbial communities than clayey or loamy soils because of their nutrient-holding capacity and moisture levels (Osman 2012). Soil pH affects microbial diversity and enzyme activity, with acidic soils favoring fungi and neutral to slightly alkaline soils supporting a broader range of bacteria (Shi et al. 2021). Changes in pH can promote or suppress the growth of specific microbial groups. Water availability also affects microbial composition. Drought or waterlogging can lead to shifts in communities (Bogati and Walczak 2022), such as an increase in spore-forming bacteria or anaerobes. Temperature fluctuations influence microbial growth and activity rates (Daunoras et al. 2024; Nguyen et al. 2021), whereas the availability of nutrients such as nitrogen, phosphorus, and potassium is crucial for their growth. Excess fertilizer use can reduce diversity and favor fast-growing microbes (Zheng et al. 2022a, b), whereas low-input systems support oligotrophic microbes. Soil organic matter enhances microbial biomass and diversity (Bastida et al. 2021). Organic inputs, such as compost and manure, enrich the rhizosphere microbiome (Yu et al. 2024). Agricultural practices, including tillage and crop rotation, significantly affect microbial dynamics, with reduced tillage and organic farming typically preserving diversity, whereas intensive practices may disrupt it (Bezboruah et al. 2024).

2.4 Rhizosphere Optimization: The Role of Microbiology in Modern-Day Farming

Modern-day agriculture relies on climate-smart technologies that improve crop health and productivity without harming the environment. These technologies include precision farming techniques that utilize data analytics and advanced sensors to optimize water usage and nutrient application. Additionally, sustainable practices, such as crop rotation and agroforestry, enhance soil health and boost biodiversity. Microbes are available as biofertilizers, which have been developed after intensive research using culture-based and culture-independent methods (Clagnan et al. 2024) (Fig. 2.2a). Innovations in agriculture, such as the adoption of microbial formulations (Mitter et al. 2021), have allowed farmers to increase crop yields while promoting ecosystem resilience in the face of climate change.

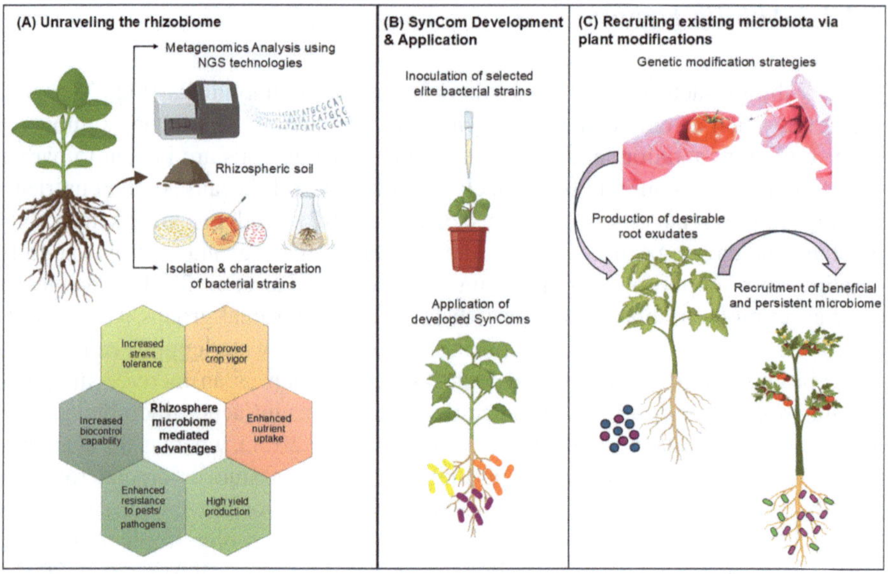

Fig. 2.2 Types of rhizobiome engineering. (**a**) Unraveling the rhizosphere microbiome via culture-based techniques and metagenomic analysis for better stress tolerance, high yield, and enhanced nutrient uptake. (**b**) "Bottom-up" rhizobiome engineering: developing synthetic microbial communities with plant growth-promoting and disease-suppressive capabilities and their application to alter the plant's metabolism. (**c**) "Top-down" rhizobiome engineering: plant genetic modification for the production of desirable root exudates/xenobiotic compounds that can recruit specific microorganisms

2.5 Rhizoengineering and Its Types

The rhizosphere can be engineered in three ways, using microbes, plants, and soil. These methods are discussed in detail in this section.

2.5.1 Rhizosphere Optimization via Microbiome Engineering

The microbiome surrounding plant roots can be modified using sustainable agricultural practices, such as synthetic community (SynCom) development (Fig. 2.2b). Synthetic microbial communities are those that have been designed in the laboratory by selecting elite bacterial isolates based on their plant growth-promoting (PGP) and disease-suppressive capabilities (Durán et al. 2025). These SynComs are then applied to plants in pot or field experiments to ensure that they withstand environmental stresses and inhabit the rhizospheric zone. Once the rhizospheric microbiome is established by the application of these selective SynCom strains, it possesses the capabilities of disease suppression (Martins et al. 2023; Wang et al.

2023), efficient nutrient solubilization (Gonçalves et al. 2023), and enhanced nutrient uptake (Xu et al. 2025). Clustered regularly interspaced short palindromic repeats (CRISPR) and other genome-editing techniques can be used to alter microbes to enhance their metabolite production and colonization efficiency (Chowdhary et al. 2022). A recent approach involves replacing the diseased microbiome with a healthy microbiome, where soil from healthy plants is taken, and the rhizosphere microbiota are transplanted into the soil of diseased plants, which helps suppress the disease in susceptible genotypes (Badar et al. 2025). Biosensors can also be used to monitor soil health conditions for optimized plant performance (Bharti et al. 2024; Gangwar 2024).

2.5.2 *Rhizosphere Optimization via Plant Engineering*

The microbial diversity in the rhizosphere can be optimized by making genetic modifications to the plant via molecular approaches or breeding strategies and using elite genotypes (Fig. 2.2c). Genetic modifications are performed to enhance root architecture, allowing plants to have an increased root surface area and a deeper, branched fine root system for beneficial microbes (Patel et al. 2025; Ragland et al. 2024). Plants can be altered for the selective release of sugars, amino acids, or secondary metabolites, which are responsible for recruiting plant growth-promoting rhizobacteria (PGPR) (Santoyo et al. 2021). Genetically engineered plants also produce certain antimicrobial compounds that kill disease-causing bacteria in the soil, allowing only beneficial microbes to thrive (Miller et al. 2022). Conditions also become favorable for mycorrhizal fungi and rhizobia, which form symbiotic associations with plants to handle environmental stress (Mitra et al. 2021).

2.5.3 *Rhizosphere Optimization via Soil Engineering*

Rhizosphere management practices utilize simple techniques such as crop rotation, reduced tillage, or organic farming approaches (Jiang et al. 2022). The sustainable agricultural approach of crop rotation ensures that pest cycles are disrupted (Shah et al. 2021). The diversity of the rhizosphere increases as different crops excrete different types of root exudates (Dhungana et al. 2023). Therefore, soil health is preserved, which in turn benefits the rhizosphere, the narrow zone where plant-microbe interactions occur. Soil organic matter and moisture are retained as the established root systems, and the residing microbes are left undisturbed due to reduced tillage practices. The utilization of compost and manure instead of chemical fertilizers engineers the soil, thus enriching microbial life and eventually helping in nutrient mobilization and disease suppression (Wei et al. 2024).

2.6 Applications: Rhizoengineering Tools

The field of rhizoengineering employs various tools, ranging from naturally occurring microbes to modern-day advancements in sequencing technologies, to improve rhizospheric microbial diversity and thus plant health. The different tools are discussed in detail in this section.

2.6.1 *Microbial Biofertilizers*

To minimize environmental damage caused by the use of chemical fertilizers, there is a need to shift towards sustainable agricultural practices that involve the utilization of microorganisms as biofertilizers. These "biofertilizers" enhance nutrient uptake; for example, in the case of nitrogen, there are many nitrogen-fixing bacteria that play a crucial role in nitrogen assimilation for plants (Li et al. 2023). Rhizobia are bacteria that form root nodules in leguminous plants and convert atmospheric nitrogen (N_2) to ammonia (NH_3), which is then assimilated by plants. There are non-nodule-forming N_2-fixing bacteria as well that exist freely in the rhizosphere, are endophytes, or form associations with certain plants. These include *Azotobacter*, *Bacillus*, *Beijierinckia*, *Burkholderia*, *Clostridium*, *Desulfovibrio*, *Derxia*, *Enterobacter*, *Klebsiella*, *Paenibacillus*, *Pseudomonas*, and *Serratia* (da Silva et al. 2011; Sanow et al. 2023; Singh et al. 2024a, b; Timmusk et al. 2017; Zhang et al. 2023; Zheng et al. 2022a, b).

Plants require micro- and macronutrients that are not readily available for uptake in the soil. Phosphate-solubilizing bacteria (PSB) and potassium-solubilizing bacteria (KSB) are involved in the conversion of P and K from insoluble to soluble forms, respectively. The presence and utilization of these bacteria reduce the need for chemical fertilizers (Soni et al. 2022). *Azotobacter*, *Bacillus*, *Baeyerlingia*, *Burkholderia*, *Pseudomonas*, *Pseudocystis*, *Enterobacter*, *Salmonella*, *Flavobacterium*, *Micrococcus*, *Pantoea*, *Serratia*, and *Thiobacillus* are PSB with *Bacillus* and *Pseudomonas* being the most abundant phosphate-solubilizing bacterial genera (Kalayu 2019; Pan and Cai 2023; Rawat et al. 2021; Saeid et al. 2018; Shome et al. 2022).

Studies suggest that by decreasing chemical fertilizer usage by 25% and replacing it with biofertilizers, which enhance nutrient uptake and provide the plant with hormones, more balanced microbial communities can be established in the rhizosphere that can better handle perturbations (Cai et al. 2017; Fasusi et al. 2021).

2.6.2 *Biocontrol Strategies*

Microbes residing in the plant rhizosphere play a crucial role in natural protection against pathogens. This is either by the direct production of such secondary metabolites that inhibit the growth of pathogens or by enhancing the plant's defense mechanism itself so that it can fight better against the disease. Antibiosis is a key mechanism used by rhizosphere microorganisms to combat pathogen infection (Ab Rahman et al. 2018). This process involves the secretion of antibiotics, organic compounds, toxins, and hydrolytic enzymes, such as beta-xylosidase, chitinase, pectin methylesterase, and β-1,3-glucanase. These substances work together to degrade the pathogen's cell wall by breaking down glycosidic linkages, resulting in its death. Rhizobia are symbiotic microbes residing in the roots of leguminous plants that play vital roles in nitrogen fixation and biocontrol (Goyal and Habtewold 2023). They promote plant growth by producing antibiotics, mycolytic enzymes, siderophores, and hydrocyanic acid (HCN), which help inhibit pathogenic fungi, such as *Fusarium* and *Rhizoctonia* (Beshah et al. 2023; Boruah and Hage 2024). Additionally, rhizobia enhance plant immunity by activating defense-related genes and systemic resistance (Tonelli et al. 2020).

Certain bacteria produce compounds such as indole-acetic acid (IAA) and salicylic acid (SA), which help in plant growth and disease suppression, respectively. For example, *Serratia marcescens*, *Bacillus pumilus*, and *Fictibacillus* have been reported as SA-producing bacteria that activate plant defense responses by activating systemic acquired resistance (SAR) in plants via the activation of non-expressor of pathogenesis-related genes 1 (*NPR1*) (Aqueel et al. 2023; Caarls et al. 2015). Salicylic acid is a key regulator of pathogenesis-related (PR) proteins through the NPR1-dependent pathway, playing a crucial role in plant defense against biotrophic pathogens and inducing SAR. PR proteins, which are produced in higher amounts during pathogen attacks or stress, belong to 17 naturally occurring families in plants (Finkina et al. 2017). *Delftia lacustris* and *Rahnella aquatilis* isolated from the rice rhizosphere exhibited plant growth promotion via IAA production (Thoa et al. 2022). Biocontrol strategies have been employed in plant disease incidence assays as monoassociation studies, SynCom experiments, or whole microbiota transplants. The main factor that leads to the success of biocontrol is the survival of the applied microbes under field conditions. In addition, it is imperative to test whether the microbes forming a consortium inhibit each other. The successful use of SynComs in suppressing plant disease has been extensively reported in wheat, tomato, and apple (Qiao et al. 2024; Yin et al. 2022; Zhou et al. 2022). Studies have reported the use of complete rhizosphere microbiota transplants from healthy plants and their application to suppress disease in susceptible plants (Badar et al. 2025; Kwak et al. 2018).

2.6.3 *Quorum Sensing*

Quorum sensing (QS) is a microbial behavior that is dependent on cell density and is guided by small signaling molecules, such as *N*-acyl-homoserine lactones (AHL) in gram-negative bacteria and peptides in gram-positive bacteria (Zeng et al. 2023). When critical microbial density is reached, they activate or suppress gene expression involved in biofilm formation (Markowska et al. 2024; Wang et al. 2022), antibiotic production (Cabuhat and Moron-Espiritu 2022), virulence or symbiosis (Loh et al. 2002), nutrient acquisition (Hartmann et al. 2021a, b), motility, and chemotaxis (Scharf et al. 2016).

QS can be used by scientists to design and guide specific microbial communities for root colonization (Shahni et al. 2025). It can also be used for the induction of specific plant hormones, such as IAA and ACC (1-aminocyclopropane-1-carboxylate) deaminase, which are involved in plant growth promotion (Hartmann et al. 2021a, b). Certain *Serratia* spp. strains increase SA production in plants by releasing the quorum-sensing compound AHL (Pang et al. 2009; Schuhegger et al. 2006). However, *Bacillus* spp. play a role in quorum quenching (QQ) and degrade the AHL signals produced by other pathogenic bacteria, preventing their biofilm formation in the rhizosphere (Hartmann et al. 2021a, b).

2.6.4 Nanotechnology

Using targeted technologies at the nanoscale (1–100 nm), the rhizospheric environment can be altered to enhance nutritional uptake, support the formation of beneficial microbial communities, and inhibit disease-causing bacteria (Poornima and Bharathiraja 2023). In rhizobiome engineering, nanoparticles can penetrate biofilms because of their small size and reach specified hotspots (Meel and Saharan 2024). Nanofertilizers, such as ZnO, FeO, and nanohydroxyapatite, are used to improve nutrient availability in the rhizosphere (Arora and Khosla 2024). Nanocarriers, mostly made of silica or clay, can protect certain bacteria and ensure their release in specific environments, such as near roots (Saleh et al. 2022). Nanoparticles, such as silver or zinc oxide, can quench certain harmful pathogens (Ahamad Khan et al. 2023). These nanoparticles may also be used to deliver antimicrobial compounds at low doses that are effective in suppressing disease-causing microbes (Farooq et al. 2021; Noman et al. 2023). Precision agriculture is based on the fact that real-time monitoring of root exudate composition, soil nutrient levels, and microbial activity can be performed using nanosensors (Beegum and Das 2022). Although nanotechnology has the benefit of precision, its unknown long-term ecological effects cannot be overlooked. In uncontrolled environments, it can also have off-target effects (Suhag et al. 2025). Therefore, it should be used under controlled conditions while minimizing chemical input and enhancing crop productivity.

2.6.5 Metagenomics

Only 1% of the bacterial population in drylands is culturable (Fazeli-Nasab et al. 2022). Next-generation sequencing (NGS) technologies and metagenomics make it possible to delve into the remaining part of the complex microbiota that inhabit the environment. These techniques enable precise microbiome profiling. This allows the identification of key microbial functions of rhizospheric microbial communities, apart from their diversity. Metagenomics is a powerful tool that has emerged in recent times for the in-depth analysis of the genetic diversity within complex microbial communities residing in the rhizosphere. Metagenomic techniques include shotgun metagenomics, which facilitates an unbiased analysis of the entire microbial genome, and targeted sequencing of the 16S rRNA gene, which is used for taxonomic profiling (Rajguru et al. 2024). Significant progress has been made in identifying core rhizospheric functional microbes that suppress plant diseases using NGS technologies, elucidating the relationship between rhizospheric interactions and resistance to pathogen invasion, and investigating how these core microbes influence plant immunity (Niu et al. 2017; Castrillo et al. 2017; Li et al. 2019). In the current era of precision agriculture, it has become mandatory that, in addition to field data analytics, machine learning utilizes soil microbiome data and provides the best possible solutions to overcome agriculture-related issues (Emmenegger et al. 2023).

2.7 Future Perspectives

With the ever-changing climatic conditions, it is imperative that the lab-developed synthetic microbial communities perform better under field conditions and do not exhibit microbial inconsistency. Therefore, soil optimization plays a key role in maintaining stable soil conditions for harboring beneficial rhizospheric microbiota. Farmers should be well educated on how to apply these developed SynComs in the field for optimal results. There is no doubt that the answer lies in "going back to the roots" (Philippot et al. 2013). The rhizosphere is where the most beneficial microbiota reside, which can be isolated, enhanced, and applied to improve crops and promote sustainable agriculture. Using advances in metagenomics, unraveling the microbial and functional diversity of rhizosphere communities can help scientists develop targeted solutions for soil microbiome communities that aid in crop improvement and promote plant health (Hakim et al. 2022).

References

Ab Rahman SFS, Singh E, Pieterse CM, Schenk PM (2018) Emerging microbial biocontrol strategies for plant pathogens. Plant Sci 267:102–111

Ahamad Khan M, Lone SA, Shahid M, Zeyad MT, Syed A, Ehtram A, Danish M (2023) Phytogenically synthesized zinc oxide nanoparticles (ZnO-NPs) potentially inhibit the bacterial pathogens: in vitro studies. Toxics 11:452

Ali MA, Naveed M, Mustafa A, Abbas A (2017) The good, the bad, and the ugly of rhizosphere microbiome. In: Probiotics plant health, vol 253, pp 253–290

Aqueel R, Badar A, Roy N, Mushtaq Q, Ali AF, Bashir A, Malik KA (2023) Cotton microbiome profiling and Cotton Leaf Curl Disease (CLCuD) suppression through microbial consortia associated with *Gossypium arboreum*. NPJ Biofilms Microbiomes 9:100

Aqueel R, Badar A, Ijaz UZ, Malik KA (2024a) Microbial influencers and cotton leaf curl disease (CLCuD) susceptibility: a network perspective. Front Microbiol 15:1381883

Aqueel R, Badar A, Roy N, Ijaz UZ, Malik KA (2024b) Disease resistance correlates with core microbiome diversity in cotton. Curr Microbiol 81:302

Arora V, Khosla B (2024) Synthesis of phosphorus nano-fertilisers: their strategic applications and effect on plant growth. Int J Environ Sci Technol 1:1–20

Backer R, Rokem JS, Ilangumaran G, Lamont J, Praslickova D, Ricci E, Subramanian S, Smith DL (2018) Plant growth-promoting rhizobacteria: context, mechanisms of action, and roadmap to commercialization of biostimulants for sustainable agriculture. Front Plant Sci 9:1473

Badar A, Aqueel R, Nawaz A, Ijaz UZ, Malik KA (2025) Microbiota transplantation for cotton leaf curl disease suppression – core microbiome and transcriptome dynamics. Commun Biol 8:380

Bakker MG, Chaparro JM, Manter DK, Vivanco JM (2015) Impacts of bulk soil microbial community structure on rhizosphere microbiomes of *Zea mays*. Plant Soil 392:115–126

Barbosa Lima A, Cannavan FS, Navarrete AA, Teixeira WG, Kuramae EE, Tsai SM (2015) Amazonian dark earth and plant species from the Amazon region contribute to shape rhizosphere bacterial communities. Microb Ecol 69:855–866

Bastida F, Eldridge DJ, García C, Kenny Png G, Bardgett RD, Delgado-Baquerizo M (2021) Soil microbial diversity–biomass relationships are driven by soil carbon content across global biomes. ISME J 15:2081–2091

Beegum S, Das S (2022) Nanosensors in agriculture. In: Ghosh S, Thongmee S, Kumar A (eds) Agricultural nanobiotechnology: biogenic nanoparticles, nanofertilizers and nanoscale biocontrol agents. Woodhead Publishing, pp 465–478

Benitez MS, Osborne SL, Lehman RM (2017) Previous crop and rotation history effects on maize seedling health and associated rhizosphere microbiome. Sci Rep 7:1570

Berendsen RL, Vismans G, Yu K, Song Y, de Jonge R, Burgman WP, Burmølle M, Herschend J, Bakker PA, Pieterse CM (2018) Disease-induced assemblage of a plant-beneficial bacterial consortium. ISME J 12:1496–1507. https://doi.org/10.1038/s41396-018-0093-1

Berlanas C (2019) New insights into the biology, ecology and control of black-foot disease in grapevine. PhD Thesis

Beshah A, Assefa F, Muleta D, Legese G (2023) Screening groundnut rhizobia for multiple plant growth promoting activities in Ethiopia. Afr J Microbiol Res 17:212–220. https://doi.org/10.5897/AJMR2022.9674

Bezboruah M, Sharma SK, Laxman T, Ramesh S, Sampathkumar T, Gulaiya S, Krishnaveni SA (2024) Conservation tillage practices and their role in sustainable farming systems. J Exp Agric Int 46:946–959. https://doi.org/10.9734/jeai/2024/v46i92768

Bharti A, Jain U, Chauhan N (2024) From lab to field: nano-biosensors for real-time plant nutrient tracking. Plant Nano Biol:100079. https://doi.org/10.1016/j.pnb.2024.100079

Bogati K, Walczak M (2022) The impact of drought stress on soil microbial community, enzyme activities and plants. Agronomy 12:189. https://doi.org/10.3390/agronomy12010189

Boruah RR, Hage A (2024) Role of rhizobial metabolites in control of soil-borne phytopathogenic fungi. In: Mishra V, Mishra J, Arora NK (eds) Bioactive microbial metabolites. Academic, pp 51–65

Bulgarelli D, Schlaeppi K, Spaepen S, van Themaat EVL, Schulze-Lefert P (2013) Structure and functions of the bacterial microbiota of plants. Annu Rev Plant Biol 64:807–838. https://doi.org/10.1146/annurev-arplant-050312-120106

Caarls L, Pieterse CM, Van Wees SC (2015) How salicylic acid takes transcriptional control over jasmonic acid signaling. Front Plant Sci 6:170. https://doi.org/10.3389/fpls.2015.00170

Cabuhat KSP, Moron-Espiritu LS (2022) Quorum sensing orchestrates antibiotic drug resistance, biofilm formation, and motility in *Escherichia coli* and quorum quenching activities of plant-derived natural products: a review. J Pure Appl Microbiol 16:1513–1529. https://doi.org/10.22207/JPAM.16.3.47

Cai F, Pang G, Li RX, Li R, Gu XL, Shen QR, Chen W (2017) Bioorganic fertilizer maintains a more stable soil microbiome than chemical fertilizer for monocropping. Biol Fertil Soils 53:861–872. https://doi.org/10.1007/s00374-017-1224-1

Castrillo G, Teixeira PJPL, Paredes SH, Law TF, De Lorenzo L, Feltcher ME et al (2017) Root microbiota drive direct integration of phosphate stress and immunity. Nature 543:513–518. https://doi.org/10.1038/nature21476

Choudhary S, Singh A, Rani M (2022) Rhizosphere: hotspot of soil-microbes-plant interaction. Agric Rev 80:1–511

Chowdhary K, Arora H, Sharma S (2022) CRISPR/Cas9-based genome editing as a way ahead for inducing production of bioactive metabolites in endophytes. Natl Acad Sci Lett 45:275–280. https://doi.org/10.1007/s40009-021-01071-0

Clagnan E, Costanzo M, Visca A, Di Gregorio L, Tabacchioni S, Colantoni E et al (2024) Culturomics- and metagenomics-based insights into the soil microbiome preservation and application for sustainable agriculture. Front Microbiol 15:1473666. https://doi.org/10.3389/fmicb.2024.1473666

Cotton TA, Pétriacq P, Cameron DD, Meselmani MA, Schwarzenbacher R, Rolfe SA, Ton J (2019) Metabolic regulation of the maize rhizobiome by benzoxazinoids. ISME J 13:1647–1658. https://doi.org/10.1038/s41396-019-0375-3

da Silva K, Nóbrega RSA, Lima AS, Barberi A, de Moreira FMS (2011) Density and diversity of diazotrophic bacteria isolated from Amazonian soils using N-free semi-solid media. Sci Agric 68:518–525. https://doi.org/10.1590/S0103-90162011000400015

Daunoras J, Kačergius A, Gudiukaitė R (2024) Role of soil microbiota enzymes in soil health and activity changes depending on climate change and the type of soil ecosystem. Biology (Basel) 13:85. https://doi.org/10.3390/biology13020085

Dhungana I, Kantar MB, Nguyen NH (2023) Root exudate composition from different plant species influences the growth of rhizosphere bacteria. Rhizosphere 25:100645

Durán P, Vailleau F, Roux F (2025) Building microbial synthetic communities: get inspired by the design of synthetic plant communities. New Phytol 246:402–405. https://doi.org/10.1111/nph.19273

Elvia JC, de Freitas RJ, Germida JJ (2021) Bacterial microbiomes associated with the rhizosphere, root interior, and aboveground plant organs of wheat and canola at different growth stages. Phytobiomes J 5:442–451. https://doi.org/10.1094/PBIOMES-02-21-0014-R

Emmenegger B, Massoni J, Pestalozzi CM, Bortfeld-Miller M, Maier BA, Vorholt JA (2023) Identifying microbiota community patterns important for plant protection using synthetic communities and machine learning. Nat Commun 14:7983. https://doi.org/10.1038/s41467-023-43766-z

Farooq T, Adeel M, He Z, Umar M, Shakoor N, da Silva W et al (2021) Nanotechnology and plant viruses: an emerging disease management approach for resistant pathogens. ACS Nano 15:6030–6047. https://doi.org/10.1021/acsnano.1c01157

Fasusi OA, Cruz C, Babalola OO (2021) Agricultural sustainability: microbial biofertilizers in rhizosphere management. Agriculture 11:163. https://doi.org/10.3390/agriculture11020163

Fazeli-Nasab B, Mahdinezhad N, Parray JA (2022) Metagenomics and microbiome engineering: identification of core microbiome and improvement of rhizosphere. In: Parray JA, Shameem N, Abd-Allah EF, Mir MY (eds) Core microbiome: improving crop quality and productivity, pp 58–100

Fernández-González AJ, Villadas PJ, Gómez-Lama Cabanás C, Valverde-Corredor A, Belaj A, Mercado-Blanco J, Fernández-López M (2019) Defining the root endosphere and rhizosphere microbiomes from the world olive germplasm collection. Sci Rep 9:20423. https://doi.org/10.1038/s41598-019-56747-8

Finkina EI, Melnikova DN, Bogdanov IV, Ovchinnikova TV (2017) Plant pathogenesis-related proteins PR-10 and PR-14 as components of innate immunity system and ubiquitous allergens. Curr Med Chem 24:1772–1787. https://doi.org/10.2174/0929867324666170127115225

Gangwar DS (2024) Recent trends in biosensor development for climate smart organic agriculture and their role in environmental sustainability. Environ Dev Sustain. https://doi.org/10.1007/s10668-024-04523-7

Giller KE, Hijbeek R, Andersson JA, Sumberg J (2021) Regenerative agriculture: an agronomic perspective. Outlook Agric 50:13–25. https://doi.org/10.1177/0030727021998063
Gonçalves OS, Creevey CJ, Santana MF (2023) Designing a synthetic microbial community through genome metabolic modeling to enhance plant–microbe interaction. Environ Microbiome 18:81. https://doi.org/10.1186/s40793-023-00487-w
Gottel NR, Castro HF, Kerley M, Yang Z, Pelletier DA, Podar M et al (2011) Distinct microbial communities within the endosphere and rhizosphere of *Populus deltoides* roots across contrasting soil types. Appl Environ Microbiol 77:5934–5944. https://doi.org/10.1128/AEM.05255-11
Goyal RK, Habtewold JZ (2023) Evaluation of legume–rhizobial symbiotic interactions beyond nitrogen fixation that help the host survival and diversification in hostile environments. Microorganisms 11:1454. https://doi.org/10.3390/microorganisms11061454
Gregory PJ (2006) Roots, rhizosphere and soil: the route to a better understanding of soil science? Eur J Soil Sci 57:2–12. https://doi.org/10.1111/j.1365-2389.2005.00778.x
Gupta S, Khan IM, Parihar N, Kumar D (2024) Introduction to the rhizosphere world. In: Kumar D, Gupta S, Hasan W, Hussain B, Ansari MJ, Singh S (eds) Rhizosphere revolution. CRC Press, pp. 1–26
Ha J, Gao Y, Zhang R, Li K, Zhang Y, Niu X et al (2021) Diversity of the bacterial microbiome associated with the endosphere and rhizosphere of different cassava (*Manihot esculenta Crantz*) genotypes. Front Microbiol 12:729022. https://doi.org/10.3389/fmicb.2021.729022
Hakim S, Nawaz MS, Siddique MJ, Hayat M, Gulzar U, Imran A (2022) Metagenomics for rhizosphere engineering. In: Dubey RC, Kumar P (eds) Rhizosphere engineering. Academic, pp 395–416. https://doi.org/10.1016/B978-0-323-90513-0.00017-6
Hartmann A, Klink S, Rothballer M (2021a) Plant growth promotion and induction of systemic tolerance to drought and salt stress of plants by quorum sensing auto-inducers of the N-acyl-homoserine lactone type: recent developments. Front Plant Sci 12:683546. https://doi.org/10.3389/fpls.2021.683546
Hartmann A, Klink S, Rothballer M (2021b) Importance of N-acyl-homoserine lactone-based quorum sensing and quorum quenching in pathogen control and plant growth promotion. Pathogens 10:1561. https://doi.org/10.3390/pathogens10121561
Hein JW, Wolfe GV, Blee KA (2008) Comparison of rhizosphere bacterial communities in *Arabidopsis thaliana* mutants for systemic acquired resistance. Microb Ecol 55:333–343. https://doi.org/10.1007/s00248-007-9283-4
Jiang Y, Zhang J, Manuel DB, De Beeck MO, Shahbaz M, Chen Y et al (2022) Rotation cropping and organic fertilizer jointly promote soil health and crop production. J Environ Manag 315:115190. https://doi.org/10.1016/j.jenvman.2022.115190
Kalayu G (2019) Phosphate solubilizing microorganisms: promising approach as biofertilizers. Int J Agron 2019:4917256. https://doi.org/10.1155/2019/4917256
Kwak MJ, Kong HG, Choi K, Kwon SK, Song JY, Lee J et al (2018) Rhizosphere microbiome structure alters to enable wilt resistance in tomato. Nat Biotechnol 36:1100–1109. https://doi.org/10.1038/nbt.4232
Li H, Su JQ, Yang XR, Zhu YG (2019) Distinct rhizosphere effect on active and total bacterial communities in paddy soils. Sci Total Environ 649:422–430. https://doi.org/10.1016/j.scitotenv.2018.08.288
Li L, Hu Z, Tan G, Fan J, Chen Y, Xiao Y et al (2023) Enhancing plant growth in biofertilizer-amended soil through nitrogen-transforming microbial communities. Front Plant Sci 14:1259853. https://doi.org/10.3389/fpls.2023.1259853
Li Y, Hong Y, Chen Y, Zhu N, Jiang S, Yao Z et al (2024) Rhizosheath formation and its role in plant adaptation to abiotic stress. Agronomy 14:2368. https://doi.org/10.3390/agronomy14102368
Loh J, Pierson EA, Pierson LS III, Stacey G, Chatterjee A (2002) Quorum sensing in plant-associated bacteria. Curr Opin Plant Biol 5:285–290. https://doi.org/10.1016/S1369-5266(02)00271-8
Markowska K, Szymanek-Majchrzak K, Pituch H, Majewska A (2024) Understanding quorum-sensing and biofilm forming in anaerobic bacterial communities. Int J Mol Sci 25:12808. https://doi.org/10.3390/ijms252312808

Marques JM, da Silva TF, Vollu RE, Blank AF, Ding GC, Seldin L, Smalla K (2014) Plant age and genotype affect the bacterial community composition in the tuber rhizosphere of field-grown sweet potato plants. FEMS Microbiol Ecol 88:424–435. https://doi.org/10.1111/1574-6941.12296

Martins SJ, Pasche J, HAO S, Selten G, Savastano N, Abreu LM et al (2023) The use of synthetic microbial communities to improve plant health. Phytopathology 113:1369–1379. https://doi.org/10.1094/PHYTO-11-22-0454-R

Meel S, Saharan BS (2024) Enhancing crop resilience towards drought: by integrating nanotechnology, microbiomes, and growth-promoting rhizobacteria. Discov Agric 2:112. https://doi.org/10.1007/s44214-024-00123-w

Miller SA, Ferreira JP, LeJeune JT (2022) Antimicrobial use and resistance in plant agriculture: a one health perspective. Agriculture 12:289. https://doi.org/10.3390/agriculture12020289

Mitra D, Djebaili R, Pellegrini M, Mahakur B, Sarker A, Chaudhary P et al (2021) Arbuscular mycorrhizal symbiosis: plant growth improvement and induction of resistance under stressful conditions. J Plant Nutr 44:1993–2028. https://doi.org/10.1080/01904167.2021.1896755

Mitter EK, Tosi M, Obregón D, Dunfield KE, Germida JJ (2021) Rethinking crop nutrition in times of modern microbiology: innovative biofertilizer technologies. Front Sustain Food Syst 5:606815. https://doi.org/10.3389/fsufs.2021.606815

Nazari M (2024) The unique rhizosphere of aerial roots. Rhizosphere 29:100867. https://doi.org/10.1016/j.rhisph.2023.100867

Nguyen J, Lara-Gutiérrez J, Stocker R (2021) Environmental fluctuations and their effects on microbial communities, populations and individuals. FEMS Microbiol Rev 45:fuaa068. https://doi.org/10.1093/femsre/fuaa068

Niu B, Paulson JN, Zheng X, Kolter R (2017) Simplified and representative bacterial community of maize roots. Proc Natl Acad Sci USA 114:E2450–E2459. https://doi.org/10.1073/pnas.1616148114

Noman M, Ahmed T, Wang J, Ijaz M, Shahid M, Islam MS et al (2023) Nano-enabled crop resilience against pathogens: potential, mechanisms and strategies. Crop Health 1:15. https://doi.org/10.1093/croph/yvad001

Olanrewaju OS, Ayangbenro AS, Glick BR, Babalola OO (2019) Plant health: feedback effect of root exudates–rhizobiome interactions. Appl Microbiol Biotechnol 103:1155–1166. https://doi.org/10.1007/s00253-018-9560-4

Ortiz A, Sansinenea E (2022) The role of beneficial microorganisms in soil quality and plant health. Sustainability 14(9):5358

Osman KT (2012) Physical properties of soil. In: Soils: principles, properties and management. Springer, Dordrecht, pp 49–65. https://doi.org/10.1007/978-94-007-5663-8_4

Pan L, Cai B (2023) Phosphate-solubilizing bacteria: advances in their physiology, molecular mechanisms and microbial community effects. Microorganisms 11:2904. https://doi.org/10.3390/microorganisms11112904

Pang Y, Liu X, Ma Y, Chernin L, Berg G, Gao K (2009) Induction of systemic resistance, root colonization and biocontrol activities of the rhizospheric strain of *Serratia plymuthica* are dependent on N-acyl homoserine lactones. Eur J Plant Pathol 124:261–268. https://doi.org/10.1007/s10658-008-9415-2

Patel R, Asthana S, Sood N (2025) Rhizosphere dynamics: the microbial influence on root architecture. In: Saraf M, Goswami D, Maheshwari DK (eds) Climate change and soil microorganisms for environmental sustainability. Springer, Singapore, pp 239–263. https://doi.org/10.1007/978-981-99-2629-6_12

Philippot L, Raaijmakers JM, Lemanceau P, van der Putten WH (2013) Going back to the roots: the microbial ecology of the rhizosphere. Nat Rev Microbiol 11:789–799. https://doi.org/10.1038/nrmicro3109

Poornima R, Bharathiraja S (2023) Nano-biofertilizers as a replacement of conventional fertilizers: a surging eco-friendly and cutting-edge approach for progressive evolution of precision and sustainable agriculture. Arch Agric Allied Sci. https://doi.org/10.5281/zenodo.8275787

Qiao R, Xu M, Jiang J, Song Z, Wang M, Yang L et al (2024) Plant growth promotion and biocontrol properties of a synthetic community in the control of apple disease. BMC Plant Biol 24:546. https://doi.org/10.1186/s12870-024-04721-6

Ragland CJ, Shih KY, Dinneny JR (2024) Choreographing root architecture and rhizosphere interactions through synthetic biology. Nat Commun 15:1370. https://doi.org/10.1038/s41467-024-45168-4

Rajguru B, Shri M, Bhatt VD (2024) Exploring microbial diversity in the rhizosphere: a comprehensive review of metagenomic approaches and their applications 3 Biotech 14(10):224

Rawat P, Das S, Shankhdhar D, Shankhdhar SC (2021) Phosphate-solubilizing microorganisms: mechanism and their role in phosphate solubilization and uptake. J Soil Sci Plant Nutr 21:49–68. https://doi.org/10.1007/s42729-020-00342-7

Richter-Heitmann T, Eickhorst T, Knauth S, Friedrich MW, Schmidt H (2016) Evaluation of strategies to separate root-associated microbial communities: a crucial choice in rhizobiome research. Front Microbiol 7:773. https://doi.org/10.3389/fmicb.2016.00773

Riseh RS, Fathi F, Vazvani MG, Tarkka MT (2025) Plant colonization by biocontrol bacteria and improved plant health: a review. Front Biosci (Landmark Ed) 30:23223. https://doi.org/10.52586/E978-1-64826-824-3.ch14

Saeid A, Prochownik E, Dobrowolska-Iwanek J (2018) Phosphorus solubilization by *Bacillus* species. Molecules 23:2897. https://doi.org/10.3390/molecules23112897

Saleh MM, Mahmoud AS, Abbas HS, Abu-Ellail FF, Kotakonda M, Salem KF (2022) Nanotechnological approaches for efficient delivery of plant ingredients. In: Faizan M, Yu F, Hayat S (eds) Sustainable agriculture reviews 53: nanoparticles: a new tool to enhance stress tolerance. Springer, Cham, pp 247–286. https://doi.org/10.1007/978-3-030-94467-2_9

Sanow S, Kuang W, Schaaf G, Huesgen P, Schurr U, Roessner U et al (2023) Molecular mechanisms of *Pseudomonas*-assisted plant nitrogen uptake: opportunities for modern agriculture. Mol Plant-Microbe Interact 36:536–548. https://doi.org/10.1094/MPMI-12-22-0235-IA

Santoyo G, Urtis-Flores CA, Loeza-Lara PD, Orozco-Mosqueda MDC, Glick BR (2021) Rhizosphere colonization determinants by plant growth-promoting rhizobacteria (PGPR). Biology (Basel) 10:475. https://doi.org/10.3390/biology10060475

Scharf BE, Hynes MF, Alexandre GM (2016) Chemotaxis signaling systems in model beneficial plant–bacteria associations. Plant Mol Biol 90:549–559. https://doi.org/10.1007/s11103-015-0393-9

Schuhegger R, Ihring A, Gantner S, Bahnweg G, Knappe C, Vogg G et al (2006) Induction of systemic resistance in tomato by N-acyl-L-homoserine lactone-producing rhizosphere bacteria. Plant Cell Environ 29:909–918. https://doi.org/10.1111/j.1365-3040.2005.01477.x

Shah KK, Modi B, Pandey HP, Subedi A, Aryal G, Pandey M et al (2021) Diversified crop rotation: an approach for sustainable agriculture production. Adv Agric 2021:8924087. https://doi.org/10.1155/2021/8924087

Shahni YS, Singh KP, Pandey S, Panigrahi CK, Das N, Panotra N et al (2025) Role of quorum sensing molecules in plant-microbe interaction for sustainable agriculture: a review. J Adv Biol Biotechnol 28:889–903. https://doi.org/10.9734/jabb/2025/v28i51456

Shi Y, Li Y, Yang T, Chu H (2021) Threshold effects of soil pH on microbial co-occurrence structure in acidic and alkaline arable lands. Sci Total Environ 800:149592. https://doi.org/10.1016/j.scitotenv.2021.149592

Shome S, Barman A, Solaiman ZM (2022) *Rhizobium* and phosphate solubilizing bacteria influence the soil nutrient availability, growth, yield, and quality of soybean. Agriculture (Basel) 12:1136. https://doi.org/10.3390/agriculture12081136

Singh P, Singh RK, Li HB, Guo DJ, Sharma A, Verma KK et al (2024a) Nitrogen fixation and phytohormone stimulation of sugarcane plant through plant growth-promoting diazotrophic *Pseudomonas*. Biotechnol Genet Eng Rev 40:15–35. https://doi.org/10.1080/02648725.2023.2186182

Singh M, Kumar S, Singh DK, Ranjan T, Pandey AK (2024b) Isolation and characterization of a novel nitrogen fixer *Beijerinckia fluminensis* strain BAUMS11 from litchi (Litchi chinensis L.) rhizosphere. J Appl Nat Sci 16. https://doi.org/10.31018/jans.v16i3.5022

Soni SK, Manhas R, Jakhar Y, Sharma A, Soni R (2022) Biofertilizers for sustainable agriculture: current trends and future perspective. In: Sobti RC, Mukesh M, Sobti A (eds) Genomic, proteomics, and biotechnology. CRC Press, Boca Raton, pp 331–356. https://doi.org/10.1201/9781003212582-17

Suhag D, Thakur P, Thakur A (2025) Future perspectives of nanobiology. In: Advancements in nanobiology. CRC Press, pp 345–361. https://doi.org/10.1201/9781003322519-18

Thoa NTK, Mai DTH, Hiu BL, Duong CA, Chau NNB, Nghiep NM et al (2022) Roles of β-indole acetic acid (IAA) producing endophytic bacteria on the recovery of plant growth and survival ability of sugarcane infected with white leaf disease (SWLD). Curr Microbiol 79:389. https://doi.org/10.1007/s00284-022-02958-4

Timmusk S, Behers L, Muthoni J, Muraya A, Aronsson AC (2017) Perspectives and challenges of microbial application for crop improvement. Front Plant Sci 8:49. https://doi.org/10.3389/fpls.2017.00049

Tonelli ML, Figueredo MS, Rodríguez J, Fabra A, Ibañez F (2020) Induced systemic resistance-like responses elicited by rhizobia. Plant Soil 448:1–14. https://doi.org/10.1007/s11104-019-04388-y

Trivedi P, Leach JE, Tringe SG, Sa T, Singh BK (2020) Plant–microbiome interactions: from community assembly to plant health. Nat Rev Microbiol 18(11):607–621

Wang Y, Bian Z, Wang Y (2022) Biofilm formation and inhibition mediated by bacterial quorum sensing. Appl Microbiol Biotechnol 106:6365–6381. https://doi.org/10.1007/s00253-022-12171-0

Wang Z, Hu X, Solanki MK, Pang F (2023) A synthetic microbial community of plant core microbiome can be a potential biocontrol tool. J Agric Food Chem 71:5030–5041. https://doi.org/10.1021/acs.jafc.2c08627

Wei X, Xie B, Wan C, Song R, Zhong W, Xin S, Song K (2024) Enhancing soil health and plant growth through microbial fertilizers: mechanisms, benefits, and sustainable agricultural practices. Agronomy 14:609. https://doi.org/10.3390/agronomy14030609

Xu X, Dinesen C, Pioppi A, Kovács ÁT, Lozano-Andrade CN (2025) Composing a microbial symphony: synthetic communities for promoting plant growth. Trends Microbiol. https://doi.org/10.1016/j.tim.2024.12.010

Xun W, Liu Y, Ma A, Yan H, Miao Y, Shao J et al (2024) Dissection of rhizosphere microbiome and exploiting strategies for sustainable agriculture. New Phytol. https://doi.org/10.1111/nph.19567

Yin C, Hagerty CH, Paulitz TC (2022) Synthetic microbial consortia derived from rhizosphere soil protect wheat against a soilborne fungal pathogen. Front Microbiol 13:908981

Yu Y, Zhang Q, Kang J, Xu N, Zhang Z, Deng Y et al (2024) Effects of organic fertilizers on plant growth and the rhizosphere microbiome. Appl Environ Microbiol 90:e01719-23. https://doi.org/10.1128/aem.01719-23

Zeng X, Zou Y, Zheng J, Qiu S, Liu L, Wei C (2023) Quorum sensing-mediated microbial interactions: mechanisms, applications, challenges and perspectives. Microbiol Res 273:127414. https://doi.org/10.1016/j.micres.2023.127414

Zhang J, Zhang N, Liu YX, Zhang X, Hu B, Qin Y et al (2018) Root microbiota shift in rice correlates with resident time in the field and developmental stage. Sci China Life Sci 61:613–621. https://doi.org/10.1007/s11427-018-9300-2

Zhang Y, Xu Q, Wang G, Shi K (2023) Mixed *Enterobacter* and *Klebsiella* bacteria enhance soybean biological nitrogen fixation ability when combined with rhizobia inoculation. Soil Biol Biochem 184:109100. https://doi.org/10.1016/j.soilbio.2023.109100

Zheng Y, Liu X, Cai Y, Shao Q, Zhu W, Lin X (2022a) Combined intensive management of fertilization, tillage, and organic material mulching regulate soil bacterial communities and functional capacities by altering soil potassium and pH in a Moso bamboo forest. Front Microbiol 13:944874. https://doi.org/10.3389/fmicb.2022.944874

Zheng J, Liu C, Liu J, Zhuang JY (2022b) Study of the effect of bacterial-mediated legume plant growth using bacterial strain *Serratia marcescens* N1.14 X-45. Front Microbiol 13:988692. https://doi.org/10.3389/fmicb.2022.988692

Zhou X, Wang J, Liu F, Liang J, Zhao P, Tsui CK, Cai L (2022) Cross-kingdom synthetic microbiota supports tomato suppression of *Fusarium* wilt disease. Nat Commun 13:7890. https://doi.org/10.1038/s41467-022-35512-z

Chapter 3
Practical Challenges in Developing Climate-Resilient Beneficial Bioinocula to Address Biotic and Abiotic Stresses in Sustainable Agriculture

Diyan Herdiyantoro, Fairus Hisanah Hibatullah, Fiqriah Hanum Khumairah, Irwandhi, Debora D. M. Ambarita, Juli Sinaga, Diyah Sri Utami, Anne Nurbaity, Nadia N. Kamaluddin, and Tualar Simarmata

Abstract The quest for sustainable agriculture necessitates innovative solutions to enhance crop resilience against various stresses. Beneficial bioinocula have emerged as a promising solution, crucial for improving soil and plant health, as well as crop productivity. These microorganisms mitigate abiotic and biotic stresses through mechanisms such as promoting nutrient uptake, improving soil structure, protecting plants from pathogens and pests, and helping plants cope with drought and salinity. By boosting natural plant resilience, bioinocula reduce the need for chemical fertilizers and pesticides, thereby contributing to more sustainable and resilient agricultural practices. This chapter explores the dual threats of biotic stresses, such as pathogens and pests, and abiotic stresses, including low nutrient availability, drought, salinity, and extreme temperatures, detailing how bioinocula can mitigate these issues. Developing resilient bioinocula faces several challenges, including scientific and technical obstacles, environmental and climatic variability, economic and logistical constraints, and regulatory and policy barriers. Advancements in technology and innovative approaches are highlighted as potential solutions to these challenges. Through lessons learned and success stories, this chapter showcases real-world applications and the potential for scaling up these solutions. Finally, future perspectives and research directions are outlined, emphasizing emerging

D. Herdiyantoro (✉) · F. H. Hibatullah · Irwandhi · J. Sinaga · D. S. Utami · A. Nurbaity
N. N. Kamaluddin · T. Simarmata (✉)
Faculty of Agriculture, Department of Soil Sciences and Land Resources, Universitas Padjadjaran, Bandung, West Java, Indonesia
e-mail: d.herdiyantoro@unpad.ac.id; tualar.simarmata@unpad.ac.id

F. H. Khumairah
Research Center of Applied Microbiology, National Research and Innovation Agency, Jakarta, Indonesia

D. D. M. Ambarita
Faculty of Agriculture, Universitas Bangka Belitung, West Java, Indonesia

S. Mehnaz, I. Shahid (eds.), *Frontiers in Rhizoengineering*, Microorganisms for Sustainability 65, https://doi.org/10.1007/978-981-92-3665-7_3

trends, research gaps, and strategies for broader adoption. This chapter aims to provide a thorough understanding of the complexities involved in developing climate-resilient beneficial bioinocula, offering valuable insights for researchers, practitioners, and policymakers committed to advancing sustainable agriculture.

Keywords Beneficial microbes · Biofertilizers · Biological agents · Climate change · Soil health · Marginal soils

3.1 Introduction

Global agricultural systems are at a critical juncture, challenged by the dual pressures of accelerating food demand and the escalating effects of climate change. The urgent need to sustainably feed a projected global population of over 9.7 billion by 2050 continues to place immense strain on agricultural productivity and the planet's finite natural resources (FAO 2017). Concurrently, climate change is intensifying the frequency and severity of both biotic (such as pests and pathogens) and abiotic stresses (such as drought, salinity, soil acidity, flooding, and heatwaves), which directly threaten crop yields, farming livelihoods, and global food security (IPCC 2022).

Recent studies have indicated that soil acidification leads to an average crop yield reduction of 13.7%, with vegetables being particularly sensitive. This reduction is attributed to a 25.4% decrease in root length and the diminished availability of essential nutrients such as calcium, magnesium, and potassium (Li et al. 2024). Salinity stress, which is increasingly prevalent due to factors such as rising sea levels and improper irrigation practices, disrupts plant water uptake, causes ion toxicity, and impairs metabolic functions, leading to reduced germination, growth, and photosynthesis (Parihar et al. 2020). Flooding events exacerbate soil oxygen depletion, alter microbial communities, and damage crop root systems, further increasing agricultural vulnerability under a changing climate (Francioli et al. 2021). These stressors also degrade soil security, reduce agroecosystem resilience, and accelerate biodiversity loss, undermining the ecological foundations required for long-term sustainable agriculture.

Traditional agricultural intensification, characterized by the excessive use of synthetic fertilizers and chemical pesticides, has led to short-term yield improvements at the expense of long-term ecological sustainability. These practices have significantly contributed to soil degradation, biodiversity loss, water contamination, and elevated greenhouse gas emissions (FAO 2021; Mahmud et al. 2023). Additionally, they disrupt the natural soil microbiome and diminish microbial diversity, thereby weakening the intrinsic stress-buffering capacity of agroecosystems (Zhou et al. 2022). Given these escalating environmental costs, there is an urgent need to transition to more ecologically sound and regenerative agricultural practices that enhance resilience and sustainability. One promising strategy is the application of beneficial bioinocula, which are microbial inputs composed of plant growth-promoting bacteria, mycorrhizal fungi, and microbial consortia. These have gained increasing attention for their capacity to improve plant nutrient uptake and enhance tolerance to both biotic and abiotic stresses under changing climatic conditions (Backer et al. 2023; Sinha et al. 2022).

Bioinocula, including plant growth-promoting rhizobacteria (PGPR), arbuscular mycorrhizal fungi (AMF), and nitrogen-fixing symbionts, are natural or formulated microbial agents that colonize plant roots and confer multiple benefits. These include enhancing nutrient acquisition, improving plant hormone regulation, increasing tolerance to abiotic stressors such as drought and salinity, and providing biocontrol against pathogens (Backer et al. 2018; Compant et al. 2019). Beyond direct plant benefits, bioinocula play a crucial role in improving soil health, restoring degraded soils, and enhancing soil security by stabilizing soil structure and promoting microbial biodiversity (Lehmann et al. 2020).

Despite the promising potential of bioinocula in sustainable agriculture, their widespread adoption remains limited because of several critical practical challenges. Chief among these is the inconsistent field performance of microbial inoculants under diverse climatic and edaphic conditions, which is a major concern. Although many strains exhibit strong efficacy in controlled laboratory or greenhouse environments, their effectiveness often diminishes in open-field applications owing to complex and dynamic interactions among plant genotypes, native soil microbiomes, and fluctuating environmental stressors (Malusá and Grządziel 2023; Vassileva et al. 2022). Additionally, numerous indigenous and potentially climate-resilient microbial strains remain underutilized, primarily because of technological constraints in their isolation, functional characterization, and large-scale production. Advancing high-throughput screening, formulation stability, and adaptive consortia design will be critical for overcoming these barriers and enhancing the reliability of bioinocula under real-world agricultural conditions.

Another major constraint limiting the large-scale adoption of bioinocula is the challenge of their formulation and delivery. Effective formulations must preserve microbial viability, ensure shelf stability, and maintain compatibility with diverse soil types and cropping systems. Achieving these goals requires advanced carrier materials and encapsulation technologies, which often demand significant research investments and technical precision (Vassileva et al. 2023). Moreover, regulatory barriers, lack of standardized quality control protocols, and limited farmer awareness or access to extension services further impede their commercialization and widespread use (Mitter et al. 2021). Bridging these gaps is essential for integrating bioinocula into climate-resilient and sustainable farming systems in the future.

In the face of increasing climate volatility, developing climate-resilient bioinocula and microbial strains capable of maintaining functional stability across diverse and stress-prone agroecosystems has become a strategic imperative for sustainable agriculture. These bioinoculants must tolerate environmental extremes, such as heat, drought, and salinity, while functioning synergistically with plant physiology to reduce stress impacts and optimize resource use efficiency. Recent advancements in multi-omics technologies, synthetic biology, and rhizosphere microbiome engineering have enabled the development of next-generation bioinocula tailored to dynamic field conditions (Ali et al. 2023; Zhu et al. 2023). These tools provide powerful means for designing microbial consortia with enhanced adaptability, resilience, and functional diversity.

This chapter critically examines the development and deployment of climate-resilient beneficial bioinocula as a sustainable strategy for mitigating biotic and abiotic stresses in agriculture. It outlines the essential traits and functional attributes required for bioinocula to enhance crop resilience and soil health under climate-induced stress. This chapter further explores the agricultural stress landscape, strategies for microbial strain selection, formulation and delivery challenges, field-level variability, and socio-regulatory limitations that hinder widespread adoption. Emphasis is placed on recent innovations in microbial technologies, including multi-omics, synthetic consortia, and digital farming. By integrating bioinocula into climate-smart agronomy and enabling policy frameworks, this chapter aims to inform researchers, practitioners, and policymakers about advancing sustainable agriculture, food security, and soil security in a changing climate.

By effectively addressing these multifaceted challenges, the full potential of bioinocula as a foundational component of sustainable agriculture can be unlocked. Their successful integration into farming systems can enhance food security, improve crop resilience against climate-induced stresses, and restore soil health. In the context of a rapidly warming world, leveraging bioinocula offers a promising pathway for building more climate-resilient agroecosystems that support long-term productivity and environmental sustainability.

3.2 Functional Mechanisms of Bioinocula

3.2.1 Definition and Types of Bioinocula

Bioinocula are preparations containing beneficial microorganisms, such as bacteria, fungi, or their consortia, which are introduced into soil, seeds, or plants to enhance growth, nutrient uptake, and stress tolerance (Malusá et al. 2012; Singh et al. 2022). These microbial formulations play a vital role in sustainable agriculture by improving soil fertility, suppressing pathogens, and mitigating abiotic stresses, such as drought and salinity (Vassileva et al. 2023). They function by modifying the native soil microbiome, increasing microbial diversity, and promoting plant health and resilience in the face of environmental challenges (Ali et al. 2023). Recognized as key biological inputs, bioinocula contribute significantly to crop productivity and soil security while reducing the reliance on chemical fertilizers and pesticides (Mitter et al. 2021). The types of bioinocula, based on their functions and roles in agriculture, are briefly presented in Table 3.1.

Table 3.1 Types of bioinocula, their functions, and contributions to sustainable agriculture

Type of bioinocula	Primary functions	Contribution to sustainable agriculture	Quantitative contribution	References
Symbiotic nitrogen fixers	Fix atmospheric nitrogen in association with legumes	Improve soil N status and reduce synthetic N fertilizer dependency	Reduce N fertilizer use by 30–60%; increase yield by 10–25%	Kumar et al. (2023b) and Singh et al. (2022)
Nonsymbiotic nitrogen fixers	Fix nitrogen independently in the rhizosphere	Enhance soil fertility in cereals and nonlegumes	Increase soil N content by 15–25%; yield improvement up to 20%	Ali et al. (2023)
Phosphate-solubilizing microbes	Solubilize insoluble phosphorus compounds	Improve phosphorus availability and uptake	Reduce P fertilizer input by 25–50%; increase P uptake by 30–60%	Li et al. (2022) and Vassileva et al. (2023)
Potassium-solubilizing microbes	Release K from minerals and soil particles	Improve K availability and strengthen stress tolerance	Increase K uptake by 20–40%; improve yield by 10–18%	Kumar et al. (2022)
Mycorrhizal fungi (AMF)	Enhance nutrient and water absorption through root symbiosis	Improve drought/salinity tolerance and nutrient uptake	Reduce irrigation needs by 20–30%; increase P uptake by 40%; improve yield by 15–25%	Zhao et al. (2023)
Biocontrol agents	Suppress pathogens via antibiosis, competition, and induced systemic resistance (ISR)	Reduce chemical pesticide use and promote plant health	Decrease disease incidence by 40–70%; reduce pesticide use by 30–50%	Li et al. (2022)
Plant growth-promoting rhizobacteria (PGPR)	Produce phytohormones, enhance nutrients, and induce resistance	Improve plant growth, stress tolerance, and nutrient efficiency	Increase crop yield by 10–35%; improve nutrient use efficiency by 20–50%	Singh et al. (2022) and Trivedi et al. (2020)
Decomposers	Degrade organic matter to release nutrients	Enhance soil organic matter turnover and nutrient cycling	Increase soil organic carbon by 15–30%; enhance nutrient mineralization by 20–40%	Sharma and Sharma (2022)

3.2.2 *Mechanisms of Bioinocula in Addressing Stresses*

Bioinocula mitigate biotic and abiotic stresses through diverse mechanisms that promote plant health and resilience. These include nitrogen fixation, phosphate and potassium solubilization, phytohormone production, and enhancement of root architecture for improved nutrient and water uptake by the plant. They also suppress

Table 3.2 Mechanisms of bioinocula in addressing stress across different ecosystems in food and tree crops

Agroecosystem/ type of stress	Mechanisms	Microbes	Contribution	References
Saline stress/soil salt toxicity, osmotic stress	Osmolyte production, ACC deaminase activity, Na^+ exclusion, antioxidant enzyme induction	Halotolerant PGPR, *Pseudomonas*, *Bacillus*, *Azospirillum*	Increase grain yield of rice by 18–25% under moderate salinity	Kumar et al. (2023a, b)
Acid stress/soil low pH, Al^{3+} toxicity	Acid tolerance, organic acid secretion, P-solubilization, Al-chelation	*Penicillium*, *Trichoderma*, *Burkholderia*, *Rhizobium*	Increase soybean yield by 20% and P availability by 30%	Rao (2022)
Drought/water deficit	EPS production, deeper rooting, ABA modulation, ROS scavenging	*Azospirillum*, *Bacillus subtilis*, *Enterobacter*	Increase maize biomass by 15–22%; WUE by 12%	Kumar et al. (2023a)
Temperate agroforestry systems seasonal temperature stress	Cold/heat shock protein expression, membrane stabilization	*Pseudomonas fluorescens*, Mycorrhizae	Increase sapling survival rate in forest nurseries by 30%	Singh et al. (2022)
Degraded upland soil/low fertility, erosion	Nutrient solubilization, mycorrhizal associations, rhizosheath formation	Arbuscular mycorrhizal fungi (AMF), phosphate-solubilizing bacteria (PSB), Actinomycetes	Increase cassava and maize yield by 20–35% on marginal lands	Gamalero and Glick (2022)

PGPR plant growth-promoting rhizobacteria, *EPS* exopolysaccharide, *ROS* reactive oxygen species, *WUE* water use efficiency

pathogens through antibiosis, competition, and the induction of systemic resistance. Under drought, salinity, or heat stress, bioinocula help to maintain cellular homeostasis and reduce oxidative damage. By modulating rhizosphere microbial communities, they improve crop performance while reducing reliance on chemical inputs, contributing to sustainable and climate-resilient agriculture (Trivedi et al. 2020; Kumar et al. 2023a, b; Gamalero and Glick 2022). A summary of the mechanisms and contributions of bioinocula, along with their potential applications, is provided in Table 3.2.

3.3 Microbial Resource Development and Characterization

3.3.1 Isolation and Screening for Stress Tolerance

The development of bioinoculants that are resistant to environmental stress (both biotic and abiotic) in the initial stages is crucial. Development begins by isolating microbes from extreme environments, such as saline, acidic, or polluted soil. These environments can be a source of microbes that have adapted to environmental stress. Microbes can be taken from the rhizosphere of healthy plants and can thrive in stressed land (Pirapak et al. 2022). This is in line with recent studies that have successfully isolated microbes from the rhizosphere of food crops and weeds in saline land (Aini et al. 2022) and from legume root nodules (Seidu et al. 2025). Sultana et al. (2023) revealed that *Azotobacter* and *Azospirillum* isolates from dry land were capable of surviving at high temperatures (50 °C), high salt concentrations (7% NaCl), and drought stress while also exhibiting potential for biocontrol against plant pathogens. Microbes obtained from stressed environments were screened.

Microbial screening is conducted by assessing microbial tolerance to biotic and abiotic stresses. The ability of microbes to withstand environmental stress is assessed by salinity stress in the form of a NaCl tolerance test (Irwandhi et al. 2024), drought stress using a medium with PEG 6000 (Sharath et al. 2021), acidity stress with a graded pH medium (Kamaluddin et al. 2025), and degraded upland soil by determining their ability to stimulate growth, such as nitrogen fixation, production of indole-3-acetic acid (IAA), and phosphate solubilization (Pirapak et al. 2022). Recent studies have indicated that *Klebsiella variicola* isolated from the wheat rhizosphere exhibits high antioxidant activity, produces indole-3-acetic acid (IAA), and dissolves phosphate under high-salinity conditions (Kusale et al. 2021). Inoculation with plant growth-promoting bacteria, such as *Pseudomonas*, *Azospirillum*, and *Rhizobium*, has been shown to increase nitrogen uptake and promote plant growth in saline soils. This condition occurs because microbes can dissolve phosphate, potassium, and zinc and produce ACC deaminase, IAA, siderophores, and EPS (Tiwari et al. 2025). Kankariya et al. (2025) revealed that *Pseudomonas fluorescens* isolates are resistant to salinity stress and can inhibit the growth of pathogenic fungi. Therefore, the selection of microbes that are resistant to multiple abiotic and biotic stresses is crucial for producing practical and durable bioinoculants in the field while supporting sustainable agriculture amid climate change.

3.3.2 Omics and Systems Biology Approaches

Omics refers to a field of biological sciences that involves the comprehensive study and analysis of large-scale biological data to understand the structure, function, and interactions of molecules within a biological system. It encompasses various disciplines, such as genomics (study of genes), transcriptomics (RNA expression), proteomics (proteins), metabolomics (metabolites), epigenomics (epigenetic modifications), and microbiomics (microbial communities). By adopting a holistic approach, omics technologies provide insights into complex biological processes and are widely used in fields such as medicine, agriculture, and environmental science to identify biomarkers, improve disease diagnosis, develop targeted therapies, and enhance crop productivity and sustainability (Rafique et al. 2024).

The exponential growth in omics technologies, such as genomics, transcriptomics, proteomics, and metabolomics, has revolutionized the development and characterization of microbial resources, particularly in the fields of microbial biotechnology, agriculture, and environmental engineering. By integrating high-throughput data from various omics layers, systems biology enables a holistic understanding of microbial physiology, ecology, and functional potential, which are crucial for precise microbial engineering and resource optimization.

The integration of omics technologies and systems biology with synthetic biology has fundamentally transformed microbial resource development and characterization over the past 5 years. Omics approaches, including genomics, transcriptomics, proteomics, and metabolomics, offer comprehensive and high-resolution insights into microbial physiology, diversity, and functional potential. For instance, metagenomics enables the reconstruction of microbial genomes directly from environmental samples, revealing the presence and roles of uncultured and rare microbial species. This is particularly valuable for mining novel enzymes, biosynthetic gene clusters, and stress adaptation mechanisms that are otherwise inaccessible through traditional cultivation-based methods (Carruthers and Lee 2022; Nazir et al. 2024).

3.3.3 Genomics and Metagenomics as Baselines for Resource Exploration

Whole-genome sequencing and metagenomic approaches have dramatically expanded our ability to explore and catalog microbial diversity in a variety of environments, including extreme, nutrient-poor, and contaminated ecosystems. These techniques enable the identification of novel microbial taxa and functional genes, which can be further developed into microbial consortia or individual strains for agricultural, industrial, and environmental applications (de Abreu et al. 2021; Pasolli et al. 2020). In particular, metagenome-assembled genomes (MAGs) have allowed the discovery of previously uncultivable microorganisms, providing access to underexplored microbial *dark matter* and unlocking a vast reservoir of genetic potential (Bowers et al. 2017).

3.3.4 Multi-omics Integration and Systems-Level Modeling

The integration of multi-omics data, including genomic, transcriptomic, proteomic, and metabolomic data, through systems biology approaches allows for the comprehensive modeling of microbial systems and their interactions with hosts or environments. For instance, genome-scale metabolic models (GEMs) are widely used to simulate metabolic fluxes, optimize metabolic pathways, and predict system responses to genetic modifications or environmental perturbations (Machado et al.

2018). This systems-level understanding is crucial for engineering microbial strains with enhanced capabilities for nitrogen fixation, phosphate solubilization, and pollutant degradation.

3.3.5 Engineering and Synthetic Biology for Microbial Enhancement

Synthetic biology, coupled with omics insights, provides powerful tools for the rational design and construction of microbial strains with tailored functions. CRISPR-Cas9-based genome editing, synthetic circuits, and modular pathway assembly have been employed to enhance beneficial traits, such as biofertilizer efficiency, stress tolerance, and antimicrobial metabolite production (Liao et al. 2019). For example, engineered rhizobia with synthetic nitrogenase pathways have demonstrated the potential to improve symbiotic nitrogen fixation in nonleguminous crops. Moreover, microbial consortia design has become a key synthetic biology frontier. Rather than focusing on single strains, systems biology aids in constructing synthetic communities with division-of-labor strategies to optimize resource use, enhance resilience, and perform complex biotransformations (Kehe et al. 2019).

3.3.6 Functional Characterization and Bioinformatics Frameworks

Functional annotation of microbial genomes and metagenomes relies on advanced bioinformatics pipelines that integrate omics data with machine learning (ML) and artificial intelligence (AI). Tools such as antiSMASH, Prokka, and MetaErg assist in identifying biosynthetic gene clusters and metabolic potential, while others model host-microbe or microbe-microbe interactions to predict community dynamics (Blin et al. 2021). These approaches are essential for evaluating the biotechnological potential of microbial isolates and communities.

3.3.7 Applications in Agriculture and Environmental Sustainability

In agriculture, omics-driven microbial resource development supports the production of next-generation biofertilizers, biopesticides, and plant growth-promoting rhizobacteria (PGPR). Microbes characterized using systems biology approaches can be tailored to specific soil or crop types, improving nutrient uptake and stress resistance (Mendes et al. 2022). Similarly, in environmental applications, engineered microbial strains are used for the bioremediation of heavy metals, plastics, and organic pollutants, benefiting from omics-guided metabolic optimization (Zhang et al. 2020).

3.4 Bioinocula Formulation and Production

Effective and efficient bioinoculants play a key role in the agricultural sector. A good bioinoculant is determined by selecting the carrier media, formulation, and production system. Effective and efficient bioinoculants require a well-established production method and play crucial roles in the agricultural sector. The designation of appropriate carrier materials and formulations is crucial for the success of bioinoculants (Baliyan et al. 2022; Khan et al. 2023). Simultaneously, standardization and quality control are crucial for ensuring effective utilization of bioinoculants in the field.

3.4.1 Carrier Selection and Formulations

One of the key components in developing bioinoculants is the selection of carriers or materials used to deliver beneficial microorganisms into the soil or plants in the field. These carriers must allow microorganisms to remain alive and active during storage, distribution, and application. Several factors need to be considered when selecting a carrier, including:

1. *Ability to Support Microbial Growth:* Studies have shown that an appropriate carrier may provide a good environment that supports microbial growth and maintains viability by providing nutrition, moisture, and physical protection of microbes during storage time (Bolan et al. 2023; Soniari and Sutari 2022).
2. *Availability and Cost:* The carrier used should be readily available and have a low production cost. Some commonly used natural carriers include soil, compost, peat, and other organic materials.
3. *Stability Under Field Conditions:* The carrier must also protect microorganisms from extreme environmental conditions such as high temperatures, low humidity, or excessive ultraviolet radiation (Gong et al. 2022; Velloso et al. 2023).
4. *Ease of Application:* The best bioinoculant should be tailored to the specific needs of the plants or soil type, with a formulation that is easy to apply in the field, whether in powder, liquid, or granule form, and adaptable to various application methods (e.g., spraying or direct soil application) (Rojas-Sánchez et al. 2022).

3.4.2 Production System

The production system of bioinoculants involves various technical and industrial aspects crucial for ensuring the quality, efficiency, and sustainability of the final product. The production process must consider several critical steps, such as microbial inoculation, culture maintenance, and packaging, which play a significant role in ensuring the successful application of bioinoculants in the field. The production system of bioinoculants involves several key technical and industrial aspects that are essential to ensuring the quality, efficiency, and sustainability of the final product:

1. *Microbial Inoculation:* The microbial inoculation process in bioinoculant production begins with the selection and isolation of effective microbial strains, followed by cultivation in suitable media, typically under controlled conditions to ensure optimal growth. Additionally, testing the compatibility and effectiveness of the selected strains is crucial to ensure that the resulting bioinoculants are of high quality, efficient, and ready for field application (Negi et al. 2024; Wang et al. 2024).
2. *Maintenance:* Maintaining microbial cultures in bioinoculant production requires controlling environmental conditions, selecting the appropriate carrier, and continuously monitoring the quality and quantity of microorganisms. This strategy is crucial to ensure that the resulting bioinoculants remain effective and of high quality until they are applied in the field. During this phase, microorganisms must be maintained under proper conditions to ensure that the quantity and quality of the microbial culture remain high (Khan et al. 2023; Suman et al. 2022).
3. *Fermentation:* Many bioinoculants are produced using large-scale fermentation processes. Fermentation enables the production of microorganisms in large quantities, which can be applied on an agricultural scale in the future. Factors such as temperature, pH, and nutrients are crucial for ensuring the survival and activity of microorganisms (Bianco et al. 2023; Ganeshan et al. 2021).
4. *Packaging and Storage:* Bioinoculants must be packaged after production to ensure the stability of microorganisms during distribution and storage. Packaging should also consider minimizing exposure to oxygen, moisture, and extreme temperatures (Härdter et al. 2021; Park et al. 2024; Sun et al. 2022).

3.4.3 Standard and Quality Control

The application of bioinoculants or microbial inoculants in sustainable agriculture has gained momentum globally because of their potential to enhance crop productivity, soil health, and environmental safety. However, the success and consistency of bioinoculant performance depend heavily on rigorous standardization and quality control (QC) during formulation and production. Without well-defined quality criteria and regulatory benchmarks, the efficacy, stability, and credibility of bioinoculants are compromised, hindering their widespread adoption by farmers and agricultural stakeholders.

3.4.4 The Importance of Standardization in Bioinoculant Development

Standardization refers to the establishment of specific benchmarks related to the identity, viability, efficacy, and purity of microbial formulations. One of the major challenges in the bioinoculant industry is the lack of global harmonization of quality parameters, as each country or region often develops its own guidelines based on local microbial strains and cropping systems (Vassilev et al. 2015). The most critical standards for bioinoculants include the following:

1. *Viable Cell Count:* Generally, a minimum count of 10^7–10^9 colony-forming units (CFU) g^{-1} or CFU mL^{-1} is required at the time of manufacture, depending on the strain and formulation type (Herrmann and Lesueur 2013).
2. *Microbial Identity Confirmation:* Molecular tools such as 16S rRNA sequencing, rep-PCR, and MALDI-TOF should be used to confirm strain identity to avoid contamination or strain substitution (Bhattacharyya and Jha 2012).
3. *Absence of Contaminants:* Bioinoculants must be free from human or plant pathogens to ensure biosafety during application (Keswani et al. 2020).

Despite guidelines established by organizations such as the Food and Agriculture Organization (FAO) and Bureau of Indian Standards (BIS), enforcement at the manufacturing and distribution levels remains inconsistent, especially in low- and middle-income countries.

3.4.5 Critical Aspects of Quality Control in Bioinoculant Formulation

Quality control must be implemented across the entire production pipeline, from strain selection and culture maintenance to formulation, packaging, and storage. Critical aspects of quality control in bioinoculant formulation must be addressed at every stage of production to ensure product consistency and efficacy:

1. *Strain Authentication and Maintenance:* QC begins with authentic and genetically stable strains. Genetic drift and plasmid loss can result in performance variability. Therefore, cryopreservation or lyophilization in culture collections and periodic revalidation using molecular tools are essential (Compant et al. 2019).
2. *Formulation and Shelf-Life Stability:* Formulation affects microbial viability and delivery efficacy. Carrier-based formulations (e.g., peat, lignite, charcoal, and

talc) are still widely used because of their low cost, but liquid- and polymer-based formulations offer better stability, shelf life (up to 1–2 years), and field performance (Malusá and Vassilev 2014). However, each formulation requires tailored QC parameters, such as pH, moisture content, carrier sterility, compatibility testing with other bioagents or agrochemicals, and shelf-life assays under varying storage conditions.

3. *Bioefficacy and Field Validation:* Laboratory and greenhouse assays are insufficient proxies for field performance because of environmental variability. Therefore, multilocation field trials must be standardized using a randomized block design (RBD) and proper controls. Parameters such as root colonization, nodulation efficiency, and nutrient solubilization should be routinely evaluated (Choudhary et al. 2022).

3.4.6 Regulatory Frameworks and Certification Systems

Robust regulatory systems are critical for ensuring the quality and trustworthiness of bioinoculants. However, many developing countries lack national certification systems for microbial products, resulting in the circulation of substandard and counterfeit formulations. For example, a survey in India reported that over 50% of commercial biofertilizers did not meet the prescribed microbial load standards (Sinha et al. 2017). In contrast, countries such as Brazil have implemented MAPA regulations requiring manufacturers to register strains and demonstrate consistent QC outcomes. Similarly, the European Union's new Fertilizing Products Regulation (EU 2019/1009) categorizes microbial plant biostimulants under CE marking, requiring compliance with purity, safety, and efficacy standards.

3.5 Advanced Technologies and Innovations

3.5.1 Advances in Bioinocula Formulation and Application

Currently, developments in bioinocula technologies are addressing the challenges of efficacy and efficiency, shelf life, and environmental adaptability. The key technologies and innovations are as follows:

1. *Carrier Formulation and Optimization:* Use of biodegradable polymers (e.g., alginate and starch blends) and nanomaterials as carriers enhances microbial survival and controlled release in the soil. Significant improvements in the delivery and efficacy of microbial inoculants have been achieved through advancements in carrier material development. The use of biodegradable polymers, such as alginate, starch blends, and cellulose derivatives, offers a protective matrix that enhances the viability of microbes during storage and application. The car-

rier retains moisture, buffers against environmental stress, and facilitates the gradual release of microbes into the rhizosphere. Furthermore, incorporating nanomaterials (e.g., nanoclays or silica nanoparticles) enables precise control over the release of nutrients and microbes, improves adhesion to the root surface, and can extend the shelf life and field performance of bioinoculants. In the specific case of encapsulated plant probiotics utilizing nanomaterials as biofertilizers, it is imperative to address prevailing technological and knowledge gaps. This includes understanding the interactions between nanomaterials and encapsulated microbes and optimizing the timing and site-specific delivery of these formulations. It is imperative that these challenges be addressed to optimize the efficacy of nano-biofertilizers and facilitate the transition towards more sustainable agricultural practices. The advancement of sustainable agriculture is contingent on emerging technologies such as next-generation sequencing, microbial genetic engineering, in silico modelling, and synthetic biology, particularly in the development of innovative bioinoculant formulations (Kumar et al. 2022).

2. *Inoculant Consortia:* Modern bioinoculant formulations are increasingly being developed using a multi-strain consortium approach, which combines various types of microbes with complementary functions. No longer relying on a single type of microbe, these formulations now utilize a combination of microbes, such as plant growth-promoting rhizobacteria (PGPR), that play a role in stimulating plant growth through the production of growth hormones, mycorrhizal microbes that help expand root tissue, thereby increasing nutrient and water absorption, and nitrogen-fixing, phosphate solubilizing, and potassium solubilizing bacteria that provide nutrient availability for plants. The synergy between these microbes has been proven to increase the efficiency of nitrogen uptake by plants, strengthen resistance to abiotic stresses such as drought and salinity, and contribute to increased crop productivity. This approach is key to delivering more effective and sustainable biological solutions in the agricultural sector. The application of multiple compatible microbial species or strains, known as microbial consortia, has been demonstrated to enhance plant health and disease control. Furthermore, these consortia represent a significant advancement in synthetic biology because of their adaptive complexity and broad enzymatic capabilities. However, further field studies are required to optimize their applications (Sadvakasova et al. 2023).
3. *Encapsulation and Microencapsulation:* Encapsulation involves encasing microbes in a protective micro-sized material. This material maintains microbial survival by protecting them from extreme temperatures, ultraviolet light exposure, and drought conditions. Additionally, this technique enables the slow, targeted release of microbes around the root zone or rhizosphere, so they can more effectively support plant growth. This protection also increases the shelf life of bioinoculants and improves the efficiency of field applications, making encapsulation an important innovation in modern biofertilizer development. Microbial bioengineering provides an alternative to pesticides for enhancing plant immunity by utilizing natural polysaccharides, particularly alginates, as carriers for beneficial microbes. However, industrial-scale encapsulation still necessitates

technological advancements, with polysaccharide composites enabling the customization of bioproduct properties (Ali et al. 2023).

4. *Smart Application Systems:* Smart application systems in agriculture, particularly for food and vegetable crops, integrate IoT sensors, artificial intelligence (AI), and data analytics to optimize resource use and maximize yields. These systems monitor real-time data on soil moisture, nutrient levels, weather, and crop health, thereby enabling precise irrigation, fertilization, and pest management. In rice and maize cultivation, smart systems have increased yield productivity by 15–25%, whereas in vegetable crops, such as tomatoes and spinach, yield increments of up to 30% have been reported owing to timely and targeted interventions (Singh and Kumar 2023). Additionally, these systems reduce water and fertilizer use by 20–40%, lower production costs, and minimize environmental impacts. The benefits include improved crop quality, enhanced food safety, efficient input utilization, and higher profitability, which support both sustainable agriculture and food security.

3.5.2 Use of Biotechnology and Genomics

The development of molecular biotechnology and genomics is currently booming in the development of beneficial bioinocula in agricultural practices, especially to increase microbial resistance to biotic (such as pathogens and pests) and abiotic (such as drought, salinity, and extreme temperatures) stresses (Chen et al. 2022a, b; Minchev et al. 2021). For example, through whole-genome sequencing, researchers can identify important genes in microbes involved in the synthesis of anti-stress compounds, such as exopolysaccharides (EPS), trehalose, or antioxidant enzymes (Cao et al. 2025; Kocaefe-Özşen et al. 2022; Nishu et al. 2022). Genetic characterization can also be assessed, such as *Pseudomonas fluorescens* and *Bacillus subtilis*, two genera of bacteria known to produce antibiotic compounds, siderophores, and lysozyme enzymes that can suppress the growth of plant pathogens (Albert et al. 2024; Baard et al. 2023). Genomics allows researchers to select the most promising strains based on the presence and expression of these genes. Some key roles of biotechnology include microbial genetic engineering, screening and selection of superior strains, and stable bioinoculum formulation. Genomics provides deep insights into stress resistance mechanisms and data-driven bioinoculum development.

Reports on the application of biotechnology in the development of innovative and sustainable bioinoculum, including the genetic modification of bacteria such as *Pseudomonas* and *Bacillus* to increase the production of phosphate-solubilizing enzymes or nitrogen fixation (Khumairah et al. 2019, 2022), as well as the use of CRISPR-Cas9 to edit microbial genes to be more resistant to high temperatures or salinity, show how genetic engineering can strengthen microbial resilience (Rottinghaus et al. 2022). In addition, high-throughput screening methods allow for the more efficient identification of microbes with plant growth-promoting (PGPR) properties, and the development of synergistic microbial consortia improves root colonization. On the formulation side, nanomaterial- or biopolymer-based encapsulation technologies protect microbes from extreme conditions, and the use of organic carriers extends their viability in the field (Ajeng et al. 2023; Yen et al. 2025). These innovations prove the crucial role of biotechnology in creating agricultural solutions that are adaptable to climate change.

Genomics has significantly contributed to the development of climate-resilient bioinocula using various advanced approaches. Genomic analysis has enabled the identification of key genes involved in drought and heat tolerance in symbiotic microbes, whereas soil metagenomic studies have revealed the complexity of microbial communities supporting plant resilience (Azri et al. 2024; Bierenbroodspot et al. 2024). Furthermore, microbial genome mapping has facilitated the design of superior strains with high precision, and synthetic biology approaches have opened up opportunities to create microbes with new capabilities such as more efficient production of plant growth hormones (IAA, ACC deaminase) (Kumar et al. 2021; Yu et al. 2021; Zaid et al. 2022). These advances in genomics provide a strong foundation for developing new-generation bioinocula that are more adaptive to environmental challenges.

In dealing with abiotic stresses, such as saline soil, halotolerant bacteria have been studied using a transcriptomic approach to identify gene expression responses under high-salinity conditions. Genes involved in osmotic pressure regulation, ion transport, and the biosynthesis of protective compounds are the main targets for increasing microbial resistance through genetic engineering or adaptive selection approaches (Liu et al. 2022a, b). In addition, metagenomics allows the exploration of soil microbes that cannot be cultured in the laboratory (Simarmata et al. 2023). This opens up opportunities to identify new microorganisms that support plant growth in marginal lands, such as drylands or lands with extreme pH. Metagenomic data can be used to design consortium-based bioinocula. Machine learning and bioinformatics are also used to model microbe-plant-environment interactions (Imran and Gul 2025; Wani et al. 2024). This helps predict bioinocula performance in various climate and soil conditions and supports the formulation of more stable and functionally effective end products.

Field implementation has been carried out, for example, in the development of bioinoculants based on *Rhizobium leguminosarum* and *Bradyrhizobium japonicum*, which have been modified to increase nitrogen fixation and resistance to high temperatures in soybean farming systems in tropical areas (Wheatley et al. 2020; Szczerba et al. 2024). Overall, the integration of biotechnology and genomics in bioinocula development accelerates the innovation process of functional microbial products that are more climate-resilient and support agricultural sustainability through reduced chemical inputs and increased resilience to extreme environments.

3.5.3 Synthetic Community (SynCom)

Synthetic microbial communities (SynComs) have emerged as a powerful innovation in sustainable agriculture, enabling the rational design of compatible microbial consortia to enhance plant growth and stress tolerance in crops. Unlike single-strain inoculants, SynComs are authentically assembled based on the properties and functional interactions of microbes, often derived from core or key taxa in the plant microbiome. Zhuang et al. (2021) successfully developed a six-strain *Pseudomonas* SynCom from the garlic rhizosphere, which significantly improved plant growth and health. In another study, a SynCom derived from the wheat rhizosphere suppressed the soil-borne pathogen *Rhizoctonia solani*, although its effectiveness depended on the specific community composition and soil context (Yin et al. 2022). These findings underscore the importance of selecting functionally complementary strains and understanding their ecological dynamics to optimize SynCom performance in the field environment.

Technological advances, such as high-throughput sequencing, synthetic biology, and computational modeling, have accelerated the development and optimization of SynComs. These tools facilitate the identification of microbial traits important for plant interactions and the prediction of the outcomes of microbial cocultures (Nazir et al. 2024). Additionally, SynCom applications are being tested in controlled environment systems, such as hydroponics and vertical farming, which have been shown to enhance nutrient uptake and plant growth under stressful conditions (Huang 2024). Despite these advances, challenges remain regarding SynCom stability, colonization efficiency, and persistence under fluctuating conditions. Future research should focus on the ecological robustness and metabolomic interactions of SynComs to ensure consistent field-level benefits.

3.6 Attributing Factors of Biofertilizer Effectivity

3.6.1 Factors Affecting Efficacy in the Field

Biofertilizers, composed of beneficial microorganisms, are increasingly recognized for their roles in sustainable agriculture. However, their inconsistent performance under field conditions necessitates a comprehensive understanding of the various influencing factors.

3.6.1.1 Specificity to Soil and Environmental Conditions

The efficacy of biofertilizers is often contingent on specific soil types and environmental parameters. Microbial strains exhibit optimal activity within certain pH ranges, salinity levels, and temperature ranges. For instance, halotolerant bacteria have demonstrated enhanced performance in saline soils, underscoring the importance of aligning microbial characteristics with environmental conditions (Masmoudi et al. 2023).

3.6.1.2 Biofertilizer Specificity Toward Plant Species

Biofertilizers may exhibit specificity towards particular plant species because of unique root exudates and rhizosphere interactions. Studies have shown that certain microbial strains promote the growth of specific crops, such as eggplant, when formulated with appropriate carriers. Therefore, evaluating the compatibility of biofertilizers across various plant species is crucial for their broader applicability (Tripti et al. 2022).

3.6.1.3 Inadequate Population Density Compared to Indigenous Microbes

The microbial populations introduced in biofertilizers often compete with the native soil microbiota. To establish dominance and exert beneficial effects, the inoculum density must exceed that of indigenous microbes. Research has indicated that bio-organic fertilizers can stimulate native *Pseudomonas* populations, thereby enhancing disease suppression (Tao et al. 2020).

3.6.1.4 Unsuitable Carrier Materials for Microbial Formulations

Carrier materials play a pivotal role in maintaining microbial viability during storage and applications. Materials such as biochar and fly ash have been identified as effective carriers that provide a conducive environment for microbial survival. The choice of carrier affects not only the shelf life but also the efficacy of the biofertilizer in the field (Tripti et al. 2022).

3.6.1.5 Preservation and Storage of Biofertilizer Isolates

The method of preserving microbial strains significantly impacts their longevity and effectiveness. Advanced preservation techniques, such as lyophilization and cryopreservation, are commonly used to maintain microbial viability over extended periods. Additionally, outsourcing microbial storage to specialized culture collections, such as the World Federation of Culture Collections (WFCC), provides an efficient and reliable solution. The WFCC represents a global network of culture collections that offers long-term preservation services for microbial strains, ensuring their viability and availability for future use. Similarly, the European Culture Collections' Organization (ECCO) provides microbial storage services that support the conservation of microbial diversity for research and commercial applications.

The use of services from these organizations helps mitigate the risk of genetic drift and ensures that microbial strains retain their functional properties over time (Puchkov 2023).

The performance of biofertilizers in agricultural settings is influenced by the complex interplay of environmental conditions, plant specificity, microbial competition, carrier materials, and preservation methods. Addressing these factors through targeted research and development can enhance the reliability and effectiveness of biofertilizers, thereby contributing to sustainable agricultural practices.

3.6.2 Scaling Laboratory Innovations to Field Applications

Recent advances in biofertilizer research have emphasized their vital role in sustainable agriculture, particularly in enhancing nutrient efficiency, soil health, and crop resilience. Innovations span microbial consortia design, formulation technologies, and precision delivery systems, focusing on translating laboratory successes into field-scale benefits (Kumar et al. 2024a, b). The following key developments highlight the current trends and innovations in biofertilizer research that are driving their transition from laboratory breakthroughs to impactful field applications:

1. *Microbial Consortia Outperform Single Strains:* Mixed inoculants, combining nitrogen fixers, phosphate solubilizers, arbuscular mycorrhizal fungi (AMF), and plant growth-promoting rhizobacteria (PGPR), offer multiple mechanisms of action and consistently outperform single-strain biofertilizers in field trials (Kumar et al. 2024a, b).
2. *Culturomics and Omics-Informed Strain Selection*: High-throughput culturomics, combined with soil metagenomics and multi-omics analysis, enables the isolation of previously uncultivable beneficial microbes and personalized biofertilizer consortia tailored for specific soil-crop systems (Kumar et al. 2024a, b).
3. *Biofilm-Based formulations:* Encapsulating bacteria and fungi within biofilms (EPS-rich matrices) enhances survival, competitiveness, stress resistance, and nutrient solubilization in the field. Studies have shown up to 50% reduction in chemical fertilizer use with biofilm biofertilizers (Kumar et al. 2024a, b).
4. *Nano-biofertilizers and Encapsulation*: Nanotechnology enables the controlled release of nutrients and microbial protection. Nano-encapsulated phosphorus carriers deliver a steady nutrient supply, reducing leaching and supporting field-level plant growth, whereas microencapsulation enhances formulation durability.
5. *Carrier Materials and Shelf-Life Improvements*: Innovative carriers, biochar, agrowaste matrices, and granular organic media help stabilize microbial viability during storage. Advances in fermentation and liquid/solid-state carrier design have extended the shelf life of probiotics and reduced costs for farmers.
6. *Precision Delivery and Real-Time Monitoring:* The integration of soil sensing, GIS (Geographic Information System)/GPS (Global Positioning System) map-

ping, and precision agriculture allows site-specific biofertilizer application. Farmers can dynamically adjust the doses and timing, thereby enhancing ecological efficiency (dos Reis et al. 2024).

7. *Standardizing Field Trials and Regulatory Evolution*: Robust standardized trial protocols (inoculum quality and effectiveness of biofertilizer) are now emerging, along with clearer regulatory frameworks for bioinoculants (e.g., Regulation of the Minister of Agriculture of the Republic of Indonesia No. 01/2019 and Decree of the Minister of Agriculture of the Republic of Indonesia No. 261/2019), helping reduce inconsistency between laboratory and field efficacy.

Innovations in microbial selection, biofilm and nano formulation, carrier design, field delivery technologies, and standardized trials together form a robust translation pipeline from the lab to the field. By combining consortia-based inoculants with precision application and supportive regulatory frameworks, biofertilizers can sustainably replace a substantial portion of chemical fertilizers while improving soil health and crop resilience.

Integrated strategies are needed for the success of biofertilizer applications in the field, including the isolation and selection of superior microbes, biological fertilizer formulations, applications on greenhouse and field scales, monitoring, and even multilocation tests. Some of these strategies include the following:

1. *Laboratory Phase:* Use culturomics and microfluidic testing to screen for complementary, root-colonizing strains with robust traits.
2. *Formulation:* Develop biofilm-encapsulated microbial consortia with nano-enhanced nutrient carriers blended into organic/waste-derived matrices.
3. *Pilot Trials:* Employ precision mapping tools and mobile diagnostics for localized application, following standardized trial protocols.
4. *Monitoring and Scaling:* Real-time sensing and robotics are applied for dynamic feedback, deployment adjustment, and replication of field trials across climates and soil types.

This innovation chain, spanning microbial discovery, advanced formulation, precision agronomy, and standardized trials, forms a robust pipeline to elevate biofertilizer effectiveness from controlled laboratory success to reliable field application.

3.6.3 Best Practices (Case Studies in Indonesia)

Biofertilizers play a crucial role in sustainable agricultural practices by enhancing soil fertility and promoting plant growth. In Indonesia, where soil fertility often limits crop yields, particularly in Inceptisols and Andisols, biofertilizers offer a promising solution to improve soil fertility. Table 3.3 summarizes recent research conducted in Indonesia on the application of various biofertilizers in agriculture. These studies demonstrate the significant benefits of biofertilizers in improving nutrient uptake, soil health, and crop productivity.

Table 3.3 Application of biofertilizers and their benefits

Type of biofertilizer	Benefits achieved from the experiment	References
Nitrogen fixing endophytic bacteria	Improved tomato plant height and endophytic bacterial population in Inceptisols, with a combination of seed and soil treatments	Setiawati et al. (2024)
Phosphate solubilizing bacteria (PSB)	Enhanced growth and yield of red chili with nutrient solutions; increased PSB population, phosphorus availability, and phosphorus uptake	Fitriatin et al. (2024)
Potassium solubilizing biofertilizer (KSB)	Increased K uptake, maize growth (stem diameter), and KSB population on Inceptisols with a dose of 4 kg ha^{-1}	Herdiyantoro et al. (2022)
Bacillus-coated NPK fertilizer	Enhanced tomato growth (height, number of leaves, root length, biomass) using *Bacillus*-coated NPK (BCN) fertilizer	Hindersah et al. (2025)
Arbuscular mycorrhizal fungi (AMF)	Increased root colonization, phosphate uptake, and yield of kangkong in Andisols, with potential reduction of NPK fertilizer usage by 50%	Nurbaity and Mbusango (2024)
Regenerative technologies (e.g., SOBARI, algal biofertilizers)	Enhanced soil health, increased fertilizer and water efficiency, reduced inorganic inputs (25–50%), and increased rice productivity (25–50%) in coastal areas of Indonesia	Simarmata et al. (2025)

A comprehensive summary of various biofertilizers and their applications in Indonesian agriculture, showcasing their effectiveness in enhancing soil fertility and plant growth, is presented in Table 3.3. A wide range of biofertilizers, such as nitrogen-fixing/endophytic bacteria, phosphate-solubilizing bacteria (PSB), potassium-solubilizing bacteria (KSB), *Bacillus* spp., and arbuscular mycorrhizal fungi (AMF), have specific benefits for crops such as maize, tomato, red chili, and kangkong. Each biofertilizer is linked to particular improvements in nutrient uptake (e.g., potassium and phosphorus), plant growth (e.g., height and biomass), and yield, demonstrating the positive role these biofertilizers play in increasing agricultural productivity and sustainability.

A study on regenerative technologies, particularly the System of Organic-Based Aerobic Rice Intensification (SOBARI), which focuses on improving rice farming in the coastal areas of Indonesia, has been added to Table 3.3. This study highlights the benefits of regenerative practices, such as reduced water and fertilizer usage, improved soil health, and increased rice yield, thus contributing to the national goals of sustainable agriculture and climate change mitigation. The incorporation of regenerative technologies emphasizes the growing importance of environmentally friendly practices in Indonesia's agricultural landscape.

The references reflect a diverse range of biofertilizers and innovative agricultural techniques, underscoring the significant strides made toward more sustainable and efficient farming systems in Indonesia. Overall, the studies demonstrate that biofertilizers and regenerative technologies offer promising solutions for improving crop yields while minimizing environmental impact. The use of biofertilizers in agricultural practices is an effective and environmentally friendly strategy for enhancing crop productivity while maintaining soil health. The studies reviewed in this table illustrate that biofertilizers, when applied using appropriate techniques and doses, can improve nutrient availability, support sustainable farming, and reduce the dependency on chemical fertilizers. As the agricultural sector in Indonesia continues to evolve, further research and application of biofertilizers will be essential to address the challenges posed by soil degradation and the need for increased food production.

3.7 Challenges and Future Perspectives

3.7.1 Challenges to the Application of Biofertilizers in Indonesia

The challenges faced in implementing the use of biofertilizers in Indonesia include low application rates by farmers, inconsistency in effectiveness due to variations in product quality, and limited access and distribution of biofertilizers, particularly in remote areas. Despite the recognized benefits of biofertilizers for sustainable agriculture, several key challenges hinder their widespread adoption in Indonesia:

1. *Low Application of Biofertilizers by Farmers:* Research shows that the level of biofertilizer adoption among Indonesian farmers is still low. This is due to several factors, including a lack of understanding of the benefits of biofertilizers, limited information and education, and the inherent habit of using chemical fertilizers (Herdiyantoro et al. 2026). In addition, farmers' perception that biofertilizers are less effective also affects their interest in using these products.
2. *Inconsistent Effectiveness and Variation in Product Quality:* The quality of biofertilizers in the market varies widely, leading to inconsistent application results. Many biofertilizer products are not well standardized; therefore, the effectiveness of the microorganisms contained in them can decrease during distribution and storage. This variation is also influenced by production technology, which is still simple and has not met international quality standards (Herdiyantoro et al. 2026).
3. *Limited Product Access and Distribution:* The distribution of biofertilizers is still concentrated in certain areas and has not optimally reached remote areas. Limited distribution networks, logistics, and product prices, which are relatively expensive compared to chemical fertilizers, are the main obstacles for farmers in remote areas (Herdiyantoro et al. 2026). In addition, the lack of support from the government and businesses in expanding access also slows down the spread of this product.

3.7.2 Emerging Trends and Technologies

Recent advances in biotechnology are redefining sustainable agriculture and environmental management through the convergence of cutting-edge tools, such as nanotechnology, encapsulation, biofilm engineering, genome editing, isotope tracing, consortia-based bioinoculants, and artificial intelligence (AI). These integrated approaches offer promising solutions for improving soil fertility, crop productivity, and environmental resilience while reducing reliance on chemical inputs.

In nanotechnology, engineered nanoparticles, such as liposomes and dendrimer-coated silica, enhance the delivery of antimicrobials and growth-promoting compounds into microbial biofilms, overcoming the barriers posed by extracellular polymeric substances (Avula et al. 2023; Mirhadi et al. 2020). These systems have broad applications in managing soil-borne pathogens and improving microbial efficacy in the rhizosphere.

Single-cell nanoencapsulation further protects microbial bioinoculants from environmental stressors, such as UV radiation, desiccation, and acidity. This enables controlled release and improves the colonization of plant roots under hostile conditions (Liu et al. 2023a, b). This is especially effective when paired with bioinoculant consortia, which include synergistic strains such as *Rhizobium*, *Azospirillum*, *Pseudomonas*, and mycorrhizal fungi. These microbial mixtures improve nutrient uptake, enhance disease resistance, and boost tolerance to abiotic stress (Glick et al. 2021; Kumar et al. 2024a, b).

Advances in modern biotechnology, particularly CRISPR/Cas9, have enabled precise and nontransgenic genome editing in crops and microbes. When delivered via nanoparticles, CRISPR can enhance microbial traits, such as nitrogen fixation, phosphate solubilization, and stress adaptation (Liang et al. 2024). Coupled with genome sequencing, scientists can rapidly identify target genes and optimize beneficial plant-microbe interactions (Koyama et al. 2020).

Isotope tracing techniques are critical for evaluating the real-time performance of these engineered microbes. Stable isotopes such as ^{15}N and ^{13}C have long been used to trace nitrogen fixation and carbon cycling in soil ecosystems (Buss et al. 2022; Lee et al. 2022). In addition, the radioisotope ^{32}P offers a powerful method for studying phosphorus dynamics, particularly phosphate solubilization by bioinoculants. By labeling phosphate sources with ^{32}P, researchers can quantitatively assess how efficiently inoculated microbes mobilize phosphorus and transfer it to host plants (Parikh et al. 2021; Spohn and Wanek 2024). This technique provides direct evidence of microbial contributions to phosphorus-use efficiency in nutrient-deficient or highly weathered tropical soils. Biofilm engineering enables the creation of living materials that serve ecological functions, such as pollutant degradation, water filtration, and soil stabilization. Engineered curli-fiber biofilms can be designed to host enzymes or biosensors, thereby expanding the role of microbes from nutrient cycling to environmental remediation (Liu et al. 2023a, b).

A transformative addition to this technological convergence is the implementation of AI in biofertilizer technologies. AI models, particularly machine learning and neural networks, are being used to optimize microbial strain selection, predict biofertilizer performance under varying conditions, and recommend site-specific consortia formulations. AI also supports decision-making in microbial formulations by analyzing large datasets, including genomic, soil, climate, and crop data (Shi et al. 2024).

3.7.3 Interdisciplinary Collaborations and Institutional Support

The development of climate-resilient bioinocula necessitates the integration of multidisciplinary knowledge spanning microbiology, plant physiology, climate science, agronomy, and soil science to formulate robust microbial consortia capable of adapting to dynamic environmental conditions. Collaboration with experts in genomics and bioinformatics further enhances this process by enabling the identification and characterization of key microbial genes and metabolic pathways that confer tolerance to abiotic stresses, such as drought and salinity, as well as biotic stresses, including pathogenic attacks. This integrated approach ensures the design of effective bioinoculants tailored to sustain plant health and productivity under variable climatic conditions.

Advancing effective bioinocula requires extensive multiyear and multilocation field trials that depend heavily on robust institutional infrastructure, sustainable funding mechanisms, and supportive policy frameworks. Strategic partnerships between academic institutions and industries play a crucial role in facilitating technology transfer, scaling up production, and navigating the registration process for microbial products to ensure their practical application. Moreover, the establishment of comprehensive policy and regulatory frameworks through public-private cooperation is vital for enabling the large-scale adoption and successful implementation of bioinoculant technologies in agricultural systems.

Interdisciplinary training and workshops are essential for building the capacity of stakeholders, such as researchers, extension agents, and farmers, to effectively utilize bioinocula in varying climatic conditions. By enhancing knowledge and practical skills, education is a pivotal driver in accelerating the adoption of microbial technologies, thereby advancing sustainable agriculture and improving resilience to environmental stress. Continuous capacity building ensures that innovations are translated into actionable practices that can be adapted to local contexts, thereby maximizing their impact on crop productivity and ecosystem health.

3.7.4 Research Gaps and Opportunities (Bridging the Laboratory to Land Gap)

Despite decades of research, biofertilizers have yet to consistently fulfill their potential as sustainable alternatives to chemical fertilizers under diverse field conditions. A critical bottleneck lies in the persistent gap between laboratory-based discoveries and their reproducible success in farmers' fields. While laboratory studies often demonstrate promising microbial traits, such as phosphate solubilization, nitrogen fixation, and phytohormone production, these benefits are frequently lost or diminished under field conditions due to environmental variability, microbial competition, and inconsistent formulations (Backer et al. 2018). This underscores the urgent need to refine biofertilizer innovations with real-world agronomic constraints.

A key research gap lies in the reliance on single microbial strains in these formulations. Although effective under controlled conditions, single-strain inoculants often lack resilience in the face of field heterogeneity. In response, there is a growing shift towards microbial consortia, which are formulations that combine arbuscular mycorrhizal fungi (AMF), plant growth-promoting rhizobacteria (PGPR), nitrogen-fixing bacteria, and potassium/phosphate solubilizers. These consortia provide a broader spectrum of functions, such as nutrient mobilization, stress mitigation, and biocontrol. Studies have shown that these multi-strain combinations outperform single-strain inoculants in improving yield and stress tolerance across crops, such as maize and wheat (Sharma et al. 2013).

Another gap lies in the delivery efficiency and survival of microbial inoculants under field conditions. Harsh soil environments and competition from native microbes reduce the survival of the introduced strains. To address this, biofilm-based formulations, in which beneficial microbes are encapsulated in extracellular polymeric substances, enhance rhizosphere colonization, resilience, and longevity (Ambarita et al. 2025). Field trials in maize have demonstrated that biofilm-formulated inoculant application reduced NP fertilizer dependency by 40% without compromising yield, demonstrating enhanced fertilizer-use efficiency.

Furthermore, nutrient delivery at the microbe-root interface remains inefficient in several formulations. The emergence of nanobiofertilizers represents a breakthrough in precision delivery systems. These technologies use nanocarriers to protect microbial viability and enable the gradual release of nutrients, such as phosphorus and nitrogen. Nano-encapsulated biofertilizers have demonstrated improved solubility, reduced nutrient loss, and increased uptake by plants (Sharma et al. 2023). These innovations show promise in minimizing environmental impacts while boosting crop productivity.

Storage and scalability remain pressing challenges. Biofertilizers often suffer from a short shelf life and temperature sensitivity. Innovations in carrier materials, such as biochar, peat alternatives, and granulated compost, have improved cell viability during storage and application. Low-cost, agrowaste-derived carriers used in both liquid and solid formulations have shown extended shelf lives and improved field performance in several tropical and subtropical trials (Kumar et al. 2022).

At the field application level, site-specific delivery systems are gaining traction. Precision agriculture tools, such as GPS, GIS, soil moisture sensors, and robotic samplers, now enable farmers to apply biofertilizers at the right dose, location, and time. These tools reduce waste and ensure microbial compatibility with in situ soil conditions (Fauziah et al. 2024; Janratchakool et al. 2024).

A final yet crucial issue is the lack of standardized protocols for evaluating the performance of biofertilizers across regions. Variability in strain selection, inoculation rates, and environmental reporting hinders reproducibility and comparisons. Comprehensive guidelines for the quality of biofertilizers and field-level trials, including standard metrics for evaluating plant-microbe interactions across agro-ecological zones, are urgently needed. The adoption of these protocols can accelerate regulatory approvals and farmer adoption (Ibáñez et al. 2023).

While the laboratory to land a gap in biofertilizer development persists, a convergence of innovations in microbial consortia design, delivery systems, nanoformulations, culturomics, and digital agriculture offers unprecedented opportunities. Addressing these research gaps through interdisciplinary collaboration and field validation can enable biofertilizers to become reliable components of sustainable agriculture.

3.8 Conclusion

The development of climate-resilient beneficial bioinocula is essential for advancing sustainable agriculture and ensuring long-term food security, particularly in the face of increasing biotic and abiotic stress. Biofertilizers play a central role in this transformation by enhancing soil fertility, promoting nutrient-use efficiency, and improving plant tolerance to environmental stressors. Field studies have indicated that the application of bioinoculants results in yield increases of 15–30% in food crops (e.g., rice, wheat, and maize), 20–40% in vegetable crops (e.g., tomato, chili, and cabbage), and 10–25% in perennial and tree crops (e.g., citrus, oil palm, and coffee), depending on strain compatibility and soil conditions. In terms of fertilizer efficiency, biofertilizers have been shown to increase nutrient uptake efficiency by 20–50%, particularly for nitrogen and phosphorus. This improvement allows for a reduction in synthetic fertilizer application by 25–50% without compromising productivity. Moreover, microbial consortia can reduce organic fertilizer use by up to 30% while maintaining or improving soil health and crop quality.

Despite these benefits, several practical challenges remain, including inconsistent field performance, limited formulation stability, and a lack of site-specific microbial products. However, innovative technologies, such as clustered regularly interspaced short palindromic repeats (CRISPR), are now being used to enhance the genetic performance of microbial strains. CRISPR-based genome editing enables precise, nontransgenic improvements in traits, such as stress tolerance and nutrient cycling efficiency. When combined with nanoencapsulation, AI-driven consortia design, and real-time monitoring using isotopes such as ^{15}N and ^{32}P, these technologies have the potential to revolutionize biofertilizer development. Moving forward, coordinated research, regulatory support, and farmer engagement are essential to scale these solutions. With continued innovation, biofertilizers will remain a cornerstone of climate-smart and sustainable agriculture.

Acknowledgments The authors sincerely thank the Academic Leadership Grant (ALG) from Universitas Padjadjaran (UNPAD) for its generous support in funding this study. Deep appreciation is also extended to the Faculty of Agriculture and all collaborators for their sincere contributions to the advancement of sustainable agricultural biotechnology through innovation and interdisciplinary collaboration.

References

Aini LQ, Aini N, Yamika WSD, Setiawan A (2022) Screening of plant growth-promoting halotolerant bacteria isolated from weeds rhizosphere grown in saline soil. AGRIVITA J Agric Sci 44(2):322–331

Ajeng AA, Abdullah R, Ling TC (2023) Biochar-*Bacillus* consortium for a sustainable agriculture physicochemical and soil stability analyses. Biochar 5(1):17

Albert D, Zboralski A, Ciotola M et al (2024) Identification and genomic characterization of *Pseudomonas* spp. displaying biocontrol activity against *Sclerotinia sclerotiorum* in lettuce. Front Microbiol 15:1304682

Ali MA, Shahid M, Nadeem SM (2023) Engineering plant microbiomes for climate-resilient agriculture: advances and prospects. Environ Microbiome 18(1):12. https://doi.org/10.1186/s40793-023-00480-y

Ambarita DDM, Jingga A, Fitriatin BN, Simarmata T (2025) Synergistic effect of biofilm-forming biofertilizers and ameliorants for enhancing fertilizer efficiency and maize agronomic traits on acidic soils. J Ecol Eng 26(7):171–184. https://doi.org/10.12911/22998993/203137

Avula S, Ahmed R, Mishra A et al (2023) Nanotechnology in combating biofilm: a smart and promising therapeutic strategy. Front Microbiol 13. https://doi.org/10.3389/fmicb.2022.1028086

Azri R, Lamine M, Bensalem-Fnayou A, Hamdi Z, Mliki A, Ruiz-Lozano JM, Aroca R (2024) Genotype-dependent response of root microbiota and leaf metabolism in olive seedlings subjected to drought stress. Plants 13(6):857

Baard V, Bakare OO, Daniel AI, Nkomo M, Gokul A, Keyster M, Klein A (2023) Biocontrol potential of *Bacillus subtilis* and *Bacillus tequilensis* against four Fusarium species. Pathogens 12(2):254

Backer R, Rokem JS, Ilangumaran G, Lamont J, Praslickova D, Ricci E, Subramanian S, Smith DL (2018) Plant growth-promoting rhizobacteria: context, mechanisms of action, and roadmap to commercialization of biostimulants for sustainable agriculture. Front Plant Sci 9:1473. https://doi.org/10.3389/fpls.2018.01473

Backer R, Saeed W, Khan A, Smith DL (2023) The future of microbial inoculants in sustainable agriculture: recent advances and emerging applications. Front Microbiol 14:1122335. https://doi.org/10.3389/fmicb.2023.1122335

Baliyan N, Qureshi K, Jaremko M, Rajput M, Singh M, Dhiman S, Maheshwari D, Kant C, Kumar A (2022) Bioformulation containing cohorts of *Ensifer adhaerens* MSN12 and *Bacillus cereus* MEN8 for the nutrient enhancement of *Cicer arietinum* L. Plants 11. https://doi.org/10.3390/plants11223123

Bhattacharyya PN, Jha DK (2012) Plant growth-promoting rhizobacteria (PGPR): emergence in agriculture. World J Microbiol Biotechnol 28(4):1327–1350. https://doi.org/10.1007/s11274-011-0979-9

Bianco A, Zara G, Garau M, Castaldi P, Atzori A, Deroma M, Coronas R, Budroni M (2023) Microbial community assembly and chemical dynamics of raw brewers' spent grain during inoculated and spontaneous solid-state fermentation. Waste Manag 174:518–527. https://doi.org/10.1016/j.wasman.2023.12.021

Bierenbroodspot MJ, Darienko T, de Vries S, Fürst-Jansen JM, Buschmann H, Pröschold T, Irisarri I, de Vries J (2024) Phylogenomic insights into the first multicellular streptophyte. Curr Biol 34(3):670–681

Blin K, Shaw S, Kloosterman AM, Charlop-Powers Z, van Wezel GP, Medema MH, Weber T (2021) AntiSMASH 6.0: improving cluster detection and comparison capabilities. Nucleic Acids Res 49(W1):W29–W35. https://doi.org/10.1093/nar/gkab335

Bolan S, Hou D, Wang L et al (2023) The potential of biochar as a microbial carrier for agricultural and environmental applications. Sci Total Environ:163968. https://doi.org/10.1016/j.scitotenv.2023.163968

Bowers R, Kyrpides N, Stepanauskas R et al (2017) Minimum information about a single amplified genome (MISAG) and a metagenome-assembled genome (MIMAG) of bacteria and archaea. Nat Biotechnol 35:725–731. https://doi.org/10.1038/nbt.3893

Buss H, Chiewattanakul M, Evershed R, Reay M, McAleer A, Griffiths R (2022) Compound-specific amino acid 15N-stable isotope probing for the quantification of biological nitrogen fixation in soils. Soil Biol Biochem. https://doi.org/10.1016/j.soilbio.2022.108654

Cao H, Peng T, Zhao W, Huang H, Yu S, Zhu Y (2025) Whole-genome sequencing uncovers the plant growth-promoting potential of *Bacillus licheniformis* G41, isolated from the rhizosphere soil of Gannan navel orange. Ann Microbiol 75(1):8

Carruthers DN, Lee TS (2022) Translating advances in microbial bioproduction to sustainable biotechnology. Front Bioeng Biotechnol 10:968437. https://doi.org/10.3389/fbioe.2022.968437

Chen X, Li Y, Wang J (2022a) Role of bioinoculants in improving acid-stressed soils and enhancing crop productivity. Agric Ecosyst Environ 334:107994. https://doi.org/10.1016/j.agee.2022.107994

Chen J, Yang Y, Ke Y, Chen X, Jiang X, Chen C, Xie S (2022b) Sulfonamide-metabolizing microorganisms and mechanisms in antibiotic-contaminated wetland sediments revealed by stable isotope probing and metagenomics. Environ Int 165:107332

Choudhary S, Sharma M, Kumawat KC, Meena RK (2022) Bioefficacy of novel insecticide molecules against sucking insect pests of Indian bean. Pharma Innov J Spec 11(5):833–838

Compant S, Samad A, Faist H, Sessitsch A (2019) A review on the plant microbiome: ecology, functions, and emerging trends in microbial application. J Adv Res 19:29–37. https://doi.org/10.1016/j.jare.2019.03.004

de Abreu VAC, Perdigão J, Almeida S (2021) Metagenomic approaches to analyze antimicrobial resistance: an overview. Front Genet 11:575592. https://doi.org/10.3389/fgene.2020.575592

dos Reis GA, Martínez-Burgos WJ, Pozzan R, Pastrana PY, Ocán-Torres D, de Queiroz FMP, Rodrigues C, Lima SJ, Scapini T, Karp SG, Soccol CR (2024) Comprehensive review of microbial inoculants: agricultural applications, technology trends in patents, and regulatory frameworks. Sustainability 16(19):8720. https://doi.org/10.3390/su16198720

FAO (2017) The future of food and agriculture: trends and challenges. Food and Agriculture Organization of the United Nations

FAO (2021) Assessment of agricultural plastics and their sustainability: a call for action. Food and Agriculture Organization of the United Nations

Fauziah NO, Fitriatin BN, Fakhrurroja H, Simarmata T (2024) Enhancing soil nutritional status in smart farming: the role of IoT-based management for meeting plant requirements. Int J Agron. https://doi.org/10.1155/2024/8874325

Fitriatin BN, Ghifari RFH, Sofyan ET, Widiantini F, Fakhrurroja H, Simarmata T (2024) The role of nutrient solutions on phosphate-solubilizing bacteria population, phosphorus availability, phosphorus uptake, growth and yield of red chili (*Capsicum annuum* L.). Jurnal Kultivasi 23(2):354–363. https://doi.org/10.24198/kultivasi.v23i3.58764

Francioli D et al (2021) Flooding causes dramatic compositional shifts and depletion of putative beneficial bacteria on the spring wheat microbiota. Front Microbiol 12:711. https://doi.org/10.3389/fmicb.2021.711

Gamalero E, Glick BR (2022) Recent advances in bacterial amelioration of plant drought and salt stress. Biology 11(437):1–26. https://doi.org/10.3390/biology11030437

Ganeshan S, Kim S, Vujanovic V (2021) Scaling-up production of plant endophytes in bioreactors: concepts, challenges and perspectives. Bioresour Bioprocess 8. https://doi.org/10.1186/s40643-021-00417-y

Glick B, Santos-Villalobos S, Parra-Cota F, Santoyo G, Guzmán-Guzmán P, Orozco-Mosqueda M (2021) Plant growth stimulation by microbial consortia. Agronomy 11:219. https://doi.org/10.3390/AGRONOMY11020219

Gong Y, Niu Q, Liu Y, Dong J, Xia M (2022) Development of multifarious carrier materials and impact conditions of immobilised microbial technology for environmental remediation: a review. Environ Pollut:120232. https://doi.org/10.1016/j.envpol.2022.120232

Härdter N, Menzen T, Winter G (2021) Minimizing oxidation of freeze-dried monoclonal antibodies in polymeric vials using a smart packaging approach. Pharmaceutics 13. https://doi.org/10.3390/pharmaceutics13101695

Herdiyantoro D, Simarmata T, Setiawati MR, Nurlaeny N, Joy B, Arifin M, Hamdani JS, Handayani I (2022) Selection of application technique and dose of potassium solubilizing biofertilizer to increase potassium uptake and growth of maize on Inceptisols. Jurnal Kultivasi 21(1):51–59. https://doi.org/10.12911/22998993/203137

Herdiyantoro D, Suryatmana P, Setiawan A, Sandrawati A, Sara DS, Kamaluddin NN, Ismail A, Odim C (2026) Effect of farmers' knowledge on interest in biofertilizers and organic fertilizers in Mekarasih Village, Jatigede, Sumedang. Agrimasta: Jurnal Pengabdian Kepada Masyarakat 3(2):45–50. https://doi.org/10.24198/agrimasta.v3i2.70294

Herrmann L, Lesueur D (2013) Challenges of formulation and quality of biofertilizers for successful inoculation. Appl Microbiol Biotechnol 97(20):8859–8873. https://doi.org/10.1007/s00253-013-5228-8

Hindersah R, Fitriatin BN, Setiawati MR, Suryatmana P, Risanti RR, Dewi YW (2025) Enhancing tomato (*Lycopersicon esculentum* Mill.) growth in a greenhouse using NPK fertilizer coated with endospore-forming bacillus. Jurnal Agrikultura 36(1):128–137. https://doi.org/10.24198/agrikultura.v36i1.62354

Huang DD (2024) Optimizing engineered SynComs for controlled environment agriculture (CEA): from theory to commercialization. International Journal of Horticulture 14(3):195–206. https://hortherbpublisher.com/index.php/ijh/article/view/3926/2952

Ibáñez A, Garrido-Chamorro S, Vasco-Cárdenas MF, Barreiro C (2023) From lab to field: biofertilizers in the 21st century. Horticulturae 9(12):1306. https://doi.org/10.3390/horticulturae9121306

Imran H, Gul A (2025) Computational approaches to understand rhizomicrobiome community and its future implications. In: Rhizomicrobiome in sustainable agriculture and environment. Academic Press, pp 391–402

IPCC (2022) Climate change 2022: impacts, adaptation and vulnerability. Contribution of Working Group II to the sixth assessment report of the Intergovernmental Panel on Climate Change. Cambridge University Press

Irwandhi I, Prihatiningsih N, Abraham S, Isroni M, Sativa RG, Kamaluddin NN, Simarmata T (2024) Diversity of bacterial isolates as biocontrol agents against *Fusarium oxysporum* f. sp. lycopersici. Biodiversitas 25(10):3403

Janratchakool W, Jansengrat P, Boonmee K et al (2024) IoT-enabled bio-organic fertilizer monitoring for residual palm materials for sustainable agriculture. Int J Eng Trends Technol 72(4):67–73. https://doi.org/10.14445/22315381/IJETT-V72I4P107

Kamaluddin NN, Irwandhi I, Arum P, Setiawan Y, Khumairah FH, Prihatiningsih N, Simarmata T (2025) Assessment of phosphate-solubilizing microbes isolated from indigenous organic-biofertilizers for enhancing plant growth in acid-stressed. Biodiversitas 26(3):1325–1333

Kankariya RA, Jape PV, Patil RP, Chaudhari AB, Dandi ND (2025) Bioprospecting of multi-stress tolerant *Pseudomonas* sp. antagonistic to *Rhizoctonia solani* for enhanced wheat growth promotion. Int Microbiol 28(1):17–35

Kehe J, Kulesa A, Ortiz A, Ackerman CM, Thakku SG, Sellers D, Kuehn S, Gore J, Friedman J, Blainey PC (2019) Massively parallel screening of synthetic microbial communities. Proc Natl Acad Sci USA 116(26):12804–12809. https://doi.org/10.1073/pnas.1900102116

Keswani C, Singh HB, García-Estrada C, Caradus J, He YW, Mezaache-Aichour S, Glare TR, Borriss R, Sansinenea E (2020) Antimicrobial secondary metabolites from agriculturally important bacteria as next-generation pesticides. Appl Microbiol Biotechnol 104(3):1013–1034. https://doi.org/10.1007/s00253-019-10300-8

Khan A, Singh A, Gautam S, Agarwal A, Punetha A, Upadhayay V, Kukreti B, Bundela V, Jugran A, Goel R (2023) Microbial bioformulation: a microbial assisted biostimulating fertilization technique for sustainable agriculture. Front Plant Sci 14. https://doi.org/10.3389/fpls.2023.1270039

Khumairah FH, Nurbaity A, Setiawati MR, Fitriatin BN, Simarmata T (2019) The ability of Phosphorhizobacteria isolates to produce organic acid and promote phosphatase activity to increase the growth of maize (*Zea mays* L.) in selected medium. IOP Conf Ser Earth Environ Sci 334(1):012023

Khumairah FH, Setiawati MR, Fitriatin BN, Simarmata T, Alfaraj S, Ansari MJ, Enshasy HAE, Sayyed RZ, Najafi S (2022) Halotolerant plant growth-promoting rhizobacteria isolated from saline soil improve nitrogen fixation and alleviate salt stress in rice plants. Front Microbiol 13:905210

Kocaefe-Özşen N, Yilmaz B, Alkım C, Arslan M, Topaloğlu A, Kısakesen HL, Gülsev E, Çakar ZP (2022) Physiological and molecular characterization of an oxidative stress-resistant *Saccharomyces cerevisiae* strain obtained by evolutionary engineering. Front Microbiol 13:822864

Koyama H, Panda S, Agrahari R, Singh P (2020) Plant-microbe interactions for sustainable agriculture in the post-genomic era. Curr Genomics 21:168–178. https://doi.org/10.2174/1389202921999200505082116

Kumar A, Maleva M, Bruno LB, Rajkumar M (2021) Synergistic effect of ACC deaminase producing *Pseudomonas* sp. TR15a and siderophore producing *Bacillus aerophilus* TR15c for enhanced growth and copper accumulation in *Helianthus annuus* L. Chemosphere 276:130038

Kumar S, Sindhu SS, Kumar R (2022) Biofertilizers: an ecofriendly technology for nutrient recycling and environmental sustainability. Curr Res Microb Sci 3:100094. https://doi.org/10.1016/j.crmicr.2021.100094

Kumar R, Singh J, Bhardwaj A (2023a) Seed bio-priming with plant growth-promoting rhizobacteria enhances germination, seedling vigor, and drought resilience in maize. Rhizosphere 27:100639. https://doi.org/10.1016/j.rhisph.2023.100639

Kumar S, Singh R, Singh H (2023b) Symbiotic and non-symbiotic nitrogen fixers and their role in enhancing crop yield under stress conditions. Agric Ecosyst Environ 334:107957. https://doi.org/10.1016/j.agee.2022.107957

Kumar A, Saharan BS, Parshad J et al (2024a) Revolutionizing Indian agriculture: the imperative of advanced biofertilizer technologies for sustainability. Discov Agric 2. https://doi.org/10.1007/s44279-024-00037-y

Kumar K, Singh S, Kour D, Chowdhury S, Ahmed N, Kaur T, Rai A, Kapoor M, Yadav A, Sharma B, Shreaz S, Kumar A, Jan T, Rustagi S, Negi R (2024b) Microbial consortia: promising tool as plant bioinoculants for agricultural sustainability. Curr Microbiol 81:222. https://doi.org/10.1007/s00284-024-03755-0

Kusale SP, Attar YC, Sayyed RZ, Malek RA, Ilyas N, Suriani NL, El Enshasy HA (2021) Production of plant beneficial and antioxidants metabolites by *Klebsiella variicola* under salinity stress. Molecules 26(7):1894

Lee J, Lazicki P, Mayes M, Jagadamma S, Neupane A (2022) The use of stable carbon isotopes to decipher global change effects on soil organic carbon: present status, limitations, and future prospects. Biogeochemistry 160:315–354. https://doi.org/10.1007/s10533-022-00963-3

Lehmann A, Veresoglou SD, Leifheit EF, Rillig MC (2020) Arbuscular mycorrhizal influence on soil structure and function. Fungal Ecol 45:100867

Li Y, He L, Zhang X (2022) Biofertilizers for sustainable agriculture: types, mechanisms, and applications. Front Plant Sci 13:852528. https://doi.org/10.3389/fpls.2022.852528

Li Y et al (2024) Heterogeneous impact of soil acidification on crop yield reduction and its regulatory variables: a global meta-analysis. Field Crop Res 319:109643. https://doi.org/10.1016/j.fcr.2024.109643

Liang W, Lou Z, Frugier F, Jiang S, Gao J, Murray J, Xu P, Su Y (2024) Applying conventional and cell-type-specific CRISPR/Cas9 genome editing in legume plants. aBIOTECH. https://doi.org/10.1007/s42994-024-00190-4

Liao C, Ttofali F, Slotkowski RA, Denny SR, Cecil TD, Leenay RT, Beisel CL (2019) Modular one-pot assembly of CRISPR arrays enables library generation and reveals factors influencing crRNA biogenesis. Nat Commun 10(1). https://doi.org/10.1038/s41467-019-10747-3

Liu Y, Cao X, Yue L, Wang C, Tao M, Wang Z, Xing B (2022a) Foliar-applied cerium oxide nanomaterials improve maize yield under salinity stress: reactive oxygen species homeostasis and rhizobacteria regulation. Environ Pollut 299:118900

Liu Y, Zhao H, Liu F (2022b) Bioinoculants in sustainable agriculture: mechanisms for alleviating soil salinity and improving tree crop health. J Appl Microbiol 132(4):1897–1910. https://doi.org/10.1111/jam.15567

Liu S, Li J, Huang Y, Zhou L, Zhong C, Guo X, Xue B, Zhong J (2023a) Effective disposal and remediation of chemical agents with designer living biofilm materials in soil and water. Fundam Res. https://doi.org/10.1016/j.fmre.2023.04.007

Liu L, Liang W, Zhang Y, Fu Q (2023b) Nanoencapsulation in polymeric materials: weaving magical coats for microorganisms. Nano Today 52:101973. https://doi.org/10.1016/j.nantod.2023.101973

Machado D, Andrejev S, Tramontano M, Patil KR (2018) Fast automated reconstruction of genome-scale metabolic models for microbial species and communities. Nucleic Acids Res 46(15):7542–7553. https://doi.org/10.1093/nar/gky537

Mahmud K, Missaoui A, Lee K (2023) Impacts of agrochemical overuse on soil microbial communities and ecosystem functioning. Soil Biol Biochem 180:108941. https://doi.org/10.1016/j.soilbio.2023.108941

Malusá E, Grządziel J (2023) The field performance gap of microbial inoculants in agriculture: causes and perspectives. Appl Soil Ecol 186:104866. https://doi.org/10.1016/j.apsoil.2023.104866

Malusá E, Vassilev N (2014) A contribution to set a legal framework for biofertilisers. Appl Microbiol Biotechnol 98:6599–6607. https://doi.org/10.1007/s00253-014-5828-y

Malusá E, Sas-Paszt L, Ciesielska J (2012) Technologies for beneficial microorganisms inocula used as biofertilizers. Sci World J 2012:491206. https://doi.org/10.1100/2012/491206

Masmoudi F, Alsafran M, Jabri HA et al (2023) Halobacteria-based biofertilizers: a promising alternative for enhancing soil fertility and crop productivity under biotic and abiotic stresses—a review. Microorganisms 11(5):1248. https://doi.org/10.3390/microorganisms11051248

Mendes CR, Dilarri G, Forsan CF et al (2022) Antibacterial action and target mechanisms of zinc oxide nanoparticles against bacterial pathogens. Sci Rep 12(1):2658. https://doi.org/10.1038/s41598-022-06657-y

Minchev Z, Kostenko O, Soler R, Pozo MJ (2021) Microbial consortia for effective biocontrol of root and foliar diseases in tomato. Front Plant Sci 12:756368. https://doi.org/10.3389/fpls.2021.756368

Mirhadi E, Malaekeh-Nikouei B, Khameneh B, Tajani A, Bazzaz B (2020) The role of nanotechnology in combating biofilm-based antibiotic resistance. J Drug Deliv Sci Technol 60:101880. https://doi.org/10.1016/j.jddst.2020.101880

Mitter EK, Tosi M, Gaiero JR, Dunfield KE (2021) Research perspectives on the regulation and use of microbial inoculants in agriculture. Front Sustain Food Syst 5:626340. https://doi.org/10.3389/fsufs.2021.626340

Nazir A, Hussain FHN, Raza A (2024) Advancing microbiota therapeutics: the role of synthetic biology in engineering microbial communities for precision medicine. Front Bioeng Biotechnol 12:1511149. https://doi.org/10.3389/fbioe.2024.1511149

Negi R, Sharma B, Jan T et al (2024) Microbial consortia: promising tool as plant bioinoculants for agricultural sustainability. Curr Microbiol 81(8):222. https://doi.org/10.1007/s00284-024-03755-0

Nishu SD, No JH, Lee TK (2022) Transcriptional response and plant growth promoting activity of *Pseudomonas fluorescens* DR397 under drought stress conditions. Microbiol Spectrum 10(4):e00979-22. https://doi.org/10.1128/spectrum.00979-22

Nurbaity A, Mbusango A (2024) Mycorrhizal root colonization, p uptake, and yield of *Ipomea reptans* poir. as affected by arbuscular mycorrhizal fungi and NPK fertilizer grown on andisols. J Trop Mycorrhiza 3(1):39–46. https://journal.ami-ri.org/index.php/JTM/article/view/57

Parihar P et al (2020) Salinity induced physiological and biochemical changes in plants: an omic approach towards salt stress tolerance. Environ Exp Bot 178:104186. https://doi.org/10.1016/j.envexpbot.2020.104186

Parikh S, Li C, Manavalan L, Rippner D (2021) Evaluation of Bacillus seed coatings on soybean phosphorus uptake in an oxisol fertilized with 32P-labeled hydroxyapatite. Plant Soil 464:273–287. https://doi.org/10.1007/s11104-021-04941-w

Park J, Liu R, Kim A et al (2024) Sweet corn phytoglycogen dendrimers as a lyoprotectant for dry-state protein storage. J Biomed Mater Res 112(12):2026–2041. https://doi.org/10.1002/jbm.a.37761

Pasolli E, De Filippis F, Mauriello IE (2020) Large-scale genome-wide analysis links lactic acid bacteria from food with the gut microbiome. Nat Commun 11(1):2610. https://doi.org/10.1038/s41467-020-16438-8

Pirapak W, Siriphap A, Inprasit T, Phuangsri C, Kraivuttinun P (2022) Isolation and characterization of plant growth promoting rhizobacteria from rhizospheric soil of selected pulses and their effect on Coriandrum sativum plants. Biodiversitas 23(11):5765–5770. https://doi.org/10.13057/biodiv/d231129

Puchkov EO (2023) Preservation of viable microorganisms in the laboratory: an overview of basics, methods and practical recommendations for beginners. Austin J Biotechnol Bioeng 10(1):1119. https://doi.org/10.26420/austinjbiotechnolbioeng.2023.1119

Rafique N, Khalil S, Cardinale M, Rasheed A, Zhao F, Abideen Z (2024) A comprehensive evaluation of the potential of plant growth-promoting rhizobacteria for applications in agriculture in stressed environments. Pedosphere 35(1):229–248. https://doi.org/10.1016/j.pedsph.2024.02.005

Rao DLN (2022) Diversity of root nodule bacteria: strengthening R&D and inoculant supply. Indian J Plant Genet Resour 35(3):343–345. https://ispgr.nbpgr.ernet.in/SpecialVol/05/03-DLN%20Rao-Rev.pdf

Rojas-Sánchez B, Guzmán-Guzmán P, Morales-Cedeño L et al (2022) Bioencapsulation of microbial inoculants: mechanisms, formulation types and application techniques. Appl Biosci 1(2):198–220. https://doi.org/10.3390/applbiosci1020013

Rottinghaus AG, Ferreiro A, Fishbein SR, Dantas G, Moon TS (2022) Genetically stable CRISPR-based kill switches for engineered microbes. Nat Commun 13(1):672. https://doi.org/10.1038/s41467-022-28163-5

Sadvakasova AK, Bauenova MO, Kossalbayev BD et al (2023) Synthetic algocyanobacterial consortium as an alternative to chemical fertilizers. Environmental Research 233(116418). https://doi.org/10.1016/j.envres.2023.116418

Seidu OA, Githiri SM, Wesonga JM, Ngumi VW (2025) Isolation, screening and in-vitro characterization of plant growth-promoting *Bradyrhizobium* isolates from the nodules of bambara groundnut (*Vigna subterranean*) for potential use as bioinoculants. Front Sustain Food Syst 8:1506346. https://doi.org/10.3389/fsufs.2024.1506346

Setiawati MR, Azkiannissa SZ, Herdiyantoro D et al (2024) The nitrogen-fixing endophytic bacteria and the application techniques for the growth of tomato plants (*Solanum lycopersicum* L.) in inceptisols from Jatinangor. AGROLOGIA 13(2):131–141. https://doi.org/10.30598/ajibt.v13i2

Sharath S, Triveni S, Nagaraju Y, Latha PC, Vidyasagar B (2021) The role of phyllosphere bacteria in improving cotton growth and yield under drought conditions. Front Agron 3:680466. https://doi.org/10.3389/fagro.2021.680466

Sharma A, Sharma R (2022) Decomposer microorganisms in soil nutrient cycling and their impact on soil health. Soil Biol Biochem 171:108717. https://doi.org/10.1016/j.soilbio.2022.108717

Sharma SB, Sayyed RZ, Trivedi MH et al (2013) Phosphate solubilizing microbes: sustainable approach for managing phosphorus deficiency in agricultural soils. Springerplus 2:587. https://doi.org/10.1186/2193-1801-2-587

Sharma B, Tiwari S, Kumawat KC, Cardinale M (2023) Nano-biofertilizers as bio-emerging strategies for sustainable agriculture development: potentiality and their limitations. Sci Total Environ 860. https://doi.org/10.1016/j.scitotenv.2022.160476
Shi T, Huang J, Guo Q, Li X (2024) A comprehensive review on the application of neural network model in microbial fermentation. Bioresour Technol 416:131801. https://doi.org/10.1016/j.biortech.2024.131801
Simarmata R, Widowati T, Sylvia JRL et al (2023) Rhizosphere bacteriome of *Allium cepa* after the application of chemical and endophyte-based fertilizer. Acta Ecol Sin 43(6):1138–1148. https://doi.org/10.1016/j.chnaes.2023.08.009
Simarmata T, Irwandhi I, Khumairah F et al (2025) Implementation of regenerative technologies for sustainable and net-zero rice farming in adapting to climate change in Indonesian coastal areas. BIO Web Conf:176, 01007. https://doi.org/10.1051/bioconf/202517601007
Singh P, Kumar V (2023) Biological approaches to manage acid soil stress in crops: a review on bioinoculants and their mechanisms. Agronomy 13(2):345. https://doi.org/10.3390/agronomy13020345
Singh D, Yadav AN, Kumar M (2022) Plant microbiome: a sustainable tool for improving crop productivity under stress environments. Arch Microbiol 204:112. https://doi.org/10.1007/s00203-021-02547-3
Sinha R, Abu-Ali G, Vogtmann E (2017) Assessment of variation in microbial community amplicon sequencing by the microbiome quality control (MBQC) project consortium. Nat Biotechnol 35(11):1077–1086. https://doi.org/10.1038/nbt.3981
Sinha R, Fraceto LF, Singh HB (2022) Nanobiotechnology and bioinoculants: a synergistic approach for stress management in agriculture. Environ Res 212:113198. https://doi.org/10.1016/j.envres.2022.113198
Soniari N, Sutari N (2022) *Azotobacter* spp. viabilities on some combinations of carrier media. Agrotrop J Agric Sci 12(1):76–84. https://doi.org/10.24843/ajoas.2022.v12.i01.p07
Spohn M, Wanek W (2024) Quantifying element fluxes using radioisotopes. New Phytol 245(2):443–445. https://doi.org/10.1111/nph.20203
Sultana U, Desai S, Pinisetty S (2023) In vitro screening for abiotic stress tolerance and biocontrol ability of plant growth promoting strains of *Azotobacter* and *Azospirillum* spp. Curr Agric Res J 11(3). https://doi.org/10.12944/CARJ.11.3.14
Suman A, Govindasamy V, Ramakrishnan et al (2022) Microbial community and function-based synthetic bioinoculants: a perspective for sustainable agriculture. Front Microbiol 12. https://doi.org/10.3389/fmicb.2021.805498
Sun M, Liao J, Jing Z, Gao H, Shen B, Xu Y, Fang W (2022) Effects of polyol excipient stability during storage and use on the quality of biopharmaceutical formulations. J Pharm Anal 12:774–782. https://doi.org/10.1016/j.jpha.2022.03.003
Szczerba A, Płażek A, Kopeć P, Wójcik-Jagła M, Dubert F (2024) Effect of different Bradyrhizobium japonicum inoculants on physiological and agronomic traits of soybean (*Glycine max* (L.) Merr.) associated with different expression of nodulation genes. BMC Plant Biol 24(1):1201. https://doi.org/10.1186/s12870-024-05911-x
Tao C, Li R, Xiong W et al (2020) Bio-organic fertilizers stimulate indigenous soil Pseudomonas populations to enhance plant disease suppression. Microbiome 8(1):137. https://doi.org/10.1186/s40168-020-00892-z
Tiwari PK, Srivastava AK, Singh R, Srivastava AK (2025) Stress-relieving plant growth-promoting bacterial co-inoculation enhances nodulation and nitrogen uptake in black gram under nitrogen-free saline conditions. Front Microbiol 15:1516748. https://doi.org/10.3389/fmicb.2024.1516748
Tripti, Kumar A, Kumar V, Anshumali, Bruno LB, Rajkumar M (2022) Synergism of industrial and agricultural waste as a suitable carrier material for developing potential biofertilizer for sustainable agricultural production of eggplant. Horticulturae 8(5):444. https://doi.org/10.3390/horticulturae8050444

Trivedi P, Leach JE, Tringe SG, Sa T, Singh BK (2020) Plant–microbiome interactions: from community assembly to plant health. Nat Rev Microbiol 18(11):607–621. https://doi.org/10.1038/s41579-020-0412-1
Vassilev N, Vassileva M, Lopez A et al (2015) Unexploited potential of some biotechnological techniques for biofertilizer production and formulation. Appl Microbiol Biotechnol 99:4983–4996. https://doi.org/10.1007/s00253-015-6656-4
Vassileva M, Mendes GDO, Vassilev N (2022) Biotechnology of microbial inoculants: overcoming bottlenecks for practical applications in agriculture. Crit Rev Biotechnol 42(7):1126–1141. https://doi.org/10.1080/07388551.2022.2070664
Vassileva M, Mendes GDO, Vassilev N (2023) Formulation of microbial inoculants and delivery systems: innovations and bottlenecks. Trends Microbiol 31(4):324–335. https://doi.org/10.1016/j.tim.2023.01.002
Velloso C, Lopes M, Badino A, Farinas C (2023) Exploring the roles of starch for microbial encapsulation through a systematic mapping review. Carbohydr Polym 306:120574. https://doi.org/10.1016/j.carbpol.2023.120574
Wang Y, Dall'Agnol R, Bertani I, Bez C, Venturi V (2024) Identification of synthetic consortia from a set of plant-beneficial bacteria. Microb Biotechnol 17. https://doi.org/10.1111/1751-7915.14330
Wani AK, Rahayu F, Alkahtani AM et al (2024) Metagenomic profiling of rhizosphere microbiota: unraveling the plant-soil dynamics. Physiol Mol Plant Pathol 133:102381. https://doi.org/10.1016/j.pmpp.2024.102381
Wheatley RM, Ford BL, Li L et al (2020) Lifestyle adaptations of rhizobium from rhizosphere to symbiosis. Proc Natl Acad Sci 117(38):23823–23834. https://doi.org/10.1073/pnas.2009094111
Yen LT, McLean JE, Anderson AJ, Norton JM, Britt DW, Lin YT, Jacobson AR (2025) Synthesis of nano-sized metavivianite and nano-sized chitosan-coated metavivianite and their solubility in a calcareous soil saturated paste extract with soil microbe inoculation. ACS Agric Sci Technol 5(5):822–839. https://doi.org/10.1021/acsagscitech.4c00826
Yin C, Hagerty CH, Paulitz TC (2022) Synthetic microbial consortia derived from rhizosphere soil protect wheat against a soilborne fungal pathogen. Front. Microbiol. 13(908981):1–15. https://doi.org/10.3389/fmicb.2022.908981
Yu P, He X, Baer M et al (2021) Plant flavones enrich rhizosphere oxalobacteraceae to improve maize performance under nitrogen deprivation. Nat Plants 7(4):481–499. https://doi.org/10.1038/s41477-021-00897-y
Zaid DS, Cai S, Hu C, Li Z, Li Y (2022) Comparative genome analysis reveals phylogenetic identity of *Bacillus velezensis* HNA3 and genomic insights into its plant growth promotion and biocontrol effects. Microbiol Spectrum 10(1):e02169-21. https://doi.org/10.1128/spectrum.02169-21
Zhang H, Yuan X, Xiong T, Wang H, Jiang L (2020) Bioremediation of co-contaminated soil with heavy metals and pesticides: influence factors, mechanisms and evaluation methods. Chem Eng J 398:125657. https://doi.org/10.1016/j.cej.2020.125657
Zhao L, He Y, Huang Q (2023) Mechanistic insights into bioinocula-mediated mitigation of biotic and abiotic stresses in agroecosystems. Microb Biotechnol 16(1):101–115. https://doi.org/10.1111/1751-7915.14107
Zhou X, Guo Z, Chen C, Wei W (2022) Long-term chemical fertilization alters soil microbial diversity and functions critical for crop productivity. Sci Total Environ 838:156335. https://doi.org/10.1016/j.scitotenv.2022.156335
Zhu Y, Kumar A, Trivedi P (2023) Harnessing multi-omics approaches for designing climate-resilient plant–microbe interactions. Trends Biotechnol 41(8):947–960. https://doi.org/10.1016/j.tibtech.2023.03.005
Zhuang L, Li Y, Wang Z, Yu Y, Zhang N, Yang C, Zeng Q, Wang Q (2021) Synthetic community with six Pseudomonas strains screened from garlic rhizosphere microbiome promotes plant growth. Microbial Biotechnology 14(2):647–659. https://doi.org/10.1111/1751-7915.13640

Part II
Comparative Analysis of Bioinoculants Developed Through Contemporary Rhizoengineering Strategies

Chapter 4
Genomic Perspective of Rhizoengineering for Sustainable Agricultural Systems

Farah Deeba, Roomina Mazhar, Raheela Waheed, Tahira Younis, and Iqra Jabeen

Abstract The specific biomanagement of root-associated microbial communities, referred to as rhizoengineering, is a growing technique that aims to improve plant yield and resilience in sustainable agriculture. The advancement of high-throughput sequencing and omics technologies has provided new insights into the genomic perspective of rhizosphere microbiomes. This chapter discusses how genomic techniques, such as metagenomics, metatranscriptomics, and genome editing, can be used to untangle microbial diversity, recognize key functional genes, and manipulate valuable plant-microbe interactions. In particular, focus is placed on the delineation of core optimizations for PGPR, alteration of root exudates, and design of synthetic consortia for target crops personalized to particular environmental conditions. Additionally, issues concerning certainty and stability, along with regulatory concerns about real-world applications within agricultural settings, are critically analyzed from the perspective of rhizoengineered systems. Coupling genomics with systems biology and precision agriculture enables a scenario in which rhizoengineering can be employed as a centerpiece in next-generation approaches for sustainable farming.

Keywords Rhizosphere engineering · Genome editing · Metagenomics · Sustainable agriculture · Root exudates · Bioinocula

F. Deeba (✉) · T. Younis · I. Jabeen
Department of Biochemistry & Biotechnology, The Women University Multan, Multan, Pakistan
e-mail: farah.9003@wum.edu.pk

R. Mazhar
Department of Botany, Banazir Bhutto Women University, Peshawar, Pakistan

R. Waheed
Department of Biosciences and Technology, Emerson University Multan, Multan, Pakistan

S. Mehnaz, I. Shahid (eds.), *Frontiers in Rhizoengineering*, Microorganisms for Sustainability 65, https://doi.org/10.1007/978-981-92-3665-7_4

4.1 Rhizosphere Engineering

Rhizosphere engineering is an advanced alignment technique that can be used to adjust the biophysical features of the soil to alter the effects of plant, soil, and microbial associations. This method aims to enhance agricultural performance by increasing the effectiveness of resource exploitation, specifically water and nutrients (Hakim et al. 2021). The primary elements that comprise the engineering of the rhizosphere are the plant, the surrounding soil environment, and the related microorganisms. By controlling these elements, one can manipulate and maximize the nutritional and microbial processes at the root interface. Despite the promising results of genetic engineering of plants and microorganisms, the response has been held back by factors regarding environmental sustainability and public health, which have barred the widespread acceptance of these results. The rhizosphere can be altered agronomically, which is now regarded as the most feasible and sustainable option for enhancing crop production and soil health (Velmourougane and Blaise 2021).

The greatest goal of rhizosphere engineering is to guide the relationship between plants and microbes to achieve diverse beneficial outcomes. They enhance mineralization and the breakdown of organic matter; optimize nutrient cycles; increase resistance to abiotic stresses, including drought, salinity, and high temperature; and improve plant pathogen resistance (Hakim et al. 2021). These three components of the rhizosphere (plants, microbes, and soil) can be modified individually or in unison to strengthen plant functions. Plant engineering techniques have advantages, including optimum plant growth, nutritive content, and pest resistance. However, it is not widely used because of social concerns regarding biosecurity and the environment (Hakim et al. 2021).

In addition to plant-based interventions, promising microbial interventions have been developed. For example, genetic changes in some rhizobacteria have enabled them to inhibit stress hormones, such as ethylene, causing root elongation inhibition. These engineered microbes can also be used to produce antibiotics and lytic enzymes that can inhibit pathogens that pass through the soil (Bano et al. 2021). Scientists have attempted to establish a positive symbiosis between plants and microbes by selectively introducing helpful species of microbes or by prompting indigenous microbial communities. The general objective of rhizosphere engineering is to promote positive interactions between plants and microbes while mitigating harmful effects, which leads to sustained improvements in crop productivity.

Beyond its agricultural importance, rhizosphere engineering is being explored for its potential environmental applications, such as phytoremediation and ecological rehabilitation. By intentionally altering the rhizosphere environment, plants may be better equipped to tolerate and detoxify pollutants, which in turn helps restore soil and develop resistant ecosystems in degraded landscapes. Rhizosphere engineering can be classified into three principal methodologies (Fig. 4.1).

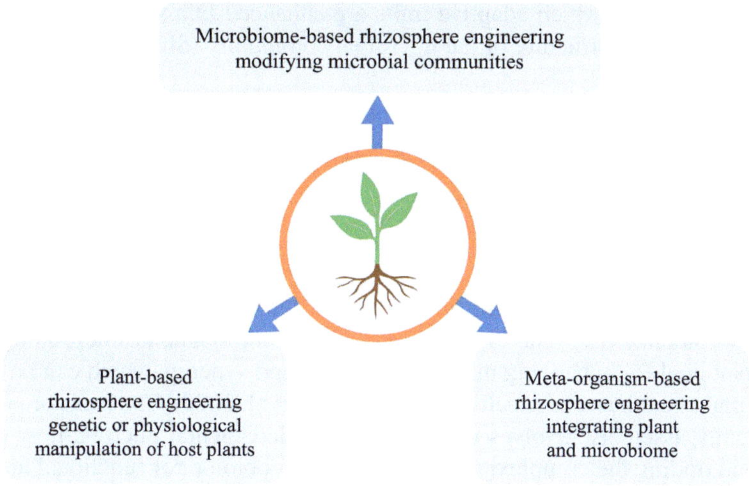

Fig. 4.1 Classification of rhizosphere engineering methodologies

4.1.1 Microbiome-Based Approach to Rhizosphere Engineering

A well-structured and diverse microbial community in the rhizosphere is often linked to improved plant vigor and resilience to stress. In this context, rhizosphere engineering efforts frequently target the enrichment of microbial diversity by fostering the proliferation of beneficial microbial populations. One of the most effective and ecologically sustainable methods for influencing the microbiome is the introduction of artificially cultured microbial inoculants (del Carmen Orozco-Mosqueda et al. 2022). Numerous commercial formulations containing one or more strains of beneficial bacteria or fungi are available as biofertilizers, which have shown promise in promoting plant growth and contributing to sustainable farming systems.

Most microorganisms, such as plant growth-promoting rhizobacteria, are either in a symbiotic or beneficial connection with hosts, assisting their growth and providing them with a greater chance to endure various environmental conditions (Kumar and Dubey 2020). These PGPR are essential for making the soil more fertile and the plants productive by aiding in the absorption and utilization of nutrients, blocking phytopathogens, and improving soil structure. In addition, they play a role in the bioremediation of polluted soils and assist in the removal of organic pollutants (Bibi et al. 2018).

To improve this area, it is essential to determine the specific components of root exudates that play a crucial role in suppressing pathogens and encouraging plant growth. Instead of just adding biofertilizers, there is now a growing focus on more sophisticated methods that shape and influence the soil's natural microbiome. In this regard, two methods have emerged: the application of synthetic microbial communities (SMC) and artificial selection, facilitated by either the plant or the multiple passage method, in which adaptive traits are enhanced in a series of plant growth cycles to create the structure of the microbial community (Shayanthan et al. 2022).

4.1.2 Modernized Approach: Constructing a Synthetic Microbiome

Synthetic communities (SynComs) are emerging communities with high development potential that address the pitfalls and instability of conventional microbial consortia inoculations. The creation of a functional synthetic microbiome is an important goal for enhancing microbial stability and synergy, which can be beneficial to plant health and productivity (Shayanthan et al. 2022). The basis of synthetic community assembly involves maintaining the fundamental microbe-host interactions and trading the complexity of the native microbiome for functional and manageable units. The typical and currently most pursued methods for developing SynComs involve a top-down approach with modification of plants to produce certain chemicals that influence the rhizobiome and a bottom-up approach with direct manipulation of the permanently present natural microbial communities (Shayanthan et al. 2022). Based on the functional screening of 1893 microbial strains isolated from root-associated compartments of soybean plants, Wang et al. (2021) built three synthetic communities (SynComs). Significant plant growth and nutrient uptake were facilitated by the functional assembly of SynComs in both nitrogen and phosphorus nutrient deficiency and sufficiency scenarios. According to additional field trials, the application of SynComs significantly and steadily increased plant growth, improved the uptake of nitrogen and phosphorus, and ultimately increased soybean yield (Wang et al. 2021).

4.1.3 Metaorganism-Based Rhizosphere Engineering

The methodology of rhizosphere engineering is also based on the concept of metaorganisms because the term describes the holobiont concept, denoting multicellular eukaryotic organisms along with their microbiomes (Compant et al. 2019). Regarding plants, microbial communities may be vertically inherited in seeds or horizontally gained from the surrounding soil. These microbes (portrayed as the second genome of the plant) have been shown to play vital roles in determining the functional traits of a plant and its health.

In this strategy, plants and their surrounding microbial communities will be co-designed to promote specific metabolic functions. For example, the genomes of the host plants might be manipulated to synthesize a particular type of metabolite, whereas the associated microorganisms would be modified so that they would use or break the products down to form a tightly regulated functional relationship between the plant and the microorganisms in its rhizosphere.

4.2 Role of Genomics in Understanding Plant-Microbe Interactions

Advances in genomics have largely deepened our understanding of the complexity of plant-microbe interactions, including beyond mutualistic and symbiotic types (Carper et al. 2022; Chiquito-Contreras et al. 2024). Advancements that have been made in the field of next-generation sequencing (NGS) allow for the acquisition of complete genome sequences, de novo assembly, and comparative sequencing at a reasonable cost (Crandall et al. 2020). This genomic research has facilitated the study of genes involved in N fixation, phytohormone production, and other beneficial agronomic characteristics. Metagenomic studies provide a thorough description of microbial community composition and dynamics. They also provide a comprehensive understanding of the functions and roles involved in plant ecology and general health.

The importance of completing transcriptomic and proteomic analyses, which are complementary, provides further insights into gene and protein activity dynamics during interactions between microbes and plants, such that the molecular basis for the relationships can be understood. The converging field of genomic tools and synthetic biology can be used to engineer a wide range of beneficial traits into microbial strains and host plants. Furthermore, microbiome engineering frameworks seek to maximize microbial assemblages, subsequently promoting plant navigation, production, and stress toughness (Chandok et al. 2022).

4.2.1 *Metagenomics*

Metagenomics is a method for obtaining the sequences of genes for whole microbial communities with respect to the traditional method of culture (Ghosh et al. 2019). This includes functional and structural methods, which have implications for particular microorganisms, as well as for ecology at a greater scope. Metagenomic studies began with Sanger sequencing. This led to the development of NGS, as the discovery of new genes is dependent on the use of 16S rRNA genes as well as random shotgun sequences (Regalado et al. 2020). These methods have been applied across a variety of conditions, such as the identification of nitrogen-fixing genes within rhizobium-legume symbioses and mycotoxin biosynthesis genes within *Alternaria alternata*, as well as the host response in apples (Bhargava et al. 2019).

Metagenomics allows the discovery of microbial taxa by cloning and counting universal genetic markers, which are rRNA genes found in sequenced environmental samples that consist of many unculturable species (Zhang et al. 2021). Targeted amplicon sequencing and random shotgun metagenomics are the two main strategies used to study the composition of microbiomes and their functions within plant-associated biomes. Diazotrophs, such as *Azotobacter vinelandii*, *Klebsiella pneumoniae*, *Rhodobacter capsulatus*, and *Rhodospirillum rubrum*, contain operons called nif (e.g., nifHDK, nifRLA, nifENB, nifUSVM, nifJ, and NIFWF) and encode enzymes essential for the fixation of nitrogen in biological systems (Usman and Wali 2024).

The emergence of omics methods, such as genetics and gene editing technology, is highly likely to increase the productivity of crops. The rhizosphere microbiome also plays a role in rendering plants resilient to environmental stress through the release of ACC (1-aminocyclopropane-1-carboxylate) deaminase that degrades ACC into O-ketobutyrate and ammonia, reducing the stress that ethylene causes on plants. The gene responsible for ACC deaminase (AcdS) is found in rhizospheric bacterial species, symbiotic rhizobia, endophytes, and nitrogen-fixing species, which are controlled by the *NifA2* gene and s54 sigma factors (Larekeng et al. 2024).

The complete genome sequencing process with platforms such as Illumina and Nanopore and assembly tools like MaSuRCA has identified metabolic genes as well as biosynthetic pathways within rhizobacteria, such as *Rhizobium* sp. and *Hydrogenophaga* sp., bolstering their ability to promote growth in plants (Ilangumaran et al. 2024). Under non-sterile conditions, *Pseudomonas sivasensis* significantly enhanced canola growth, attributed to the presence of genes associated with ACC deaminase (acdA), indole-3-acetic acid (IAA) biosynthesis (trpF, trpG), and siderophore production (fbpA, mbtH, acrB) (Świątczak et al. 2024).

Additionally, rhizobacteria trigger phytodefense responses by initiating pathogenesis-related (PR) proteins, such as chitinases. A genome-wide study of soybean discovered GmChi01, GmChi02, and GmChi16, as transgenic Arabidopsis expressed GmChi02 and GmChi16 showed resistance to *Fusarium* oxysporum. Notably, GmChi02 expression was also stimulated by *Burkholderia ambifaria* (Chen et al. 2024). Furthermore, the transgenic expression of antimicrobial proteins of *Bacillus thuringiensis* (Bt) in maize could be a viable alternative for managing Fusarium ear decay (FER). The migration of beneficial microbes under the conditions described above is triggered by the genotype of the host plant and demonstrates the importance of the interactions between microbial species in controlling FER.

4.2.2 Epigenomics

Epigenomics, as a framework for exploring plant-microbe interactions, involves the comprehensive examination of epigenetic modifications and heritable changes in gene expression that do not involve changes in DNA sequence (Samantara et al. 2021). The field includes DNA methylation, histone modification (e.g., histone methylation and acetylation), regulatory noncoding RNAs (such as microRNAs), and chromatin remodeling dynamic processes that influence stress response and adaptability in microbes and plants (Ali et al. 2023).

Epigenetic markers are crucial in influencing plant responses to biotic and environmental stresses. In particular, techniques such as EpiEffectors engineered proteins composed of transcription activator-like effectors (TALEs) and CRISPR/Cas9 that have been deactivated (dCas9) with epigenetic enzymes permit the precise manipulation to control the epigenome (Doddavarapu et al. 2024). The degree of active demethylation (part of the plant defense against infections caused by *Pseudomonas syringae*) (Masenya et al. 2024).

The proline-alanine-valine (*PAV*) gene, potentially linked to epigenomic adaptations in fungal pathogens, may also inform predictive models of host-pathogen dynamics and guide improved disease management strategies (Joubert and Krasileva 2024). Epigenomic research has revealed that *Neobacillus* species, prevalent within Naxos tubers, alter host gene expression via the hypermethylation of certain proteins, such as xyloglucan endotransglycosylase (Boutsika et al. 2023). Consequently, epigenomics provides a valuable lens through which key microbial taxa and regulatory mechanisms governing plant-microbe interactions can be identified.

4.2.3 Proteomics and Metaproteomics

Proteomics research has made immense contributions to gene expression research because of the elucidation of the functional role of proteins in the metabolism of microbes under various environmental settings. Proteins are essential in the development of cellular balance, signal transduction, protection pathways, and defense responses in medicinal plants. Proteomic technologies are known to be of great use in most instances, particularly in the study of interactions with microorganisms that cannot be cultured. For example, Khatabi et al. (2019) used proteomic strategies to examine the effectors of pathogens and host proteins during phytopathogen infestation (Khatabi et al. 2019). The same is true for Jain et al. (2021), who used proteomic approaches to determine secretory proteins involved in the development of the fire blight infection in apple by *Erwinia amylovora* (Jain et al. 2021). These methods provide great insights into microbial pathogens, their responses, and antioxidant interactions, contributing to our overall understanding of plant physiology and the interaction between pathogens and plants.

Protein-protein interactions (PPI) can be investigated using functional proteomics. The yeast two-hybrid (Y2H) model (bait and dam proteins) was used to investigate the dynamism of these interactions. Studies on the proteomic impact rely on analytical instruments made using mass spectrometry, such as PeptIdent, MultiIdent, MASCOT SEQUEST, and Sherpa, which facilitate the complete characterization and identification of proteins. Specific software, such as Found MS-fit, MOWSE, PepSea, PepFrag, and MS-Tag, enhance the analysis of curated data and, more importantly, enhance the resolution and precision (Chandok et al. 2022). Gel-based methodologies, such as two-dimensional electrophoresis (DEG) and differential electrophoresis (DIGE), when coupled with mass spectrometry, allow the identification and quantification of the expression patterns that differ between proteins. The analysis performed by 2D-LC/MS/MS revealed nodular proteins within *Medicago truncatula* and changes in rice proteomes following exposure to *S. meliloti*.

Comparative proteomic studies have revealed different protein secretion profiles between symbiotic organisms, such as *S. meliloti*, and pathogens, such as *Pseudomonas syringae* DC3000, in response to varied root exudates. Research on almost isogenic line alleles of FHB1 has revealed molecular components that confer resistance to *Fusarium graminearum*. Proteomic responses in pea curls to infection by the necrotrophic fungus *Sclerotinia sclerotiorum* have further illustrated the dynamic nature of the beneficial associations of the plant. This research collectively underlines the relevance of proteomic methods in deciphering the complex interactions of plants (Jain et al. 2021).

The biochemical profile of *Botrytis cinerea* reveals modifications in the secreted proteins, particularly pectinases, which are crucial for cell wall destruction. Serine is one of the most abundant metalloproteases (Sarim et al. 2020). Furthermore, ITRAQ-based proteomic research has identified various regulatory pathways between susceptible and resistant rice cultivars battling *Magnaporthe oryzae* disease.

In contrast, metaproteomics serves as a vital tool for investigating the functions of microbial communities related to soil health, plant interactions, nutrient cycling, and ecological restoration. In a historical study, Priya et al. (2021) performed a metaproteogenomic analysis of bacterial populations inhabiting the phyllosphere of several plant species, revealing preserved bacterial taxa and patterns of associated protein expression (Priya et al. 2021). Rane et al. (2022) utilized the chance to determine the proteins that are associated with the effectiveness of *Trichoderma harzianum*'s biocontrol. This was even in the same way that Priya et al. (2021) discovered proteins implicated within the pathogenicity *Fusarium oxysporum* as well as the modulation of immune function of the host within the watermelon (Rane et al. 2022). Additionally, metaproteomic analysis revealed a connection between protein and biochemical use by microbial organisms of certain carbon substrates found in the phyllosphere.

4.2.4 Transcriptomics and Metatranscriptomics

Transcriptomics and next-generation sequencing can be employed to measure and analyze the transcripts of RNA found in biological samples, providing complete information regarding the activities of gene expression in specific environmental settings. This is done using microarrays and laughing and solid analysis that permit the analysis of transcript profile, analysis of mutations, differential expression studies, and alternative joint event detection (Weidemüller et al. 2021; Jeon et al. 2022).

This technique has been applied to explore the genes involved in the transition between mutual relationships and pathogenic nature in the plant microbiome (Rathnasamy et al. 2023). For example, Iavicoli et al. (2003) reported that root inoculation with *Pseudomonas fluorescens* CHA0 induced systemic resistance in *Arabidopsis thaliana*, thereby activating plant defense-related responses, and Jeon et al. (2022) studied the transcriptional response of the host to infection by *Penicillium expansum* of blueberry (Jeon et al. 2022; lavicoli et al. 2003). In addition, RNA-Seq has elucidated the interaction mechanisms between the bacteria *Delftia acidovorans* Ray209, which stimulates plant growth and canola and soybean roots during the colonization periods (Gamalero et al. 2022).

Metatranscriptomics extends transcriptomic profiling of complete microbial populations to the in situ description of whole-population functional gene expression. The metatranscriptomics approach provides real-time dynamics of gene expression in environmental samples when assessing total RNA and enhances our understanding of microbial functional responses (Dubey et al. 2020). This approach assists in the identification of active microbial genes involved in plant interactions and enables the functional profiling of microbial consortia (Gamalero et al. 2022).

4.3 Genomic Approaches for Rhizoengineering

The introduction of genomic technologies has exponentially increased the scope of research, enabling the modification and discovery of genes involved in root growth and functions. Rhizoengineering is gaining attention because of its capability to modify crop performance and sustainability. Genetic tools are at the forefront of this development, permitting the identification and manipulation of genes that control root structure and function. Modern genome editing technologies, such as CRISPR/Cas9, which was the first instance, have transformed crop breeding through the precisely controlled manipulation of genes in the development of the root. Recent research highlights the possibilities of genome editing to enhance plant root structures and nutrient uptake. Tao et al. (2025), as a result of an instance, employed CRISPR/Cas9 technology to alter the *Os*NRT2.3 of rice, which enhanced the efficiency of nitrogen use. Transcriptomics is a method used to study patterns in gene expression and develop tools to assist in determining the molecular processes that regulate the growth of root tissue as well as its interactions with environmental factors (Tao et al. 2025). Transcriptomic analysis has produced a highly precise mapping of Arabidopsis roots in temporal and spatial dimensions. This has led to the development of a new understanding of developmental processes, as well as the distinction of root structures (Brady et al. 2007). Genomics is a field of research that is focused on studying the genome holistically as well as genomes and has resulted in the emergence of genetic variants that are like roots with obvious characteristics. Recent research using genome-wide association studies (GWAS) has led to the discovery of genetic variations that control roots with optimal architecture and nutrient absorption. Zhang et al. (2020) employed GWAS to determine the genetic changes that affect the structure of rice root systems (Zhang et al. 2020).

4.3.1 Third-Generation Sequencing Technologies

High-throughput third-generation sequencing (HTS) platforms can read hundreds of thousands or millions of DNA molecules in a single read while regaining specific taxa in low amounts (Bik et al. 2012; Nilsson et al. 2019). Library preparation and running of HTS instruments typically take several days. There could be waits of up to months for commercial service suppliers to respond to requests. Furthermore,

Table 4.1 Comparison of PacBio SMRT and Oxford Nanopore sequencing technologies

Technology	Benefits	Limitations
PacBio SMRT sequencing	Extensive read lengths, typically over 10 K base pairs High accuracy with error rates of less than 1% Appropriate for applications requiring high accuracy and long read lengths, such as the classification of epigenetic modifications and de novo assembly	High cost per base, which makes it the most expensive option for most applications Lower quantity compared to other technologies, with fewer reads generated in a single run
Oxford nanopore	Long read lengths, classically over 10 K base pairs Higher accuracy for longer read lengths, with error rates of less than 1% Transportable and flexible, allowing sequencing in remote locations or in real time	High cost per base compared to other technologies, although lower than PacBio SMRT sequencing Lower throughput compared to other technologies, with only a small number of reads generated in a single run Higher unpredictability in accuracy than other technologies, with a higher error rate for shorter read lengths

sequencing is expensive, making it impractical for the rapid identification of pathogens (Tedersoo et al. 2019). Despite millions of output reads, these second-generation SOLiD, Roche 454, Illumina, and Ion Torrent platforms have shorter read lengths, which makes it difficult to identify the exact species of microorganisms because of the poor taxonomic resolution of 100–500-bp markers (Mosher et al. 2014). Third-generation sequencing platforms from Pacific Biosciences (PacBio; RSII and Sequel instruments) and Oxford Nanopore Technologies (ONT; MinION, GridION, and PromethION instruments) can accommodate an average length of sequences of approximately 220,000 bases. A comparison of PacBio SMRT and Oxford Nanopore Sequencing Technologies is presented in Table 4.1. However, this comes with an error rate ranging between 5% and 20% (Tedersoo et al. 2018). PacBio devices utilize circular consensus sequences and produce multiple duplicates of the same DNA fragment to produce the most precise consensus sequence. In the end, long high-fidelity readings proved to be effective in assembling complex genomes from scratch (Rhoads and Au 2015) and in the barcoding of DNA (Hebert et al. 2018). In comparison to the short-read second-generation high-throughput sequencing (HTS), metabarcoding based on PacBio offers more resolution, specifically for bacteria (Singer et al. 2016) as well as fungi (Tedersoo et al. 2018), including plant pathogens (Walder et al. 2017). In contrast to other HTS platforms, which typically require expensive and bulky devices, MinION is a more compact device. The MinION device is small, roughly the size of a smartphone, and is reasonably priced. Therefore, it is an ideal choice for government laboratories and researchers working in the field and small-scale organizations (Mikheyev and Tin 2014). The small size of MinION and its low energy requirements allow it to be used almost anywhere and in any remote environment, such as tropical rainforests,

Antarctic deserts, and even space-based missions. MinION can generate more than one million sequences daily, with reading lengths in the range of 20,000 bases, and occasionally reaching more than one million bases. Owing to the low quality of sequences, MinION has been used for whole-genome sequencing studies to remove long repeats, bridge contigs, and sequence genomes (Quick et al. 2014). The read error rate can be decreased from 10–15% to 1–5% through the sequencing of the complementary strand (1D2 procedure) or the preparation of the tandem repeat molecules (concatemers); however, these methods are labor-intensive and provide a low level of sequencing and are, therefore, rarely employed (Volden et al. 2018). However, MinION has been used to generate lengthy DNA barcodes that identify particular human pathogens based on consensus sequences that can be easily distinguished and recognized in reference sequence databases (Kilianski et al. 2015). Although several reports suggest that species-level taxonomic resolution can be obtained from nanopore sequencing of complex environmental samples, the excessive error rate remains an issue when analyzing exploratory metabarcoding of the natural community. In contrast, metagenomics has become popular for the detection of human-related pathogens to avoid the complete PCR procedure and errors. Recently, Bronzato Badial et al. (2018) showed that pathogenic plant virus and bacteria can be detected with MinION, whereas Hu et al. (2019) extended this in the field to fungal pathogens from cereals.

4.3.2 *Transcriptomics and Epigenomics for Studying Microbe-Plant Interactions*

Plants do not grow independently and are home to a range of microbes that live in their habitats. Plants can benefit certain organisms, and others can directly affect plants, which assists both. Understanding how microbes and plants interact is a crucial component of understanding the ways in which microbes impact plants, both positively and negatively. Microbes can also indirectly affect plants by altering their environment. The rapid development of sequencing technologies within the genomics age and various OMICs tools has greatly improved the field of biological sciences. Examining the complex interplay between microbes and plants could yield different ways to sustainably increase plant productivity by providing a greater understanding of this relationship.

Plants have developed a variety of cellular and molecular methods to resist extended periods of exposure to different pathogens. Plant defense against pathogens includes the host's transcriptional reconfiguration. Recent research has linked DNA methylation and demethylation and chromatin remodeling via posttranslational modifications to histones, such as acetylation, methylation, and ubiquitination, with changes in the levels of expression of defense genes in response to pathogen attack. It is interesting to note that inducible defense mechanisms may be prepared prior to pathogen attack through epigenetic alterations. The increased resistance can later be handed down to subsequent generations via the transfer of these traits.

4.3.2.1 Molecular Mechanisms Underlying Plant-Microbe Interactions

Complex molecular signaling pathways that are crucial for the development of diseases, symbiosis, and the ability to adapt to changing environments are involved in the interactions between microbial organisms and host plants. By signaling through plant receptors, phytohormones such as auxins and cytokinin, produced by microorganisms such as *Rhizobium leucaminosarum*, are able to stimulate plant growth as well as nutrient absorption. Microbes, such as *Pseudomonas fluorescens*, produce exopolysaccharides, which assist in the formation of biofilm adhesion on plant surfaces and defense against environmental stress.

Bradyrhizobium japonicum and other rhizobia release nod factors that induce nodulation signaling cascades and create symbiotic associations in leguminous plants. Pathogen effectors, such as *Pseudomonas syringae*, alter the cellular functions of hosts to modify plant defenses. Although lipopolysaccharides derived from bacteria, such as *Pseudomonas aeruginosa*, trigger plant defense mechanisms, quorum-sensing bacteria from microbes, such as *Agrobacterium tumefaciens*, can trigger plant reactions. In addition, microorganisms such as *Trichoderma harzianum* produce volatile organic compounds that trigger reactions in the plant system (White and Winans 2007).

4.3.2.2 Challenges in Translating Genomic Advancements into Field Applications

- Genomic data can be heterogeneous, which requires modern analytical instruments (Zhang et al. 2020).
- Cross-validation of the results obtained from genetic analysis across diverse species and different environments could be a challenge (Ioannidis et al. 2008).
- Regulatory models for genomics can be unambiguous or inconsistent (Kwon et al. 2024).
- The acceptance of genetic technologies could be affected by questions about security, ethics, and privacy (Caulfield and McGuire 2012).
- Successful translation requires collaboration among experts in various disciplines, such as bioinformatics, genomics, and field-based science (Gevers et al. 2012).

4.3.2.3 Solutions

- Encouraging collaboration between experts from different fields is essential.
- Standardized methods for genomic analysis and validation should be developed.
- Establish clear regulatory frameworks for genomic applications.
- Work with the public to raise awareness and increase understanding of genetic technologies.

- They indicate a constant requirement for cooperation, research, and clarity about rules to connect genomic advancements and practical applications within the area.

4.3.3 Genome-Wide Association Studies (GWAS) for Linking Microbial Traits to Plant Performance

Genome-wide association studies (GWAS) have also been employed to investigate the connections between microbial and plant traits, eventually revealing connections with respect to the performance of plants and the microbiome. By comparing the genetic variations between a group of plant species and GWAS, it is possible to identify the genes that are relevant to several characteristics of plants, including those associated with interactions with harmful or beneficial microbes in the rhizosphere. These datasets are then used to determine how plant genotypes affect the composition and function of plant-related microbiomes, in addition to how microbiome function affects plant productivity. In the general sense, GWAS is an excellent method for identifying genetic variations (e.g., SNPs) that are relevant to certain plant characteristics. When comparing different plant genome variants from a large population, GWAS can determine the genetic factors associated with plant performance, resistance to disease, and other characteristics. This can help in the analysis of how genes affect plant characteristics. Plants interact in a complex way with their microbes inside their rhizosphere (the part of the soil that surrounds their roots), which can significantly affect the growth, development, and overall health of the plant. Beneficial microbes provide nutrients, protect against pathogens, and enhance root growth. They can also harm microbes that cause disease or impair performance. Understanding the impact of plant genotypes on rhizosphere microbiome function and composition is crucial for optimizing plant health and productivity. It can be used with GWAS to determine how genetic variation affects microbiome composition and activity in the rhizosphere. This will require the study of plant genotypes along with microbial factors such as community structure, diversity, and function.

4.3.3.1 Benefits of this Approach

- Understanding the genetic basis of interactions between microbes and plants could lead to the development of innovative ways to boost plant health and the capacity of plants to withstand stress.
- The insights gained from these studies can be utilized to inform the development of sustainable and environmentally friendly farming practices, such as the use of beneficial microbes to increase plant efficiency.

- Utilizing GWAS or microbial trait studies provides researchers with an overview of the aspects that impact the effectiveness of plants, which, in turn, leads to better and longer-lasting plants.

4.3.4 CRISPR-Cas Genome Editing in Beneficial Rhizosphere Microbes

The rhizosphere is a thriving city for microbes around plant roots. There are three important types of beneficial microbes: fertilizer makers, such as *Rhizobium*, which absorb nitrogen from the air and provide it to plants. Bodyguard microbes, such as *Pseudomonas*, fight off harmful fungi and bacteria. The Delivery Crew—Fungi that act like extra roots, bringing water and nutrients.

Overall, these microscopic helpers do much for plants: they feed, protect, and help them grow. With new CRISPR gene-editing tools, scientists are finding ways to make these microbes even better at their jobs. CRISPR technology helps improve these soil bacteria and what it might mean for farming in the future.

CRISPR works like molecular scissors. The components of the model are explained below:

- The GPS (gRNA): This tells the system where to cut in the DNA.
- The Scissors (Cas9): This protein makes the actual cut.
- The Repair Kit: The cell then fixes the cut, either making small changes (NHEJ) or adding new instructions (HDR).

CRISPR is special because it is precise and relatively simple compared to older genetic engineering methods. Scientists can now edit genes in microbes much more easily than before.

4.3.4.1 Editing Helpful Soil Bacteria

Researchers are using CRISPR to improve soil microbes in several ways, including editing nitrogen-fixing bacteria to work with more crops and helping phosphate-solubilizing bacteria work in poor soils. They act as stronger protectors by removing weak points in biocontrol bacteria and adding genes to help them fight more types of plant diseases. It also serves as a survival upgrade by helping microbes handle drought or salty soils. These changes could lead to the development of microbes that help crops grow better with less chemical fertilizers.

4.3.4.2 Challenges

While CRISPR editing sounds promising, there are real concerns, such as:

- Some soil bacteria are hard to edit.
- Changes might not last in real field conditions.
- Edited microbes might not compete well with natural ones.

Overall, future demand is as follows:

- More precise editing tools beyond basic CRISPR
- Custom microbes designed for specific crops or soils
- Better testing to ensure safety
- Possible combination with organic farming methods

The key is to balance innovation with caution to avoid harming natural systems.

4.4 Engineering Microbial Bioinocula Through Genomics

Bioinoculants from beneficial microorganisms are essential for sustainable agriculture because they facilitate an increase in the intake of nutrients, enhance plant growth, and reduce environmental and biotic stressors. The genomics period and related molecular tools have transformed bioinoculant development because one can identify, functionally analyze, and genetically manipulate high-quality strains. This assignment addresses the developments in genomic-based microbial bioinoculant engineering over the last 2 years and highlights key research in this rapidly growing field.

4.4.1 Genome Mining in **Paenibacillus** *for Antimicrobial Compounds*

In a recent study, Kim et al. (2024) applied genome mining strategies to investigate biosynthetic gene clusters (BGCs) of *Paenibacillus* species. The use of this research methodology enabled the authors to discern the genes implicated in producing novel antimicrobial metabolites (Kim et al. 2024). The identification of these gene sets can enable research on producing fresh microbial inoculants with enhanced disease suppression efficacy, especially for use in agricultural contexts in which microbial biocontrol is an important tool.

4.4.2 Genetic Enhancement of Rhizobia for Iron Uptake

Geetha and Joshi (2013) published a study on the engineering of rhizobial strains that enhance iron acquisition by increasing the expression of genes associated with siderophore manufacturing. Increased iron levels enhance nitrogen fixation, eventually leading to higher plant biomass accumulation. This approach illustrates how the genomic toolkit can be used to improve the delivery of nutrients by microbes (Geetha and Joshi 2013).

4.4.3 *Overexpression of Mycosubtilin in* Bacillus subtilis

The effectiveness of microbes for biocontrol is derived from their generation of antimicrobial compounds. Leclère et al. (2005) created engineered strains from *Bacillus subtilis* that have increased the biosynthesis of mycosubtilin, which is a powerful lipopeptide that has antifungal properties. Strain engineering has significantly improved the ability of phytopathogens to fight off pathogens, with an optimistic trend towards creating strains of microbes with enhanced biocontrol effectiveness (Leclère et al. 2005).

4.4.4 Functional Genomics of Plant Growth-Promoting Bacteria

Identifying the genetic factors that determine the traits that promote plant growth is crucial for identifying successful bioinoculants. Adedayo and Babalola (2023) provided a genomic overview of the beneficial bacteria associated with leguminous crops. This study identified genes responsible for critical functions, such as nitrogen fixation, hormone production (e.g., auxins), and abiotic stress tolerance. Such insights pave the way for the rational engineering and application of targeted microbial consortia (Adedayo and Babalola 2023).

4.4.4.1 Genomic Characterization of Tomato Core Microbiome

In 2024, a research group studied ten bacterial isolates of the tomato core microbiome using whole-genome sequencing. Genomic exploration revealed genes associated with plant colonization, nutrient mobilization, and resistance to pathogens. Such high-resolution genomic clarity makes it possible to design strain-specific inoculants tailored specifically at the crop system level, promoting inoculant efficacy and strength under realistic field conditions (Nicotra et al. 2024).

4.4.4.2 Genomics-Based Development of Sustainable Inoculants

Díaz-Rodríguez et al. (2025) reviewed the advancements in microbial inoculants with an emphasis on sustainable agriculture. The authors emphasized how genomics, and specifically omics, are changing the way we identify and improve microbial functional characteristics. They also highlighted the significance of environmental and regulatory surveillance to ensure the safe use of genetically engineered microbial products (Díaz-Rodríguez et al. 2025).

4.4.5 Identification and Selection of Superior Microbial Strains Based on Genomic Data

The development of genomics has significantly improved the way we analyze and enhance microbial strains. With the help of comprehensive genomic information and analysis, scientists have been able to identify microorganisms with traits suitable for agricultural, industrial, and therapeutic purposes. The genomics-informed method is a far more efficient and precise alternative to the traditional phenotypic screening (Zhou et al. 2020).

Current high-throughput sequencing methods, such as whole-genome sequencing (WGS), provide a complete view of the genomics of microbial organisms. Comparative genomics aids in discovering the functional genes that are responsible for significant traits, including resistance to stress, production of antibiotics, and efficacy of metabolic processes (Medema et al. 2014). Methods such as average nucleotide identification (ANI) and phylogenomic reconstruction play an important role in confirming species-specific recognition and determining genetic connections between different species (Richter and Rosselló-Móra 2009). Analyses of the pangenome revealed gene sets that shared commonality between strains based on associated genes that code for idiosyncratic roles. Such a separation is useful for identifying genes that correspond to significant attributes. Marker-assisted selection employs these types of gene markers to identify strains that are capable of enhancing productivity or resistance to environmental stress. For example, identifying genes that participate in the biosynthetic pathway of secondary metabolites may help select the most effective antibiotic-producing strains, especially in actinomycetes (Blin et al. 2019).

4.4.5.1 Applications and Future Directions

Genetic-based selection has proven to be efficient in various sectors. In agriculture, strains such as *Rhizobium* and *Bacillus* have been selected based on their ability to improve plant development and deter pathogens. In industrial applications, strains of *E. coli* and *S. cerevisiae* are often engineered and optimized for biofuel and bioproduct production. As multiomics strategies that combine genomics with transcriptomics and metabolomics continue to mature, they will inevitably extend strain optimization strategies even further (Palazzotto and Weber 2018).

The introduction of genomic analysis in microbial strain creation has opened a new phase in precision biology. By utilizing advanced sequencing techniques, bioinformatics, and multiomics analysis, scientists can rationally choose and create highly efficient microbial strains, paving the way for advancements across a wide range of scientific disciplines.

4.4.6 Engineering Microbial Consortia for Enhanced Stress Tolerance and Nutrient Mobilization

Metabolic engineering presents novel strategies for enhancing the biosynthetic pathways of these substances at the microbial level for industrial production. It is a process of optimizing cell-based genetic and regulatory processes to enhance the production of substances. It involves:

- Gene editing
- Pathway reconstruction
- Flux redirection
- Regulatory modifications

4.4.6.1 Strategies in Metabolic Engineering for Enhancing Secondary Metabolite Gene Overexpression and Knockout

- Enhancing the expression of rate-limiting enzymes
- Knocking out competitive pathways to divert precursors

1. *Heterologous Expression:* Transferring biosynthetic gene clusters to a more manageable host, such as *E. coli* or *Saccharomyces cerevisiae*.
2. *Promoter Engineering:* Using strong and inducible promoters to precisely control gene expression.
3. *CRISPR/Cas9 and Synthetic Biology Tools:* Editing genes with high precision is important for optimal strain development in biotechnological applications.
4. *Flux Balance Analysis and Omics Integration:* Using computational models and omics data (genomics, transcriptomics, and proteomics) for pathway optimization.

4.4.6.2 Case Studies

Artemisinin Production in Yeast: A notable example is the engineering *of Saccharomyces cerevisiae* to produce artemisinic acid, a precursor of the antimalarial drug artemisinin.
Lovastatin Production in Aspergillus terreus: Lovastatin biosynthesis was enhanced by gene cluster overexpression, which increased the yield in industrial fermenters.

4.4.6.3 Applications

- *Pharmaceutical Industry:* Antibiotics and anticancer drugs
- *Agriculture:* Biopesticides and growth regulators
- *Food Industry:* Flavor compounds and preservatives

4.4.6.4 Challenges and Future Prospects

- *Genetic Stability:* Maintenance of engineered traits over generations
- *Metabolic Burden:* Balancing growth and production
- *Regulatory Hurdles:* Biosafety and approval concerns

The future lies in the integration of AI-driven metabolic modeling and synthetic biology for precision metabolic engineering (Fig. 4.2).

Genomic methodologies have significantly advanced rhizoengineering, unlocking new possibilities for crop enhancement. The integration of genomics, transcriptomics, and genome editing has enabled the development of crops with superior root systems, thereby improving nutrient acquisition and stress resilience. To fully exploit these genomic strategies and translate laboratory findings into field-level agricultural improvements, much work is required. In addition, metabolic engineering can provide a high-throughput platform for enhancing secondary metabolite production to meet industrial demands and support sustainable biomanufacturing. As genomic data informatics, synthetic biology, and CRISPR-mediated strategies continue to develop, microbial cell factories can be tailor-made for global use in medicine and industrial biotechnology. The use of genomic technologies has greatly enhanced the discovery of efficient and effective microbial bioinoculants. By means of genome mining, genome editing, and comparative genomics, high-value key traits of improved plant-microbe interactions can be ascertained and enhanced. As these technologies continue to develop, their integration into crop systems will support increased crop productivity, resilience, and environmental sustainability.

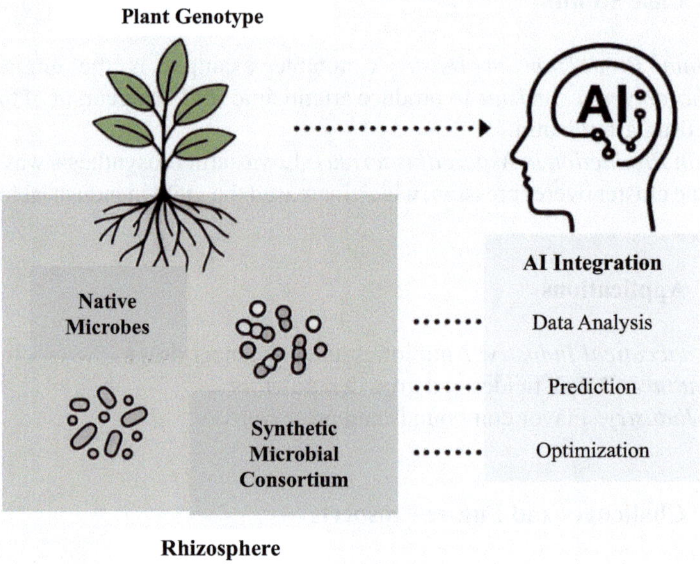

Fig. 4.2 Conceptual diagram of synthetic microbial consortia and AI integration in rhizosphere engineering

4.5 Host Plant Genomics and Its Role in Shaping the Rhizomicrobiome

Plant roots are in contact with various microbes within the rhizosphere (Wang et al. 2024). The majority of these interactions have some impact on plant growth (Cheng et al. 2019; Carrión et al. 2019). The outcomes of plant root-microbiome interactions are shaped by interdependent variables, including the host's genetics, climate, edaphic factors, and rhizosphere characteristics (Escudero-Martinez et al. 2022; Liu et al. 2023; Edwards et al. 2023). Recent research has revealed that the primary factor influencing the microbiome composition of the rhizosphere is host genes, which regulate root structure exudates, immunity, and nutrient intake (Wang et al. 2022; Yu et al. 2021). In particular, the genes involved in the formation of hair roots and root growth of vine species of plants could affect the structure and function of the microbial community by altering the exudates from roots. In *Arabidopsis,* the two genes are involved in the development of root hairs, *RHD6* and *AXR2*, and boost the area of root surfaces and alter root exudation, which forms the rhizosphere microbial population (Mendoza and Alvarez-Buylla 2000; Yi et al. 2010). Furthermore, in maize, genes such as *RTH6*, *LRT1*, *RUM1*, and *RTC*s can also change the composition of microbial communities related to roots (Yu et al. 2021; Coudert et al. 2010).

4.5.1 Genetic Determinants of Plant-Microbe Specificity

The microecosystem hybrid rhizomicrobiome comprises microbes found within the rhizosphere. It plays a role in nutrition, plant health, and stress tolerance to environmental stresses (Chai et al. 2024). The type of interaction that a plant has with rhizosphere microbes and the specificity of these interactions are primarily determined by the genetic system of the host plant. Genomic control affects not only the physical and biochemical features of root tissue but also the signaling pathways that manage symbiosis or defensive mechanisms.

Therefore, the microbes engaged in potential habitats are determined by the root exudates produced and genetically controlled features of the roots, including root hair density and lateral root branching (Upadhyay et al. 2022). For example, resistance (*R*) genes in plants for immune receptors that recognize microbe-associated molecular patterns have mechanisms to distinguish between symbionts and pathogens, which enables them to differentiate symbionts from pathogens (Tian et al. 2020). This form of molecular recognition may either promote further colonization, which is beneficial, as in the case of nitrogen-fixing bacteria residing in legumes, or initiate immune responses that block the cellular pathways of intruding pathogens.

Additionally, the dimensions of genome-wide association research (GWAS) and quantitative trait loci (QTLs) are being studied and have identified specific regions of the genome that are relevant to the ability to capture certain microbes. In maize, certain QTLs are associated with the diversity of fungal and bacterial populations in relation to the rhizosphere of maize, establishing the transmission characteristics of the development of microbial communities (Li et al. 2022).

4.5.2 The Function of Root Exudates in Microbial Selection and Recruitment

Root exudates mainly serve as low-signaling molecules during the recruitment and activity of microorganisms in the rhizomicrobiome and supply nutrients to microorganisms. Root exudates consist of short sugars, amino acids, organic acids, and phenolics, some of which vary with the plant's genotype, developmental phase, and prevailing environment (Tian et al. 2020). Different classes of these substances exist, each with distinct ecologically interesting functions. Flavonoids washed out of leguminous roots, for example, induce nodulation signaling in rhizobia and bring about a series of events leading progressively to the formation of nitrogen-fixing nodules (Singla and Garg 2017). Similar to these processes, several organic acids, such as malic, citric, and succinic acids, have been proven to boost the recruitment of plant rhizobacteria (PGPR), such as *Pseudomonas fluorescens and Azospirillum brasilense*, via certain regulatory routes (Panichikkal and Edayileveetil Krishnankutty 2020).

Table 4.2 Examples of root exudate compounds and their microbial interactions

Plant species	Root exudate compounds	Recruited microbes	Functional outcome	References
Groundnut (*Arachis hypogaea*)	Naringenin, citric acid, lactic acid	*Bradyrhizobium*, *Azospirillum brasilense*	Enhanced root exudation and PGPR interaction under water deficit	Cesari et al. (2019)
Wheat (*Triticum aestivum*)	2,4-Diacetylphloroglucinol (DAPG)	*Pseudomonas* spp.	PGPR colonization and increased yield	Raaijmakers and Weller (2001)
Rice (*Oryza sativa*)	Carbohydrates, amino acids	*Azospirillum brasilense* and *Bacillus* spp.	Attraction of endophytes and nutrient acquisition	Bacilio-Jiménez et al. (2003)
Tomato (*Solanum lycopersicum*)	Citric, succinic, malic acids	*Bacillus amyloliquefaciens*	Biocontrol and increased biomass	Tan et al. (2013)
Maize (*Zea mays*)	Humic acid	*Herbaspirillum seropedicae*	Enhanced colonization and nitrogen fixation	da Silva Lima et al. (2014)

The amount and quality of exudates from the roots can be affected by abiotic factors. For instance, a condition that is deficient in phosphate can be illustrated in which plants release more organic acids to release bound phosphate and, at the same time, draw the phosphate-solubilizing bacteria (Sarkar et al. 2014). Some examples of root exudate compounds and their microbial interactions are presented in Table 4.2. These adaptational changes highlight the significance of root exudates in maintaining a vibrant and supportive microbial community in the rhizosphere. Furthermore, phenolic exudates can impact diversity by promoting beneficial microorganisms and inhibiting opportunistic pathogens. Some phenolic compounds such as coumarins and benzoxazinoids have been documented to possess phyto-pathogen antifungal properties while simultaneously aiding beneficial microbial colonization (Lombardi et al. 2018).

4.5.3 Engineering Plants for Improved Microbial Interactions and Stress Tolerance

Biotechnological approaches are now used to modify host plant genomes to improve rhizomicrobiome interactions. Genetic engineering and synthetic biology techniques allow the alteration of the root exudate neutral immune system, exudate composition, responders, and stress tolerance mechanisms to enhance microbial recruitment and efficiency under environmental stress (Zhang et al. 2024). To enhance the recruitment of symbiotic nitrogen-fixing bacteria in legumes and cereals, the overexpression of genes involved in flavonoid biosynthesis is a promising strategy. For instance, transgenic plants modified to enhance naringenin production yield greater quantities of *Bradyrhizobium* and *Azospirillum* (Nouwen et al. 2019). This enhances nitrogen absorption while lowering dependence on chemical fertilizers.

Similarly, directing microbial exudates towards pathogenic microbes while attracting beneficial ones can be achieved by engineering specific plant exudate compounds. In tomato and wheat, the secretion of certain organic acids has been linked to the selective clustering of biocontrol agents, such as *Pseudomonas fluorescens* (Ma et al. 2021; Tan et al. 2013).

Entirely designing plants for tolerance to abiotic stresses also affects the rhizomicrobiome indirectly. Enhanced drought or salinity tolerance is often linked to modified root exudation profiles, which attract many microbes that aid in further alleviating stress (Wang et al. 2024). For instance, drought-tolerant transgenic lines may start secreting higher quantities of lactic acid and some amino acids, aiding in the colonization of these sites by drought-resilient PGPR (Bouremani et al. 2023).

Understanding the interplay at the molecular level between microbes and plants is enhanced by the application of multiomics technologies, such as proteomics, transcriptomics, and metabolomics. These approaches have assisted researchers in using the assembly processes of microbial community structures to determine plant genes that control crosstalk with microbes (Kumar et al. 2023). This understanding is critical when developing strategies for designing next-generation crops capable of enduring the productivity challenges posed by a changing climate.

4.6 Practical Limitations and Bottlenecks

4.6.1 Challenges in Scaling Up Genomically Optimized Bioinocula

The use of genomics in developing microbial bioinoculants has enhanced the accuracy and effectiveness of the interactions between crops and microbes. There is no doubt that the industry has not yet mastered the scaling-up challenges associated with genetically optimized bioinocula in the laboratory. First, a laboratory setting usually does not capture the unpredictable nature of heterogeneous field environments, and as a result, microbes do not perform as expected (Compant et al. 2025). Some beneficial features exhibited in cultures may not manifest when placed under dynamic pH, temperature, or moisture flux conditions.

In addition, the long-term retention of genomic traits as functional viability within bioinoculants makes mass production technologically challenging. Fermentation-based production systems must guarantee high yields in terms of

Table 4.3 Common limitations in the field application of genomically engineered bioinoculants

Limitation	Description	References
Environmental heterogeneity	Variability in pH, moisture, and soil type reduces bioinoculant efficacy	Compant et al. (2010)
Native microbiome competition	Endogenous microbes outcompete or neutralize introduced strains	Johnston-Monje and Raizada (2011)
Regulatory delays	Extensive approval processes delay product commercialization	Backer et al. (2018)
Public perception	Skepticism about GMOs affects market acceptance	Eckerstorfer et al. (2025)
Poor storage and delivery systems	Loss of microbial viability during transport and field application	Bashan and de-Bashan (2010)

identity, stability, and survivability during storage and transport of inocula while preserving minimal genetic change or loss of traits, which is often referred to as the "genetic drift trap" (Backer et al. 2018).

Achieving stable microbial colonization and persistence in the rhizosphere across different soil types is particularly challenging. Competition among soil microorganisms can lead to competitive exclusion, restricting the survival and growth of the introduced strains (Mitter et al. 2013). Therefore, spatial variability and interference from the native microbiome are important limitations. The common limitations in the field application of genomically engineered bioinoculants are summarized in Table 4.3.

4.6.2 Regulatory and Biosafety Issues Related to the Use of Genetically Modified Microbes

The application of genetically modified (GM) organisms in agriculture is accompanied by complex layers of societal, ethical, biosafety, and regulatory concerns. In most regions, the application of GM organisms in agriculture is highly regulated by the government. Many regions face burdensome, fragmented, expensive, and cumbersome approval processes (Ahmad et al. 2024).

Biosafety concerns tend to focus on the concept of horizontal gene transfer (HGT), which raises the possibility that foreign engineered genes may assimilate into native microbes, potentially disrupting ecosystems or giving rise to new pathogenic organisms (Alamnie 2020). There is also the concern of gene escape from non-targeted plants or microbial entities.

In addition, societal perceptions are critical for the acceptance of GM technology in agriculture. There has been long-standing skepticism about the safety and ecological impacts of GM microbes, even within some segments of the scientific community (Mahdizade Ari et al. 2024). Such societal resistance tends to obstruct policy initiatives, even when strong and rational cost-benefit analyses favor the proposed change.

4.6.3 Performance Gaps and Environmental Constraints

Regardless of meeting all regulatory requirements and increasing production capacity, bioinoculants often fail to perform as expected in real-world field conditions. In addition to weather conditions, soil chemistry, and biology, the specific crop cultivated and the interactions of these with soil- and plant-associated microbes are important variables that can greatly impact the performance and efficacy of bioinoculants (Backer et al. 2018).

Abiotic stresses, such as drought, salinity, and low nutrient availability, can negatively affect both beneficial microbes and the services they provide to the host plant. Over time, the first inoculant strains may lose ground because native organisms swarm in and compete or even absorb them. In addition, current application techniques can scatter inoculants unevenly or place them far from the roots. This problem is exacerbated by insufficient training of farmers and infrastructure for microbial product application.

4.7 Future Directions and Innovations

4.7.1 Incorporating Big Data and Artificial Intelligence in Rhizosphere Genomics

Recent advances in computational and data science biology have opened new pathways in rhizosphere genomics. Owing to artificial intelligence (AI) and machine learning (ML), which can manage massive multiomics datasets, scientists can predict the relationships between microbes and plants or even pinpoint the best microbial mixtures (Fadiji et al. 2023). For example, an AI model might merge metagenomic, transcriptomic, and phenotypic signatures to highlight plant genotypes that excel in attracting helpful microbes. Similar algorithms also predict the behavior of microbial communities under shifting environmental conditions, thereby guiding practical field management (Xu et al. 2021).

Big-data platforms that pool results from numerous trials across different sites can further deliver on-demand bioinoculant advice tailored to local soil, climate, and crop needs. Such targeted recommendations boost the accuracy of inoculant performance in commercial farming, instilling confidence in growers.

4.7.2 Synthetic Biology's Potential in Engineering Microbial Communities

Advanced gene-editing tools, such as CRISPR/Cas9, along with synthetic gene circuits and purpose-built microbial chassis, are now being harnessed to craft microbes that sense, interpret, and adapt to environmental changes (Teng et al. 2024). Research teams can now assemble microbial consortia that pair nitrogen fixers, phosphate-solubilizing organisms, and stress-tolerant plant growth-promoting rhizobacteria (PGPR) into a single synergistic blend that supports plants on multiple fronts. Stability and field deliverability are then boosted through encapsulation methods that protect the microbes during storage and release them gradually to the roots. Furthermore, engineered strains can be programmed to self-monitor and deactivate, or even die, when environmental triggers signal that their job is done, easing safety concerns. If widely adopted, these approaches could dramatically reduce the use of chemical fertilizers and signal the opening of a new genomics-led era in sustainable farming.

4.7.3 Policy Frameworks for Commercializing Genomically Engineered Biofertilizers

As public and private stakeholders invest resources in microbial biofertilizers, the need for balanced rules that allow new products to reach the market while protecting ecosystems and maintaining trust is increasing. Merging risk assessment with the push to innovate is the chief test for regulators (Mitter et al. 2021).

Streamlined approvals for moving bioinoculants across borders require clearer and aligned rules among nations. Any new framework should promise open information, avenues for citizen input, and postlaunch checks.

Co-investments from the government and industry can accelerate progress by supplying funds, test facilities, and education for farmers. Backers commit capital if laws shield intellectual property and spell out who is liable when products fail.

4.7.4 Bioinoculant Genome Design and Climate-Smart Technology: A Comprehensive Overview

1. Articulate AI algorithms to predict plant and microbe interactions and their responses to environmental stressors.
2. Construct synthetic consortia through gene editing for multitrait advantages.
3. Establish policies that enable safe and open market access to targeted bioproducts.

4. Enhance agricultural field test networks for optimal inoculant performance in real-world conditions.
5. Promote public confidence through outreach initiatives and education targeted at farmers.

Unlocking the full potential of rhizosphere genomics for sustainable agriculture requires a global commitment that is scientifically sound and policy-aligned.

4.8 Conclusion

Plant genetics largely determines which microbes anchor to its roots, guiding the traits it expresses and the chemical signals it leaks. Fine-tuning this partnership presents a novel path to sharper nutrient absorption, greater stress tolerance, and higher harvests. However, real-world hurdles, such as scalability, shifting weather, public trust, and tangled regulations, still slow the spread of bioinoculants built through genome-guided design. Variable performance in the field and the complex interactions among soil, microbes, and roots pose major obstacles. Addressing these challenges requires a broad toolkit that combines smart policies with agronomy, microbiology, data science, and plant genomics. A bold approach merges artificial intelligence, large datasets, and synthetic biology to build rational, purpose-driven inoculants. Future work must map the entire farming ecosystem and tailor products to the needs of each agroecological zone. Close dialogue with farmers, regulators, and the public will be vital to gain approval and secure the enabling laws.

References

Adedayo AA, Babalola OO (2023) Genomic mechanisms of plant growth-promoting bacteria in the production of leguminous crops. Front Genet 14:1276003

Ahmad J, Grunden A, Kuzma J (2024) Biotechnology executive order opens door for regulatory reform and social acceptance of genetically engineered microbes in agriculture. GM Crops Food 15:248–261. https://doi.org/10.1080/21645698.2024.2381294

Alamnie G (2020) Biosafety issues of unintended horizontal transfer of recombinant DNA. In: To KY (ed) Genetic transformation in crops. IntechOpen, Rijeka. https://doi.org/10.5772/intechopen.93170

Ali MH, Mandal S, Ghorai M, Lal MK, Tiwari RK, Kumar M, Ghosh A, Al-Tawaha AR, Gopalakrishnan AV, Shekhawat MS (2023) Perspectives of omics and plant microbiome. In: Genomics, transcriptomics, proteomics and metabolomics of crop plants. Elsevier, pp 131–144

Bacilio-Jiménez M, Aguilar-Flores S, Ventura-Zapata E, Pérez-Campos E, Bouquelet S, Zenteno E (2003) Chemical characterization of root exudates from rice (*Oryza sativa*) and their effects on the chemotactic response of endophytic bacteria. Plant Soil 249(2):271–277. https://doi.org/10.1023/A:1022888900465

Backer R, Rokem JS, Ilangumaran G, Lamont J, Praslickova D, Ricci E, Subramanian S, Smith DL (2018) Plant growth-promoting rhizobacteria: context, mechanisms of action, and roadmap to commercialization of biostimulants for sustainable agriculture. Front Plant Sci 9:1473

Bano S, Wu X, Zhang X (2021) Towards sustainable agriculture: rhizosphere microbiome engineering. Appl Microbiol Biotechnol 105:1–20

Bashan Y, de-Bashan LE (2010) How the plant growth-promoting bacterium *Azospirillum* promotes plant growth—a critical assessment. In: Sparks DL (ed) Advances in agronomy, vol 108. Academic Press, pp 77–136. https://doi.org/10.1016/S0065-2113(10)08002-8

Bhargava P, Khan M, Verma A, Singh A, Singh S, Vats S, Goel R (2019) Metagenomics as a tool to explore new insights from plant-microbe interface. In: Kumar A, Singh J, de Azevedo LCB (eds) Plant microbe interface. Springer, pp 271–289

Bibi N, Hamayun M, Khan SA, Iqbal A, Islam B, Shah F, Khan MA, Lee I-J (2018) Anthracene biodegradation capacity of newly isolated rhizospheric bacteria *Bacillus cereus* S13. PLoS One 13(8):e0201620

Bik HM, Porazinska DL, Creer S, Caporaso JG, Knight R, Thomas WK (2012) Sequencing our way towards understanding global eukaryotic biodiversity. Trends Ecol Evol 27(4):233–243

Blin K, Kim HU, Medema MH, Weber T (2019) Recent development of antiSMASH and other computational approaches to mine secondary metabolite biosynthetic gene clusters. Brief Bioinform 20(4):1103–1113

Bouremani N, Cherif-Silini H, Silini A, Bouket AC, Luptakova L, Alenezi FN, Baranov O, Belbahri L (2023) Plant growth-promoting rhizobacteria (PGPR): a rampart against the adverse effects of drought stress. Water 15(3):418

Boutsika A, Michailidis M, Ganopoulou M, Dalakouras A, Skodra C, Xanthopoulou A, Stamatakis G, Samiotaki M, Tanou G, Moysiadis T (2023) A wide foodomics approach coupled with metagenomics elucidates the environmental signature of potatoes. Iscience 26(1):105917

Brady SM, Orlando DA, Lee J-Y, Wang JY, Koch J, Dinneny JR, Mace D, Ohler U, Benfey PN (2007) A high-resolution root spatiotemporal map reveals dominant expression patterns. Science 318(5851):801–806

Bronzato Badial A, Sherman D, Stone A, Gopakumar A, Wilson V, Schneider W, King J (2018) Nanopore sequencing as a surveillance tool for plant pathogens in plant and insect tissues. Plant Dis 102(8):1648–1652

Carper DL, Appidi MR, Mudbhari S, Shrestha HK, Hettich RL, Abraham PE (2022) The promises, challenges, and opportunities of omics for studying the plant holobiont. Microorganisms 10(10):2013

Carrión VJ, Perez-Jaramillo J, Cordovez V, Tracanna V, de Hollander M, Ruiz-Buck D, Mendes LW, van Ijcken WFJ, Gomez-Exposito R, Elsayed SS, Mohanraju P, Arifah A, van der Oost J, Paulson JN, Mendes R, van Wezel GP, Medema MH, Raaijmakers JM (2019) Pathogen-induced activation of disease-suppressive functions in the endophytic root microbiome. Science 366(6465):606–612. https://doi.org/10.1126/science.aaw9285

Caulfield T, McGuire AL (2012) Direct-to-consumer genetic testing: perceptions, problems, and policy responses. Annu Rev Med 63(1):23–33

Cesari A, Paulucci N, López-Gómez M, Hidalgo-Castellanos J, Plá CL, Dardanelli MS (2019) Restrictive water condition modifies the root exudates composition during peanut-PGPR interaction and conditions early events, reversing the negative effects on plant growth. Plant Physiol Biochem 142:519–527. https://doi.org/10.1016/j.plaphy.2019.08.015

Chai YN, Qi Y, Goren E, Chiniquy D, Sheflin AM, Tringe SG, Prenni JE, Liu P, Schachtman DP (2024) Root-associated bacterial communities and root metabolite composition are linked to nitrogen use efficiency in sorghum. Msystems 9(1):e01190-01123

Chandok IK, Afreen H, Afreen R, Haider S, Moharana DP, Hussain T, Aamir M, Shanmugam V, Malik MZ (2022) Functional genomics tools for studying microbe-mediated stress tolerance in plants. In: Santoyo G, Kumar A, Aamir M, Uthandi S (eds) Mitigation of plant abiotic stress by microorganisms: applicability and future directions. Academic Press, London, pp 175–204

Chen JY, Sang H, Chilvers MI, Wu CH, Chang HX (2024) Characterization of soybean chitinase genes induced by rhizobacteria involved in the defense against *Fusarium oxysporum*. Front Plant Sci 15:1341181

Cheng YT, Zhang L, He SY (2019) Plant-microbe interactions facing environmental challenge. Cell Host Microbe 26:183–192. https://doi.org/10.1016/j.chom.2019.07.009

Chiquito-Contreras CJ, Meza-Menchaca T, Guzmán-López O, Vásquez EC, Ricaño-Rodríguez J (2024) Molecular insights into plant–microbe interactions: a comprehensive review of key mechanisms. Front Biosci-Elite 16:9

Compant S, Van Der Heijden MG, Sessitsch A (2010) Climate change effects on beneficial plant–microorganism interactions. FEMS Microbiol Ecol 73(2):197–214

Compant S, Samad A, Faist H, Sessitsch A (2019) A review on the plant microbiome: ecology, functions, and emerging trends in microbial application. J Adv Res 19:29–37

Compant S, Cassan F, Kostić T, Johnson L, Brader G, Trognitz F, Sessitsch A (2025) Harnessing the plant microbiome for sustainable crop production. Nat Rev Microbiol 23(1):9–23. https://doi.org/10.1038/s41579-024-01079-1

Coudert Y, Perin C, Courtois B, Khong NG, Gantet P (2010) Genetic control of root development in rice, the model cereal. Trends Plant Sci 15(4):219–226

Crandall SG, Gold KM, Jiménez-Gasco MM, Filgueiras CC, Willett DS (2020) A multi-omics approach to solving problems in plant disease ecology. PLoS One 15(9):e0237975

da Silva Lima L, Olivares FL, Rodrigues de Oliveira R, Vega MRG, Aguiar NO, Canellas LP (2014) Root exudate profiling of maize seedlings inoculated with *Herbaspirillum seropedicaeand* humic acids. Chem Biol Technol Agric 1(1):23. https://doi.org/10.1186/s40538-014-0023-z

del Carmen Orozco-Mosqueda M, Fadiji AE, Babalola OO, Glick BR, Santoyo G (2022) Rhizobiome engineering: unveiling complex rhizosphere interactions to enhance plant growth and health. Microbiol Res 263:127137

Díaz-Rodríguez AM, Parra Cota FI, Cira Chávez LA, García Ortega LF, Estrada Alvarado MI, Santoyo G, de Los Santos-Villalobos S (2025) Microbial inoculants in sustainable agriculture: advancements, challenges, and future directions. Plants 14(2):191

Doddavarapu B, Lata C, Shah JM (2024) Epigenetic regulation influenced by soil microbiota and nutrients: paving road to epigenome editing in plants. Biochim Biophys Acta Gen Subj 1868(4):130580

Dubey RK, Tripathi V, Prabha R, Chaurasia R, Singh DP, Rao CS, El-Keblawy A, Abhilash PC (2020) Metatranscriptomics and metaproteomics for microbial communities profiling. In: Dubey RK, Tripathi V, Prabha R, Chaurasia R, Singh DP, Rao CS, El-Keblawy A, Abhilash PC (eds) Unravelling the soil microbiome: perspectives for environmental sustainability. Springer, Cham, pp 51–60

Eckerstorfer MF, Dolezel M, Miklau M, Greiter A, Heissenberger A, Kastenhofer K, Schulz F, Hagen K, Otto M, Engelhard M (2025) Environmental applications of GM microorganisms: tiny critters posing huge challenges for risk assessment and governance. IJMS 26(7):3174

Edwards JA, Saran UB, Bonnette J, MacQueen A, Yin J, Uyen Nguyen T, Schmutz J, Grimwood J, Pennacchio LA, Daum C (2023) Genetic determinants of switchgrass-root-associated microbiota in field sites spanning its natural range. Curr Biol 33(10):1926–1938.e1926

Escudero-Martinez C, Coulter M, Alegria Terrazas R, Foito A, Kapadia R, Pietrangelo L, Maver M, Sharma R, Aprile A, Morris J (2022) Identifying plant genes shaping microbiota composition in the barley rhizosphere. Nat Commun 13(1):3443

Fadiji AE, Barmukh R, Varshney RK, Singh BK (2023) Exploring the connectivity between rhizosphere microbiomes and the plant genes: a way forward for sustainable increase in primary productivity. J Sustain Agric Environ 2(4):424–443

Gamalero E, Bona E, Glick BR (2022) Current techniques to study beneficial plant-microbe interactions. Microorganisms 10(7):1380

Geetha S, Joshi SJ (2013) Engineering rhizobial bioinoculants: a strategy to improve iron nutrition. Sci World J 2013(1):315890

Gevers D, Knight R, Petrosino JF, Huang K, McGuire AL, Birren BW, Nelson KE, White O, Methé BA, Huttenhower C (2012) The human microbiome project: a community resource for the healthy human microbiome. PLoS Biol 10:e1001377. https://doi.org/10.1371/journal.pbio.1001377

Ghosh A, Mehta A, Khan AM (2019) Metagenomic analysis and its applications. In: Ranganathan S, Gribskov M, Nakai K, Schönbach C (eds) Encyclopedia of bioinformatics and computational biology. Elsevier, Amsterdam, pp 184–193

Hakim S, Naqqash T, Nawaz MS, Laraib I, Siddique MJ, Zia R, Mirza MS, Imran A (2021) Rhizosphere engineering with plant growth-promoting microorganisms for agriculture and ecological sustainability. Front Sustain Food Syst 5:617157

Hebert PD, Braukmann TW, Prosser SW, Ratnasingham S, DeWaard JR, Ivanova NV, Janzen DH, Hallwachs W, Naik S, Sones JE (2018) A sequel to sanger: amplicon sequencing that scales. BMC Genomics 19:1–14

Hu Y, Green GS, Milgate AW, Stone EA, Rathjen JP, Schwessinger B (2019) Pathogen detection and microbiome analysis of infected wheat using a portable DNA sequencer. Phytobiomes J 3(2):92–101

Iavicoli A, Boutet E, Buchala A, Métraux JP (2003) Induced systemic resistance in Arabidopsis thaliana in response to root inoculationwith Pseudomonas fluorescens CHA0. Mol Plant Microbe Interact 16(10):851–858

Ilangumaran G, Subramanian S, Smith DL (2024) Complete genome sequences of *Rhizobium* sp. strain SL42 and *Hydrogenophaga* sp. strain SL48, microsymbionts of *Amphicarpaea bracteata*. Front Microbiomes 3:1309947

Ioannidis JP, Boffetta P, Little J, O'Brien TR, Uitterlinden AG, Vineis P, Balding DJ, Chokkalingam A, Dolan SM, Flanders WD (2008) Assessment of cumulative evidence on genetic associations: interim guidelines. Int J Epidemiol 37:120–132. https://doi.org/10.1093/ije/dym159

Jain A, Singh HB, Das S (2021) Deciphering plant-microbe crosstalk through proteomics studies. Microbiol Res 242:126590. https://doi.org/10.1016/j.micres.2020.126590

Jeon J, Kim KT, Choi J, Cheong K, Ko J, Choi G, Lee H, Lee GW, Park SY, Kim S (2022) Alternative splicing diversifies the transcriptome and proteome of the rice blast fungus during host infection. RNA Biol 19:373–386. https://doi.org/10.1080/15476286.2022.2043040

Johnston-Monje D, Raizada MN (2011) Conservation and diversity of seed associated endophytes in *Zea* across boundaries of evolution, ethnography and ecology. PLoS One 6(6):e20396

Joubert PM, Krasileva KV (2024) Distinct genomic contexts predict gene presence–absence variation in different pathotypes of *Magnaporthe oryzae*. Genetics 226(4):iyae012

Khatabi B, Gharechahi J, Ghaffari MR, Liu D, Haynes PA, McKay MJ, Mirzaei M, Salekdeh GH (2019) Plant–microbe symbiosis: what has proteomics taught us? Proteomics 19:1800105

Kilianski A, Haas JL, Corriveau EJ, Liem AT, Willis KL, Kadavy DR, Rosenzweig CN, Minot SS (2015) Bacterial and viral identification and differentiation by amplicon sequencing on the MinION nanopore sequencer. Gigascience 4:12. https://doi.org/10.1186/s13742-015-0051-z

Kim MS, Jeong DE, Jang JP, Jang JH, Choi SK (2024) Mining biosynthetic gene clusters in *Paenibacillus* genomes to discover novel antibiotics. BMC Microbiol 24:226

Kumar A, Dubey A (2020) Rhizosphere microbiome: engineering bacterial competitiveness for enhancing crop production. J Adv Res 24:337–352

Kumar U, Raj S, Sreenikethanam A, Maddheshiya R, Kumari S, Han S, Kapoor KK, Bhaskar R, Bajhaiya AK, Gahlot DK (2023) Multi-omics approaches in plant–microbe interactions hold enormous promise for sustainable agriculture. Agronomy 13:1804

Kwon S, Youtie J, Porter A, Newman N (2024) How does regulatory uncertainty shape the innovation process? Evidence from the case of nanomedicine. J Technol Transf 49:262–302

Larekeng SH, Ngadiman N, Khairina Y, Simarmata R, Christita M (2024) Unraveling the potential of ACC deaminase-producing microbes in various agricultural stresses: current status, limitations, and recommendations. Pak J Bot 56(2):791–805

Leclère V, Béchet M, Adam A, Guez J-S, Wathelet B, Ongena M, Thonart P, Gancel F, Chollet-Imbert M, Jacques P (2005) Mycosubtilin overproduction by *Bacillus subtilis* BBG100 enhances the organism's antagonistic and biocontrol activities. Appl Environ Microbiol 71(8):4577–4584

Li Y, Qu Z, Xu W, Chen W, Hu Y, Wang Z (2022) Maize (*Zea mays* L.) genotypes induce the changes of rhizosphere microbial communities. Arch Microbiol 204(6):321

Liu Q, Cheng L, Nian H, Jin J, Lian T (2023) Linking plant functional genes to rhizosphere microbes: a review. Plant Biotechnol J 21(5):902–917
Lombardi N, Vitale S, Turrà D, Reverberi M, Fanelli C, Vinale F, Marra R, Ruocco M, Pascale A, d'Errico G (2018) Root exudates of stressed plants stimulate and attract *Trichoderma* soil fungi. Mol Plant-Microbe Interact 31(10):982–994
Ma L, Wang W-Q, Shi R, Zhang X-M, Li X, Yang Y-S, Mo MH (2021) Effects of organic acids on the chemotaxis profiles and biocontrol traits of antagonistic bacterial endophytes against root-rot disease in *Panax notoginseng*. Antonie Van Leeuwenhoek 114:1771–1789
Mahdizade Ari M, Dadgar L, Elahi Z, Ghanavati R, Taheri B (2024) Genetically engineered microorganisms and their impact on human health. Int J Clin Pract 2024:6638269. https://doi.org/10.1155/2024/6638269
Masenya K, Manganyi MC, Dikobe TB (2024) Exploring cereal metagenomics: unravelling microbial communities for improved food security. Microorganisms 12(3):510
Medema MH, Cimermancic P, Sali A, Takano E, Fischbach MA (2014) A systematic computational analysis of biosynthetic gene cluster evolution: lessons for engineering biosynthesis. PLoS Computational Biol 10(12):e1004016
Mendoza L, Alvarez-Buylla ER (2000) Genetic regulation of root hair development in *Arabidopsis thaliana*: a network model. J Theor Biol 204(3):311–326. https://doi.org/10.1006/jtbi.2000.2014
Mikheyev AS, Tin MM (2014) A first look at the Oxford Nanopore MinION sequencer. Mol Ecol Resour 14:1097–1102. https://doi.org/10.1111/1755-0998.12324
Mitter B, Brader G, Afzal M, Compant S, Naveed M, Trognitz F, Sessitsch A (2013) Advances in elucidating beneficial interactions between plants, soil, and bacteria. In: Sparks DL (ed) Advances in agronomy, vol 121. Academic Press, pp 381–445
Mitter EK, Tosi M, Obregón D, Dunfield KE, Germida JJ (2021) Rethinking crop nutrition in times of modern microbiology: innovative biofertilizer technologies. Front Sustain Food Syst 5:606815. https://doi.org/10.3389/fsufs.2021.606815
Mosher JJ, Bowman B, Bernberg EL, Shevchenko O, Kan J, Korlach J, Kaplan LA (2014) Improved performance of the PacBio SMRT technology for 16S rDNA sequencing. J Microbiol Methods 104:59–60. https://doi.org/10.1016/j.mimet.2014.06.012
Nicotra D, Ghadamgahi F, Ghosh S, Anzalone A, Dimaria G, Mosca A, Massimino ME, Vetukuri RR, Catara V (2024) Genomic insights and biocontrol potential of ten bacterial strains from the tomato core microbiome. Front Plant Sci 15:1437947
Nilsson RH, Anslan S, Bahram M, Wurzbacher C, Baldrian P, Tedersoo L (2019) Mycobiome diversity: high-throughput sequencing and identification of fungi. Nat Rev Microbiol 17:95–109
Nouwen N, Gargani D, Giraud E (2019) The modification of the flavonoid naringenin by *Bradyrhizobium* sp. strain ORS285 changes the nod genes inducer function to a growth stimulator. Mol Plant-Microbe Interact 32(11):1517–1525
Palazzotto E, Weber T (2018) Omics and multi-omics approaches to study the biosynthesis of secondary metabolites in microorganisms. Curr Opin Microbiol 45:109–116
Panichikkal J, Edayileveetil Krishnankutty R (2020) Rhizobacterial biofilm and plant growth promoting trait enhancement by organic acids and sugars. Biofouling 36(8):990–999
Priya P, Aneesh B, Harikrishnan K (2021) Genomics as a potential tool to unravel the rhizosphere microbiome interactions on plant health. J Microbiol Methods 185:106215
Quick J, Quinlan AR, Loman NJ (2014) A reference bacterial genome dataset generated on the MinION™ portable single-molecule nanopore sequencer. Gigascience 3(1):2047-2217X-2043-2022
Raaijmakers JM, Weller DM (2001) Exploiting genotypic diversity of 2,4-diacetylphlor oglucinol-producing *Pseudomonas* spp.: characterization of superior root-colonizing *P. fluorescens* strain Q8r1-96. Appl Environ Microbiol 67:2545–2554. https://doi.org/10.1128/AEM.67.6.2545-2554.2001
Rane NR, Tapase S, Kanojia A, Watharkar A, Salama ES, Jang M, Yadav KK, Amin MA, Cabral-Pinto MM, Jadhav JP (2022) Molecular insights into plant–microbe interactions for sustainable remediation of contaminated environment. Bioresour Technol 344:126246

Rathnasamy SA, Gothandapani S, Chellamuthu S, Chakraborty A, Gurusamy D, Roy A (2023) Omics technologies unravelling the plant–pathogen interaction and stress response. In: Mani A, Kushwaha S (eds) Genomics of plant–pathogen interaction and the stress response. CRC Press, Boca Raton, pp 74–110

Regalado J, Lundberg DS, Deusch O, Kersten S, Karasov T, Poersch K, Shirsekar G, Weigel D (2020) Combining whole-genome shotgun sequencing and rRNA gene amplicon analyses to improve detection of microbe–microbe interaction networks in plant leaves. ISME J 14(8):2116–2130

Rhoads A, Au KF (2015) PacBio sequencing and its applications. Genomics Proteomics Bioinformatics 13:278–289. https://doi.org/10.1016/j.gpb.2015.08.002

Richter M, Rosselló-Móra R (2009) Shifting the genomic gold standard for the prokaryotic species definition. Proc Natl Acad Sci 106(45):19126–19131

Samantara K, Shiv A, de Sousa LL, Sandhu KS, Priyadarshini P, Mohapatra SR (2021) A comprehensive review on epigenetic mechanisms and application of epigenetic modifications for crop improvement. Environ Exp Bot 188:104479

Sarim KM, Srivastava R, Ramteke PW (2020) Next-generation omics technologies for exploring complex metabolic regulation during plant-microbe interaction. In: Singh JS, Vimal SR (eds) Microbial services in restoration ecology. Elsevier, Amsterdam, pp 123–138

Sarkar S, Bhaduri D, Chakraborty K (2014) Plant adaptation mechanisms in phosphorus-deprived soil: mitigation of stress and way to balanced nutrition. Adv Plant Physiol 15:254–282

Shayanthan A, Ordoñez PAC, Oresnik IJ (2022) The role of synthetic microbial communities (SynCom) in sustainable agriculture. Front Agron 4:896307

Singer E, Bushnell B, Coleman-Derr D, Bowman B, Bowers RM, Levy A, Gies EA, Cheng J-F, Copeland A, Klenk H-P (2016) High-resolution phylogenetic microbial community profiling. ISME J 10(8):2020–2032

Singla P, Garg N (2017) Plant flavonoids: key players in signaling, establishment, and regulation of rhizobial and mycorrhizal endosymbioses. In: Varma A, Prasad R, Tuteja N (eds) Mycorrhiza – function, diversity, state of the art, 4th edn. Springer, Cham, pp 133–176

Świątczak J, Kalwasińska A, Brzezinska MS (2024) Plant growth–promoting rhizobacteria: *Peribacillus frigoritolerans* 2RO30 and *Pseudomonas Sivasensis* 2RO45 for their effect on canola growth under controlled as well as natural conditions. Front Plant Sci 14:1233237

Tan S, Yang C, Mei X, Shen S, Raza W, Shen Q, Xu Y (2013) The effect of organic acids from tomato root exudates on rhizosphere colonization of *Bacillus amyloliquefaciens* T-5. Appl Soil Ecol 64:15–22. https://doi.org/10.1016/j.apsoil.2012.10.011

Tao Y, Chen Z, Xu Y, Wang F, Jiang Y, Fan F, Li W, Zhu J, Li X, Wang J (2025) CRISPR-mediated targeted mutagenesis for improving nitrogen use efficiency in japonica rice. Plant Commun 6:101189

Tedersoo L, Tooming-Klunderud A, Anslan S (2018) PacBio metabarcoding of fungi and other eukaryotes: errors, biases and perspectives. New Phytol 217(3):1370–1385

Tedersoo L, Drenkhan R, Anslan S, Morales-Rodriguez C, Cleary M (2019) High-throughput identification and diagnostics of pathogens and pests: overview and practical recommendations. Mol Ecol Resour 19:47–76

Teng Y, Jiang T, Yan Y (2024) The expanded CRISPR toolbox for constructing microbial cell factories. Trends Biotechnol 42(1):104–118

Tian T, Reverdy A, She Q, Sun B, Chai Y (2020) The role of rhizodeposits in shaping rhizomicrobiome. Environ Microbiol Rep 12:160–172

Upadhyay SK, Srivastava AK, Rajput VD, Chauhan PK, Bhojiya AA, Jain D, Chaubey G, Dwivedi P, Sharma B, Minkina T (2022) Root exudates: mechanistic insight of plant growth promoting rhizobacteria for sustainable crop production. Front Microbiol 13:916488

Usman NI, Wali MM (2024) Nitrogen fixation by rhizobacterial nif mechanism: an advanced genetic perspective. In: Gull M (ed) Updates on rhizobacteria. IntechOpen, London. https://doi.org/10.5772/intechopen.1004087

Velmourougane K, Blaise D (2021) Rhizoengineering: a strategy to enhance soil and crop productivity. In: Pudake RN, Sahu BB, Kumari M, Sharma AK (eds) Omics science for rhizosphere biology. Springer, Singapore, pp 235–262. https://doi.org/10.1007/978-981-16-0889-6_13

Volden R, Palmer T, Byrne A, Cole C, Schmitz RJ, Green RE, Vollmers C (2018) Improving nanopore read accuracy with the R2C2 method enables the sequencing of highly multiplexed full-length single-cell cDNA. Proc Natl Acad Sci USA 115:9726–9731. https://doi.org/10.1073/pnas.1806447115

Walder F, Schlaeppi K, Wittwer R, Held AY, Vogelgsang S, van der Heijden MG (2017) Community profiling of *Fusarium* in combination with other plant-associated fungi in different crop species using SMRT sequencing. Front Plant Sci 8:2019

Wang C, Li Y, Li M, Zhang K, Ma W, Zheng L, Xu H, Cui B, Liu R, Yang Y (2021) Functional assembly of root-associated microbial consortia improves nutrient efficiency and yield in soybean. J Integr Plant Biol 63(6):1021–1035

Wang Y, Wang X, Sun S, Jin C, Su J, Wei J, Luo X, Wen J, Wei T, Sahu SK (2022) GWAS, MWAS and mGWAS provide insights into precision agriculture based on genotype-dependent microbial effects in foxtail millet. Nat Commun 13(1):5913

Wang L, Zhang T, Li C, Zhou C, Liu B, Wu Y, He F, Xu Y, Li F, Feng X (2024) Overexpression of wild soybean expansin gene GsEXLB14 enhanced the tolerance of transgenic soybean hairy roots to salt and drought stresses. Plants 13(12):1656

Weidemüller P, Kholmatov M, Petsalaki E, Zaugg JB (2021) Transcription factors: bridge between cell signaling and gene regulation. Proteomics 21(23–24):2000034

White CE, Winans SC (2007) Cell–cell communication in the plant pathogen *Agrobacterium tumefaciens*. Philos Trans R Soc Lond Ser B Biol Sci 362:1135–1148. https://doi.org/10.1098/rstb.2007.2040

Xu L, Pierroz G, Wipf HML, Gao C, Taylor JW, Lemaux PG, Coleman-Derr D (2021) Holo-omics for deciphering plant-microbiome interactions. Microbiome 9:1–11. https://doi.org/10.1186/s40168-021-01114-4

Yi K, Menand B, Bell E, Dolan L (2010) A basic helix-loop-helix transcription factor controls cell growth and size in root hairs. Nat Genet 42(3):264–267

Yu P, He X, Baer M, Beirinckx S, Tian T, Moya YA, Zhang X, Deichmann M, Frey FP, Bresgen V (2021) Plant flavones enrich rhizosphere Oxalobacteraceae to improve maize performance under nitrogen deprivation. Nat Plants 7(4):481–499

Zhang N, Wang M, Fu J, Shen Y, Ding Y, Wu D, Shu X, Song W (2020) Identifying genes for resistant starch, slowly digestible starch, and rapidly digestible starch in rice using genome-wide association studies. Genes Genomics 42:1227–1238

Zhang L, Chen F, Zeng Z, Xu M, Sun F, Yang L, Bi X, Lin Y, Gao Y, Hao H (2021) Advances in metagenomics and its application in environmental microorganisms. Front Microbiol 12:766364

Zhang Y, Ku Y-S, Cheung T-Y, Cheng S-S, Xin D, Gombeau K, Cai Y, Lam H-M, Chan T-F (2024) Challenges to rhizobial adaptability in a changing climate: genetic engineering solutions for stress tolerance. Microbiol Res 288:127886

Zhou Z, Alikhan N-F, Mohamed K, Fan Y, Achtman M, Brown D, Chattaway M, Dallman T, Delahay R, Kornschober C (2020) The Entero Base user's guide, with case studies on *Salmonella transmissions*, *Yersinia pestis* phylogeny, and *Escherichia* core genomic diversity. Genome Res 30(1):138–152

Chapter 5
Evaluation of Traditional and Contemporary Rhizoengineering Approaches to Alleviate Abiotic Stresses

Ayesha Shehzadi and Saima Majeed

Abstract Growing environmental and food safety problems have given momentum to the finding and implementation of nature-based solutions (NbS) as modern agroecological tools or biological potent strategies that can better deal with these currently emerging agroeconomic challenges at a global scale, specifically caused by climatic shifts, and are constantly being highlighted by the United Nations Intergovernmental Panel on Climate Change (IPCC) and Sustainable Development Goals (SDGs) at the international level. The historical development of agriculture has provided us with the opportunity to evaluate major milestones in the evolution of worldwide agriculture from obsolete and traditional to modern and contemporary rhizoengineering approaches to alleviate abiotic stresses with rising biotechnological inventions and advancements. Evaluations and recommendations via interdisciplinary crosstalk will assist in proposing future research directions by recognizing novel pathways for climatic mitigation and the restoration of agroecosystems vital for sustainable agronomy and biodiversity preservation. Rationalization of contemporary rhizoengineering strategies is significantly contributing to modern synthetic biotechnological tools and multiomics technologies, for instance, knowing how the functions of microbes and plant gene expression will work in concert to regulate plant stress resistance and adaptability with profitable yield.

Keywords Nature-based solutions (NbS) · Sustainable Development Goals (SDGs) · Abiotic stresses · Climate mitigation · Rhizoengineering · Multiomics

A. Shehzadi · S. Majeed (✉)
Institute of Biochemistry, Biotechnology, Bioinformatics (IBBB), The Islamia University of Bahawalpur, Bahawalpur, Pakistan
e-mail: saima.majeed@iub.edu.pk

S. Mehnaz, I. Shahid (eds.), *Frontiers in Rhizoengineering*, Microorganisms for Sustainability 65, https://doi.org/10.1007/978-981-92-3665-7_5

5.1 Background

Traditional rhizoengineering approaches, including past strategies during the Green Revolution era, have boosted food production (Kenneth et al. 2019). However, the excessive use of agrochemicals in the form of chemical fertilizers and pesticides has caused serious harm to humans, animals, aquatic life, land, and the environment (Kumar et al. 2022; Shahwar et al. 2023). Similarly, unsustainable agricultural practices, grazing, deforestation, industrialization, mining, and anthropogenic activities have led to several problems, such as environmental degradation, health issues, and deteriorated soil physicochemical properties, over the last few decades (Xun et al. 2021). Industrial activities and mining have led to the accumulation of heavy metals in agricultural soils, reducing soil microbial diversity and organic vitality and disrupting nutrient absorption, metabolism, respiration, and photosynthesis in plants (Akbar et al. 2024; Okereafor et al. 2020). These difficulties, coupled with increased pesticide resistance, reduced crop yields, and the growing global population from 7.6 billion in 2017 to an estimated 11.2 billion by 2100, pose a significant challenge to food security (UN DESA 2017).

For the rapidly growing human population, an almost 50% increase in agricultural yield is considered necessary by 2029 (Nawaz et al. 2024b), and according to a report by the Food and Agricultural Organization (FAO), it must be increased by 60% by 2050 to meet global food demands (FAO Report 2016). Approaches to expanding farmland to obtain larger areas with more growth through deforestation have also proven insufficient to fulfill needs on a large scale (Dubey 2024). All of these earlier traditional approaches, strategies, techniques, and practices of the past Green Revolution era have no longer been found effective against twenty-first-century climate challenges, which are increasing the area of land unbefitting for cultivation (Li et al. 2025) and have become a pressing global issue. Climatic changes have shown a significant increase in global warming since the 1970s, coupled with more frequent extreme weather events and changes in precipitation patterns (Capua and Rahmstorf 2023). The accumulation of greenhouse gases due to industrial and anthropogenic activities has led to global warming (Fahad et al. 2017).

The Intergovernmental Panel on Climate Change (IPCC), established in 1988, has reported a 1 °C increase in global surface temperature over the past century, with projections estimating an additional rise of 1.4–5.8 °C by the end of this century (Alotaibi 2023; Giordano et al. 2021; IPCC Report 2001). Even minor shifts in temperature have been recorded to be associated with an amplified frequency of extreme weather happenings all over the world, such as droughts, desertification, salinization, heatwaves, habitat loss, and floods, which not only generate abiotic stresses on crops increasingly but also could create biotic stresses by widely expanding the range of plant disease-causing pathogens and pests (Alotaibi 2023; DeWolf and Isard 2007; Garrett et al. 2021). Overall, these facts greatly impact global socioeconomics, disturb agricultural ecosystems, reduce productivity, and threaten biodiversity worldwide. These environmental changes also negatively affect other ecosystem services vital for sectors such as forestry, health, and tourism (Bellard et al. 2012; Mace et al. 2012). Therefore, the conservation of biodiversity and the restoration of agroecosystems are essential for achieving long-term sustainability and resilience. For this purpose, not only from the last decade of the twentieth century but also from the very beginning of the twenty-first century, the focus has been shifted from traditional approaches to advance sustainable strategies to get climate-resilient and stress-tolerant crops such as by applying modern cultural practices (e.g., water-conserving methods of irrigation to combat drought stress), traditional breeding of stress-tolerant crops, and genetic engineering approaches (Fahad et al. 2017; Henry 2020; Varshney et al. 2021). However, these techniques have been found to be labor-intensive, costly, poorly manageable, and slow to implement (Aalipour et al. 2020). Therefore, there is an urgent need for more efficient, natural, affordable, environment-friendly, and sustainable methods that leverage scientific tools and modern technologies to enhance crop productivity and fight against these dangers.

During the mid-second decade of the current century, global frameworks have been designed to meet all of these key international development objectives, including the United Nations 2030 Agenda, which has been developed as Sustainable Development Goals (SDGs), Aichi Biodiversity Targets, and the Paris Agreement on Climate Change (Intergovernmental Science-Policy Platform on Biodiversity and Ecosystem Services IPBES Report), which collectively aim to address these interconnected challenges of agricultural productivity, land degradation, and emerging climatic conditions (UN SDGs 2015; Pandit et al. 2018). Nature-based solutions (NbS) are endorsed by the IUCN and UN for the preservation of diverse bioecosystems and limiting global warming to 1.5 °C (IPCC Report 2022; IUCN Report 2023). Most recently, the 16th Conference of the Parties (COP16) of the United Nations Convention to Combat Desertification (UNCCD 2024) reported that increasing global aridity projections highlight an estimated fact that over 45% of the Earth's land surface could be classified as arid or semi-arid by 2050 and is expected to pose significant threats to food safety, crop cultivation, and biodiversity. The IUCN 2024 report indicates that 38% of the world's tree species are at risk of extinction (IUCN Report 2024). Recent data have shown that crop yield is highly vulnerable to abiotic stresses compared to biotic stresses owing to the negative impacts of the current climate-shifting conditions at the global level (Bowerman et al. 2023). In response to all the reports of these mutually connected problems, biodiversity-conserving and agroecology-protecting natural bioelements are being recommended for sustainability.

Diverse rhizoengineering approaches or strategies have been implemented globally that share common goals or objectives to counter the abundantly occurring abiotic threats. Researchers have developed a wide range of techniques and management plans, spanning traditional breeding methods to modern biotechnological principles, including advanced rhizoengineering and agronomic innovations with insights into microbiology, which have collectively assisted in organizing rhizosphere or soil components in recent years. Plant rhizosphere microbiology was established in the late nineteenth century by the emergence of the field of plant-soil-microbe interactions, which was motivated by Koch's postulates (Xu et al. 2025) under the single-strain paradigm. The updating of microbiology at the beginning of the twenty-first century by the progression in agroecology and biotechnology has highlighted the essential role of rhizomicrobiomes, such as bacteria and fungi, as the most widely studied life forms promoting crop yield. However, other groups, such as archaea, viruses, nematodes, and protists, also contribute notably to plant benefits and microbiome assembly (Trivedi et al. 2020). Harnessing phyllosphere, rhizospheric, rhizoplane, and root endospheric microbiota, particularly plant growth-promoting rhizobacteria (PGPR), is being recognized as the most promising rhizoengineering approach for steadily enhancing the positive effects on plant growth and tolerance under stress. Many experiments have been conducted that support these types of beneficial microorganisms as Natural Bioelements mediating stress tolerance more effectively in an eco-friendly and cost-effective way (Ojuederie et al. 2019). Interactions between plants and microbes under abiotic stress are intricate and constantly evolving. To effectively utilize the potential of plant-associated microbiota in agriculture, it is crucial to understand how abiotic stressors affect not only plants and microorganisms individually but also their mutual interactions (Fadiji et al. 2023). The plant-associated microbiome forms a symbiotic unit known as the "holobiont," whose composition and function can be shaped by environmental conditions (Trivedi et al. 2020).

Holobionts are complex, interconnected communities of organisms that coexist in close association across diverse ecosystems (Matthews 2024). Microbial members of this system play a vital role in enhancing plant performance (Muhammad et al. 2024), but plant adaptations are often insufficient under severe stress conditions (Rahim et al. 2025). Recent rapid theoretical and technological advances have modified the global focus not only on plant growth-promoting bacteria (PGPB) but also on their bioactive compounds or secondary metabolites as modern agroecological tools that are being proposed for agrosustainability (Bhardwaj et al. 2014). The rhizosphere is the central hub that provides an environment where most beneficial soil microbes interact with plant roots, and these rhizobacteria act as biofertilizers by enabling the integration of macronutrients into the plant nutrient cycle (Thepbandit and Athinuwat 2024; Trivedi et al. 2020). PGPB, particularly rhizobacteria, are under high investigation and are being explored in the latest volume of research studies as eco-friendly alternatives to chemical fertilizers and pesticides due to their dual role as biofertilizers and biocontrol agents that can not only mitigate both biotic (pathogens) and abiotic (drought, salinity, temperature, etc.) stresses (Hassan et al. 2019) but also help in ecosystem recovery. Researchers and scientists have predominantly focused on identifying effective PGPRs and their benefits, necessitating deeper investigations into their underlying mechanisms. Climatic factors and global warming significantly alter soil parameters, which in turn influence the root exudate profiles of stressed plants and affect rhizomicrobial recruitment (Naylor et al. 2017; Schmidt et al. 2019; Simmons et al. 2020). These root exudates, comprising specific sugars and organic acids, attract specific microbial communities, which enhance plant resilience under environmental stress through metabolite production (Aslam et al. 2022; Guo et al. 2018). Emerging research underscores that these types of microbial responses to stress are not inherent but induced because stressed plants shape these rhizobacterial community compositions and endophytes, which are actually steered by root exudation and environmental stress to help in modulating defensive responses to stress and are important for plant adaptation (Dubey et al. 2021; Jiao et al. 2021; Lopes et al. 2021; Toju et al. 2018). For instance, extended drought stress periods have been shown to enhance the abundance of microbial groups such as Actinobacteria and Firmicutes (Vescio et al. 2021). Similar to the previously mentioned example, a comparative increase in Acidobacteria and Cyanobacteria was observed in the peanut rhizosphere under salt stress (Xu et al. 2020). In a recent study, Dollete et al. (2024) reported the effects of drought stress on forage legumes in the form of a deficiency in symbiotic nitrogen fixation, which impacts plant health, growth, and development. Additionally, the phosphate solubilization potential of the biocontrol fungus *Trichoderma* sp. was observed to be altered under abiotic stresses, such as extreme pH, temperature shifts, and the inclusion of heavy metals (Muhammad et al. 2024; Rawat and Tewari 2011). This stress-induced shaping of microbial community compositions and their specific mechanisms of protective responses are opening pathways for climatic resilience and restoration of agroecosystems vital for sustainable agronomy plus biodiversity preservation (Chen et al. 2022; Gaete et al. 2021; Gouda et al. 2018). In particular, the rhizomicrobiome aids in stress mitigation by regulating different plant mechanisms, such as physiological, morphological, biochemical, and molecular modifications (Haichar et al. 2021; Jabborova et al. 2021). Their symbiotic relationships with plants enhance soil fertility and organic vitality through the activation of defensive pathways against stress, modulation of phytohormones, and facilitation of nutrient acquisition (Abdelaal et al. 2021; Li et al. 2020). These microbes possess traits such as mineral transport, exopolysaccharide (EPS) production, and secretion of beneficial compounds or metabolites, enabling their application in alleviating the effects of different stressors (Barquero et al. 2022; Shaffique et al. 2022). Under stressed conditions, PGPBs enhance antioxidant defense systems, reducing oxidative stress and modulating gene expression related to cellular homeostasis, transcription factors, aquaporins, and stress signaling (Chakraborty et al. 2013; Manjunatha et al. 2022).

Overall, rhizoengineering by plant growth-promoting rhizomicrobiomes (PGPR) strongly offers a nature-based solution (NbS) to counter climatic fluctuations with better crop resilience, sustained yield, improved stress resistance, and reversed stunted growth (Glick 2014; Inbaraj 2021; Kumari et al. 2016). Genetic modification can sometimes succeed where some methods of genetic engineering fail. For example, the *HKT1;1* gene in *Arabidopsis thaliana* did not improve salt tolerance, whereas a *strain of Bacillus subtilis* was able to do so by *differentially regulating AtHKT1, downregulating it in roots and upregulating it in shoots.* This *microbe-mediated, tissue-specific regulation* enhanced *Na^+ homeostasis* under salt stress where genetic recombination failed, illustrating the powerful potential of *plant-microbe interactions to combat stresses that are increasingly disturbing crops due to climatic variability* (Maser et al. 2002; Rus et al. 2004; Schmitz et al. 2022; Zhang et al. 2008).

5.2 Abiotic Stressors

Plants in their natural habitat are continuously subjected to an ever-changing external climate and connect in vibrant or dynamic crosstalk with various environmental stimuli (both biotic and abiotic) via a complicated web of communicatory networks (Muhammad et al. 2024; Nawaz et al. 2024a). Some regions around the globe are already experiencing severely fluctuating climatic states, which are highly producing soil-based abiotic stresses for plants, such as heavy metal toxicity, salinity, water logging, nutrient imbalances, temperature, and drought, which are all nonliving environmental factors that negatively influence plant physiology, health, development, crop production, and the stability of ecological units (Muhammad et al. 2024). An analysis revealed that it has evolved as a significant global concern because a momentous rise in research on soil abiotic stresses can be seen since the 1980s, especially from 2011 onward. Such stressors are becoming more prevalent due to rapid industrial revolution, urbanization, mining, and, particularly, in the face of intensifying climate shifts threatening global food protection and environmental health (Lal 2012; Mustafa et al. 2022; Rahim et al. 2025). Each type of abiotic stress affects crops differently. Salinity and water shortages also cause osmotic stress, oxidative damage, nutrient deficiencies, and reduced germination, ultimately leading to substantial yield loss. Poorly managed irrigation systems result in salinization, which causes toxic ion accumulation and can lead to productivity declines ranging from 5% to 50% (Oyebamiji et al. 2024). High temperatures affect pollen viability and seed set. Cold stress in legumes, such as chickpeas and mungbeans, has caused yield losses of up to 70%. Almost 10–40% of crop yield is lost due to heat and cold stress (Ahmad et al. 2023; Oyebamiji et al. 2024). These abiotic pressures have significant ecological consequences, such as the degradation of biodiversity and agricultural losses, with future projections showing exacerbation due to global warming (Padhiary and Kumar 2024; Rahim et al. 2025). More than 50% of crop losses are attributable to abiotic stresses, which affect over 91% of global croplands. The frequency of abiotic stresses is projected to increase in the coming years, driven by the ongoing progressive escalation of climate change phenomena (Oyebamiji et al. 2024; Younis et al. 2020). Therefore, it is crucial to develop more productive strategies to mitigate the adverse effects of these challenges on the agricultural industry and improve crop resilience and yield.

A comprehensive compilation of data on traditional, modern, and current trends in rhizoengineering, along with innovative strategies, should be intensified to fully elucidate the mechanisms involved in the action of emerging technologies. Consequently, interdisciplinary crosstalk through collaborations, conferences, evaluations, recommendations, journal publications, and information sharing at the international level is required to help researchers incorporate knowledge and skills from many disciplines and obtain an improved and systematic understanding of the problems, resulting in comprehensive solutions. This will assist in proposing future research directions by highlighting pathways for climatic mitigation and restoration of agroecosystems vital for sustainable agronomy and biodiversity preservation via contemporary and multiomics technologies, for instance, knowing how the functions of microbes and plant gene expression will work in concert to regulate plant stress resistance and adaptability (Li et al. 2025).

5.3 Diverse Approaches to Improve Plant Health

5.3.1 Breeding Approaches

5.3.1.1 Conventional Breeding

It is a conventional approach that utilizes crop genetic variation at different levels, such as intraspecific, interspecific, and intergeneric (Table 5.1), to generate tolerant lines and cultivars by traditional selection, specifically introducing desirable root traits. It involves methods such as hybridization, backcrossing, mutation, and recurrent selection; for example, eight lines of soybean showed waterlogging tolerance and *KRL1-4* as a salt-tolerant wheat genotype (Guan et al. 2022; Seleiman et al. 2022). The limitations of this method are its time consumption and narrow genetic pools (Oyebamiji et al. 2024).

5.3.1.2 Grafting

It is a very old technique to replace a susceptible root system (scion) with a vigorous one as a stress-tolerant rootstock, which can be considered as a part of the practical application of rhizoengineering principles. For example, especially in vegetables like tomatoes and cucumbers, grafting involves nutrient uptake, biomass accumulation, hormone regulation, antioxidant activity, photosynthesis, root development, and resistance to stresses like waterlogging (Singh et al. 2020), heat (Ali et al. 2020; Muneer et al. 2016), and drought (Coskun 2023; Oyebamiji et al. 2024). Due to current advanced approaches, it is possible to monitor graft union via omics approaches that show altered physiological responses and the maintenance of several proteins for defense (Oyebamiji et al. 2024).

5.3.1.3 QTL Mapping and Marker-Assisted Selection (MAS)

Quantitative trait loci (QTL) mapping is a quick approach, and MAS is a DNA-based marker that helps recognize and incorporate alleles of desirable traits, specifically root characteristics associated with stress tolerance, into elite cultivars (Table 5.1). For example, QTLs linked to drought tolerance in barley (Ajayi et al. 2023) and chilling tolerance in sorghum have been successfully integrated into breeding programs (LaBorde et al. 2023; Oyebamiji et al. 2024). This enables the possibility of more precise breeding programs (Oyebamiji et al. 2024).

Table 5.1 Comprehensive table of modern and traditional strategies for mitigating abiotic stress in crops

Stress	Plant	Strategies	Technical approach	Characteristics improved	References
Salinity	Tomato	Edkway	Natural selection	Salt tolerance	Fageria et al. (2012)
Salinity	Rice	CSR1, Dasal, Getu	Pure line selection	Salt tolerance	Meena et al. (2016)
Salinity	Chickpea	qPHC5.2, qYPPS6.1	QTL mapping	Productivity traits	Soren et al. (2020)
Salinity	Tomato	Bet A (*E. coli*)	Genetic engineering	Biomass improves	Ashraf et al. (2008)
Salinity	Maize, wheat	Biopriming	Seed priming	Physiological/morphological characteristics	Arun et al. (2022)
Salinity	Cotton	Proline	PGR	Hormone accumulation, decreased MDA	Seleiman et al. (2022)
Salinity	Beans	10–20% biochar	Biochar application	Root/shoot growth	Wani et al. (2022)
Drought	Rice	*Tarom Mahalli*	Induced mutation	Drought tolerance	Oladosu et al. (2019)
Drought	Maize	qCH1–1, qCW2–1	QTL mapping	Cob characteristics	Sarkar et al. (2023)
Drought	Potato	BADH (spinach)	Genetic engineering	Fresh weight	Kikuchi et al. (2015)
Drought	Soybean, rye	Hydropriming	Seed priming	Germination, antioxidant activity	Marthadan et al. (2020), Langeroodi and Noora (2017)
Drought	Wheat	Melatonin	PGR	Seedling vigor	Kumar et al. (2021a, b)
Drought	Tomato	5% biochar	Biochar application	Seedling growth	Zhang et al. (2023)
Waterlogging	Soybean	NN1138-2, M8206	Selection	Tolerance	Ali et al. (2018)
Waterlogging	Rice	qAG3.1, qSF10.1	QTL mapping	Anaerobic germination	Rani et al. (2022)
Waterlogging	Rice	OsARD1	Genetic engineering	Multistress tolerance	Liang et al. (2019)
Waterlogging	Rice, okra	Phytohormone/chemical priming	Seed priming	Enzyme induction, tolerance	Rhaman et al. (2020), Kumar et al. (2021a, b)
Waterlogging	Soybean	GA3	PGR	Flooding tolerance	Guan et al. (2022)
Waterlogging	Soybean	0.1% biochar	Biochar application	Growth characteristics	Kang et al. (2021)

Temperature	Tomato	L72, Saldate	Hybridization/ selection	Heat tolerance	Hazra et al. (2007), Hall (1992)
Temperature	Tomato	HSP21	Genetic engineering	Photosystem safety	Krishna et al. (2019)
Temperature	Maize	Phytohormone priming	Seed priming	Growth, pigment, yield	Iqbal et al. (2020)
Temperature	Mustard	24-Epibrassinolide	PGR	Induced heat tolerance	Hassan et al. (2021)
Temperature	Rice, thymus	Biochar (40 g–5%)	Biochar application	Fertility, photosynthesis	Huang et al. (2021), Alharbi et al. (2023)
Cold	*Alfalfa*	dbr1, dft1	QTL mapping	Cold tolerance	Adhikari et al. (2021)
Cold	*Arabidopsis*	LcFIN1, TaMYC2	Genetic engineering	Cold tolerance, antioxidants	Adhikari et al. (2022)
Cold	Canola, rapeseed	Thermo/hydropriming + phytohormone	Seed priming	Germination, enzyme activity	Ashraf et al. (2008), Zhu et al. (2021)
Cold	Tomato, pea	Brassinosteroids, auxin	PGR	Decline in oxidative stress	Raza et al. (2023), Jankovska et al. (2023)
Cold	Cucumber	5-Methoxytrypt-amine	Genetic engineering	Cold tolerance	Mu et al. (2021)
Heavy metal	Wheat, sunflower	Biochar (5–50%)	Biochar application	Growth enhanced, heavy metal reduction	Tang et al. (2022)
Heavy metal	Rice, soybean	MTP1, Cda1	Genetic engineering	Tolerance to Zn/Cd	Ding et al. (2018), Jegadeesan et al. (2010)
Heavy metal	Pigeon pea, mustard	Phytohormone priming	Seed priming	Germination, metal accumulation	Rhaman et al. (2020), Sytar et al. (2019)
Heavy metal	Lupin, mint	GA3, SA, triacontanol	PGR	Increased enzyme activities, Cd reduction	Saini et al. (2021), Zaid et al. (2020)

5.3.1.4 Genetic Engineering

It is a modern technique that enables the direct insertion of stress-tolerant genes of desired traits into the cell genome via agrobacterium-mediated and particle bombardment techniques (Hossain et al. 2021), such as *HSP21* for heat tolerance (Krishna et al. 2019) or *MTP1* for heavy metal resistance, where genetic engineering has led to the modification of gene expression (Table 5.1) by employing microRNA (Ding et al. 2018). Despite the advantages and acceptability of genetically modified organisms (GMOs) among the public, their utilization remains limited because of potential health concerns. Therefore, transgenic plants must be tested in the field, as well as on humans and animals, to evaluate their effects (Oyebamiji et al. 2024).

5.3.1.5 Genome Editing

It is an emerging approach, particularly a multiplex genome editing strategy using CRISPR-associated Cas protein systems CRISPR-Cas9, which has shown promising results. For instance, a salt- and drought-tolerant gene in the indica mega rice genotype *MTU1010* was edited, and the resultant plant showed a decline in stomatal activity with expanded leaves and improved leaf water conservation (Kushwaha et al. 2023). This tool can be used to generate new germplasms for breeding (Oyebamiji et al. 2024).

5.3.1.6 Speed Breeding

It is a new technique to accelerate the breeding process in which we can increase the number of crop cycles using optimum temperature, prolonged light, and controlled environments by using artificial LED lights with 400–700 nm wavelength and a photoperiod of 22 h and 2 h of darkness (Kumari and Singh 2022). To reduce the time before seed maturity, an embryo rescue technique was applied to obtain more crop cycles. Crops such as wheat and chickpeas complete six growing cycles, and groundnuts are grown for up to four generations annually (Samantara et al. 2022). When combined with MAS and genome editing, this method rapidly produces high-yield, abiotic stress-resilient cultivars (Pandey et al. 2022). Drawbacks include a shortage of skilled personnel, equipment, and funding to set up a standard controlled environment for such types of culture rooms (Oyebamiji et al. 2024).

5.3.2 *Agronomic and Physiological Approaches*

5.3.2.1 Seed Priming

During the seedling phase, immersion in water, chemicals, or biological agents prepares seeds to germinate more rapidly and resist abiotic stresses (Table 5.1). Techniques such as hydropriming, osmopriming, and biopriming have been found to be effective across various crops (Arun et al. 2022; Rhaman et al. 2020).

5.3.2.2 Melatonin and Nanoparticles

New emerging agents for seed priming, such as melatonin (Rajora et al. 2022) and nanoparticles, boost antioxidant defenses and promote seedling vigor under abiotic stresses. Optimized usage of nanoparticles, such as Si-NPs, ZnO, TiO_2, and FeO, is a promising tool that can penetrate plant cells to improve plant health and safety from ecosystem damage. They are utilized in diverse forms, such as nanosensors, nanofertilizers, nanoherbicides/nanopesticides, and nanoremediators. Further research is required to avoid the negative effects of these additives on consumer health (El-Saadony et al. 2022; Oyebamiji et al. 2024). Melatonin is a tryptophan derivative that enhances tolerance to drought, salinity, heavy metals, and temperature extremes (Table 5.1). It does this by modulating gene expression, regulating hormonal crosstalk, and reducing oxidative stress (Godoy et al. 2021; Hoque et al. 2021). For example, melatonin-treated maize and cucumber showed improved growth under heat and cold conditions, respectively (Pardo-Hernández et al. 2020). Numerous field trials should be conducted in arid regions to examine its potential (Oyebamiji et al. 2024).

5.3.2.3 Microbial Seed Treatments

Biological agents, such as fungi and bacteria, especially species such as *Trichoderma*, *Bacillus*, and *Azospirillum*, improve germination, reduce reactive oxygen species (ROS) accumulation, and promote growth under drought, salinity, and cold stress (Cardarelli et al. 2022; Hassan et al. 2021; Paravar et al. 2023). These microbes have been observed as the most promising bioelements presenting nature-based solutions (NbS) to tackle emerging problems in agroecology. Extensive research is still needed to fully understand the role and fundamental mechanisms involved (Oyebamiji et al. 2024).

5.3.2.4 Plant Growth Regulators (PGRs) and Osmoprotectants

The utilization of polyols, glycine betaine (GB), trehalose, salicylic acid, jasmonic acid, gibberellic acid, abscisic acid, uniconazole, brassinolide, zeatin, ethephon, thiourea, and proline as plant growth regulators (PGRs) and osmoprotectants (Table 5.1) improves abiotic stress responses by modulating antioxidant systems and hormonal balances (Alotaibi 2023; Oyebamiji et al. 2023). These compounds enhance water-use efficiency, decrease lipid peroxidation, and increase the photosynthetic capacity of crops under stress (Oyebamiji et al. 2024). Melatonin and strigolactones are relatively new plant growth regulators. Improved tolerance has been achieved due to the positive crosstalk between melatonin and growth phytohormones (Oyebamiji et al. 2024).

5.3.2.5 Arbuscular Mycorrhizal Fungi (AMF)

To alleviate abiotic stress, arbuscular mycorrhizal fungi work as microbial plant biostimulants (MPBs), similar to plant growth-promoting rhizobacteria (PGPR), and establish mutualistic relationships with roots, improving water uptake, nutrient absorption, and heavy metal tolerance (Hassan et al. 2017). They also modulate hormone levels (e.g., ABA and jasmonic acid) and antioxidant enzyme activities, enhancing plant resilience under multiple stresses (Behrooz et al. 2019; Oyebamiji et al. 2024).

5.3.3 Cultural and Nutrient Management

5.3.3.1 Cultural Practices

The implementation of cultural farm practices, such as salinity tolerance strategies, includes leaching, excess sodium, mixing of calcareous subsoil into topsoil by deep plowing, and adding acidifying minerals such as gypsum and pyrite (Kumar et al. 2021a). Similarly, proper spacing (Kumar et al. 2021b), appropriate shading (Formisano et al. 2021), timely planting (Hassan et al. 2021), proper cutting management (Adhikari et al. 2022), irrigation management, and plastic film mulching (Hossain et al. 2021) are used to reduce abiotic stresses (Oyebamiji et al. 2024).

5.3.3.2 Nutrient Management

This is critical because almost 60% of soil has a deficiency of sufficient nutrients at the global level, which affects plant output (Hassan et al. 2021). Nutrient deficiencies exacerbate the negative effects of abiotic stress. The application of macronutrients (such as N, P, K, and Ca) and micronutrients (such as B, Zn, Fe, Cu, and Si) supports essential physiological functions, enhances antioxidant responses, and maintains the cellular integrity of the plant to withstand stresses. For example, silicon is critical for improving tolerance to drought, salinity, cold, waterlogging, and heavy metal stresses by stabilizing cell walls and increasing water retention (Kumar et al. 2021a, b). The implementation of two or more nutrients improves plants owing to crosstalk between them at the molecular and physiological levels (Oyebamiji et al. 2024).

5.3.4 Organic Amendments as Emerging Sustainable Strategies

5.3.4.1 Biochar

It is a carbon-rich charcoal that is a by-product of biomass thermal degradation (pyrolysis). It recovers soil structure, improves water or nutrient retention, and adsorbs heavy metals (Table 5.1). Its porous nature enhances microbial activity and nutrient availability, thereby increasing plant growth under drought, salinity (Wu et al. 2023), and toxic scenarios (Ali et al. 2020; Oyebamiji et al. 2024).

5.3.4.2 Nonmicrobial Biostimulants

These have shown promising effects on plant growth. For instance, kaolin is a mineral mud that dissolves quickly in water and is usually sprayed on plants to form a safe particle film to protect plants from the detrimental effects of abiotic stresses, such as heat and drought, under greenhouse conditions (Terán et al. 2023). Similarly, chitosan, a linear polysaccharide, is biodegradable. Soil amendment with chitosan reduces abiotic stress, excites antioxidant responses, improves water retention, and boosts osmotic adjustment (Aazami et al. 2023; Tan et al. 2023). However, the precise mechanisms of their action remain unexplored (Oyebamiji et al. 2024). Yeast contains essential elements such as vitamins (B1, B2, and B12), pyridoxine, thiamine, riboflavin, cytokines, minerals, and essential hormones that improve the physical and chemical properties of soil and tolerance to a wide range of abiotic stresses (Sadak and Dawood 2023). Similarly, higher plant extracts can be used as beneficial chemical fertilizers (Ahmad et al. 2022a, b). Similar to yeast and higher plant extracts, seaweeds (also known as macroalgae), which are multicellular marine algae, contain diverse beneficial compounds for botanical applications, such as phytohormones, macro- and micronutrients, proteins and peptides, carbohydrates, pigments, and complex phenolic compounds (Deolu-Ajayi et al. 2022). Superabsorbents are effective adsorbents in the form of cross-linked hydrophilic water-soluble polymers, such as hydrogels, which act as buffers to neutralize the influence of abiotic stresses (Du et al. 2023). Comprehensive information should be investigated regarding the use of biostimulants at optimum rates to achieve maximum growth (Oyebamiji et al. 2024).

5.4 Rationalization of Contemporary Rhizoengineering Strategies

Evaluations of the above-mentioned fields of rhizoengineering are still needed to confirm their stability and performance in resource-limited regions and to better comprehend the mechanisms governing nutrient uptake by plants under a wide variety of abiotic stresses. The application of nanomaterials, biostimulants, osmoprotectants, phytohormones, grafting techniques, and nutrient administration has shown promise. Recent advancements have deepened our knowledge of these mechanisms; however, further clarity is required. In addition to the process of conventional breeding, the involvement of QTL mapping, marker-assisted modern breeding, genetic recombination, and speed-breeding techniques strongly supports the improvement of the plant's ability to cope with abiotic stress problems. These programs can be employed to breed new germplasm or to develop new cultivars with better traits, such as the release of more high-yielding and tolerant genotypes. Stringent screening of more tolerant germplasm is required using new technologies under different stress conditions. Comprehensive investigations are required to test the adaptability of these new genotypes in several environments to determine not only the genotype-specific environments but also the efficient growth of these genetically modified plants in the natural environment. Natural troubles, climatic variations, and genetic erosion have led to the loss of genetic resources. Conservation methods should be applied for the long-term preservation of diverse genetic materials using both in situ (e.g., storage of seeds, gene banks, pollen, and DNA, as well as botanical parks) and ex situ (e.g., on-farm, field farms, natural gardens, gene sanctuaries, and hotspots of diverse life forms) methods.

Soil microorganisms are the chief source of nutrients and energy for plants and also affect soil fertility in extreme environments. The flora (herbs/shrubs/grasses) and their rhizomicrobiomes in extreme regions have been found to develop different ecological, biochemical, physiological, and molecular adaptations in plants to survive under stressful conditions. Research studies recommend that after determining the category and functioning capacity of beneficial rhizomicrobiomes, they can be employed as biofertilizers/biological control agents for the rhizoengineering of agricultural crops to counteract the effects of abiotic stresses. Studies have revealed that only flora and their associated bacterial communities are natural prolific constituents of the ecosystem that can prevent soil deterioration due to diverse abiotic stresses and climatic instabilities by augmenting the functioning of the soil system, and in this way, such rhizomicrobial symbionts can improve yield (Abd El-Ghani et al. 2017). These plants secrete a variety of root exudates and induce variability in the structure and composition of microbial communities and their biological properties in the rhizosphere due to stress. Plants rely heavily on *rhizosphere microbes* to withstand abiotic stress. These microorganisms are often considered the second genome of plants, broadening their genetic and metabolic capabilities (Berendsen et al. 2012). Another important aspect is that most commercially important plants that belong to the same nature or family as the successfully surviving wild plants found in extreme environments can have almost similar root architecture/few related features, which may facilitate the colonization of similar beneficial bacterial species in the rhizosphere of the important crop plants of the same group as biofertilizers. The diversified PGP bacterial communities associated with stress-tolerant diverse plants growing in an extremely dry, harsh, water-logged, cold, or heated environment need to be explored for their plant growth-promoting traits or abiotic stress-resistant capabilities/antimicrobial/antifungal properties. Research has been limited to the reporting of plant growth-promoting traits or compounds of rhizobacteria/root endosphere bacteria and their synergistic effectiveness on plant systems/mechanisms for plant growth promotion and disease control (Hussain et al. 2022). In these studies, only inadequate data illustrate the maximum metabolic potential of bacterial species by revealing their biosynthetic activities (Saharan and Nehra 2011; Shahid et al. 2024). After characterization and identification, beneficial bacterial communities can be employed as biofertilizers/biological control agents for the bioinoculation of important crops, such as in water-starved environments/stresses due to climatic variations, and for the protection of vegetation to address soil deprivation/land degradation/desertification along with poor vegetation (Ahmad et al. 2022a, b). The biosynthetic pathways and putative genes involved in the production and regulation of plant growth-promoting compounds by effective bacterial communities are still new insights that are not completely known. Genome editing of beneficial microorganisms inhabiting the rhizosphere of crops could be explored, with the plan of improving their symbiotic connection with plants (Yi et al. 2018; Wang et al. 2021). Additionally, there are not many planned investigations in the previously published literature that compare the specificity of host plant species to bacterial communities to better understand microbiome responses to environmental changes and host (plant) adaptation. Only a handful of studies have focused on these aspects and their combined effects (Ahmad et al. 2022a, b). It has been observed that PGPB inoculation results in improved leaf chlorophyll and photosynthetic pigment contents, root/shoot length, leaf number, fresh/dry weights, membrane stability index (MSI), osmolyte production, relative water content, and proline content. Plant growth is also improved by rhizobacteria due to reduced electrolyte leakage (EL) and malondialdehyde (MDA). Plant-beneficial bacteria also upregulate the antioxidants in the host plant, such as peroxidase (POD), catalase (CAT), superoxide dismutase (SOD), polyphenol oxidase (PPO), and phenylalanine (PAL), to scavenge the manufacturing of reactive oxygen species (ROS) under abiotic stresses (Li et al. 2020; Inbaraj 2021; Si et al. 2021; Fadiji et al. 2022). Many studies have analyzed the effects of single microbial strains on plants under sterile conditions. However, this does not reflect the original agroenvironments, where plants coexist with compound indigenous microbial consortia. In field trials opposite to the sterile environments, it has been observed that competitive interactions with other indigenous microbes cause failure of individual microbial strains (Baez-Rogelio et al. 2017). Consequently, the traditional “one-microbe-at-a-time” strategy is insufficient. An alternative approach is the expansion of *synthetic microbial communities (SynComs),* which are derived from natural microbiomes and promote plant growth under nonsterile conditions, particularly when abiotic stresses are present. The record of published data on SynComs has evolved since 2014 (Xu et al. 2025). Standardized methodologies for constructing functional SynComs remain to be investigated fully (Schmitz et al. 2022). Recent advancements in computational biology, bioinformatics, and multiomics technologies have enabled the exploration of this field without the need for culturing microbes. Metagenomics can provide a comprehensive overview of microbial composition, abundance, functionality, and interactions with plant health under abiotic stress conditions. The sequence of instrumental approaches on bioformulation of SynComs that has been found in experimental methodologies of latest research publications includes bottom-up and top-down approaches (Xu et al. 2025), as shown below:

Bottom-up	Isolations	Growth assays	Bioassays	Gene annotation	Bioformulation
Top-down	Relative abundance	Genome sequencing	Selection	Pot experiments	Bioformulation

Future studies should utilize *multiomics* tools to reveal plant–microbiome functional networks by exploring *root exudate–microbial interaction mechanisms* and integrating *genetic engineering* and *microbiome-based breeding* to improve stress resilience. Understanding and manipulating rhizosphere interactions and rhizoengineering of the plant microbiome will be central to *next-generation sustainable agriculture* (Berendsen et al. 2018; Trivedi et al. 2020).

5.5 Beneficial Rhizomicrobiome Optimization

Four core strategies for *rhizosphere microbiome optimization using multiomics have emerged.*

5.5.1 Soil Management

Biochar develops *soil structure*, *rhizomicrobial diversity*, and *nutrient accumulation* (Yan et al. 2022).

5.5.2 Root Exudate Regulation

Genome editing (e.g., *CRISPR*) can be used to modify root exudate profiles to favor the recruitment of beneficial microbes (Bai et al. 2015; Yang et al. 2023).

5.5.3 *Probiotic Application*

The accumulation of artificial or synthetic microbial consortia as probiotics is a promising approach in rhizoengineering, *which* can stimulate plant growth and stress resistance but requires adaptation for constant *field performance by successful colonization* (Xu et al. 2018; Sánchez-Gil et al. 2023).

5.5.4 *Use of Wild Crop Relatives*

Wild-type species related to agricultural crops harbor untapped genetic characteristics for stress tolerance and symbiosis. In combination with an understanding of their rhizomicrobiomes, they have been found to be valuable for producing *resilient cultivars or varieties* and are being used as a strategy to achieve sustainably cultivated crops with profitable yields (Berendsen et al. 2012).

5.6 Contemporary Rhizoengineering Approaches to Mitigate Abiotic Stresses

To obtain profound insights into the real-time dynamics of rhizosphere microbiomes, metagenomics presents a comprehensive perspective of the dynamics, specificities, frequencies, and functions of cultivable and noncultivable rhizomicrobiota. RFLP (restriction-fragment length polymorphism) analysis helps to determine the presence of potentially *novel microbial lineages. 16S rRNA pyrosequencing* assists in analyzing the influence of abiotic stress on local or regional microbial communities. Functional metagenomics uncovers the roles of *novel stress-responsive genes* (such as *DNA/RNA helicases*, *endonucleases,* membrane-bound branched-chain amino acid BCAA, and domains of unknown function DUF). Molecular techniques such as DNA-RNA hybridization, automated ribosomal intergenic spacer analysis (ARISA), temperature gradient gel electrophoresis (TGGE), denaturing gradient gel electrophoresis (DGGE), and ribosomal intergenic spacer analysis support the characterization *of nonculturable* PGPRs. *DNA microarray* studies have examined functional gene expression under stress. *Single-strand conformation polymorphism (SSCP)* helps reveal microbial genetic diversity. Genome mining tools help explore *biosynthetic gene clusters (BGCs). Transcriptomics and proteomics* together provide deeper insights into the production of resilient metabolites. *Differential proteomics* studies protein expression across different growth stages. *Metabolomics*, using *mass spectrometry (MS)* and nuclear magnetic resonance *(NMR)*, has identified novel metabolites involved in stress tolerance. *VarQuest* identifies novel peptide-based natural products, and *meta-omic techniques* continue to discover novel polyketides and microbial metabolites *under abiotic stress* (Shahid et al. 2024). These techniques demonstrate the power of next-generation sequencing and multiomic tools in decoding plant–microbiome interactions, especially under abiotic stress (Table 5.2). Unlike traditional single-strain bioinoculants, SynComs

Table 5.2 Techniques used to study plant rhizomicrobiome in relation to abiotic stress

Technique	Targets	Mechanism	Advantages	Approach
Stable isotope probing (SIP)	^{13}C, ^{15}N, ^{18}O, CO_2	Traces C and N flow in microbial communities by incorporating isotopic labels into the biomolecules of active microbes	This enables the functional recognition of stress-resilient microbes that participate in carbon/nutrient cycling in stressed environments	Conventional to contemporary
DNA arrays	Gene expression profiles	Hybridization of nucleic acid probes to detect gene expression patterns linked to stress responses	Allows rapid detection of stress-induced transcriptional changes but is limited by known sequences	Traditional
Bioreporters	Active microbial genes and functions	Genetic engineering of microbes to emit signals (e.g., fluorescence/luminescence) when specific stress-responsive genes are expressed	Enables noninvasive real-time tracking of gene expression and microbial behavior or traits in planta	Modern
Whole genome sequencing/genomics and metagenomics	Microbial genomes and communities	High-throughput DNA sequencing involving Illumina and PacBio was used to identify the taxonomic and functional potential of both culturable and unculturable microbes	Facilitates the discovery of novel microbial taxa, genes, and pathways involved in abiotic stress mitigation	Modern and advanced
Proteomics and metaproteomics	Functional proteins and enzymes	Mass spectrometry-based profiling of proteins to evaluate functional microbial adaptations to abiotic stress	Reveals active metabolic responses and protein networks associated with stress tolerance in the microbiome	Contemporary
Growth kinetics and metabolite profiling	PGPR metabolites and growth response	Monitoring PGPR growth and metabolite synthesis (e.g., phytohormones, VOCs, and exopolysaccharides) under stress	Identifies microbial traits and bioactive compounds vital for plant growth promotion, health, and stress adaptation	Contemporary
Metabolomics and pathway analysis	Stress-induced metabolites and biosynthetic pathways	Analytical techniques (e.g., LC-MS and GC-MS) are used to characterize primary and secondary metabolites and their biosynthetic regulation	Links specific metabolites with stress alleviation and integrates metabolome with genomics/transcriptomics	Advanced
Transcriptomics and metatranscriptomics	RNA transcripts (mRNA, sRNA)	Sequencing and quantification of RNA to identify actively expressed genes in microbes and plants under stress	Uncovers real-time microbial functional responses and helps design synthetic microbial consortia	Modern
In situ microbial community analysis	Whole microbial communities	The application of tools such as FISH, qPCR, and biosensors to assess community composition and activity directly in the rhizosphere	Provides spatial and functional insights into microbiome dynamics in real time under natural conditions	Contemporary

aim to replicate or optimize natural microbiome functions via a *reductionist and modular approach* (Vorholt et al. 2017; Xu et al. 2025). SynComs have shown benefits under diverse stress conditions. For example, a SynCom derived from *Indigofera argentea* enhanced tomato biomass by activating salt-responsive genes under salinity stress (Schmitz et al. 2022). Another SynCom improved drought tolerance in *Arabidopsis thaliana* by increasing abscisic acid production and biofilm formation (Yang et al. 2024). Similarly, tomato antibacterial SynCom mitigated salinity stress via PGP and ACC deaminase activities (Almazrouei et al. 2025). Recently, SynCom has provided significantly strong PGP effects on maize seedlings (Yu et al. 2025).

5.7 Conclusion

Evolutionary advancements have constantly deepened our knowledge. Evaluations of the above-mentioned fields of rhizoengineering, such as nanomaterials, biostimulants, osmoprotectants, phytohormones, grafting techniques, and nutrient administration, are needed, although these support the improvement of the plant's ability to cope with abiotic stress problems. In addition to the process of conventional breeding, the involvement of QTL mapping, marker-assisted modern breeding, genetic recombination, and techniques of speed breeding also contribute, but more clarity is required to confirm their stability, safety, and performance. The release of root exudates by stressed plants is known as the "cry for help" strategy. To understand the background of the interplay of plant-microbe interactions in the rhizosphere due to the rising threats of climate-induced abiotic stresses, there is a need for an integrated or novel strategy in sustainable agronomic practices by implementing SynComs in combination with or in place of current approaches. By sharing protocols or experiences and incorporating expertise across fields or laboratories, future researchers can improve the reproducibility, safety, experimental efficiency, and performance of this method. To obtain these desired applications, synthetic biology tools have been developed to engineer specific functions using contemporary high-throughput microbial culturomics with computational solutions.

References

Aalipour H, Nikbakht A, Etemadi N, Rejali F, Soleimani M (2020) Biochemical response and interactions between arbuscular mycorrhizal fungi and plant growth promoting rhizobacteria during establishment and stimulating growth of Arizona cypress (*Cupressus arizonica G.*) under drought stress. Sci Hortic 261:108923. https://doi.org/10.1016/j.scienta.2019.108923

Aazami MA, Maleki M, Rasouli F, Gohari G (2023) Protective effects of chitosan based salicylic acid nanocomposite (CS-SA NCs) in grape (*Vitis vinifera* cv.'Sultana') under salinity stress. Sci Rep 13:883. https://doi.org/10.1038/s41598-023-27618-z

Abd El-Ghani MM, Huerta-Martínez FM, Hongyan L, Qureshi R (2017) The deserts of Pakistan. In: Plant responses to hyperarid desert environments. Springer, New York, pp 547–573. https://doi.org/10.1007/978-3-319-59135-3

Abdelaal K, AlKahtani M, Attia K, Hafez Y, Király L, Künstler A (2021) The role of plant growth-promoting bacteria in alleviating the adverse effects of drought on plants. Biol 10(6):520. https://doi.org/10.3390/biology10060520

Adhikari L, Makaju SO, Lindstrom OM, Missaoui AM (2021) Mapping freezing tolerance QTL in alfalfa: based on indoor phenotyping. BMC Plant Biol 21:403. https://doi.org/10.1186/s12870-021-03182-4

Adhikari L, Baral R, Paudel D, Min D, Makaju SO, Poudel HP, Acharya JP, Missaoui AM (2022) Cold stress in plants: strategies to improve cold tolerance in forage species. Plant Stress 4:100081. https://doi.org/10.1016/j.stress.2022.100081

Ahmad A, Blasco B, Martos V (2022a) Combating salinity through natural plant extracts based biostimulants: a review. Front Plant Sci 13:862034. https://doi.org/10.3389/fpls.2022.862034

Ahmad M, Iqbal Z, Bushra Hussain A, Abdullah M, Alataway A, Dewidar AZ, Mattar MA (2022b) Seasonal changes modulate the rhizosphere of desert plant species. J Agron 13(1):57. https://doi.org/10.3390/agronomy13010057

Ahmad N, Naeem M, Ali H, Alabbosh KF, Hussain H, Khan I, Siddiqui SA, Khan AA, Iqbal B (2023) From challenges to solutions: the impact of melatonin on abiotic stress synergies in horticultural plants via redox regulation and epigenetic signaling. Sci Hortic 321:112369. https://doi.org/10.1016/j.scienta.2023.112369

Ajayi OO, Bregitzer P, Klos K, Hu G, Walling JG, Mahalingam R (2023) QTL mapping of shoot and seed traits impacted by drought in barley using a recombinant inbred line population. BMC Plant Biol 23(1):283. https://doi.org/10.1186/s12870-023-0429-x

Akbar WA, Rahim HU, Irfan M et al (2024) Assessment of heavy metal distribution and bioaccumulation in soil and plants near coal mining areas: implications for environmental pollution and health risks. Environ Monit Assess 196:97. https://doi.org/10.1007/s10661-023-12258-7

Alharbi K, Khan AA, Alhaithloul HAS, Al-Harbi NA, Al-Qahtani SM, Aloufi SS, Soliman MH (2023) Synergistic effect of B-Sitosterol and biochar application for improving plant growth of *Thymus vulgaris* under heat stress. Chemosphere 340:139832. https://doi.org/10.1016/j.chemosphere.2023.139832

Ali MJ, Yu Z, Xing G, Zhao T, Gai J (2018) Establishment of evaluation procedure for soybean seed-flooding tolerance and its application to screening for tolerant germplasm sources. Legume Res-Int J 41:34–40. https://doi.org/10.18805/lr.v0iOF.9112

Ali S, Rizwan M, Arif MS, Ahmad R, Hasanuzzaman M, Ali B, Hussain A (2020) Approaches in enhancing thermotolerance in plants: an updated review. J Plant Growth Regul 39:456–480. https://doi.org/10.1007/s00344-019-09994-x

Almazrouei LM, Purayil GP, Alnuaimi LS, Barhumi AI, Elameen AJ, Ibrahim SM, Elnahhas RB, AbuQamar SF, El-Tarabily KA (2025) Rhizosphere-competent actinobacterial consortium alleviates salinity stress in tomato through plant growth-promoting traits and ACC deaminase activity. BMC Plant Biol 1:1443. https://doi.org/10.1186/s12870-025-07446-1

Alotaibi M (2023) Climate change, its impact on crop production, challenges, and possible solutions. Not Bot Horti Agrobot Cluj-Napoca 51(1):13020. https://doi.org/10.15835/nbha51113020

Arun MN, Hebbar SS, Senthivel T, Nair AK, Padmavathi G, Pandey P, Singh A (2022) Seed priming: the way forward to mitigate abiotic stress in crops. In: Hasanuzzaman M, Nahar K, Brzozowski T (eds), vol 11. IntechOpen, London, p 173. https://doi.org/10.5772/intechopen.102033

Ashraf M, Athar HR, Harris PJC, Kwon TR (2008) Some prospective strategies for improving crop salt tolerance. Adv Agron 97:45–110. https://doi.org/10.1016/S0065-2113(07)00002-8

Aslam MM, Idris AL, Zhang QI, Weifeng XU, Karanja JK, Wei YU (2022) Rhizosphere microbiomes can regulate plant drought tolerance. Pedosphere 32(1):61–74. https://doi.org/10.1016/S1002-0160(21)60061-9

Baez-Rogelio A, Morales-Garcia YE, Quintero-Hernandez V, Munoz-Rojas J (2017) Next generation of microbial inoculants for agriculture and bioremediation. Micro Biotechnol 10:19–21. https://doi.org/10.1111/1751-7915.12448

Bai Y, Müller DB, Srinivas G, Garrido-Oter R, Potthoff E, Rott M, Dombrowski N et al (2015) Functional overlap of the *Arabidopsis* leaf and root microbiota. Nature 528:364–369. https://doi.org/10.1038/nature16192

Barquero M, Poveda J, Laureano-Marín AM, Ortiz-Liébana N, Brañas J, González-Andrés F (2022) Mechanisms involved in drought stress tolerance triggered by rhizobia strains in wheat. Front Plant Sci 13:1036973. https://doi.org/10.3389/fpls.2022.1036973

Behrooz A, Vahdati K, Rejali F, Lotfi M, Sarikhani S, Leslie C (2019) Arbuscular mycorrhiza and plant growth-promoting bacteria alleviate drought stress in walnut. Hort Sci 54:1087–1092. https://doi.org/10.21273/HORTSCI13961-19

Bellard C, Bertelsmeier C, Leadley P, Thuiller W, Courchamp F (2012) Impacts of climate change on the future of biodiversity. Ecol Lett 15(4):365–377. https://doi.org/10.1111/j.1461-0248.2011.01736.x

Berendsen RL, Pieterse CM, Bakker PA (2012) The rhizo sphere microbiome and plant health. Trends Plant Sci 17:478–486. https://doi.org/10.1016/j.tplants.2012.04.001

Berendsen RL, Vismans G, Yu K, Song Y, de Jonge R, Burgman WP et al (2018) Disease-induced assemblage of a plant beneficial bacterial consortium. ISME J 12:1496–1507. https://doi.org/10.1038/s41396-018-0093-1

Bhardwaj D, Ansari MW, Sahoo RK, Tuteja N (2014) Biofertilizers function as key player in sustainable agriculture by improving soil fertility, plant tolerance and crop productivity. Microb Cell Factories 13:1–10. https://doi.org/10.1186/1475-2859-13-66

Bowerman AF, Byrt CS, Roy SJ, Whitney SM, Mortimer JC, Ankeny RA et al (2023) Potential abiotic stress targets for modern genetic manipulation. Plant Cell 35:139–161. https://doi.org/10.1093/plcell/koac327

Capua GD, Rahmstorf S (2023) Extreme weather in a changing climate. Environ Res Lett 18:102001. https://doi.org/10.1088/1748-9326/acfb23

Cardarelli M, Woo SL, Rouphael Y, Colla G (2022) Seed treatments with microorganisms can have a biostimulant effect by influencing germination and seedling growth of crops. Plants 11:259. https://doi.org/10.3390/plants11030259

Chakraborty U, Chakraborty BN, Chakraborty AP, Dey PL (2013) Water stress amelioration and plant growth promotion in wheat plants by osmotic stress tolerant bacteria. World J Microbiol Biotechnol 29:789–803. https://doi.org/10.1007/s11274-012-1234-8

Chen Y, Yao Z, Sun Y, Wang E, Tian C, Sun Y, Liu J, Sun C, Tian L (2022) Current studies of the effects of drought stress on root exudates and rhizosphere microbiomes of crop plant species. Int J Mol Sci 23(4):2374. https://doi.org/10.3390/ijms23042374

Coskun ÖF (2023) The effect of grafting on morphological, physiological and molecular changes induced by drought stress in cucumber. Sustainability 15(1):875. https://doi.org/10.3390/su15010875

Deolu-Ajayi AO, van der Meer IM, Van der Werf A, Karlova R (2022) The power of seaweeds as plant biostimulants to boost crop production under abiotic stress. Plant Cell Environ 45:2537–2553. https://doi.org/10.1111/pce.14391

DeWolf ED, Isard SA (2007) Disease cycle approach to plant disease prediction. Annu Rev Phytopathol 45:203–220. https://doi.org/10.1146/annurev.phyto.44.070505.143329

Ding Y, Gong S, Wang Y, Wang F, Bao H, Sun J, Cai C, Yi K, Chen Z, Zhu C (2018) MicroRNA166 modulates cadmium tolerance and accumulation in rice. Plant Physiol 177:1691–1703. https://doi.org/10.1104/pp.18.00485

Dollete D, Lumactud RA, Carlyle CN, Szczyglowski K, Hill B, Thilakarathna MS (2024) Effect of drought stress on symbiotic nitrogen fixation, soil nitrogen availability and soil microbial diversity in forage legumes. Plant Soil 495:445–467. https://doi.org/10.1007/s11104-023-06348-1

Du F, Liu L, Pan Y, Wu C, Wang R, Zhao Z, Wang JA (2023) Novel biochar-based composite hydrogel for removing heavy metals in water and alleviating cadmium stress in tobacco seedlings. Sci Rep 13:15656. https://doi.org/10.1038/s41598-023-41946-0

Dubey A (2024) Sustainable agriculture. Encyclopedia Britannica. https://www.britannica.com/technology/sustainable-agriculture. Accessed 1 Dec 2025

Dubey A, Saiyam D, Kumar A, Hashem A et al (2021) Bacterial root endophytes: characterization of their competence and plant growth promotion in soybean (*Glycine max* (L.) Merr.) under drought stress. Int J Environ Res Public Health 18(3):931. https://doi.org/10.3390/ijerph18030931

El-Saadony MT, Saad AM, Soliman SM, Salem HM, Desoky ESM, Babalghith AO, AbuQamar SF (2022) Role of nanoparticles in enhancing crop tolerance to abiotic stress: a comprehensive review. Front Plant Sci 13:946717. https://doi.org/10.3389/fpls.2022.946717

Fadiji AE, Santoyo G, Yadav AN, Babalola OO (2022) Efforts towards overcoming drought stress in crops: revisiting the mechanisms employed by plant growth-promoting bacteria. Front Microbiol 13. https://doi.org/10.3389/fmicb.2022.962427

Fadiji AE, Yadav AN, Santoyo G, Babalola OO (2023) Understanding the plant-microbe interactions in environments exposed to abiotic stresses: an overview. Microbiol Res 271:127368. https://doi.org/10.1016/j.micres.2023.127368

Fageria NK, Stone LF, Santos AB (2012) Breeding for salinity tolerance. In: Fritsche-Neto R, Borém A (eds) Plant breeding for abiotic stress tolerance. Springer, Berlin, pp 103–122. https://doi.org/10.1007/978-3-642-30553-5_7

Fahad S, Bajwa AA, Nazir U, Anjum SA, Farooq A, Zohaib A, Sadia S, Nasim W, Adkins S, Saud S, Ihsan MZ (2017) Crop production under drought and heat stress: plant responses and management options. Front Plant Sci 8:1147. https://doi.org/10.3389/fpls.2017.01147

FAO (2016) The state of food and agriculture 2016 (SOFA): climate change, agriculture and food security. FAO, Rome. http://www.fao.org/3/a-i6030e.pdf. Accessed 18 Oct 2025

Formisano L, Ciriello M, Cirillo V, Pannico A, El-Nakhel C, Cristofano F, De Pascale S (2021) Divergent leaf morpho physiological and anatomical adaptations of four lettuce cultivars in response to different greenhouse irradiance levels in early summer season. Plants 10:1179. https://doi.org/10.3390/plants10061179

Gaete A, Pulgar R, Hodar C, Maldonado J, Pavez L, Zamorano D, Pastenes C, González M, Franck N, Mandakovic D (2021) Tomato cultivars with variable tolerances to water deficit differentially modulate the composition and interaction patterns of their rhizosphere microbial communities. Front Plant Sci 12:688533. https://doi.org/10.3389/fpls.2021.688533

Garrett KA, Nita M, De Wolf E, Esker PD, Gomez-Montano L, Sparks AH (2021) Plant pathogens as indicators of climate change. In: Letcher TM (ed) Climate change: observed impacts on planet earth. Elsevier, Amsterdam, pp 499–513. https://doi.org/10.1016/B978-0-12-82157 5-3.00024-4

Giordano M, Petropoulos SA, Rouphael Y (2021) Response and defence mechanisms of vegetable crops against drought, heat and salinity stress. Agriculture 11:463. https://doi.org/10.3390/agriculture11050463

Glick BR (2014) Bacteria with ACC deaminase can promote plant growth and help to feed the world. Microbiol Res 169(1):30–39. https://doi.org/10.1016/j.micres.2013.09.009

Godoy F, Olivos-Hernández K, Stange C, Handford M (2021) Abiotic stress in crop species: improving tolerance by applying plant metabolites. Plants 10:186. https://doi.org/10.3390/plants10020186

Gouda S, Kerry RG, Das G, Paramithiotis S, Shin HS, Patra JK (2018) Revitalization of plant growth promoting rhizobacteria for sustainable development in agriculture. Microbiol Res 206:131–140. https://doi.org/10.1016/j.micres.2017.08.016

Guan Y, Xie Z, Guan J, Zhao Q, Rasheed A, Hussain MI, Ali I, Zhang S, Hassan MU, Hashem M et al (2022) The intervention of classical and molecular breeding approaches to enhance flooding stress tolerance in soybean- an review. Front Plant Sci 13:1085368. https://doi.org/10.3389/fpls.2022.1085368

Guo X, Feng J, Shi Z, Zhou X, Yuan M, Tao X, Hale L, Yuan T, Wang J, Qin Y, Zhou A (2018) Climate warming leads to divergent succession of grassland microbial communities. Nat Clim Chang 8(9):813–818. https://doi.org/10.1038/s41558-018-0254-2

Haichar FE, Cernava T, Liu J, Timm CM (2021) Novel insights into the response of the plant microbiome to abiotic factors. Front Plant Sci 12:607874. https://doi.org/10.3389/fpls.2021.607874

Hall A (1992) Breeding for heat tolerance. In: Janick J (ed) Plant breeding reviews, vol 10. Wiley, New York, pp 129–168. https://doi.org/10.1002/9780470650011.CH5

Hassan ZU, Ali S, Rizwan M, Ibrahim M, Nafees M, Waseem M (2017) Role of bioremediation agents (bacteria, fungi, and algae) in alleviating heavy metal toxicity. In: Kumar V, Kumar

M, Sharma S, Prasad R (eds) Probiotics in agroecosystem. Springer, Singapore, pp 517–537. https://doi.org/10.1007/978-981-10-4059-7_27

Hassan MK, McInroy JA, Kloepper JW (2019) The interactions of rhizodeposits with plant growth-promoting rhizobacteria in the rhizosphere: a review. Agri 9(7):142. https://doi.org/10.3390/agriculture9070142

Hassan MU, Chattha MU, Khan I, Chattha MB, Barbanti L, Aamer M, Hossain J, Sarkar S, Saha S, Bhadra P et al (2021) Heat stress in cultivated plants: nature, impact, mechanisms, and mitigation strategies – a review. Plant Biosyst-Int J Deal All Asp Plant Biol 155:211–234. https://doi.org/10.1080/11263504.2020.1727987

Hazra P, Samsul HA, Sikder D, Peter KV (2007) Breeding tomato (*Lycopersicon esculentum* mill) resistant to high temperature stress. Int J Plant Breed 1:31–40

Henry RJ (2020) Innovations in plant genetics adapting agriculture to climate change. Curr Opin Plant Biol 56:168–173. https://doi.org/10.1016/j.pbi.2019.11.004

Hoque MN, Tahjib-Ul-Arif M, Hannan A, Sultana N, Akhter S, Hasanuzzaman M, Akter F, Hossain MS, Sayed MA, Hasan MT et al (2021) Melatonin modulates plant tolerance to heavy metal stress: morphological responses to molecular mechanisms. Int J Mol Sci 22:11445. https://doi.org/10.3390/ijms222111445

Hossain A, Skalicky M, Brestic M, Maitra S, Ashraful Alam M, Syed MA, Islam T (2021) Consequences and mitigation strategies of abiotic stresses in wheat (*Triticum aestivum* L.) under the changing climate. Agronomy 11:241. https://doi.org/10.3390/agronomy11020241

Huang M, Yin X, Chen J, Cao F (2021) Biochar application mitigates the effect of heat stress on rice (*Oryza sativa* L.) by regulating the root-zone environment. Front Plant Sci 12:711725. https://doi.org/10.3389/fpls.2021.711725

Hussain A, Iqbal Z, Aurangzaib M, Naeem M, Mustafa S, Shoaib M, Khalid M (2022) Comparative effectiveness of plant growth promoting rhizobacteria and various organic carriers on wheat growth, physiology, antioxidative activities and rhizosphere properties. Pak J Bot 54(1):317–324. https://doi.org/10.30848/PJB2022-1(43)

Inbaraj MP (2021) Plant-microbe interactions in alleviating abiotic stress – a mini review. Front Agron 28. https://doi.org/10.3389/fagro.2021.667903

IPCC (2001) Climate change 2001: the scientific basis. In: Houghton JT, Ding Y, Griggs DJ, Noguer M, van der Linden PJ, Dai X, Maskell K, Johnson CA (eds) Contribution of working group I to the third assessment report of the intergovernmental panel on climate change. Cambridge University Press, Cambridge, p 881

IPCC (2022) Climate change 2022: impacts, adaptation and vulnerability. In: Pörtner HO, Roberts DC, Tignor M, Poloczanska ES, Mintenbeck K, Alegría A, Craig M, Langsdorf S, Löschke S, Möller V, Okem A, Rama B (eds) Contribution of working group II to the sixth assessment report of the intergovernmental panel on climate change. Cambridge University Press, Cambridge, p 3675. https://doi.org/10.1017/9781009325844

Iqbal H, Yaning C, Waqas M, Ahmed Z, Raza ST, Shareef M (2020) Improving heat stress tolerance in late planted spring maize by using different exogenous elicitors. Chil J Agric Res 80:30–40. https://doi.org/10.4067/S0718-58392020000100030

IUCN (2023) IUCN 2023 annual report. IUCN, Gland. https://portals.iucn.org/library/sites/library/files/documents/2024-012-En.pdf

IUCN (2024) IUCN 2024 annual report. IUCN, Gland. https://portals.iucn.org/library/sites/library/files/documents/2025-012-En.pdf

Jabborova D, Kannepalli A, Davranov K, Narimanov A, Enakiev Y, Syed A, Elgorban AM, Bahkali AH, Wirth S, Sayyed RZ, Gafur A (2021) Co-inoculation of rhizobacteria promotes growth, yield, and nutrient contents in soybean and improves soil enzymes and nutrients under drought conditions. Sci Rep 11(1):22081. https://doi.org/10.1038/s41598-021-01337-9

Jankovska-Bortkevi˘c E, Katerova Z, Todorova D et al (2023) Effects of auxin-type plant growth regulators and cold stress on the endogenous polyamines in Pea plants. Horticulturae 9:244. https://doi.org/10.3390/horticulturae9020244

Jegadeesan S, Yu K, Poysa V, Gawalko E, Morrison MJ, Shi C, Cober E (2010) Mapping and validation of simple sequence repeat markers linked to a major gene controlling seed cadmium

accumulation in soybean [*Glycine max* (L.) Merr]. Theor Appl Genet 121:283–294. https://doi.org/10.1007/s00122-010-1309-6
Jiao X, Takishita Y, Zhou G, Smith DL (2021) Plant associated rhizobacteria for biocontrol and plant growth enhancement. Front Plant Sci 12:634796. https://doi.org/10.3389/fpls.2021.634796
Kang SM, Adhikari A, Khan MA, Kwon EH, Park YS, Lee IJ (2021) Influence of the rhizobacterium *Rhodobacter sphaeroides* KE149 and biochar on waterlogging stress tolerance in *Glycine max* L. Environments 8:94. https://doi.org/10.3390/environments8090094
Kenneth OC, Nwadibe EC, Kalu AU, Unah UV (2019) Plant growth promoting rhizobacteria (PGPR): a novel agent for sustainable food production. Am J Agric Biol Sci 14(35):54. https://doi.org/10.3844/ajabssp.2019.35.54
Kikuchi A, Huynh HD, Endo T, Watanabe K (2015) Review of recent transgenic studies on abiotic stress tolerance and future molecular breeding in potato. Breed Sci 65:85–102. https://doi.org/10.1270/jsbbs.65.85
Krishna R, Karkute SG, Ansari WA, Jaiswal DK, Verma JP, Singh M (2019) Transgenic tomatoes for abiotic stress tolerance: status and way ahead. 3 Biotech 9:143. https://doi.org/10.1007/s13205-019-1665-0
Kumar A, Nayak AK, Hanjagi PS, Kumari K, Vijayakumar S, Mohanty S, Panneerselvam P (2021a) Submergence stress in rice: adaptive mechanisms, coping strategies and future research needs. Environ Exp Bot 186:104448. https://doi.org/10.1016/j.envexpbot.2021.104448
Kumar S, Mahapatra S, Chakraborty S, Mukharjee S (2021b) Effect of abiotic stresses and mitigation strategy associated with their tolerance in wheat. J Cereal Res 13:16–37. https://doi.org/10.25174/25822675/2021/111211
Kumar S, Sindhu SS, Kumar R (2022) Biofertilizers: an ecofriendly technology for nutrient recycling and environmental sustainability. Curr Res Microbial Sci 3:100094. https://doi.org/10.1016/j.crmicr.2021.100094
Kumari A, Singh AP (2022) Speed breeding: accelerating crop improvement for global food security. Innov Approaches Agric 1:92–103
Kumari S, Vaishnav A, Jain S, Varma A, Choudhary DK (2016) Induced drought tolerance through wild and mutant bacterial strain *Pseudomonas simiae* in mung bean (*Vigna radiata L.).* World J Microbiol Biotechnol 32:1–10. https://doi.org/10.1007/s11274-015-1974-3
Kushwaha AK, Shiv A, Samantara K, Chand S, Majhi PK, Goyal L, Kaldate S, Kumar S, Razzaq A, Wani SH (2023) Editing genomes to modify plant response to abiotic stress. In: Ghorbanpour M, Shahid MA (eds) Plant stress mitigators. Academic, Chennai, pp 1–15. https://doi.org/10.1016/B978-0-323-89871-3.00004-5
LaBorde N, Rajewski J, Dweikat I (2023) Novel QTL for chilling tolerance at germination and early seedling stages in sorghum. Front Genet 14:1129460. https://doi.org/10.3389/fgene.2023.1129460
Lal R (2012) Climate change and soil degradation mitigation by sustainable management of soils and other natural resources. Agric Res 1:99–212. https://doi.org/10.1007/s40003-012-0031-9
Langeroodi ARS, Noora R (2017) Seed priming improves the germination and field performance of soybean under drought stress. J Anim Plant Sci 27:1611–1620
Li X, Sun P, Zhang Y, Jin C, Guan C (2020) A novel PGPR strain *Kocuria rhizophila* Y1 enhances salt stress tolerance in maize by regulating phytohormone levels, nutrient acquisition, redox potential, ion homeostasis, photosynthetic capacity and stress-responsive genes expression. Environ Exp Bot 174:104023. https://doi.org/10.1016/j.envexpbot.2020.104023
Li X, Xiao R, Tao Y (2025) Enhancing plant stress resilience and agricultural sustainability through rhizosphere microbiome optimization. Plant Soil 19:1–3. https://doi.org/10.1007/s11104-025-07359-w
Liang S, Xiong W, Yin C, Xie X, Jin YJ, Zhang S, Yang B, Ye G, Chyen S, Luan WJ (2019) Overexpression of *OsARD1* improves submergence, drought, and salt tolerances of seedling through the enhancement of ethylene synthesis in rice. Front Plant Sci 10:1088. https://doi.org/10.3389/fpls.2019.01088

Lopes M, Dias-Filho MB, Gurgel ESC (2021) Successful plant growth promoting microbes: inoculation methods and abiotic factors. Front Sustain Food Syst 5:606454. https://doi.org/10.3389/fsufs.2021.606454

Mace GM, Norris K, Fitter AH (2012) Biodiversity and ecosystem services: a multilayered relationship. Trends Ecol Evol 27:19–26. https://doi.org/10.1016/j.tree.2011.08.006

Manjunatha BS, Nivetha N, Krishna GK, Elangovan A, Pushkar S, Chandrashekar N et al (2022) Plant growth-promoting rhizobacteria *Shewanella putrefaciens* and *Cronobacter dublinensis* enhance drought tolerance of pearl millet by modulating hormones and stress-responsive genes. Plant Physiol 174:e13676. https://doi.org/10.1111/ppl.13676

Marthandan V, Geetha R, Kumutha K, Renganathan VG, Karthikeyan A, Ramalingam J (2020) Seed priming: a feasible strategy to enhance drought tolerance in crop plants. Int J Mol Sci 21:8258. https://doi.org/10.3390/ijms21218258

Maser P, Hosoo Y, Goshima S, Horie T, Eckelman B, Yamada K et al (2002) Glycine residues in potassium channel-like selectivity filters determine potassium selectivity in four-loop-per subunit *HKT* transporters from plants. Proc Natl Acad Sci USA 99:6428–6433. https://doi.org/10.1073/pnas.082123799

Matthews JL (2024) Holobiont interactions. Front Media SA 12:1382169. https://doi.org/10.3389/fevo.2024.1382169

Meena HP, Bainsla NK, Yadav DK (2016) Breeding for abiotic stress tolerance in crop plants. In: Recent advances in plant stress physiology. Daya Publishing House, New Delhi, pp 329–378

Mu J, Fu Y, Liu B, Zhang Y, Wang A, Li Y, Zhu J (2021) *SiFBA5,* a cold-responsive factor from *Saussurea involucrata* promotes cold resilience and biomass increase in transgenic tomato plants under cold stress. BMC Plant Biol 21:75. https://doi.org/10.1186/s12870-021-02851-8

Muhammad A, Kong X, Zheng S, Bai N, Li L, Khan MHU, Fiaz S, Zhang Z (2024) Exploring plant-microbe interactions in adapting to abiotic stress under climate change: a review. Front Plant Sci 15:1482739. https://doi.org/10.3389/fpls.2024.1482739

Muneer S, Ko CH, Wei H, Chen Y, Jeong BR (2016) Physiological and proteomic investigations to study the response of tomato graft unions under temperature stress. PLoS One 11:e0157439

Mustafa M, Szalai Z, Divéky-Ertsey A, Gál I, Csambalik L (2022) Conceptualizing multiple stressors and their consequences in agroforestry systems. Stress 2(3):242–255. https://doi.org/10.3390/stresses2030018

Nawaz T, Gu L, Fahad S, Saud S, Bleakley B, Zhou R (2024a) Exploring sustainable agriculture with nitrogen-fixing cyanobacteria and nanotechnology. Molecules 29:2534. https://doi.org/10.3390/molecules29112534

Nawaz T, Saud S, Gu L, Khan I, Fahad S, Zhou R (2024b) Cyanobacteria: harnessing the power of microorganisms for plant growth promotion, stress alleviation, and phytoremediation in the era of sustainable agriculture. Plant Stress 11:100399. https://doi.org/10.1016/j.stress.2024.100399

Naylor D, DeGraaf S, Purdom E, Coleman-Derr D (2017) Drought and host selection influence bacterial community dynamics in the grass root microbiome. ISME J 11(12):2691–2704

Ojuederie OB, Olanrewaju OS, Babalola OO (2019) Plant growth promoting rhizobacterial mitigation of drought stress in crop plants: implications for sustainable agriculture. Agronomy 9:712. https://doi.org/10.3390/agronomy9110712

Okereafor U, Makhatha M, Mekuto L et al (2020) Toxic metal implications on agricultural soils, plants, animals, aquatic life and human health. Int J Environ Res Pub Health 17:2204. https://doi.org/10.3390/ijerph17072204

Oladosu Y, Rafii MY, Samuel C, Fatai A, Magaji U, Kareem I, Kamarudin ZS, Muhammad I, Kolapo K (2019) Drought resistance in rice from conventional to molecular breeding: a review. Int J Mol Sci 20:3519. https://doi.org/10.3390/ijms20143519

Oyebamiji YO, Shamsudin NAA, Ikmal AM, Yusop MR (2023) Heat stress in vegetables: impacts and management strategies – a review. Sains Malays 52(7):1925–1938. https://doi.org/10.17576/jsm-2023-5207-03

Oyebamiji YO, Adigun BA, Shamsudin NAA, Ikmal AM, Salisu MA, Malike FA, Lateef AA (2024) Recent advancements in mitigating abiotic stresses in crops. Horticulture 10(2):156. https://doi.org/10.3390/horticulturae10020156

Padhiary M, Kumar R (2024) Assessing the environmental impacts of agriculture, industrial operations, and mining on agro-ecosystems. In: Azrour M, Mabrouki J, Alabdulatif A, Guezzaz A, Amounas F (eds) Smart internet of things for environment and healthcare, vol 1165. Springer Nature, Cham, pp 107–126. https://doi.org/10.1007/978-3-031-70102-3_8

Pandey S, Singh A, Parida SK, Prasad M (2022) Combining speed breeding with traditional and genomics-assisted breeding for crop improvement. Plant Breed 141:301–313. https://doi.org/10.1111/pbr.13012

Pandit R, Scholes R, Montanarella L, Brainich A, Barger N et al (2018) Summary for policymakers of the IPBES assessment report on land degradation and restoration of the Intergovernmental Science-Policy Platform on Biodiversity and Ecosystem Services. IPBES Report Bonn, Germany, p 744. https://doi.org/10.5281/zenodo.3237

Paravar A, Piri R, Balouchi H, Ma Y (2023) Microbial seed coating: an attractive tool for sustainable agriculture. Biotechnol Rep 37:e00781. https://doi.org/10.1016/j.btre.2023.e00781

Pardo-Hernández M, López-Delacalle M, Rivero RM (2020) ROS and NO regulation by melatonin under abiotic stress in plants. Antioxidants 9:1078. https://doi.org/10.3390/antiox9111078

Rahim HU, Ali W, Uddin M, Ahmad S, Khan K, Bibi H, Alatalo JM (2025) Abiotic stresses in soils, their effects on plants, and mitigation strategies: a literature review. Chem Ecol 41(4):552–585. https://doi.org/10.1080/02757540.2024.2439830

Rajora N, Vats S, Raturi G, Thakral V, Kaur S, Rachappanavar V, Deshmukh R (2022) Seed priming with melatonin: a promising approach to combat abiotic stress in plants. Plant Stress 4:100071. https://doi.org/10.1016/j.stress.2022.100071

Rani MG, Satyanarayana PV, Chamundeswari N, Rao PR, Prabhakar M, Ravikumar BNVSR, Nagakumari P, Kalpana K (2022) Mapping of QTLs for flood tolerance in rice using recombinant inbred lines of Indra and new plant genetic resource *AC39416*. A Plant Genet Resour 20:270–276. https://doi.org/10.1017/S147926212300014X

Rawat R, Tewari L (2011) Effect of abiotic stress on phosphate solubilization by biocontrol fungus *Trichoderma* sp. Curr Microbiol 62:1521–1526. https://doi.org/10.1007/s00284-011-9888-2

Raza A, Charagh S, Najafi-Kakavand S, Abbas S, Shoaib Y, Anwar S, Siddique KH (2023) Role of phytohormones in regulating cold stress tolerance: physiological and molecular approaches for developing cold-smart crop plants. Plant Stress 8:100152. https://doi.org/10.1016/j.stress.2023.100152

Rhaman MS, Tania SS, Imran S, Rauf F, Kibria MG, Ye W, Murata Y (2020) Seed priming with nanoparticles: an emerging technique for improving plant growth, development, and abiotic stress tolerance. J Soil Sci Plant Nutr 22:4047–4062. https://doi.org/10.1007/s42729-022-01007-3

Rus A, Lee BH, Munoz-Mayor A, Sharkhuu A, Miura K, Zhu JK et al (2004) *AtHKT1* facilitates Na^+ homeostasis and K^+ nutrition in planta. Plant Physiol 136:2500–2511. https://doi.org/10.1104/pp.104.042234

Sadak MS, Dawood MG (2023) Biofertilizer role in alleviating the deleterious effects of salinity on wheat growth and productivity. Gesunde Pflanz 75:1207–1219. https://doi.org/10.1007/s10343-022-00783-3

Saharan BS, Nehra V (2011) Plant growth promoting rhizobacteria: a critical review. Life Sci Med Res 21(1):30

Saini S, Kaur N, Pati PK (2021) Phytohormones: key players in the modulation of heavy metal stress tolerance in plants. Ecotoxicol Environ Saf 223:112578. https://doi.org/10.1016/j.ecoenv.2021.112578

Samantara K, Bohra A, Mohapatra SR, Prihatini R, Asibe F et al (2022) Breeding more crops in less time: a perspective on speed breeding. Biology 11:275. https://doi.org/10.3390/biology11020275

Sánchez-Gil JJ, Poppeliers SWM, Vacheron J, Zhang H, Odijk B, Keel C, de Jonge R (2023) The conserved iol gene cluster in *Pseudomonas* is involved in rhizosphere competence. Curr Biol 33:3097–3110. https://doi.org/10.1016/j.cub.2023.05.057

Sarkar B, Varalaxmi Y, Vanaja M, RaviKumar N, Prabhakar M, Yadav SK, Maheswari M, Singh VK (2023) Mapping of QTLs for morphophysiological and yield traits under water-deficit stress and well-watered conditions in maize. Front Plant Sci 14:1124619. https://doi.org/10.3389/fpls.2023.1124619

Schmidt JE, Kent AD, Brisson VL, Gaudin AC (2019) Agricultural management and plant selection interactively affect rhizosphere microbial community structure and nitrogen cycling. Microbiome 7:1–8. https://doi.org/10.1186/s40168-019-0756-9

Schmitz L, Yan Z, Schneijderberg M, de Roij M, Pijnenburg R, Zheng Q, Franken C, Dechesne A, Trindade LM, Van Velzen R, Bisseling T (2022) Synthetic bacterial community derived from a desert rhizosphere confers salt stress resilience to tomato in the presence of a soil microbiome. ISME J 16:1907–1920. https://doi.org/10.1038/s41396-022-01238-3

Seleiman MF, Aslam MT, Alhammad BA, Hassan MU, Maqbool R, Chattha MU, Khan I, Gitari HI, Uslu SO, Roy R et al (2022) Salinity stress in wheat: effects, mechanisms and management strategies. Phyton 91:667–694. https://doi.org/10.32604/phyton.2022.017365

Shaffique S, Khan MA, Imran M, Kang SM, Park YS, Wani SH, Lee IJ (2022) Research progress in the field of microbial mitigation of drought stress in plants. Front Plant Sci 13:870626. https://doi.org/10.3389/fpls.2022.870626

Shahid I, Batool S, Hassan M, Ismail H, Mehnaz S, Deeba F, Anwar M, Zulfiqar F, Iqbal R, Ali HM (2024) A decade of progress in rhizoengineering to exploit plant microbiome for salt stress amelioration. Plant Stress 11:100325. https://doi.org/10.1016/j.stress.2023.100325

Shahwar D, Mushtaq Z, Mushtaq H, Alqarawi AA, Park Y, Alshahrani TS et al (2023) Role of microbial inoculants as bio fertilizers for improving crop productivity: a review. Heliyon 9(6):e16134. https://doi.org/10.1016/j.heliyon.2023.e16134

Si J, Froussart E, Viaene T, Vázquez-Castellanos JF, Hamonts K, Tang L, Beirinckx S, De Keyser A, Deckers T, Amery F, Vandenabeele S (2021) Interactions between soil compositions and the wheat root microbiome under drought stress: from an in silico to in planta perspective. Comput Struct Biotechnol J 19:4235–4247. https://doi.org/10.1016/j.csbj.2021.07.027

Simmons T, Styer AB, Pierroz G, Gonçalves AP, Pasricha R, Hazra AB, Bubner P, Coleman-Derr D (2020) Drought drives spatial variation in the millet root microbiome. Front Plant Sci 11:599. https://doi.org/10.3389/fpls.2020.00599

Singh H, Sethi S, Kaushik P, Fulford A (2020) Grafting vegetables for mitigating environmental stresses under climate change: a review. J Water Clim Change 11:1784–1797. https://doi.org/10.2166/wcc.2019.177

Soren KR, Madugula P, Kumar N, Barmukh R, Sengar MS, Bharadwaj C, Sharma PC, Singh S, Bhandari A, Singh J et al (2020) Genetic dissection and identification of candidate genes for salinity tolerance using Axiom® Cicersnp Array in chickpea. Int J Mol Sci 21:5058. https://doi.org/10.3390/ijms21145058

Sytar O, Kumari P, Yadav S, Brestic M, Rastogi A (2019) Phytohormone priming: regulator for heavy metal stress in plants. J Plant Growth Regul 38:739–752. https://doi.org/10.1007/s00344-018-9886-8

Tan C, Li N, Wang Y, Yu X, Yang L, Cao R, Ye X (2023) Integrated physiological and transcriptomic analyses revealed improved cold tolerance in cucumber (*Cucumis sativus* L.) by exogenous chitosan oligosaccharide. Int J Mol Sci 24:6202. https://doi.org/10.3390/ijms24076202

Tang H, Wang S, Liu Y, Mu H, Song Y, Huang G, Hashem M, Alamri S, Mostafa YS (2022) Biochar: a promising soil amendment to mitigate heavy metals toxicity in plants. Not Bot Horti Agrobot Cluj-Napoca 50:12778. https://doi.org/10.15835/nbha50312778

Terán F, Vives-Peris V, López-Climent MF, Gómez-Cadenas A, Pérez-Clemente RM (2023) Palliative effects of Kaolin on citrus plants under controlled stress conditions of high temperature and high light intensity. J Plant Growth Regul:1–14. https://doi.org/10.1007/s00344-023-11103-y

Thepbandit W, Athinuwat D (2024) Rhizosphere microorganisms supply availability of soil nutrients and induce plant defense. Microorganisms 12:558. https://doi.org/10.3390/microorganisms12030558

Toju H, Peay KG, Yamamichi M, Narisawa K, Hiruma K, Naito K et al (2018) Core microbiomes for sustainable agroecosystems. Nat Plants 4:247–257. https://doi.org/10.1038/s41477-018-0139-4

Trivedi P, Leach JE, Tringe SG, Sa T, Singh BK (2020) Plant microbiome interactions: from community assembly to plant health. Nat Rev Microbiol 18:607–621. https://doi.org/10.1038/s41579-020-0412-1

UN DESA (2017) The Sustainable Development Goals report 2017. United Nations, New York. https://unstats.un.org/sdgs/report/2017/

UNCCD (2024) 16th session of the Conference of the Parties to the United Nations Convention to Combat Desertification (COP16), Riyadh, Saudi Arabia, 2–13 December 2024. https://www.unccd.int/cop16

United Nations General Assembly (2015) Transforming our world: the 2030 Agenda for Sustainable Development. A/RES/70/1. https://undocs.org/A/RES/70/1

Varshney RK, Barmukh R, Roorkiwal M, Qi Y, Kholova J, Tuberosa R, Reynolds MP, Tardieu F, Siddique KH (2021) Breeding custom-designed crops for improved drought adaptation. Adv Genet 2.3:e202100017. https://doi.org/10.1002/ggn2.202100017

Vescio R, Malacrinò A, Bennett AE, Sorgonà A (2021) Single and combined abiotic stressors affect maize rhizosphere bacterial microbiota. Rhizosphere 17:100318. https://doi.org/10.1016/j.rhisph.2021.100318

Vorholt JA, Vogel C, Carlström CI, Müller DB (2017) Establishing causality: opportunities of synthetic communities for plant microbiome research. Cell Host Microbe 22(2):142–155. https://doi.org/10.1016/j.chom.2017.07.004

Wang L, Xiao Y, Wei X, Pan J, Duanmu D (2021) Highly efficient CRISPR-mediated base editing in *Sinorhizobium meliloti*. Front Microbiol 12:686008. https://doi.org/10.3389/fmicb.2021.686008

Wani OA, Parthiban M, Bhat MA, Mahdi SS, Jan R, Bhat MA, Kumar SS (2022) Biochar: a new emerging tool to mitigate abiotic stresses and its effect on soil properties. In: Bahar FA, Anwar Bhat M, Mahdi SS (eds) Secondary agriculture: sustainability and livelihood in India. Springer, Cham, pp 97–108. https://doi.org/10.1007/978-3-031-09218-3_9

Wu Y, Wang X, Zhang L, Zheng Y, Liu X, Zhang Y (2023) The critical role of biochar to mitigate the adverse impacts of drought and salinity stress in plants. Front Plant Sci 14:1163451. https://doi.org/10.3389/fpls.2023.1163451

Xu L, Naylor D, Dong Z, Simmons T, Pierroz G, Hixson KK, Kim YM et al (2018) Drought delays development of the sorghum root microbiome and enriches for monoderm bacteria. Proc Natl Acad Sci USA 115:E4284–E4293. https://doi.org/10.1073/pnas.171730811

Xu Y, Zhang G, Ding H, Ci D, Dai L, Zhang Z (2020) Influence of salt stress on the rhizosphere soil bacterial community structure and growth performance of groundnut (*Arachis hypogaea* L.). Int Microbiol 23:453–465. https://doi.org/10.1007/s10123-020-00118-0

Xu X, Dinesen C, Pioppi A, Kovács ÁT, Lozano-Andrade CN (2025) Composing a microbial symphony: synthetic communities for promoting plant growth. Trends Microbiol 33:738–751. https://doi.org/10.1016/j.tim.2025.01.006

Xun W, Shao J, Shen Q, Zhang R (2021) Rhizosphere microbiome: functional compensatory assembly for plant fitness. Comput Struct Biotechnol J 19:5487–5493. https://doi.org/10.1016/j.csbj.2021.09.035

Yan H, Cong M, Hu Y, Qiu C, Yang Z, Tang G, Xu W, Zhu X, Sun X, Jia H (2022) Biochar-mediated changes in the microbial communities of rhizosphere soil alter the architecture of maize roots. Front Microbiol 13:1023444. https://doi.org/10.3389/fmicb.2022.1023444

Yang K, Fu R, Feng H, Jiang G, Finkel O, Sun T et al (2023) RIN enhances plant disease resistance via root exudate-mediated assembly of disease-suppressive rhizosphere micro biota. Mol Plant 16:1379–1395. https://doi.org/10.1016/j.molp.2023.08.004

Yang N, Røder HL, Wicaksono WA, Wassermann B, Russel J, Li X, Nesme J, Berg G, Sørensen SJ a BM (2024) Interspecific interactions facilitate keystone species in a multispecies biofilm that promotes plant growth. ISME J 18(1):wrae012. https://doi.org/10.1093/ismejo/wrae012

Yi Y, Li Z, Song C, Kuipers OP (2018) Exploring plant-microbe interactions of the rhizobacteria *Bacillus subtilis* and *Bacillus mycoides* by use of the CRISPR-Cas9 system. Environ Microbiol 20:4245–4260

Younis A, Ramzan F, Ramzan Y, Zulfiqar F, Ahsan M, Lim KB (2020) Molecular markers improve abiotic stress tolerance in crops: a review. Plants (Basel) 9(10):1374. https://doi.org/10.3390/plants9101374

Yu S, Mao M, Zhang H, Song H, Sun Y (2025) Growth-promoting effects and mechanisms of synthetic plant growth-promoting rhizobacteria on maize seedlings. Microorganisms 13(11):2460. https://doi.org/10.3390/microorganisms13112460

Zaid A, Mohammad F, Fariduddin Q (2020) Plant growth regulators improve growth, photosynthesis, mineral nutrient and antioxidant system under cadmium stress in menthol mint (*Mentha arvensis* L.). Physiol Mol Biol Plants 26:25–39. https://doi.org/10.1007/s12298-019-00715-y

Zhang H, Kim MS, Sun Y, Dowd SE, Shi H, Pare PW (2008) Soil bacteria confer plant salt tolerance by tissue-specific regulation of the sodium transporter *HKT1*. Mol Plant-Microbe Interact 21:737–744. https://doi.org/10.1094/MPMI-21-6-0737

Zhang W, Wei J, Guo L, Fang H, Liu X, Liang K, Siddique KH (2023) Effects of two biochar types on mitigating drought and salt stress in tomato seedlings. Agronomy 13:1039. https://doi.org/10.3390/agronomy13041039

Zhu ZH, Sami A, Xu QQ, Wu LL, Zheng WY, Chen ZP, Zhou KJ (2021) Effects of seed priming treatments on the germination and development of two rapeseed (*Brassica napus* L.) varieties under the co-influence of low temperature and drought. PLoS One 16:e0257236. https://doi.org/10.1371/journal.pone.0257236

Part III
Revisiting Traditional Approaches and Rationalizing Contemporary Targets in Rhizoengineering

Chapter 6
Role of Beneficial Soil Microorganisms in Engineering Agricultural Sustainability Under Changing Climate Scenario

Priyanka Verma, Ankita Bhattacharya, and Naveen Kumar Arora

Abstract Climate change has a significant impact on plant health and productivity. Climate change causes the intensification of environmental stresses, including flashes of extreme weather events and prolonged climate stress episodes. These stress factors affect various plant physiological processes and mechanisms at different growth stages. Climate change-induced environmental stressors, along with human population outbursts, have created huge pressure on the agricultural sector, leading to the use of excessive chemicals in the agro-ecosystems. In recent years, beneficial soil microorganisms have emerged as excellent, cost-effective, and sustainable alternatives to combat climate change-related environmental stresses. These microorganisms employ various mechanisms to reduce the effects of environmental stress on plants. Microorganisms ranging from rhizospheric bacteria and fungi to endophytic origins have gained recognition as biofertilizers and biopesticides; however, many challenges must be overcome to realize their true potential from laboratories to fields. In the future, advances in technologies will create new insights into plant–microbe interactions, combating climate change, and their deleterious effects on agriculture.

Keywords Climate change · Biotic stress · Abiotic stress · Food security · Agricultural sustainability

---

P. Verma · A. Bhattacharya · N. K. Arora (✉)
Department of Environmental Science, Babasaheb Bhimrao Ambedkar University, Lucknow, Uttar Pradesh, India

S. Mehnaz, I. Shahid (eds.), *Frontiers in Rhizoengineering*, Microorganisms for Sustainability 65, https://doi.org/10.1007/978-981-92-3665-7_6

6.1 Introduction

The human population has increased from 5 billion in 1986 to 8.2 billion by mid-2025 and is expected to expand beyond 9 billion by 2037 and 9.8 billion by 2050 (Gu et al. 2021). The twentieth century experienced the highest growth in the world population, reaching from 1.65 billion to 6 billion (Gu et al. 2021). The human population and climate change are inextricably related. An increase in population demands higher resources for consumption and ultimately results in increased greenhouse gas (GHG) emissions (Mishra et al. 2025a). Similarly, climate change is believed to impact the human population by inducing health issues, compromising food security, increasing the occurrence of extreme weather events, and causing economic losses (Arora et al. 2025). Climate change is defined as a condition in which there are increasing levels of atmospheric CO_2, which is more than 400 ppm; sudden and notable variations in daily, monthly, and interannual temperatures; and shifts in wet and dry seasons causing interplay of heavy precipitation events or protracted drought periods, severe frost, and hot and dry times (Corwin 2021). Climate change is widely accepted to impact food security in the anthropogenic era. The four dimensions of food security—food availability, food access, food utilization, and food system stability—are variably affected by climate change (Bhuyan et al. 2025). Moreover, the quest for food security through agricultural intensification and agricultural land expansion increases GHG emissions from deforestation and land-use changes. Climate change causes alterations in the seasonal patterns of crops, which in turn makes it harder for farmers to predict planting and harvesting times. Moreover, extreme weather events negatively impact food production and increase biotic stress, further reducing crop yield. Plants are sensitive to their environment, and their genetic makeup is adapted to natural variations in weather. However, abrupt and extreme fluctuations in the environment due to climate change can severely retard vegetative and reproductive growth in plants, thereby reducing productivity (Arora et al. 2018). Climate change amplifies environmental stress, generating a feedback cycle that exacerbates the problem. These stresses include extreme fluctuations in temperature, precipitation patterns, and biotic stresses. These extreme events are associated with retardation in plant growth caused by changes at both the morphological and molecular levels. The production of stress ethylene, reactive oxygen species (ROS), osmotic stress, water stress, reduced enzymatic activity related to photosynthesis and respiration, cell toxicity, decreased nutrient mineralization, and phytohormonal activities that promote growth are all products of environmental stress (Verma et al. 2024a). Plants cope with these stresses using their innate immunity, which is an inbuilt complex immune system. The intricate interactions between stress hormones, signaling pathways, and cellular reactions in this system aid plant survival and adaptation to harsh environmental conditions.

Microbes have an intricate relationship and association with plants from the germination stage. Microbes have built-in metabolic and genetic abilities to combat extreme environmental conditions. The 1000 years of relationship between plants and microbial communities has evolved over millennia, and during stressful conditions, microbes trigger various levels of local and systemic responses in plants with cellular, biochemical, and molecular mechanisms to protect them from environmental stress. Microbes surround plant systems or reside within the plant tissues, in the rhizospheric zone, or are free-living in soil, aiding in plant development. The rhizosphere is the interface region between plant roots and soil, and here, the number of microbes is comparatively high due to secretions called root exudates, which attract microbial communities. Plant endophytes are microbes that grow inside healthy plant tissues without causing harm or disease. Plant growth-promoting microorganisms (PGPMs) enhance plant growth through various direct and indirect mechanisms. Direct mechanisms for plant growth promotion include the facilitation of essential micronutrient uptake of phosphorus, potassium, zinc, and calcium in a form that plants can utilize, siderophore production for iron chelation, synthesis of various phytohormones and their regulation, and nitrogen fixation. Indirect mechanisms, on the other hand, include the protection of plants against pathogenic attacks and abiotic stressors. The popular bacterial genera known to promote plant growth and help ameliorate various environmental stresses include *Azotobacter* (Sahoo et al. 2014a, b), *Azospirillum* (Creus et al. 2004; Omar et al. 2009), *Bacillus* (Ashraf et al. 2004; Marulanda et al. 2008; Tiwari et al. 2011; Vardharajula et al. 2011), *Bradyrhizobium* (Figueiredo et al. 1999; Swaine et al. 2007), *Burkholderia* (Ait Barka et al. 2006; Oliveira et al. 2009), *Enterobacter* (Grichko and Glick 2001; Nadeem et al. 2007), *Pantoea* (Amellal et al. 1998; Egamberdiyeva and Höflich 2003; Sorty et al. 2016), *Pseudomonas* (Grichko and Glick 2001; Ali et al. 2009; Sorty et al. 2016), and *Rhizobium* (Alami et al. 2000; Remans et al. 2008; Sorty et al. 2016). In addition to bacteria, fungi are important plant growth promoters and have been extensively reported to have stress tolerance mechanisms in plants. Symbiotic fungi that promote plant growth can be categorized into endophytic and mycorrhizal fungi. Endophytic fungi completely reside in plant tissues, whereas mycorrhizal fungi are only associated with root tissues, with their extensions reaching the rhizosphere. Some of the important genera of endophytic fungi that have been reported to confer tolerance against drought, salinity, and heat stress include *Colletotrichum*, *Fusarium*, *Piriformospora,* and *Trichoderma* (Rodriguez and Redman 2008; Kavroulakis et al. 2018; Santra and Banerjee 2022; Rawal et al. 2022; Li et al. 2023a, b, c, d). The most studied fungal symbionts are AM fungi, which are found in approximately 90% of terrestrial plant species and offer their host plants several advantages. Plants colonized by mycorrhizal fungi are better nourished and more suited to their surroundings.

PGPMs employ an array of defense mechanisms to shield plants from stress, including the synthesis of osmolytes, exopolysaccharides, ROS-scavenging enzymes, ACC deaminase enzymes, phytohormone regulation, solubilization of minerals such as potassium, zinc, and phosphorus, production of siderophores, and upregulation of many genes involved in stress tolerance mechanisms (Bhattacharya et al. 2024). Moreover, PGPMs improve soil quality and aid in climate change mitigation by enhancing carbon sequestration in soil. In addition to abiotic stresses resulting from climate change, PGPMs are also involved in enhancing plant immunity against biotic stresses associated with climate change (Verma et al. 2024a).

Beneficial microorganisms cause a plant's immune system to grow more resilient to future attacks by insects and diseases, a defense process known as "induced systemic resistance (ISR)." This improved resistance results from an increase in the chemical or physical defenses of the plant.

6.2 Climate Change and Its Linkage with Environmental Stresses

Climate change is one of the most pressing global challenges, exacerbating environmental stresses that threaten ecosystems, biodiversity, and human well-being. Rising global temperatures, driven by increased greenhouse gas emissions, have intensified extreme weather events, such as hurricanes, droughts, and heatwaves (IPCC 2023). These changes disrupt water cycles, leading to altered precipitation patterns that cause both severe floods and prolonged droughts, stressing freshwater resources (UNEP 2022). Additionally, ocean warming and acidification resulting from higher CO_2 levels are degrading marine ecosystems, particularly coral reefs, which face bleaching and collapse (Hoegh-Guldberg et al. 2023). Terrestrial ecosystems are equally affected, with shifting climate zones forcing species to migrate or face extinction, thereby reducing biodiversity (WWF 2024). Climate change also interacts with land degradation, as deforestation and soil erosion reduce the carbon sequestration capacity, further amplifying global warming (FAO 2022). Air pollution worsens as higher temperatures increase ground-level ozone formation, thereby exacerbating respiratory diseases (WHO 2023). Furthermore, climate-induced stress on agriculture, such as erratic rainfall and pest outbreaks, threatens food security, particularly in vulnerable regions (FAO 2015). The compounding effects of these environmental stresses create feedback loops; for instance, melting permafrost releases stored methane, accelerating climate change (NASA 2023). Addressing these interconnected challenges requires integrated mitigation and adaptation strategies, including sustainable land use, renewable energy adoption, and the development of resilient infrastructure (IPCC 2023). Without urgent global action, the synergistic impacts of climate change and environmental degradation will intensify, posing irreversible risks to both planetary health and human societies. Many computer models also project that a further increase in the average global earth temperature will result in more frequent and intense episodes of abiotic stress (www.ipcc.ch/, Zandalinas et al. 2021).

6.2.1 Abiotic Stresses

Climate change intensifies various abiotic stresses, severely impacting ecosystems and agricultural productivity. Global warming and irregular precipitation patterns are responsible for a substantial increase in the occurrence of abiotic stresses, including temperature extremes, droughts, floods, and salinity (Tewari and Arora

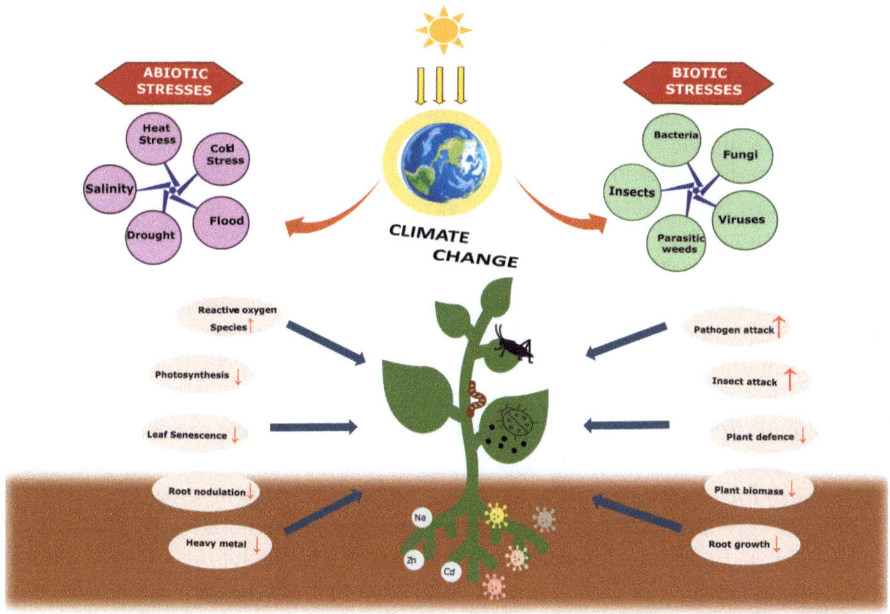

Fig. 6.1 Impact of different climate scenarios on crops

2013; IPCC 2023) (Fig. 6.1). Heatwaves exacerbate thermal stress, impairing crop yields and disrupting physiological processes in both flora and fauna (FAO 2023). Salinization of soils, worsened by rising sea levels and irrigation mismanagement, further degrades the arable land. Additionally, extreme weather events, such as floods and storms, cause soil erosion and nutrient leaching, thereby diminishing agricultural resilience. Elevated CO_2 levels, while sometimes enhancing plant growth, can reduce the nutritional quality of crops, affecting food security (Raza et al. 2024). These interconnected abiotic stresses threaten biodiversity, ecosystem stability, and global food systems, necessitating adaptive strategies such as drought-resistant crops, sustainable water management, and climate-smart agriculture (IPCC 2023). Addressing these challenges is critical for mitigating the cascading impacts of climate change on environmental and human health. Climate change and human activity have caused the degradation of approximately a quarter of the world's land area, as mentioned in the reports of the Global Assessment of Land Degradation and Improvement (GLADA). Plants are sensitive to environmental fluctuations, and the availability of optimum water and temperature is crucial for healthy plant productivity. Extreme temperature events, even for a short duration, impact various growth stages of plants, from germination to senescence. The optimum temperature for the majority of plant species is in the range of 25–30 °C (Hatfield and Prueger 2015). The IPCC (2014) has predicted that the mean temperature of the Earth will rise by 1.2–3.7 °C in the coming decades.

The most detrimental environmental stressor is drought, which significantly hampers crop yields (Chaudhry and Sidhu 2022). Global drought patterns are altered by multiple interconnected mechanisms. Recent studies have demonstrated that rising temperatures increase atmospheric evaporative demand, with each 1 °C increase in temperature estimated to intensify agricultural drought by approximately 9% in vulnerable regions (Ukkola et al. 2024). The 2020–2023 megadrought in southwestern North America, the most severe in 1200 years according to tree-ring reconstructions, has been directly attributed to anthropogenic climate change (Williams et al. 2022). Changing precipitation regimes are creating new drought hotspots, as seen in the recent unprecedented drying of the Amazon Basin (Boulton et al. 2023), whereas traditional drought-prone areas, such as the Horn of Africa, face worsening conditions, with consecutive failed rainy seasons (2020–2022) linked to Indian Ocean warming (Otto 2023). The impacts cascade across systems: agricultural droughts now develop five to ten times faster than in the twentieth century (Christian et al. 2023), whereas ecological droughts trigger mass forest mortality events globally (Hammond et al. 2024). The 2022 European drought caused €20 billion in agricultural losses and reduced hydropower generation by 20% (Rossi et al. 2023). Emerging research highlights feedback, including drought-induced reductions in terrestrial carbon uptake equivalent to 1–2% of annual anthropogenic emissions (Liu et al. 2023). These findings underscore the urgent need for improved early warning systems and adaptive water management strategies in a warming climate.

Floods have been occurring more frequently since the 1950s in most continents around the globe, affecting approximately 23% of the world's population (Rentschler et al. 2022). The frequency of severe events, particularly flood-related ones, has increased many folds over the past 10 years. Floods cause nutrient-rich topsoil to wash away, resulting in low-productive lands for years to come.

Salinity is one of the main factors limiting agricultural productivity worldwide. Soil with electrical conductivity (EC) greater than 4 dS m^{-1} is considered saline. Soil salinity is present in some areas around the world, with high evaporation rates, low water availability, and poor water drainage systems that tend to accumulate more salt ions. Naturally, some areas near oceans and seas also face salinity problems. However, anthropogenic factors such as improper irrigation practices, saltwater intrusion, and global warming have triggered the occurrence of soil salinization worldwide. Currently, 800 million hectares of land are affected by soil salinization, and it has been predicted that approximately 50% of the planet's land will experience varying degrees of salt stress over the next 35 years (Arora et al. 2020). The majority of plant species cannot tolerate salt concentrations beyond 200 mM (Munns and Tester 2008). Corwin (2021) explained how erratic rainfall events with too much or too less water availability may lead to the buildup of salt in the root zone. High rainfall causes the water table to rise, which causes salt ions to come to the surface by capillary action. During the dry season or drought-like conditions, inadequate rainfall does not leach salts from the root zone, resulting in salt patches on the soil surface. Climate change significantly affects soil dynamics and composition, directly impacting plant productivity. Climate change, through rising sea levels and altered precipitation patterns, intensifies the soil salinity (Verma et al. 2024b). Sea-level rise causes saltwater intrusion into coastal farmland. Increased evaporation and reduced rainfall concentrate salts in the soil, a process known as secondary salinization. This degrades soil health, reduces crop yields, and threatens global food security, particularly in vulnerable arid and coastal regions (Arora and Mishra 2024; FAO 2023).

6.2.2 *Biotic Stresses*

Climate change is a significant driver altering the dynamics of biotic stress in plants, profoundly impacting agricultural and natural ecosystems. Biotic stress, caused by organisms such as insects, pathogens, and weeds, is intensifying owing to direct and indirect climate-induced changes (Skendžić et al. 2021) (Fig. 6.1). Rising temperatures and altered precipitation patterns directly enhance the development, survival, and geographic distribution of many pests and pathogens. Warmer winters reduce low temperature mortality, allowing pest populations to expand into new regions and at higher altitudes (Pureswaran et al. 2018). Furthermore, increased temperatures often accelerate insect metabolic rates and reproduction, leading to more generations per year and larger outbreaks (Deutsch et al. 2018). Climate change indirectly affects plant-pest interactions by altering host plant physiology. Elevated CO_2 levels can reduce nitrogen concentrations in plant tissues, lowering nutritional quality and compelling herbivores to consume more biomass to compensate, thereby increasing damage (Zavala et al. 2017). Plants subjected to abiotic stresses, such as drought or heat, often experience weakened defensive capabilities, becoming more susceptible to infection and infestation (Fathi and Barakat 2022). These changes disrupt the ecological balance. Phenological mismatches may occur when the life cycles of plants, pests, and their natural enemies (e.g., predators or parasitoids) become desynchronized. This can reduce the effectiveness of biological control, further exacerbating pest pressure. The combined result is increased crop losses, greater economic costs for management, and heightened threats to global food security (Thakur et al. 2023). Understanding these complex linkages is essential for developing resilient agricultural systems and effective adaptation strategies.

Warmer temperatures can accelerate pathogen life cycles, thereby increasing the number of infection cycles per growing season. Furthermore, milder winters expand the overwintering range of many pathogens, allowing them to establish themselves in previously uninhabitable regions (Bebber et al. 2013). Altered rainfall and humidity are particularly critical; drought stress can weaken plant defense, whereas increased humidity and leaf wetness duration facilitate spore germination and penetration by fungal and bacterial pathogens (Velásquez et al. 2018). Elevated CO_2 levels exert complex effects in plants and their different developmental processes. While potentially enhancing plant growth, they can also lead to the dilution of defensive compounds or alter leaf canopy structure, creating a more favorable microclimate for disease development. Conversely, some studies have indicated that elevated CO_2 may induce systemic resistance in certain plant species, highlighting the context-dependent nature of these interactions (Zhang et al. 2015). Climate change can disrupt the evolutionary arms race between hosts and pathogens. Rapidly changing environmental conditions may select for new, more virulent pathogen strains or render existing plant resistance (R) genes ineffective, a phenomenon known as resistance breakdown (Fones et al. 2020). This necessitates a shift towards developing climate-resilient crops and implementing adaptive disease management strategies that account for these evolving interactions.

The complex physiological and biochemical defenses that plants use to fend off biotic stressors are jeopardized by climate change. Abiotic stressors, such as drought, heat, and climate change-induced nutrient imbalance, force plants to devote their limited energy and metabolic resources to survival and repair, frequently at the expense of creating costly defense compounds (Fathi and Barakat 2022). As a result of this resource trade-off, they become more vulnerable to diseases and pests. A higher carbon-to-nitrogen ratio in plant tissues is frequently the result of increased CO_2 levels, which further complicates the dynamics by promoting photosynthesis and growth. Although this nitrogen dilution lowers the nutritional value for herbivores, it can also lower the concentration of nitrogen-based defense compounds, such as cyanogenic glycosides and alkaloids, which may change plant resistance (Zavala et al. 2017). Moreover, heat stress can hinder the activation of induced resistance pathways by suppressing the synthesis of important defensive hormones, such as salicylic acid and jasmonic acid (Berens et al. 2019). Therefore, climate change can be a silent disruptor of plant immunity, making plants more palatable and susceptible to attack before a pest even shows up.

The ecological dynamics of agroecosystems are drastically altered by climate change because it has a direct impact on important biotic stressors, such as weed competition and insect herbivore behavior. Temperature increases the metabolism and development of insects, which frequently results in more feeding and more generations per season (voltinism), increasing crop damage (Deutsch et al. 2018). The efficacy of biological control can be diminished by altered climatic conditions that disrupt the balance between insect pests and their natural enemies (Skendžić et al. 2021). Moreover, high CO_2 levels can alter the volatile organic compound (VOC) profiles of plants, which may interfere with the way insects seek hosts and attract pollinators and other beneficial insects (Dyer et al. 2023).

Weeds, which are typically C3 plants, may have a greater competitive advantage for light, water, and nutrients than major C4 crops such as maize because they frequently show a stronger growth response to increased atmospheric CO_2 (Ziska 2022). Additionally, invasive weed species can spread poleward into new areas owing to climate change. Increases in temperature can shorten the life cycles of weeds and speed up their growth, and changes in rainfall patterns can impact the effectiveness of pre-emergent herbicides, further tipping the scales in favor of weeds (Ramesh et al. 2017). There is a synergistic interaction between these stressors. For example, weeds can change the microclimate in a crop canopy, making it more humid and ideal for certain pathogens and insect pests. Therefore, to guarantee future crop productivity and food security, integrated pest and weed management strategies must be modified to account for these climate-driven changes.

6.3 Effects of Climate Change-Based Stresses on Plants

Climate change has led to an increase in various environmental stresses that significantly impact plant growth, development, and productivity (Fig. 6.2). Key climate-based stresses include drought, extreme temperatures, salinity, and elevated CO_2 levels, each of which affects plants at the physiological, biochemical, and molecular levels. Elevated CO_2 initially boosts photosynthesis but often reduces nutritional quality in crops (Smith et al. 2024). Heat stress disrupts photosynthesis and increases oxidative damage, whereas drought limits water uptake, triggering stomatal closure and growth inhibition (IPCC 2023). Flooding induces hypoxia, activates fermentative metabolism, and reduces root function (Bailey-Serres et al. 2012). Shifting climates also alter plant-pollinator interactions and species distributions, threatening biodiversity (Martén-Rodríguez et al. 2025). These stresses typically trigger a suite of phenotypic adaptations aimed at survival, often at the expense of crop yield. Drought and salinity stress commonly result in reduced biomass, stunted growth, shorter root and shoot systems, and smaller, thicker leaves with a higher root-to-shoot ratio to conserve water (Arora et al. 2020; Ghadirnezhad Shiade et al. 2023). Elevated temperatures can cause stem elongation but reduce the leaf area and disrupt reproductive development, leading to flower abortion and poor fruit set (Tandel et al. 2025). These morphological changes are direct consequences of physiological dysfunctions, including osmotic imbalance, oxidative stress, and impaired photosynthetic efficiency (Dastogeer et al. 2020). However, the inoculation of plants with beneficial soil microorganisms, such as arbuscular mycorrhizal fungi (AMF) and plant growth-promoting rhizobacteria (PGPR), can significantly mitigate these negative effects (Bahrami-Rad and Hajiboland 2024). These microbes enhance root architecture, increasing the exploratory capacity for water and nutrients, improving

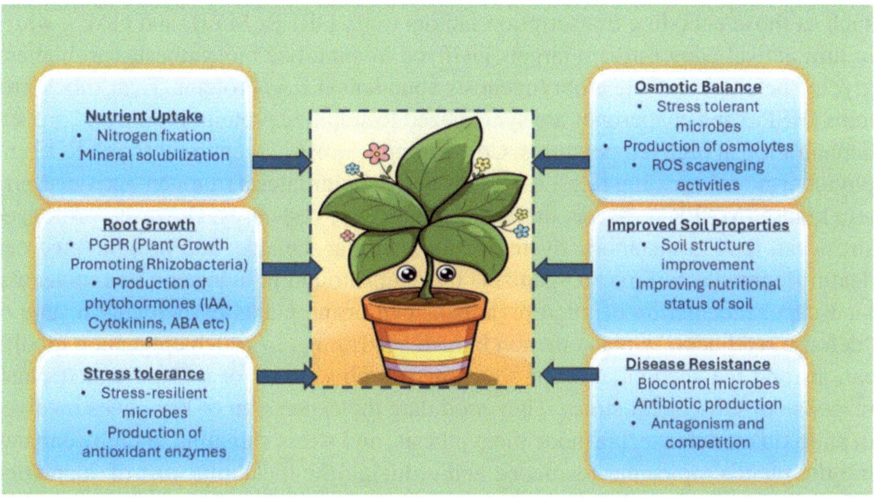

Fig. 6.2 Mechanisms of beneficial soil microorganisms for managing stress in plants

plant water status through osmotic adjustment and altered phytohormone profiles, and bolstering overall biomass accumulation under stress (Table 6.1). By essentially "rewiring" plant metabolic and developmental pathways, these microbial symbionts help maintain a more optimal morphology, thereby enhancing resilience and supporting sustainable yields in a changing climate (Backer et al. 2018).

Climate change-induced abiotic stress disrupts core physiological processes in plants, severely threatening crop performance. Drought, salinity, and heat stress induce osmotic imbalance, leading to stomatal closure, which reduces carbon dioxide influx and consequently inhibits photosynthesis. Photoinhibition is often exacerbated by the overproduction of reactive oxygen species (ROS), which cause oxidative damage to chloroplasts, lipids, proteins, and DNA. Furthermore, elevated temperatures can denature enzymes involved in photosynthesis and respiration, thereby uncoupling metabolic pathways. Salinity stress imposes dual osmotic and ionic toxicity, primarily from Na^+ and Cl^- ions, which disrupts enzyme function and nutrient homeostasis. These physiological impairments manifest as reduced photosynthetic pigment content, compromised water-use efficiency, and diminished nutrient assimilation, ultimately stunting growth and yield. Beneficial soil microorganisms, such as arbuscular mycorrhizal fungi (AMF) and PGPR, are pivotal mitigators of this physiological decline. They enhance photosynthetic efficiency by improving water and nutrient uptake (e.g., P and Zn), upregulating antioxidant defense systems to scavenge ROS, and facilitating osmotic adjustment through the accumulation of compatible solutes (Ilangumaran et al. 2024; Ma et al. 2024). By priming the physiological responses of plants, these microbes are integral to maintaining metabolic function and ensuring agricultural sustainability under a changing climate.

At the molecular level, climate change stressors trigger complex reprogramming of gene expression, metabolic pathways, and protein activity as plants attempt to acclimate. Key alterations include the rapid upregulation of stress-responsive genes, such as those encoding transcription factors (e.g., DREB, MYB, and NAC), which in turn activate downstream targets involved in osmolyte biosynthesis (proline and glycine betaine) and late embryogenesis abundant (LEA) proteins. Heat shock proteins (HSPs) and chaperones are synthesized to stabilize protein folding and prevent denaturation at high temperatures. Concurrently, stress perception through ROS and phytohormone signaling (e.g., ABA, JA, and SA) initiates a mitogen-activated protein kinase (MAPK) cascade, orchestrating defense responses. However, prolonged stress can lead to metabolic dysregulation, cellular damage, and programmed cell death. Beneficial soil microorganisms have a profound influence on this molecular landscape. Inoculation of microorganisms can prime plants by inducing a state of "defense readiness," a phenomenon known as priming, in which epigenetic modifications and the accumulation of signaling proteins allow for a faster and stronger response to subsequent stress. They modulate the expression of key genes involved in antioxidant defense, hormone biosynthesis, and stress signaling, thereby enhancing the plant's molecular resilience and reducing the metabolic cost of adaptation (Chandrasekaran et al. 2024; Javaid et al. 2025). This molecular priming is a crucial mechanism by which microbes contribute to sustainable crop production.

Table 6.1 Mechanisms of action of beneficial soil microorganisms in abiotic stress alleviation and climate change mitigation

S. no.	Soil microbes	Targeted stress	Targeted plants	Mechanism of action	Role in climate change	References
1.	*Pseudomonas atacamensis* KSS-6	Soil salinity	Rice	Increment in growth, chlorophyll content, nutrient content, and antioxidant content	Bioinoculant mitigates negative impacts of salt stress in soil and enhances the yield and nutritional properties in crop	Arora et al. (2024)
2.	*Pseudomonas taiwanensis* PWR-1	Soil salinity	Wheat	Production of osmoprotectants (proline and glycine betaine) and antioxidants	Bioremediation of saline soil with the help of ST-PGPR strain	Mishra et al. (2025b)
3.	*Pseudomonas hunanensis* SPT26	Soil salinity	Tomato	Antioxidant activity and ROS scavenging activity	Enhancing crop resilience and yield stability	Verma et al. (2024a)
4.	*Pseudomonas argentinensis* and *Pseudomonas azotoformans*	Soil salinity	Mustard	Production of aminolevulinic acid, indole acetic acid, 1-aminocyclopropane-1-carboxylate deaminase, and solubilization of potassium and phosphorus	Salt tolerance and crop productivity	Phour and Sindhu (2020)
5.	*Bacillus pumilus* JPVS11	Soil salinity	Rice	Promoting photosynthetic pigments, proline, and antioxidant production	Improved microbial count, soil enzyme activity, and soil properties	Kumar et al. (2021)
6.	*Bacillus flexus*	Soil salinity	Mustard	Antioxidant enzymes such as catalase (CAT), peroxidase (POX), and ascorbate peroxidase (APX) activities	Improved mustard growth in challenged soils	Singh and Prasad (2025)
7.	*Paenibacillus* sp. SPR11 and *Bradyrhizobium yuanmingense* PR3	Soil salinity	Black gram	Improved membrane stability, elevated uptake of nitrogen, phosphorus, potassium, and key micronutrients (Zn, Fe)	Enhancing legume productivity and resilience in salt-affected soils	Tiwari et al. (2025)
8.	*Bradyrhizobium japonicum*	Soil salinity	Soybean	Enhancing antioxidant enzyme activities to mitigate oxidative damage and soil microbial activities	Enhancing plant tolerance and soil health under salt stress conditions	Sarioğlu (2025)
9.	*B. japonicum*	Drought	Soybean	Enhancing plant growth and N_2 fixation	Mitigating the adverse effects of soil drought on yield	Szczerba et al. (2025)

(continued)

Table 6.1 (continued)

10.	*Burkholderia seminalis* 869T2	Soil salinity	Bananas, *Arabidopsis*, and several leafy vegetables	Phytohormones signaling and lowering of accumulation levels of hydrogen peroxide (H_2O_2), electrolyte leakage (EL), and malondialdehyde (MDA)	Increased salt and drought stress tolerances of host plants	Hwang et al. (2025)
11.	*P. azotoformans* JRBHU5 and *Burkholderia seminalis* JRBHU6	Drought	Wheat	EPS production and strengthening of proline and lignin contents and antioxidant defense system	Inducing osmotic tolerance	Prasad et al. (2025)
12.	*Pseudomonas palleroniana* GZNU148	Drought	*Themeda japonica*	Decreased MDA and H_2O_2 content, SOD and CAT activities, and regulated 1158 differentially expressed genes (DEGs)	Recruiting beneficial rhizosphere bacteria and increasing secondary metabolite content in rhizosphere soil	Chen et al. (2025)
13.	*Pseudomonas fluorescens* and *Pseudomonas aeruginosa*	Drought	Maize	Enhancing osmotic adjustment and antioxidant enzyme activity	Enhancing maize drought tolerance	Khaledi et al. (2025)
14.	*Bacillus subtilis*	Drought	Cotton	Activities of urease, sucrase, and alkaline phosphatase increased, and malondialdehyde content decreased	Alleviating drought stress and increasing yield in arid region	Ren et al. (2025)
15.	*Azospirillum brasilense*	Drought	Wheat	Improved plant biomass, photosynthetic rate, stomatal conductance, and soil nutrient uptake	Enhancing physiological resilience, nutrient uptake, and soil health	Ikram et al. (2025)
16.	*Enterobacter* sp. J49	Drought	Peanut	Phosphate solubilization and osmotic adjustments	Enhancing peanut resilience and growth under combined drought and phosphorus deficiency stresses, supporting sustainable agriculture in challenging environments	Lopez et al. (2025)
17.	*Enterobacter* sp. SA187	Heat stress	Tomato	Increase in ion homeostasis stress-related genes and higher expression of thermotolerance proteins	Biostimulant for improving crop tolerance and productivity	Rahman et al. (2025)

18.	*Pseudomonas putida*	Heat stress and intense light stress	Citrus	Reduced malondialdehyde content and enhanced catalase activity, improving ROS scavenging and stress mitigation	Improved photosynthesis and PSII efficiency could be a sustainable strategy for climate change adaptation	Fátima et al. (2025)
19.	*Staphylococcus epidermidis* CK9 and *Bacillus australimaris* CK11	Salinity, heat, and drought stress	Tomato	Restored normal TBARS levels, POD activity, and O_2^- levels and increased the expression of *HSP90*, *HSP70*, *CAT*, and *GR* genes	Ameliorating combined abiotic stress resistance in plants	Jan et al. (2024)
20.	*Bacillus licheniformis* BE-L60	Heat stress	Spinach	Secretes indole-3-acetic acid, dissolves phosphorus, and produces amylases and proteases	Heat tolerance by modulating plants	Li et al. (2023a, b, c, d)
21.	*Bacillus* spp. AH-08, AH-67, SH-16 and *Pseudomonas* spp. SH-29	Heat stress	Maize	Upregulation of heat shock proteins	Cope with the heat stress due to climate change and induce plant-heat tolerance	Ahmad et al. (2023)
22.	*Sphingomonas* spp.	Heavy metal stress	Rice	Arsenic redox transformation and plant growth promotion	Microbe-assisted remediation of the arsenic and other heavy metal-contaminated soil	Sultana et al. (2023)
23.	*P. putida* KNP9	Heavy metal stress	Soybean	Bioaccumulation of Pb and Cd	Promoting soybean growth, which could be used for biodiesel production	Tripathi et al. (2023)
24.	*Bipolaris maydis* AF7	Heavy metal stress	Rice	Heavy metal tolerance and higher expression of the *OsLSi6*, *PHT1*, and *OsHKT1* genes	Nutrient availability to the crops could be improved, and heavy metal toxicity and environmental hazards could be minimized	Adhikari et al. (2024)

(continued)

Table 6.1 (continued)

25.	*Metarhizium anisopliae* (MetA1)	Salinity	Rice	Antioxidant enzymes catalase, glutathione S-transferase, ascorbate peroxidase, and peroxidase, as well as the nonenzymatic antioxidants phenol and flavonoids, were significantly enhanced	Potential salt stress mitigation strategy	Chowdhury et al. (2024)
26.	*Rhizoglomus irregulare*	Salinity	Wheat	Enhancing plant photosynthesis ability and upregulating the antioxidant metabolism and osmolyte accumulation	Mediate salt tolerance and enhance wheat production in degraded areas	Ikan et al. (2023)
27.	*Bacillus cereus* JYZ-SD2 and *Peniophora cinerea* XC	Salinity	Willow	Expression of *vp2* gene	Improving the salt tolerance and growth of willow plants	Wu et al. (2023)
28.	*Rhizophagus clarus* and PGPB	Salinity	Maize	Increased plant biomass through osmoregulation and protection of the photosynthetic apparatus	AMF-PGPB consortium can minimize damages caused by saline stress	de Carvalho Neta et al. (2024)
29.	*Talaromyces* sp. GS1, *Pseudeurotium* sp. GRs12, *Penicillium* sp. *GR19*, and *Trichoderma* sp. GR21	Drought	Mulberry	Increased production of catalase, soluble sugar, and chlorophyll	Productivity enhancement in agriculture under stressed conditions	Ou et al. (2023)
30.	*Aspergillus oryzae* and *Aspergillus fumigatus*	Drought	Maize	Expression of the *ZmBSK1*, *ZmAPX*, and *ZmCAT1* genes and phytohormonal signaling pathway leading to the antioxidant upregulation	Physio-hormonal attributes and antioxidant potential to alleviate drought stress	Niaz et al. (2024)

Drought stress is one of the most critical abiotic constraints on plant growth and productivity, particularly under climate change scenarios. Recent studies have elucidated sophisticated plant responses to water deficits, spanning from immediate physiological adjustments to long-term molecular adaptations (Takahashi et al. 2020). The primary effect of drought is reduced turgor pressure, which triggers rapid stomatal closure mediated by abscisic acid (ABA) signaling through guard cell receptors (PYR/PYL/RCAR) and downstream effectors (SnRK2 kinases) (Chen et al. 2022). This response, while conserving water, simultaneously limits CO_2 uptake, leading to photoinhibition and reactive oxygen species (ROS) accumulation in chloroplasts (Zandalinas et al. 2021).

At the cellular level, drought induces profound metabolic reprogramming. Advanced omics technologies have revealed increased synthesis of osmoprotectants, such as proline (via P5CS enzymes) and trehalose (through TPS/TPP pathways), which stabilize membranes and proteins (Fàbregas and Fernie 2021). Plants also activate a sophisticated antioxidant system involving superoxide dismutase (SOD), ascorbate peroxidase (APX), and glutathione reductase (GR), with recent studies identifying thioredoxin systems as crucial ROS scavengers (Choudhary et al. 2023). Molecular responses involve extensive transcriptional reprogramming through drought-responsive elements (DREBs, NAC, and WRKY TFs) and epigenetic modifications. Cutting-edge research demonstrates that histone variants (e.g., H2A.Z) and DNA methylation patterns dynamically regulate stress memory (Kim et al. 2022). Small RNAs (miR169 and miR398) have been identified as key post-transcriptional regulators of drought responses (Li et al. 2023a).

Morphological adaptations include root architecture remodeling mediated by auxin redistribution, with newly discovered peptide signals (CLE25) coordinating root-to-shoot drought responses (Takahashi et al. 2020). Remarkably, the plant microbiome contributes significantly to drought tolerance, and specific rhizobacteria enhance water-use efficiency through ACC deaminase activity and exopolysaccharide production (Niu et al. 2023). Emerging biotechnological approaches leverage CRISPR-Cas9 to edit drought-responsive genes (e.g., *OsPYL9* in rice) and engineer synthetic promoters for precise modulation of stress responses (Waqas et al. 2023).

Salinity stress affects plants through complex physiological and molecular mechanisms, which have been increasingly elucidated in recent studies. The primary effects occur through osmotic stress, ionic toxicity, and oxidative damage (Van Zelm et al. 2020). Modern research has revealed that salt stress immediately triggers osmotic stress perception through root-based mechanosensitive channels (MSLs and OSCAs), activating calcium signaling within seconds (Jiang et al. 2019). This rapid signaling cascade involves the Salt Overly Sensitive (SOS) pathway, where SOS3–SOS2 complexes phosphorylate SOS1, a plasma membrane Na^+/H^+ antiporter crucial for ion homeostasis (Zhao et al. 2021). Recent transcriptomic studies have shown that salinity induces massive reprogramming of gene expression, particularly through ABA-dependent (SnRK2s-mediated) and ABA-independent (MAPK cascade) signaling pathways (Julkowska and Testerink 2015). Novel regulatory networks involving microRNAs (miR393 and miR169) and long noncoding RNAs have been identified as key modulators of salt stress responses (Yuan et al. 2021). At the cellular level, advanced imaging techniques have demonstrated that plants compartmentalize Na^+ into vacuoles via NHX exchangers while maintaining cytosolic K^+ levels through AKT1 channels (Wu et al. 2020). Cutting-edge research highlights the crucial role of root microbiome interactions in salinity tolerance, with specific bacterial strains enhancing plant growth under stress through ACC deaminase activity (Etesami and Beattie 2018). Furthermore, recent metabolic studies have revealed that polyamine biosynthesis and GABA shunt pathways are critical for salt stress mitigation (Alcázar et al. 2020). Emerging technologies, such as CRISPR-Cas9, are being employed to engineer salt-tolerant crops by targeting key regulators, such as HKT transporters (Zhang et al. 2022). These advancements provide new avenues for developing climate-resilient crops through precision breeding and microbiome manipulation.

Heat stress has emerged as a critical constraint on plant productivity, with rising global temperatures causing significant yield losses in major crops. When exposed to supra-optimal temperatures, plants experience immediate physiological disruptions, including alterations in membrane fluidity, protein denaturation, and damage to the photosynthetic apparatus (Zhao et al. 2023). Recent studies using thermal imaging have revealed that heat stress triggers rapid (within minutes) stomatal opening as a cooling mechanism, paradoxically leading to water loss while reducing the leaf temperature (Chavan et al. 2022). This response is mediated by intricate calcium signaling cascades involving cyclic nucleotide-gated channels (CNGCs) and calmodulin-binding transcription activators (CAMTAs) (Wu et al. 2023). At the molecular level, plants activate sophisticated heat shock responses through the conserved HSF-HSP pathway. Cutting-edge research has identified 21 heat shock transcription factors (HSFs) in *Arabidopsis* and 25 in rice, with HSFA1s serving as master regulators that orchestrate the expression of heat shock proteins (HSP70, HSP90, and sHSPs) (Wang et al. 2023). These molecular chaperones prevent protein aggregation and facilitate the refolding of denatured proteins. Novel findings have demonstrated that small HSPs form phase-separated condensates to protect client proteins during extreme heat (Li et al. 2023b).

Metabolically, plants undergo significant reprogramming during heat stress. Advanced metabolomic studies have revealed the accumulation of thermoprotective metabolites, including branched-chain amino acids, polyamines, and anthocyanins (Qu et al. 2023). The GABA shunt pathway has been identified as critical for maintaining TCA cycle flux and reducing ROS production under heat stress (Wang et al. 2025). Plants also activate antioxidant systems, and recent studies have highlighted the role of alternative oxidase (AOX) in preventing mitochondrial ROS overproduction (Vanlerberghe et al. 2002). Epigenetic mechanisms contribute significantly to the response to heat stress. Chromatin remodellers (SWI/SNF) and histone variants (H2A. Z) regulate accessibility of heat-responsive genes (Liu et al. 2023). Novel research has demonstrated that m6A RNA methylation modulates the mRNA stability of key heat tolerance genes (Zhou et al. 2025). The plant microbiome also confers thermotolerance, with specific endophytes producing ACC deaminase and volatile organic compounds that enhance heat resistance (Cui et al. 2024). Emerging approaches include CRISPR-based editing of HSFs and the engineering of thermostable Rubisco activase (Cavanagh et al. 2023). High-throughput phenotyping platforms combined with machine learning have accelerated the development of heat-resilient crop varieties.

Waterlogging conditions severely disrupt plant growth and survival by limiting oxygen availability (hypoxia) and altering soil redox potential. Recent studies have revealed that oxygen deprivation triggers rapid metabolic shifts, suppressing aerobic respiration and activating fermentative pathways (Bailey-Serres et al. 2025). Plants respond through specialized mechanisms, including the formation of aerenchyma tissue, a process mediated by ethylene-induced programmed cell death (PCD) and regulated by transcription factors such as *HRA1* (Striker et al. 2025). Hypoxia activates the *ERF-VII* transcription factor family (e.g., *RAP2.12*), which stabilizes anaerobic genes under low-oxygen conditions (Gibbs et al. 2015). The *N-end rule pathway* degrades these factors upon reoxygenation, ensuring dynamic stress adaptation (Hartman et al. 2019). Additionally, plants enhance glycolysis and ethanolic fermentation (via *PDC* and *ADH*) to sustain ATP production. This stress condition also induces ROS accumulation, triggering antioxidant defense systems, including SOD and peroxidase (POD) activation (Lin et al. 2025). Recent findings have highlighted the role of *RBOHD*-mediated ROS signaling in promoting adventitious root formation (Shukla et al. 2024). Submergence-tolerant rice varieties (e.g., *SUB1A*-expressing lines) exhibit restrained elongation to conserve energy, whereas deep-water rice employs *SNORKEL* genes (e.g., *SK1/SK2*) for rapid internode elongation (Kuroha and Ashikari 2020).

Additionally, climate stress alters plant-microbe interactions, affecting symbiotic nitrogen fixation and mycorrhizal associations, further influencing nutrient uptake (Compant et al. 2010). Understanding these mechanisms is crucial for developing stress-resilient crops through genetic engineering and breeding strategies to sustain agricultural productivity under changing climatic conditions.

6.4 Beneficial Soil Microorganisms for Managing Stresses in Plants

Microbial interactions with plants modulate molecular mechanisms to activate metabolic networks at the gene, enzyme, and metabolite levels. This works in parallel to enhance the plant's intrinsic strength to support stress mitigation. Bioformulations developed from PGPMs, especially PGPR, plant growth-promoting endophytic bacteria (PGPEB), AMF, and ectomycorrhizal fungi (ECM), are gaining popularity as more sustainable and environmentally sound options for the management of stressed agroecosystems (Dodd and Rez-Alfocea 2012; Tewari and Arora 2016; Etesami and Beattie 2018). Bioformulations can be defined as products containing microbes or their metabolites, which can be used to provide nutrients or bioregulators or combat phytopathogens to promote plant growth (Arora and Mishra 2016).

Most microbes present in the environment do not tolerate environmental extremes. A very thin population of high-tolerance microbes exists, which can not only thrive under stress but also enhance plant growth; thus, these natural entities need to be identified and used as bioinoculants in stressed agro-ecosystems. In general, the mechanism of action of PGPMs can be categorized into direct and indirect mechanisms. The direct mechanism that aids plant growth involves the synthesis and regulation of phytohormones, nitrogen fixation, mineral solubilization, siderophore and organic acid production, and the production of enzymes such as ACC deaminase, which can lower plant ethylene levels (Verma et al. 2021, Khare et al. 2018). Indirect mechanisms of plant growth promotion include the production of secondary metabolites, such as antibiotics, hydrogen cyanide (HCN), enzymes, and antioxidants that degrade the cell wall, inhibit pathogen quorum sensing, induce systemic resistance, and sequester iron by bacterial siderophores to inhibit pathogens (Whipps 2001; Ali et al. 2014).

6.4.1 Salinity

PGPMs, including rhizobacteria, endophytic bacteria, and AMF, enhance plant growth under salinity stress in numerous ways, including the regulation and production of plant hormones, solubilizing minerals such as phosphate (P), zinc (Zn), potassium (K), calcium (Ca), and magnesium (Mg), and fixing nitrogen (N). The production of beneficial antioxidant enzymes, such as superoxide dismutase (SOD), catalase (CAT), peroxidase (POX), and enzymes associated with the ascorbate (ASC)—glutathione cycle, such as monodehydroascorbate dehydrogenase (MDHAR), ASC peroxidase (APX), dehydroascorbate reductase (DHAR), and glutathione reductase (GR), as well as nonenzymatic antioxidants such as ascorbic acid (vitamin C), glutathione (GSH), and proline (Pro), plays a significant role in reducing oxidative stress induced by abiotic stress (Mohanty et al. 2021; Orozco-Mosqueda et al. 2023; Acharya et al. 2024). Under salinity stress, the soil becomes nutrient deficient, or the nutrients are present in a form that is unavailable to plants for uptake. PGPMs help plants in nutrient uptake and their utilization. Under saline conditions, excess sodium and chloride ions hinder the uptake of essential micro- and macronutrients, such as N, K, Zn, P, Ca, and Mg, which cause early senescence in plant tissues, restricting their growth. PGPR such as *Pseudomonas*, *Bacillus*, *Burkholderia*, *Achromobacter*, *Azospirillum*, and *Azotobacter* have been reported as excellent mineral solubilizers and growth promoters even under salinity-stressed ecosystems (Upadhyay and Singh 2015; Abdel Latef et al. 2020; Mishra et al. 2022). Under saline stress, plant immune responses help them produce certain osmolytes to fight against salt ions and maintain cell electrolyte equilibrium. These osmolytes include mannitol, inositol, trehalose, glycine, betaine, and proline. PGPMs have been reported to aid osmolyte production under salinity stress. Beneficial microorganisms also protect plants from salt stress by maintaining ionic equilibrium and reducing the buildup of toxic Na and Cl ions. Under such conditions, the potassium ions decrease in the cytosol. Certain beneficial microbes have the ability to enhance K^+/Na^+ ratios. The SOS signaling pathway, high-affinity K^+ transporter 1 (HKT1), and Na^+/H^+ antiporters are crucial for lowering intracellular Na^+ (Kumari et al. 2021). Under salinity stress, biofilms of exopolysaccharides (EPS) produced by microbes around the root system act as a protective shield against salt ions and protect plant roots against drought-like conditions produced by salinity stress (Fátima et al. 2025).

6.4.2 Heat Stress

Heat stress can be defined as an increase in the temperature of the environment from the optimum temperature to 10–15 °C above it or a temperature greater than 75 °F (Shaffique et al. 2022). *Bacillus amyloliquefaciens* UCMB5113, when applied to *Triticum aestivum* and exposed to short-term heat stress, showed HSP expression. Exopolysaccharides produced by microorganisms around the root system or even within plant tissues by endophytes act in the retention of water up to 97% in their mesh-like structure, which helps plants in regulating temperature to a lower side during heat stress. The endophytic bacterium *Enterobacter* sp. SA187 induced thermotolerance in *Arabidopsis thaliana* and wheat plants in a greenhouse experiment using thermopriming (Akhtar et al. 2022). In another experiment, soybean plants inoculated with the PGPR *B. cereus* SA1 under heat stress showed thermoregulation. *Bacillus cereus*, *Pseudomonas* spp., *Serratia liquefaciens*, *P. fluorescens*, and *Pseudomonas putida*, which are hosted in *Solanum lycopersicum* L, *Glycine max, and Triticum* spp., showed thermotolerance by aiding in plant growth-promoting activities, such as phytohormone, ACC deaminase, and antioxidant enzyme production (Kang et al. 2019). Heat stress affects the thylakoid and photosystem structures of plants, reducing the efficiency of plants; however, some specialized PGPMs like cyanobacteria are known to efficiently perform photosynthesis for plants and improve their growth (Hasanuzzaman et al. 2013; Scafaro et al. 2021). Certain microbes have been reported to enhance stomatal closure in plants through the production of abscisic acid (ABA), which is also known as a stress hormone that helps plants combat water loss through transpiration and helps in osmoregulation under heat stress (Iqbal et al. 2021; Li et al. 2022).

6.4.3 Droughts and Floods

It has been established by many omics approaches along with metagenomic analysis that rhizosphere microbial communities can significantly lessen agricultural drought stress by adjusting their structural, functional, and genomic makeup to water constraints (Shaffique et al. 2022; Kour and Yadav 2022). Stressful conditions give rise to oxidative stress by accumulating ROS molecules, which impart deleterious effects on cellular components. Small-molecular-weight metabolites, such as phenolics, tocopherol, phenylpropanoids, and carotenoids, and antioxidative enzymes produced by beneficial microbes help maintain the redox state in cells during oxidative damage during drought stress (Sachdev et al. 2021). In addition to these mechanisms, microbes have been reported to induce PAL gene transcripts, resulting in the activation of the PAL enzyme, which further facilitates the accumulation of phenylpropanoids (Chandrasekaran et al. 2017). Various endophytic microbes associated with plant tissues reduce the negative effects of drought-induced stress by inducing biochemical and molecular changes in plants. *Trichoderma harzianum* TH-56 in a pot trial with rice showed better drought tolerance by upregulating the transcription of certain drought tolerance-related genes, *AQU* for aquaporin and *DHN* for dehydrin, which help in water uptake (Pandey et al. 2016). Dehydrins (DHNs) play a key role in responding to adaptation against abiotic stresses. Microbial inoculation upregulated the expression of *DHN* genes in rice plants grown under stressed and non-stressed conditions (Singh et al. 2020). Dehydration responsive element binding (DREB) transcription factors improve abiotic stress tolerance in plants by regulating stress-inducible gene expression. The fungal endophyte also enhanced proline and dehydrin content in the leaf tissue of rice plants under drought stress, along with enhancing superoxide dismutase (SOD) activity. Plant responses to *Trichoderma* inoculation as a biocontrol agent are manifested by early escape from abiotic stresses through the activation of antioxidant machinery (Pacheco-Trejo et al. 2022). Inoculation with *T. harzianum* helped alleviate water deficit in tomato (Khoshmanzar et al. 2020) and rice (Singh et al. 2023a, b) through enhanced activation of ascorbate- and glutathione-related defense enzymes. Enzymatic and nonenzymatic antioxidant reactions in plants grown with microbial inoculation under non-drought and drought conditions were improved. The expression of defense-related genes that help plants regulate ROS as key steps in microbe-mediated stress mitigation processes was explored. Under drought stress, antioxidant enzyme stressors postpone wilt and drought-induced changes, such as maintaining transpiration rates by regulating stomatal conductance, maintaining photosynthesis rates, delaying drought-induced leaf senescence, and boosting amino acid and protein production, thus improving enzymatic (mono-dehydroascorbate reductase (MDHAR), glutathione reductase (GR), dehydroascorbate reductase (DHAR), and APX) activities. These activities play a vital role in improving nitrogen fixation and mineral and water uptake, reducing oxidative stress by lowering ROS and H_2O_2 content, reducing the loss of electrolytes, maintaining leaf relative water content, and improving water-use efficiency in plants. Likewise, ACC deaminase is produced to lower stress ethylene content, increase total phenolics and flavonoids in plants, and produce exopolysaccharides (EPS) to retain water in root zones (Poudel et al. 2021; Bogati and Walczak 2022). Inoculation with the *Piriformospora indica* fungal endophyte also demonstrated drought resistance by upregulating antioxidant enzymes, drought-regulated genes, and CAS mRNA levels in drought-challenged leaves. Pandey et al. (2016) demonstrated the role of *T. harzianum* in stress mitigation in rice genotypes due to the upregulation of aquaporin, dehydrin, and malonialdehyde genes, along with various other physiological parameters. Rhizobacteria-induced drought endurance and resilience (RIDER), which includes changes in the levels of phytohormones, defense-related proteins and enzymes, antioxidants, and exopolysaccharides, has been observed for microbe-mediated plant responses. Such strategies make plants more resilient to abiotic stress (Kaushal and Wani 2016). Polyphenolics in plant leaves are shown to have a protective role against stresses through antioxidation and ROS deactivation.

In contrast to drought stress, plant roots under waterlogged or flood conditions face hypoxic conditions. Flooding significantly alters the physical, chemical, and biological characteristics of soil and promotes the growth of specialized microbial communities that can only endure in environments with low oxygen levels. Features such as endospore development, osmoprotectant synthesis, and specific cell wall architectures are examples of microbial adaptations that improve microbial tolerance and resilience to hydrological stress. The abundance of anaerobic taxa, such as Firmicutes, Desulfobacterota, methanogens, and Geobacteraceae, is mostly increased by soil flooding, and the number of beneficial microflora that help in showing plant growth-promoting traits is reduced (Das et al. 2025). However, under waterlogged conditions, the number of AMF has been found to increase in some ecosystems. AMF in these ecosystems have been found to suppress pathogenic fungal growth, thus enhancing plant growth. AMF have the potential to improve nutrient uptake and strengthen antioxidant responses in plants (Tuheteru and Wu 2017). Several PGPMs isolated from wet greenhouse soil have been found in recent model research to have characteristics that promote plant growth, such as the solubilization of phosphate and potassium, the formation of ammonia and organic acids, and the activity of amylase and cellulase (Senko et al. 2024).

6.4.4 Biotic Stresses

Climate change causes the emergence of new diseases and modifies the severity of existing diseases (Ghini et al. 2011). Beneficial soil microorganisms (BSMs) are a cornerstone of sustainable agriculture, offering a powerful and eco-friendly arsenal against the myriad biotic stresses that threaten global crop production. In the face of a changing climate, which can alter pathogen distribution and host susceptibility, harnessing these microbial allies is becoming increasingly critical. BSMs, including PGPR, AMF, and filamentous fungi such as *Trichoderma* spp., deploy sophisticated and multi-layered defense systems to protect plants. PGPMs, such as *Bacillus, Pseudomonas*, and *Trichoderma*, induce resistance or priming in plants without directly interacting with the targeted pathogen (Köhl et al. 2019; Poveda et al. 2019). Some AMF, *Bacillus*, and *Pseudomonas* species have been reported to show excellent responses in defense against phytopathogens (Vafadar et al. 2014). Their mechanisms of action are rarely singular; instead, they function through a complex interplay of direct antagonism, competition, induction of systemic resistance in the host plant, and modulation of the broader rhizosphere microbiome to create a suppressive environment for the pathogen. This holistic approach not only controls specific pathogens but also enhances the overall resilience of the plant-soil system, reduces dependence on chemical pesticides, and contributes to agricultural sustainability.

Upon subsequent challenge by a pathogen or pest, ISR-primed plants exhibit faster and more robust activation of defense responses. These include strengthening of cell walls through lignification, accumulation of antimicrobial phytoalexins, and production of defense-related enzymes and pathogenesis-related (PR) proteins. *Bacillus subtilis* has proven to be an excellent biotic stress-tolerant species. They induce various stress-responsive genes, stress-related metabolites, and phytohormones, conferring disease resistance against bacteria, viruses, and fungi in plants (Jayaraj et al. 2004; Gupta et al. 2021). A study conducted by Manzar et al. (2023) demonstrated that seed treatment with the PGPR *Bacillus velezensis* JB0319 effectively primed ISR in cucumber plants, leading to a significant reduction in the severity of gray mold caused by *Botrytis cinerea*. The study linked this protection to the enhanced expression of key JA/ET pathway genes and a subsequent increase in defensive enzyme activities. Similarly, AMF colonization has been shown to induce a primed state in plants, providing broad-spectrum resistance against foliar pathogens like powdery mildew (Shang et al. 2025) and root pathogens like *Fusarium* wilt (Pu et al. 2022).

In addition to priming the host, BSMs directly antagonize pathogens through antibiosis, competition, and parasitism. Antibiosis involves the production of a diverse array of antimicrobial secondary metabolites that inhibit or kill competing microorganisms. PGPR from genera such as *Bacillus* and *Pseudomonas* are prolific producers of antibiotics, including lipopeptides (e.g., surfactin and iturin), polyketides, phenazines, and hydrogen cyanide. Recently, Zhao et al. (2024) isolated a novel strain of *Pseudomonas chlororaphis* from the soybean rhizosphere that exhibited potent biocontrol activity against the fungal pathogen *Sclerotinia sclerotiorum* through the production of a specific phenazine antibiotic, which inhibited both mycelial growth and the formation of durable sclerotia. *Pseudomonas protegens* produces a suite of antimicrobial compounds, including 2,4-diacetylphloroglucinol (DAPG), pyoluteorin, and pyrrolnitrin, which provide broad-spectrum activity against fungal pathogens like *Pythium* and *Fusarium* (Andreolli et al. 2019; Vick et al. 2021; Pellicciaro et al. 2022). Similarly, a study performed by Verma et al. (2024a) found that the strain *Pseudomonas hunanensis* SPT26 produced antifungal compounds to biocontrol Fusarium oxysporum infection in tomato crops growing under saline conditions. *Bacillus velezensis* strain FZB42 produces lipopeptides, including surfactin, iturin, and fengycin, which disrupt the membrane integrity of fungal pathogens (Borriss et al. 2019; Han et al. 2021). Manzar et al. (2023) showed that *Bacillus velezensis* JB0319 not only directly antagonized *Botrytis cinerea* but also potently induced systemic resistance (ISR) in cucumber plants, priming JA/ET-dependent defense pathways for enhanced protection.

Mycoparasitism is a direct attack strategy employed by fungi, such as *Trichoderma*. These biocontrol agents exhibit a remarkable ability to sense, grow toward, coil around, and penetrate the hyphae of pathogenic fungi. They then secrete a cocktail of cell wall-degrading enzymes, including chitinases, glucanases, and proteases, to lyse the pathogen cells and utilize them as a nutrient source (Poveda 2021). *Trichoderma asperellum* has been effectively used to manage *Rhizoctonia solani*, the causative agent of damping-off in numerous crops (Das et al. 2019; Herrera et al. 2020; Nurhayati et al. 2021; Almaghasla et al. 2023). Furthermore, the role of BSMs extends beyond individual interactions to shape the entire soil microbial community. Successful colonization by beneficial microbes can alter the rhizosphere microbiome, enriching it with other beneficial taxa and creating consistently disease-suppressive soil. This suppressive state is an emergent property of a complex microbial network that prevents pathogen establishment via collective antagonism. For instance, the application of consortia rather than single strains is gaining traction owing to their superior efficacy and stability. A meta-analysis performed by Li et al. (2024) concluded that microbial consortia consistently outperformed individual strains in suppressing soil-borne diseases across multiple crop systems, highlighting synergistic interactions between different microbial functional groups. The roles of beneficial soil microorganisms in managing biotic stress are multifaceted and integral to the development of climate-resilient agricultural systems. Through a combination of induced resistance, direct antagonism, and community manipulation, BSMs provide effective, durable, and sustainable pest and disease management strategies. They also produce stress-responsive antimicrobial compounds, such as antibiotics and lytic enzymes, which remain effective under stress conditions (Arora et al. 2020). Bhattacharya et al. (2025) observed that the bacterial endophyte *Pseudomonas azotoformans* efficiently inhibited the growth of *Alternaria brassicae*, causing blight disease in mustard plants through the secretion of antifungal metabolites. Integrating these microbial solutions into mainstream agricultural practices is not merely an alternative but a necessity for ensuring food security under the challenging conditions of a changing climate.

6.5 Future Prospects

Integrating beneficial soil microorganisms into agricultural systems presents a paradigm shift toward holistic soil health management, where improved nutrient cycling, soil structure, and carbon sequestration contribute to the creation of robust, self-sustaining agroecosystems capable of buffering climatic extremes. However, the transition from promising research to widespread on-farm application is a critical next step (Arora and Mishra 2024). Although they rely on deliberate research and development, there are numerous opportunities to optimize the impact of BSMs in the future. Priority should be given to developing specialized microbial consortia in addition to single-strain inoculants (Singh et al. 2023a, b; Sharma et al. 2023; Negi et al. 2024; Zhong et al. 2025). By formulating an efficient consortium of beneficial microbes with different capabilities, such as nutrient-mobilizing bacteria with ISR-inducing fungi and stress-tolerant actinomycetes, synergistic blends that offer broader and more reliable efficacy across diverse soils and crops can be utilized (Santoyo et al. 2021; Arora and Fatima 2023). Advanced technologies, such as microbiome engineering and machine learning-driven analysis of soil metagenomes, will be crucial for designing next-generation bio-inoculants specific to regional climates and cropping systems (Suman et al. 2022; Clagnan et al. 2024; Sharma et al. 2025). Furthermore, overcoming formulation and delivery challenges is essential for encapsulating microbes in protective, nutrient-rich carriers that ensure their survival, colonization, and activity in the field will determine commercial success (Bhattacharjee et al. 2023; Fadiji et al. 2024). Concurrently, farming practices must evolve to support these microbes through reduced tillage, organic amendments, and cover cropping, which foster a conducive environment for inoculant establishment and the proliferation of indigenous beneficial communities (Zhang et al. 2025). Ultimately, realizing the full potential of BSMs requires concerted efforts among researchers, industries, policymakers, and farmers (Hubballi et al. 2023). By investing in this green technology, we can cultivate climate-resilient agriculture that ensures food security, minimizes environmental impact, and aligns with the urgent need for sustainable development in the face of global changes.

6.6 Conclusion

Global population expansion, along with climate change scenarios, puts tremendous pressure on the agricultural sector to meet the food demand of the population. The exploration of BSMs underscores their indispensable role as keystones of agricultural sustainability in the era of climate change. They are not simply additives but fundamental components that reprogram plant responses to drought, salinity, heat, and pathogens, thereby safeguarding productivity and reducing reliance on unsustainable chemical inputs in agriculture. In the face of a rapidly changing climate, BSMs have emerged as indispensable allies in engineering agricultural sustainability. Soil microorganisms form a natural synergistic defense system by enhancing plant resilience to abiotic stresses, such as drought and salinity, and simultaneously suppressing soil-borne pathogens. Their ability to improve nutrient-use efficiency and sequester carbon directly mitigates the drivers of climate change while bolstering food security. Therefore, transitioning to management practices that protect and harness this microbial potential is not merely an option but a critical imperative. Incorporating these biological powerhouses into future agricultural frameworks offers a viable, ecocentric pathway toward resilient and productive agroecosystems for future generations.

References

Abdel Latef AAH, Zaid A, Abo-Baker ABAE, Salem W, Abu Alhmad MF (2020) Mitigation of copper stress in maize by inoculation with *Paenibacillus polymyxa* and *Bacillus circulans*. Plants 9(11):1513

Acharya BR, Gill SP, Kaundal A, Sandhu D (2024) Strategies for combating plant salinity stress: the potential of plant growth-promoting microorganisms. Front Plant Sci 15:1406913. https://doi.org/10.3389/fpls.2024.1406913

Adhikari A, Kwon EH, Khan MA, Shaffique S, Kang SM, Lee IJ (2024) Enhanced use of chemical fertilizers and mitigation of heavy metal toxicity using biochar and the soil fungus *Bipolaris maydis* AF7 in rice: genomic and metabolomic perspectives. Ecotoxicol Environ Saf 271:115938. https://doi.org/10.1016/j.ecoenv.2024.115938

Ahmad M, Imtiaz M, Nawaz MS, Mubeen F, Sarwar Y, Hayat M et al (2023) Thermotolerant PGPR consortium B3P modulates physio-biochemical and molecular machinery for enhanced heat tolerance in maize during early vegetative growth. Ann Microbiol 73(1):34. https://doi.org/10.1186/s13213-023-01732-x

Ait Barka E, Nowak J, Clément C (2006) Enhancement of chilling resistance of inoculated grapevine plantlets with a plant growth-promoting rhizobacterium, *Burkholderia phytofirmans* strain PsJN. Appl Environ Microbiol 72(11):7246–7252

Akhtar N, Wani AK, Dhanjal DS, Mukherjee S (2022) Insights into the beneficial roles of dark septate endophytes in plants under challenging environment: resilience to biotic and abiotic stresses. World J Microbiol Biotechnol 38:79. https://doi.org/10.1007/s11274-022-03266-1

Alcázar R, Bueno M, Tiburcio AF (2020) Polyamines: small amines with large effects on plant abiotic stress tolerance. Cells 9(11):2373. https://doi.org/10.3390/cells9112373

Ali SZ, Sandhya V, Grover M, Kishore N, Rao LV, Venkateswarlu B (2009) Pseudomonas sp. strain AKM-P6 enhances tolerance of sorghum seedlings to elevated temperatures. Biol Fertil Soils 46(1):45–55

Ali S, Charles TC, Glick BR (2014) Amelioration of high salinity stress damage by plant growth-promoting bacterial endophytes that contain ACC deaminase. Plant Physiol Biochem 80:160–167

Almaghasla MI, El-Ganainy SM, Ismail AM (2023) Biological activity of four Trichoderma species confers protection against *Rhizoctonia solani*, the causal agent of cucumber damping-off and root rot diseases. Sustainability 15(9):7250. https://doi.org/10.3390/su15097250

Amellal N, Burtin G, Bartoli F, Heulin T (1998) Colonization of wheat roots by an exopolysaccharide-producing *Pantoea agglomerans* strain and its effect on rhizosphere soil aggregation. Appl Environ Microbiol 64(10):3740–3747

Andreolli M, Zapparoli G, Angelini E, Lucchetta G, Lampis S, Vallini G (2019) *Pseudomonas protegens* MP12: a plant growth-promoting endophytic bacterium with broad-spectrum antifungal activity against grapevine phytopathogens. Microbiol Res 219:123–131. https://doi.org/10.1016/j.micres.2018.11.008

Alami Y, Achouak W, Marol C, Heulin T (2000) Rhizosphere soil aggregation and plant growth promotion of sunflowers by an exopolysaccharide-producing Rhizobium sp. strain isolated from sunflower roots. Appl Environ Microbiol 66(8):3393–3398

Arora NK, Fatima T, Mishra I, Verma M, Mishra J, Mishra V (2018) Environmental sustainability: challenges and viable solutions. Environ Sustain 1(4):309–340

Arora NK, Fatima T (2023) Development and application of consortia-based microbial bioinoculants for sustainable agriculture. In: Singh DP, Gupta VK, Prabha R (eds) Sustainable agrobiology: design and development of microbial consortia. Springer Nature, Singapore, pp 401–405

Arora NK, Mishra J (2024) Next generation microbe-based bioinoculants for sustainable agriculture and food security. Environ Sustain 7(1):1–4. https://doi.org/10.1007/s42398-024-00300-0

Arora NK, Fatima T, Mishra J, Mishra I, Verma S, Verma R et al (2020) Halo-tolerant plant growth promoting rhizobacteria for improving productivity and remediation of saline soils. J Adv Res 26:69–82. https://doi.org/10.1016/j.jare.2020.07.003

Arora P, Fatima T, Arora NK (2025) SDGs beyond 2030: shaping the future with lessons from the past. Environ Sust 8:389–394. https://doi.org/10.1007/s42398-025-00388-2

Arora NK, Mishra J (2016) Prospecting the roles of metabolites and additives in future bioformulations for sustainable agriculture. Appl Soil Ecol 107:405–407

Arora NK, Mishra J, Singh P, Fatima T (2024) Salt-tolerant plant growth-promoting *Pseudomonas atacamensis* KSS-6 in combination with organic manure enhances rice yield, improves nutrient content and soil properties under salinity stress. J Basic Microbiol 64(6):2300767. https://doi.org/10.1002/jobm.202300767

Ashraf M, Hasnain S, Berge O, Mahmood T (2004) Inoculating wheat seedlings with exopolysaccharide-producing bacteria restricts sodium uptake and stimulates plant growth under salt stress. Biol Fertil Soils 40(3):157–162

Backer R, Rokem JS, Ilangumaran G, Lamont J, Praslickova D, Ricci E et al (2018) Plant growth-promoting rhizobacteria: context, mechanisms of action, and roadmap to commercialization of biostimulants for sustainable agriculture. Front Plant Sci 9:1473. https://doi.org/10.3389/fpls.2018.01473

Bahrami-Rad S, Hajiboland R (2024) Arbuscular mycorrhizal fungi improve growth and physiological status of *Alkanna tinctoria* under drought and salinity stresses. J Soil Sci Plant Nutr. https://doi.org/10.1007/s42729-024-01679-z

Bailey-Serres J, Fukao T, Gibbs DJ, Holdsworth MJ, Lee SC, Licausi F et al (2012) Making sense of low oxygen sensing. Trends Plant Sci 17(3):129–138. https://doi.org/10.1016/j.tplants.2011.12.004

Bailey-Serres J, Geigenberger P, Perata P, Sasidharan R, Schwarzländer M (2025) Hypoxia as challenge and opportunity: from cells to crops, to synthetic biology. Plant Physiol 197(1):kiae640. https://doi.org/10.1093/plphys/kiae640

Bebber DP, Ramotowski MA, Gurr SJ (2013) Crop pests and pathogens move polewards in a warming world. Nat Clim Chang 3(11):985–988. https://doi.org/10.1038/nclimate1998

Berens ML, Berry HM, Mine A, Argueso CT, Tsuda K (2019) Evolution of hormone signaling networks in plant defense. Annu Rev Phytopathol 57:25.1–25.23. https://doi.org/10.1146/annurev-phyto-082718-100035

Bhattacharjee A, Dubey S, Sharma S (2023) Next-generation bioformulations for plant growth promotion and stress mitigation: a promising approach for sustainable agriculture. J Plant Growth Regul 42(10):6741–6759. https://doi.org/10.1007/s00344-023-11035-9

Bhattacharya A, Mishra P, Mishra I, Arora P, Arora NK (2024) Microbe-based biostimulants: latest developments and future perspectives. In: Arora NK, Bouizgarne B (eds) Microbial biotechnology for sustainable agriculture, Microorganisms for sustainability, vol 34, vol 2. Springer, Singapore. https://doi.org/10.1007/978-981-97-2355-3_2

Bhattacharya A, Verma P, Bharti C, Arora NK (2025) Characterization of antifungal metabolites produced by plant growth promoting bacterial endophyte *pseudomonas azotoformans* SS7 from *Brassica juncea* as a biocontrol tool for *Alternaria brassicae*. Arch Phytopathol Plant Protect 58(10):604–629. https://doi.org/10.1080/03235408.2025.2465385

Bhuyan B, Mohanty RK, Patra S (2025) Impact of climate change on food security in India: an evidence from autoregressive distributed lag model. Environ Dev Sustain 27:6349–6369. https://doi.org/10.1007/s10668-023-04139-3

Bogati K, Walczak M (2022) The impact of drought stress on soil microbial community, enzyme activities and plants. Agronomy 12(1):189. https://doi.org/10.3390/agronomy12010189

Borriss R, Wu H, Gao X (2019) Secondary metabolites of the plant growth promoting model rhizobacterium *Bacillus velezensis* FZB42 are involved in direct suppression of plant pathogens and in stimulation of plant-induced systemic resistance. In: Sharma V, Salwan R, Al-Ani LKT (eds) Secondary metabolites of plant growth promoting rhizomicroorganisms: discovery and applications. Springer, Singapore, pp 147–168

Boulton CA, Lenton TM, Boers N (2023) Reply to: little evidence that Amazonian rainforests are approaching a tipping point. Nat Clim Chang 13(12):1321–1323. https://doi.org/10.1038/s41558-023-01853-6

Cavanagh AP, Slattery R, Kubien DS (2023) Temperature-induced changes in *Arabidopsis* Rubisco activity and isoform expression. J Exp Bot 74(2):651–663. https://doi.org/10.1093/jxb/erac428

Chandrasekaran M, Belachew ST, Yoon E, Chun SC (2017) Expression of β-1,3-glucanase (GLU) and phenylalanine ammonia-lyase (PAL) genes and their enzymes in tomato plants induced after treatment with *Bacillus subtilis* CBR05 against *Xanthomonas campestris* pv. *vesicatoria*. J Gen Plant Pathol 83(1):7–13. https://doi.org/10.1007/s10327-016-0697-8

Chandrasekaran U, Park S, Kim K, Byeon S, Han AR, Lee YS, et al (2024) Energy deprivation affects nitrogen assimilation and fatty acid biosynthesis leading to leaf chlorosis under waterlogging stress in the endangered *Abies koreana*. Tree Physiol 44(6):tpae055

Chaudhry S, Sidhu GPS (2022) Climate change regulated abiotic stress mechanisms in plants: a comprehensive review. Plant Cell Rep 41(1):1–31. https://doi.org/10.1007/s00299-021-02789-5

Chavan SG, Duursma RA, Tausz M, Ghannoum O (2022) Moderate heat stress prevented the observed biomass and yield stimulation caused by elevated CO_2 in two well-watered wheat cultivars. Plant Mol Biol 110(4):365–384. https://doi.org/10.1007/s11103-022-01278-3

Chen X, Li N, Liu C, Wang H, Li Y, Xie Y et al (2022) Exogenous GABA improves the resistance of apple seedlings to long-term drought stress by enhancing GABA shunt and secondary cell wall biosynthesis. Tree Physiol 42(12):2563–2577. https://doi.org/10.1093/treephys/tpac071

Chen L, Gong J, Jin J, Wang L, Chen X, Wang C et al (2025) Intrinsic and extrinsic regulatory mechanisms of *Pseudomonas palleroniana* GZNU148 for enhancing *Themeda japonica* tolerance to drought stress. Plant Soil 508(1):589–612. https://doi.org/10.1007/s11104-024-06678-2

Choudhary S, Wani KI, Naeem M, Khan MMA, Aftab T (2023) Cellular responses, osmotic adjustments, and role of osmolytes in providing salt stress resilience in higher plants: polyamines and nitric oxide crosstalk. J Plant Growth Regul 42(2):539–553

Chowdhury MZH, Mostofa MG, Mim MF, Haque MA, Karim MA, Sultana R et al (2024) The fungal endophyte *Metarhizium anisopliae* (MetA1) coordinates salt tolerance mechanisms of rice to enhance growth and yield. Plant Physiol Biochem 207:108328. https://doi.org/10.1016/j.plaphy.2024.108328

Christian JI, Martin ER, Basara JB, Furtado JC, Otkin JA, Lowman LE et al (2023) Global projections of flash drought show increased risk in a warming climate. Commun Earth Environ 4(1):165. https://doi.org/10.1038/s43247-023-00810-7

Clagnan E, Costanzo M, Visca A, Di Gregorio L, Tabacchioni S, Colantoni E et al (2024) Culturomics- and metagenomics-based insights into the soil microbiome preservation and application for sustainable agriculture. Front Microbiol 15:1473666. https://doi.org/10.3389/fmicb.2024.1473666

Compant S, Clément C, Sessitsch A (2010) Plant growth-promoting bacteria in the rhizo- and endosphere of plants: their role, colonization, mechanisms involved and prospects for utilization. Soil Biol Biochem 42(5):669–678. https://doi.org/10.1016/j.soilbio.2009.11.024

Corwin DL (2021) Climate change impacts on soil salinity in agricultural areas. Eur J Soil Sci 72(2):842–862. https://doi.org/10.1111/ejss.13010

Creus CM, Sueldo RJ, Barassi CA (2004) Water relations and yield in Azospirillum-inoculated wheat exposed to drought in the field. Can J Bot 82(2):273–281

Cui J, Nie F, Zhao Y, Zhang D, Zhou D, Wu J et al (2024) A review on plant endophytes in response to abiotic stress. Environ Pollut Bioavailab 36(1):2323123. https://doi.org/10.1080/26395940.2024.2323123

Das AK, Lee DS, Woo YJ, Sultana S, Mahmud A, Yun BW (2025) The impact of flooding on soil microbial communities and their functions: a review. Stresses 5(2):30

Das MM, Haridas M, Sabu A (2019) Biological control of black pepper and ginger pathogens, *Fusarium oxysporum*, *Rhizoctonia solani* and *Phytophthora capsici*, using *Trichoderma* spp. Biocatal Agric Biotechnol 17:177–183. https://doi.org/10.1016/j.bcab.2018.11.019

Dastogeer KM, Tumpa FH, Sultana A, Akter MA, Chakraborty A (2020) Plant microbiome – an account of the factors that shape community composition and diversity. Curr Plant Biol 23:100161. https://doi.org/10.1016/j.cpb.2020.100161

de Carvalho Neta SJ, Araújo VLV, Fracetto FJC, da Silva CCG, de Souza ER, Silva WR et al (2024) Growth-promoting bacteria and arbuscular mycorrhizal fungus enhance maize tolerance to saline stress. Microbiol Res 284:127708. https://doi.org/10.1016/j.micres.2024.127708

Deutsch CA, Tewksbury JJ, Tigchelaar M, Battisti DS, Merrill SC, Huey RB, Naylor RL (2018) Increase in crop losses to insect pests in a warming climate. Science 361(6405):916–919. https://doi.org/10.1126/science.aar3465

Dodd IC, Rez-Alfocea P (2012) F. microbial amelioration of crop salinity stress. J Exp Bot 63:3415–3428

Dyer LA et al (2023) Climate change and tritrophic interactions: a review of empirical studies. Curr Opin Insect Sci:101049. https://doi.org/10.1016/j.cois.2023.101049

Egamberdiyeva D, Höflich G (2003) Influence of growth-promoting bacteria on the growth of wheat in different soils and temperatures. Soil Biol Biochem 35(7):973–978

Etesami H, Beattie GA (2018) Mining halophytes for plant growth-promoting halotolerant bacteria to enhance the salinity tolerance of non-halophytic crops. Front Microbiol 9:148. https://doi.org/10.3389/fmicb.2018.00148

Fàbregas N, Fernie AR (2021) The interface of central metabolism with hormone signalling in plants. Curr Biol 31(23):R1535–R1548. https://doi.org/10.1016/j.cub.2021.09.062

Fadiji AE, Xiong C, Egidi E, Singh BK (2024) Formulation challenges associated with microbial biofertilizers in sustainable agriculture and paths forward. J Sustain Agric Environ 3(3):e70006. https://doi.org/10.1002/sae2.70006

FAO (2015) Climate change and food security: risk and responses. https://openknowledge.fao.org/server/api/core/bitstreams/a4fd8ac5-4582-4a66-91b0-55abf642a400/content. Accessed 27 Apr 2026

FAO (2022) The state of the world's forests. https://openknowledge.fao.org/items/4c8bd12f-d6b8-4755-a82f-1284c41bf012. Accessed 27 Apr 2026

FAO (2023) The state of food and agriculture 2023. https://doi.org/10.4060/cc7724en. Accessed 27 Apr 2026

Fathi A, Barakat MM (2022) Integrated omics techniques for crop plant biotic stress resistance. In: Aftab T, Roychoudhury A (eds) Omics technologies for sustainable agriculture and global food security, vol II. Springer, Singapore, pp 1–29

Fátima T, Vicente VP, Aurelio GC (2025) Enhancing citrus resilience to high temperature and intense light stress with *Pseudomonas putida* and *Novosphingobium* sp. Environ Exp Bot:106167. https://doi.org/10.1016/j.envexpbot.2025.106167

Figueiredo MVB, Vilar JJ, Burity HA, De França FP (1999) Alleviation of water stress effects in cowpea by *Bradyrhizobium* spp. inoculation. Plant Soil 207(1):67–75

Fones HN, Bebber DP, Chaloner TM, Kay WT, Steinberg G, Gurr SJ (2020) Threats to global food security from emerging fungal and oomycete crop pathogens. Nat Food 1:332–342. https://doi.org/10.1038/s43016-020-0068-7

Ghadirnezhad Shiade SR, Fathi A, Taghavi Ghasemkheili F, Amiri E, Pessarakli M (2023) Plants' responses under drought stress conditions: effects of strategic management approaches—a review. J Plant Nutr 46(9):2198–2230. https://doi.org/10.1080/01904167.2022.2155552

Ghini R, Bettiol W, Hamada E (2011) Diseases in tropical and plantation crops as affected by climate changes: current knowledge and perspectives. Plant Pathol 60:122–132. https://doi.org/10.1111/j.1365-3059.2010.02392.x

Gibbs DJ, Conde JV, Berckhan S, Prasad G, Mendiondo GM, Holdsworth MJ (2015) Group VII ethylene response factors coordinate oxygen and nitric oxide signal transduction and stress responses in plants. Plant Physiol 169(1):23–31. https://doi.org/10.1104/pp.15.00338

Grichko VP, Glick BR (2001) Amelioration of flooding stress by ACC deaminase-containing plant growth-promoting bacteria. Plant Physiol Biochem 39(1):11–17

Gu D, Andreev K, Dupre ME (2021) Major trends in population growth around the world. China CDC Wkly 3(28):604. https://doi.org/10.46234/ccdcw2021.158

Gupta A, Bano A, Rai S, Dubey P, Khan F, Pathak N, Sharma S (2021) Plant growth promoting rhizobacteria (PGPR): a sustainable agriculture to rescue the vegetation from the effect of biotic stress: a review. Lett Appl NanoBiosci 10:2459–2465. https://doi.org/10.33263/LIANBS103.24592465

Hammond J, Putman A, Barnhart T, Sexstone G, McCabe G, Wolock D et al (2024) Streamflow timing and magnitude during snow drought depend on snow drought type and regional hydroclimate. Hydrol Sci J 69(13):1702–1716. https://doi.org/10.1080/02626667.2024.2386051

Han X, Shen D, Xiong Q, Bao B, Zhang W, Dai T et al (2021) The plant-beneficial rhizobacterium *Bacillus velezensis* FZB42 controls the soybean pathogen *Phytophthora sojae* due to bacilysin production. Appl Environ Microbiol 87(23):e01601-21. https://doi.org/10.1128/AEM.01601-21

Hartman S, Liu Z, Van Veen H, Vicente J, Reinen E, Martopawiro S et al (2019) Ethylene-mediated nitric oxide depletion pre-adapts plants to hypoxia stress. Nat Commun 10(1):4020. https://doi.org/10.1038/s41467-019-12032-3

Hasanuzzaman M, Nahar K, Alam MM, Roychowdhury R, Fujita M (2013) Physiological, biochemical, and molecular mechanisms of heat stress tolerance in plants. Int J Mol Sci 14:9643–9684. https://doi.org/10.3390/ijms14059643

Hatfield JL, Prueger JH (2015) Temperature extremes: effect on plant growth and development. Weather Clim Extrem 10:4–10. https://doi.org/10.1016/j.wace.2015.08.001

Herrera W, Valbuena O, Pavone-Maniscalco D (2020) Formulation of *Trichoderma asperellum* TV190 for biological control of *Rhizoctonia solani* on corn seedlings. Egypt J Biol Pest Control 30(1):44. https://doi.org/10.1186/s41938-020-00260-x

Hoegh-Guldberg O, Skirving W, Dove SG, Spady BL, Norrie A, Geiger EF et al (2023) Coral reefs in peril in a record-breaking year. Science 382(6676):1238–1240. https://doi.org/10.1126/science.adk5183

Hubballi M, Rajamanickam S, Mawar R, Tuladhar R, Singh A, Sayyed RZ, Nakkeeran S (2023) PGPM: fundamental, bioformulation, commercialization, and success at farmer's field. In: Sharma SK, Singh UB (eds) Plant growth promoting microorganisms of arid region. Springer Nature, Singapore, pp 257–301

Hwang HH, Huang YT, Chien PR, Huang FC, Wu CL, Chen LY et al (2025) A plant endophytic bacterium *Burkholderia seminalis* strain 869T2 increases plant growth under salt stress by affecting several phytohormone response pathways. Bot Stud 66(1):7. https://doi.org/10.1186/s40529-025-00444-4

Ikan C, Ben-Laouane R, Ouhaddou R, Anli M, Boutasknit A, Lahbouki S et al (2023) Interactions between arbuscular mycorrhizal fungus and indigenous compost improve salt stress tolerance in wheat (*Triticum durum*). S Afr J Bot 158:417–428. https://doi.org/10.1016/j.sajb.2023.04.032

Ikram K, Mansha MZ, Mahmood K, Hameed A, Omar MM, Hassan RM et al (2025) Integrative effects of different mulching practices and *Azospirillum brasilense* on wheat growth, physiology, and soil health under drought stress. Sci Rep 15(1):33435. https://doi.org/10.1038/s41598-025-16948-5

Ilangumaran G, Schwinghamer TD, Smith DL (2024) Rhizobacteria and arbuscular mycorrhizal fungi mitigate drought stress in corn by modulating phytohormones and antioxidant enzymes. J Plant Growth Regul. https://doi.org/10.1007/s00344-024-11300-3

IPCC (2014) Agriculture, forestry and other land use (AFOLU). In: Edenhofer O et al (eds) Climate change 2014: mitigation of climate change. Contribution of working group III to the fifth assessment report of the Intergovernmental Panel on Climate Change. Cambridge University Press, Cambridge, United Kingdom and New York, NY, USA, pp 811–921

IPCC (2023) Sixth assessment report. Climate change 2023. https://www.ipcc.ch/report/sixth-assessment-report-cycle/. Accessed 27 Apr 2026

Iqbal N, Umar S, Khan NA, Corpas FJ (2021) Crosstalk between abscisic acid and nitric oxide under heat stress: exploring new vantage points. Plant Cell Rep 40:1429–1450. https://doi.org/10.1007/s00299-021-02722-w

Jan SS, Khan NA, Asaf S, Shahzad R, Lubna, Imran M et al (2024) Consortium of endophytic *Bacillus australimaris* CK11 and *Staphylococcus epidermidis* CK9 from *Commiphora gileadensis* mediates tomato resilience to combined salinity, heat, and drought stresses. J Plant Growth Regul:1–24. https://doi.org/10.1007/s00344-024-11234-w

Javaid MH, Ali B, Neelam A, Bukhari SAH, Munir R, Haider Z, et al (2025) Green-engineered calcium-doped carbon nanospheres enhance salt tolerance and growth in maize by modulating gene expressions and antioxidant defense mechanisms. J Environ Chem Eng 13(3):117061

Jayaraj J, Yi H, Liang GH, Muthukrishnan S, Velazhahan R (2004) Foliar application of *Bacillus subtilis* AUBS1 reduces sheath blight and triggers defense mechanisms in rice. J Plant Dis Prot 111:115–125. https://doi.org/10.1007/BF03356137

Jiang Z, Zhou X, Tao M, Yuan F, Liu L, Wu F et al (2019) Plant cell-surface GIPC sphingolipids sense salt to trigger Ca2+ influx. Nature 572(7769):341–346. https://doi.org/10.1038/s41586-019-1449-z

Julkowska MM, Testerink C (2015) Tuning plant signaling and growth to survive salt. Trends Plant Sci 20(9):586–594. https://doi.org/10.1016/j.tplants.2015.06.008

Kang SM, Khan AL, Waqas M, Asaf S, Lee KE, Park YG, Kim AY, Khan MA, You YH, Lee IJ (2019) Integrated phytohormone production by the plant growth-promoting rhizobacterium *Bacillus tequilensis* SSB07 induced thermotolerance in soybean. J Plant Interact 14:416–423. https://doi.org/10.1080/17429145.2019.1643815

Kaushal M, Wani SP (2016) Rhizobacterial-plant interactions: strategies ensuring plant growth promotion under drought and salinity stress. Agric Ecosyst Environ 231:68–78

Kavroulakis N, Doupis G, Papadakis IE, Ehaliotis C, Papadopoulou KK (2018) Tolerance of tomato plants to water stress is improved by the root endophyte *Fusarium solani* FsK. Rhizosphere 6:77–85. https://doi.org/10.1016/j.rhisph.2018.04.002

Khaledi F, Balouchi H, Movahhedi Dehnavi M, Salehi A, Dedicova B (2025) Mitigating drought stress in maize: synergistic effects of zinc sulfate and *Pseudomonas* spp. on physiological and biochemical responses. Plants 14(10):1483. https://doi.org/10.3390/plants14101483

Khare E, Mishra J, Arora NK (2018) Multifaceted interactions between endophytes and plant: developments and prospects. Front Microbiol 9:2732. https://doi.org/10.3389/fmicb.2018.02732

Khoshmanzar E, Aliasgharzad N, Neyshabouri MR, Khoshru B, Arzanlou M, Asgari Lajayer B (2020) Effects of *Trichoderma* isolates on tomato growth and inducing its tolerance to water-deficit stress. Int J Environ Sci Technol 17(2):869–878. https://doi.org/10.1007/s13762-019-02523-3

Kim J, Lee H, Yi SJ, Kim K (2022) Gene regulation by histone-modifying enzymes under hypoxic conditions: a focus on histone methylation and acetylation. Exp Mol Med 54(7):878–889. https://doi.org/10.1038/s12276-022-00819-6

Köhl J, Kolnaar R, Ravensberg WJ (2019) Mode of action of microbial biological control agents against plant diseases: relevance beyond efficacy. Front Plant Sci 10:845. https://doi.org/10.3389/fpls.2019.00845

Kour D, Yadav AN (2022) Bacterial mitigation of drought stress in plants: current perspectives and future challenges. Curr Microbiol 79(9):248. https://doi.org/10.1007/s00284-022-02952-1

Kumar A, Singh S, Mukherjee A, Rastogi RP, Verma JP (2021) Salt-tolerant plant growth-promoting *Bacillus pumilus* strain JPVS11 to enhance plant growth attributes of rice and improve soil health under salinity stress. Microbiol Res 242:126616. https://doi.org/10.1016/j.micres.2020.126616

Kumari S, Chhillar H, Chopra P, Khanna RR, Khan MIR (2021) Potassium: a track to develop salinity tolerant plants. Plant Physiol Biochem 167:1011–1023. https://doi.org/10.1016/j.plaphy.2021.09.019

Kuroha T, Ashikari M (2020) Molecular mechanisms and future improvement of submergence tolerance in rice. Mol Breed 40(4):41. https://doi.org/10.1007/s11032-020-01127-x

Li X, Palta JA, Liu F (2022) Modulation of stomatal response by elevated CO_2 in plants under drought and heat stress. Front Plant Sci 13:843999. https://doi.org/10.3389/fpls.2022.843999

Li L, Feng Y, Qi F, Hao R (2023a) Research progress of *Piriformospora indica* in improving plant growth and stress resistance to plant. J Fungi 9(10):965. https://doi.org/10.3390/jof9100965

Li SS, Li SY, Zhang ML, Wang D, Li S, Zhai Y (2023b) *Bacillus licheniformis* (BE-L60) improved tolerance of spinach seedlings against heat stress. Russ J Plant Physiol 70(4):65. https://doi.org/10.1134/S1021443723601001

Li B, Zhang M, Sun W, Yue D, Ma Y, Zhang B et al (2023c) N6-methyladenosine RNA modification regulates cotton drought response in a Ca2+ and ABA-dependent manner. Plant Biotechnol J 21(6):1270–1285. https://doi.org/10.1111/pbi.14062

Li J, Li Z, Li X, Tang X, Liu H, Li J, Song Y (2023d) Effects of spraying KH_2PO_4 on flag leaf physiological characteristics and grain yield and quality under heat stress during the filling period in winter wheat. Plants 12(9):1801. https://doi.org/10.3390/plants12091801

Li Y, Wang M, Liu X (2024) The efficacy of microbial consortia for biocontrol of soil-borne plant diseases: a meta-analysis. Appl Soil Ecol 195:105237. https://doi.org/10.1016/j.apsoil.2023.105237

Lin Y, Sun Z, Wang X, Cao S, Liu Y, Liu S et al (2025) Exogenous proanthocyanidins improved postharvest quality and resistance to *Botrytis cinerea* infection of strawberry via enhanced non-enzymatic antioxidants accumulation and salicylic acid signaling pathway. Postharvest Biol Technol 230:113732. https://doi.org/10.1016/j.postharvbio.2025.113732

Liu H, Zeng B, Zhao J, Yan S, Wan J, Cao Z (2023) Genetic research progress: heat tolerance in rice. Int J Mol Sci 24(8):7140. https://doi.org/10.3390/ijms24087140

Lopez A, Anzuay MS, Loser UA, Taurian T, Furlan AL (2025) The combined effects of drought stress and phosphorus deficiency on peanut (*Arachis hypogaea* L.) plants are mitigated by drought tolerant phosphate-solubilizing bacteria. Symbiosis 97(1):53–70. https://doi.org/10.1007/s13199-025-01041-4

Ma Y, Dias MC, Freitas H (2024) The role of arbuscular mycorrhizal fungi in plant photosynthesis and water use efficiency under drought stress: a meta-analysis. Plant Cell Environ 47(2):660–675. https://doi.org/10.1111/pce.14769

Manzar N, Kashyap AS, Goutam RS, Rajawat MVS, Sharma PK, Sharma SK, Singh HV (2023) *Bacillus velezensis* JB0319 promotes plant growth and induces resistance against *Botrytis cinerea* in cucumber via altering the phytohormonal and transcriptional landscape. Biol Control 177:105127. https://doi.org/10.1016/j.biocontrol.2022.105127

Martén-Rodríguez S, Cristobal-Pérez EJ, de Santiago-Hernández MH, Huerta-Ramos G, Clemente-Martínez L, Krupnick G et al (2025) Untangling the complexity of climate change effects on plant reproductive traits and pollinators: a systematic global synthesis. Glob Change Biol 31(2):e70081. https://doi.org/10.1111/gcb.70081

Marulanda-Aguirre A, Azcón R, Ruiz-Lozano JM, Aroca R (2008) Differential effects of a *Bacillus megaterium* strain on *Lactuca sativa* plant growth depending on the origin of the arbuscular mycorrhizal fungus coinoculated: physiologic and biochemical traits. J Plant Growth Regul 27(1):10–18

Mishra P, Bhattacharya A, Verma P, Bharti C, Arora NK (2022) Plant growth-promoting bacteria as biostimulants of crops in saline agroecosystems. In: Arora NK, Bouizgarne B (eds) Microbial biotechnology for sustainable agriculture, Microorganisms for sustainability, vol 33, vol 1. Springer, Singapore

Mishra S, Ahmad R, Srivastava A (2025a) Bridging renewable energy sources with non-isolated DC-DC converters: challenges and innovations. Discov Electron 2(1):69. https://doi.org/10.1007/s44306-025-00069-4

Mishra P, Mishra J, Bharti C, Arora NK (2025b) Salt-tolerant *Pseudomonas taiwanensis* PWR-1 mediated organic acid production for biofortification of zinc and reducing fertilizer dependency in wheat under saline conditions. J Plant Growth Regul:1–21. https://doi.org/10.1007/s00344-025-11654-4

Mohanty P, Singh PK, Chakraborty D, Mishra S, Pattnaik R (2021) Insight into the role of PGPR in sustainable agriculture and environment. Front Sustain Food Syst 5:667150

Munns R, Tester M (2008) Mechanisms of salinity tolerance. Annu Rev Plant Biol 59(1):651–681. https://doi.org/10.1146/annurev.arplant.59.032607.092911

Nadeem SM, Zahir ZA, Naveed M, Arshad M (2007) Preliminary investigations on inducing salt tolerance in maize through inoculation with rhizobacteria containing ACC deaminase activity. Can J Microbiol 53(10):1141–1149

NASA (2023) NASA helps find thawing permafrost adds to near-term global warming. https://www.nasa.gov/earth/nasa-helps-find-thawing-permafrost-adds-to-near-term-global-warming/. Accessed 27 Apr 2026

Negi R, Sharma B, Jan T, Kaur T, Chowdhury S, Kapoor M et al (2024) Microbial consortia: promising tool as plant bioinoculants for agricultural sustainability. Curr Microbiol 81(8):222. https://doi.org/10.1007/s00284-024-03752-5

Niaz K, Rauf M, Arif M, Hamayun M, Gul H, Hashem A et al (2024) Drought-tolerant fungal microbes, *Aspergillus oryzae* and *Aspergillus fumigatus*, elevate physiohormonal and antioxidant responses of maize under drought stress. Front Microbiol 15:1488639. https://doi.org/10.3389/fmicb.2024.1488639

Niu X, Lai Z, Wang L, Ma R, Ren Y, Wang X et al (2023) Co-expression of JcNAC1- and JcZFP8-improved agronomic traits of tobacco and enhanced drought resistance through NbbHLH1 and NbbHLH2. Plants 12(17):3029. https://doi.org/10.3390/plants12173029

Nurhayati Y, Suryanti S, Wibowo A (2021) In vitro evaluation of *Trichoderma asperellum* isolate UGM-LHAF against *Rhizoctonia solani* causing sheath blight disease of rice. J Perlindungan Tanaman Indones 25(1):64–73. https://doi.org/10.22146/jpti.100633

Oliveira ALM, Stoffels M, Schmid M, Reis VM, Baldani JI, Hartmann A (2009) Colonization of sugarcane plantlets by mixed inoculations with diazotrophic bacteria. Eur J Soil Biol 45(1):106–113

Omar MNA, Osman MEH, Kasim WA, Abd El-Daim IA (2009) Improvement of salt tolerance mechanisms of barley cultivated under salt stress using *Azospirillum brasilense*. In Salinity and water stress: improving crop efficiency, Dordrecht: Springer Netherlands, pp 133–147

Orozco-Mosqueda MDC, Santoyo G, Glick BR (2023) Recent advances in the bacterial phytohormone modulation of plant growth. Plants (Basel) 12:606. https://doi.org/10.3390/plants12030606

Otto FE (2023) Attribution of extreme events to climate change. Annu Rev Environ Resour 48(1):813–828. https://doi.org/10.1146/annurev-environ-112620-092421

Ou T, Zhang M, Gao H, Wang F, Xu W, Liu X et al (2023) Study on the potential for stimulating mulberry growth and drought tolerance of plant growth-promoting fungi. Int J Mol Sci 24(4):4090. https://doi.org/10.3390/ijms24044090

Pacheco-Trejo J, Aquino-Torres E, Reyes-Santamaría MI, Islas-Pelcastre M, Pérez-Ríos SR, Madariaga-Navarrete A, Saucedo-García M (2022) Plant defensive responses triggered by *Trichoderma* spp. as tools to face stressful conditions. Horticulturae 8(12):1181. https://doi.org/10.3390/horticulturae8121181

Pandey V, Ansari MW, Tula S, Yadav S, Sahoo RK, Shukla N et al (2016) Dose-dependent response of *Trichoderma harzianum* in improving drought tolerance in rice genotypes. Planta 243(5):1251–1264. https://doi.org/10.1007/s00425-015-2425-6

Pellicciaro M, Padoan E, Lione G, Celi L, Gonthier P (2022) Pyoluteorin produced by the biocontrol agent *Pseudomonas protegens* is involved in the inhibition of *Heterobasidion* species present in Europe. Pathogens 11(4):391. https://doi.org/10.3390/pathogens11040391

Phour M, Sindhu SS (2020) Amelioration of salinity stress and growth stimulation of mustard (*Brassica juncea* L.) by salt-tolerant *Pseudomonas* species. Appl Soil Ecol 149:103518. https://doi.org/10.1016/j.apsoil.2020.103518

Poudel M, Mendes R, Costa LA, Bueno CG, Meng Y, Folimonova SY et al (2021) The role of plant-associated bacteria, fungi, and viruses in drought stress mitigation. Front Microbiol 12:743512. https://doi.org/10.3389/fmicb.2021.743512

Poveda J (2021) Trichoderma as biocontrol agent against pests: new uses for a mycoparasite. Biol Control 159:104634. https://doi.org/10.1016/j.biocontrol.2021.104634

Poveda J, Hermosa R, Monte E, Nicolás C (2019) The *Trichoderma harzianum* Kelch protein ThKEL1 plays a key role in root colonization and the induction of systemic defense in Brassicaceae plants. Front Plant Sci 10:1478. https://doi.org/10.3389/fpls.2019.01478

Prasad JK, Dey R, Pandey P, Raghuwanshi R (2025) Exploring drought tolerance in PGPR-protected wheat plants through physiological and molecular expression. J Plant Growth Regul:1–16. https://doi.org/10.1007/s00344-025-11655-3

Pu C, Ge Y, Yang G, Zheng H, Guan W, Chao Z et al (2022) Arbuscular mycorrhizal fungi enhance disease resistance of *Salvia miltiorrhiza* to *Fusarium* wilt. Front Plant Sci 13:975558. https://doi.org/10.3389/fpls.2022.975558

Pureswaran DS, Roques A, Battisti A (2018) Forest insects and climate change. Curr For Rep 4(2):35–50. https://doi.org/10.1007/s40725-018-0078-4

Qu Y, Mueller-Cajar O, Yamori W (2023) Improving plant heat tolerance through modification of Rubisco activase in C3 plants to secure crop yield and food security in a future warming world. J Exp Bot 74(2):591–599. https://doi.org/10.1093/jxb/erac392

Rahman S, Rahmani AA, Aziz MA, Ahmad M, Sivasankaran SK, Hirt H, Masmoudi K (2025) The *Enterobacter* sp. SA187 stimulates stress-responsive genes and promotes salt and heat stress tolerance in tomato plants. Sci Hortic 342:114038. https://doi.org/10.1016/j.scienta.2025.114038

Ramesh K, Matloob A, Aslam F, Florentine SK, Chauhan BS (2017) Weeds in a changing climate: vulnerabilities, consequences, and implications for future weed management. Front Plant Sci 8:95. https://doi.org/10.3389/fpls.2017.00095

Rawal R, Scheerens JC, Fenstemaker SM, Francis DM, Miller SA, Benitez MS (2022) Novel *Trichoderma* isolates alleviate water deficit stress in susceptible tomato genotypes. Front Plant Sci 13:869090. https://doi.org/10.3389/fpls.2022.869090

Raza A, Safdar M, Adnan Shahid M, Shabir G, Khil A, Hussain S, et al (2024) Climate change impacts on crop productivity and food security: an overview. In: Kanga S, Singh SK, Shevkani K, Pathak V, Sajan B (eds) Transforming agricultural management for a sustainable future. World Sustainability Series. Springer, Cham, pp 163–186. https://doi.org/10.1007/978-3-031-63430-7_8

Ren P, Zhou B, Bi Y, Chen X, Yao S (2025) Study on the microbial mechanism of *Bacillus subtilis* in improving drought tolerance and cotton yield in arid areas. Agronomy 15(8):1932. https://doi.org/10.3390/agronomy15081932

Rentschler J, Salhab M, Jafino BA (2022) Flood exposure and poverty in 188 countries. Nat Commun 13(1):3527

Remans R, Ramaekers L, Schelkens S, Hernandez G, Garcia A, Reyes JL, et al (2008) Effect of *Rhizobium–Azospirillum* coinoculation on nitrogen fixation and yield of two contrasting *Phaseolus vulgaris* L. genotypes cultivated across different environments in Cuba. Plant Soil 312(1):25–37

Rodriguez R, Redman R (2008) More than 400 million years of evolution and some plants still can't make it on their own: plant stress tolerance via fungal symbiosis. J Exp Bot 59(5):1109–1114

Rossi L, Wens M, DE M, Cotti D, SABINO SAS, Toreti A, et al (2023) European drought risk atlas. Publications Office of the European Union, Luxembourg

Sachdev S, Ansari SA, Ansari MI, Fujita M, Hasanuzzaman M (2021) Abiotic stress and reactive oxygen species: generation, signaling, and defense mechanisms. Antioxidants 10(2):277. https://doi.org/10.3390/antiox10020277

Sahoo RK, Ansari MW, Dangar TK, Mohanty S, Tuteja N (2014a) Phenotypic and molecular characterisation of efficient nitrogen-fixing Azotobacter strains from rice fields for crop improvement. Protoplasma 251(3):511–523

Sahoo RK, Ansari MW, Pradhan M, Dangar TK, Mohanty S, Tuteja N (2014b) A novel *Azotobacter vinellandii* (SRI Az 3) functions in salinity stress tolerance in rice. Plant Signal Behav 9(7):511–23

Santoyo G, Guzmán-Guzmán P, Parra-Cota FI, Santos-Villalobos SDL, Orozco-Mosqueda MDC, Glick BR (2021) Plant growth stimulation by microbial consortia. Agronomy 11(2):219. https://doi.org/10.3390/agronomy11020219

Santra HK, Banerjee D (2022) Production, optimization, characterization and drought stress resistance by β-glucan-rich heteropolysaccharide from an endophytic fungi *Colletotrichum alatae* LCS1 isolated from clubmoss (*Lycopodium clavatum*). Front Fungal Biol 2:796010. https://doi.org/10.3389/ffunb.2021.796010

Sarioğlu A (2025) Growth of soybean plants under saline conditions: the role of potassium and *Bradyrhizobium japonicum* inoculation. BMC Plant Biol 25(1):473. https://doi.org/10.1186/s12870-025-06444-1

Scafaro AP, Fan Y, Posch BC, Garcia A, Coast O, Atkin OK (2021) Responses of leaf respiration to heatwaves. Plant Cell Environ 44:2090–2101. https://doi.org/10.1111/pce.14038

Senko H, Kajić S, Hud A, Palijan G, Petek M, Rajnović I et al (2024) Will the beneficial properties of plant-growth promoting bacteria be affected by waterlogging predicted in the wake of climate change: a model study. Appl Soil Ecol 198:105379. https://doi.org/10.1016/j.apsoil.2024.105379

Shaffique S, Khan MA, Imran M, Kang SM, Park YS, Wani SH, Lee IJ (2022) Research progress in the field of microbial mitigation of drought stress in plants. Front Plant Sci 13:870626. https://doi.org/10.3389/fpls.2022.870626

Shang P, Zheng R, Li Y, Han S, Tang S, Wu J, Duan T (2025) Effect of AM fungi on the growth and powdery mildew development of *Astragalus sinicus* L. under water stress. Plant Physiol Biochem 219:109422. https://doi.org/10.1016/j.plaphy.2024.109422

Sharma S, Rathod ZR, Jain R, Goswami D, Saraf M (2023) Strategies to evaluate microbial consortia for mitigating abiotic stress in plants. In: Maheshwari DK, Dheeman S (eds) Sustainable agrobiology, Microorganisms for sustainability, vol 43. Springer, Singapore. https://doi.org/10.1007/978-981-19-9570-5_9

Sharma V, Kumari A, Meena AK, Sharma AK, Meena PK, Kalpana, Gambhir AY (2025) Advancing precision farming through microbial genomics and biotechnological tools: innovations for improving crop productivity and soil health. Int J Agric Food Sci 7(7):694–706. https://doi.org/10.33545/2664844X.2025.v7.i7a.123

Shukla A, Srivastava S, Mishra V, Kumar K, Suprasanna P (2024) Arsenic induced toxicity and regulation through metabolomic, hormonal and signaling hubs in plants. In: Singh VP, Singh S, Prasad SM, Sharma P (eds) Metals and metalloids in plant signaling. Springer Nature, Cham, pp 171–200

Singh A, Prasad V (2025) Mitigation of salinity induced negative impacts by salt tolerant plant growth promoting rhizobacteria *Bacillus flexus* in mustard (*Brassica juncea* L.). Front Microbiol 16:1638366. https://doi.org/10.3389/fmicb.2025.1638366

Singh DP, Singh V, Gupta VK, Shukla R, Prabha R, Sarma BK, Patel JS (2020) Microbial inoculation in rice regulates antioxidative reactions and defense related genes to mitigate drought stress. Sci Rep 10(1):4818. https://doi.org/10.1038/s41598-020-61742-8

Singh P, Singh R, Madhu GS, Singh VP (2023a) Seed biopriming with *Trichoderma harzianum* for growth promotion and drought tolerance in rice (*Oryza sativus*). Agric Res 12(2):154–162. https://doi.org/10.1007/s40003-022-00646-3

Singh A, Yadav VK, Chundawat RS, Soltane R, Awwad NS, Ibrahium HA et al (2023b) Enhancing plant growth promoting rhizobacterial activities through consortium exposure: a review. Front Bioeng Biotechnol 11:1099999. https://doi.org/10.3389/fbioe.2023.1099999

Skendžić S, Zovko M, Živković IP, Lešić V, Lemić D (2021) The impact of climate change on agricultural insect pests. Insects 12(5):440. https://doi.org/10.3390/insects12050440

Smith NG, Zhu Q, Keenan TF, Riley WJ (2024) Acclimation of photosynthesis to CO_2 increases ecosystem carbon storage due to leaf nitrogen savings. Glob Change Biol 30(11):e17558. https://doi.org/10.1111/gcb.17558

Sorty AM, Meena KK, Choudhary K, Bitla UM, Minhas PS, Krishnani KK (2016) Effect of plant growth promoting bacteria associated with halophytic weed (*Psoralea corylifolia* L) on germination and seedling growth of wheat under saline conditions. Appl Biochem Biotechnol 180(5):872–882

Striker GG, Yamauchi T, Mollard FP (2025) Maize physiology under waterlogging: shoot and root adaptive responses to mitigate low oxygen stress. J Exp Bot. https://doi.org/10.1093/jxb/eraf424

Sultana R, Islam SMN, Sultana T (2023) Arsenic and other heavy metals resistant bacteria in rice ecosystem: potential role in promoting plant growth and tolerance to heavy metal stress. Environ Technol Innov 31:103160. https://doi.org/10.1016/j.eti.2023.103160

Suman A, Govindasamy V, Ramakrishnan B, Aswini K, SaiPrasad J, Sharma P et al (2022) Microbial community and function-based synthetic bioinoculants: a perspective for sustainable agriculture. Front Microbiol 12:805498. https://doi.org/10.3389/fmicb.2021.805498

Swaine EK, Swaine MD, Killham K (2007) Effects of drought on isolates of *Bradyrhizobium elkanii* cultured from *Albizia adianthifolia* seedlings of different provenances. Agrofor Syst 69(2):135–145

Szczerba A, Płażek A, Kopeć P, Surówka E, Dubert F (2025) Mitigating soil drought effects in soybean with *Bradyrhizobium japonicum* inoculants. BMC Plant Biol 25(1):1533. https://doi.org/10.1186/s12870-025-06561-x

Takahashi F, Kuromori T, Urano K, Yamaguchi-Shinozaki K, Shinozaki K (2020) Drought stress responses and resistance in plants: from cellular responses to long-distance intercellular communication. Front Plant Sci 11:556972. https://doi.org/10.3389/fpls.2020.556972

Tandel YN, Zala VR, Koti S (2025) Impact of high-temperature on growth and development of fruit crops. In: Nandwani D, Karki S (eds) Growth and development in plants and their medicinal and environmental impact. IntechOpen, London. https://doi.org/10.5772/intechopen.1009830

Tewari S, Arora NK (2016) Exopolysaccharides based bioformulation from P*seudomonas aeruginosa* combating saline stress. Recent Trends PGPR Res Sustain Crop Product 93

Tewari S, Arora NK (2013) Plant growth promoting rhizobacteria for ameliorating abiotic stresses triggered due to climatic variability. Clim Chang Environ Sustain 1(2):95–103

Thakur M, Kumar R, Kumar S (2023) Assessing the impact of changing climate on insect pest management and food security. In: Singh P, Sharma SK, Choudhary SP (eds) Climate change and agriculture. Elsevier, Amsterdam, pp 213–229

Tittabutr P, Piromyou P, Longtonglang A, Noisa-Ngiam R, Boonkerd N, Teaumroong N (2013) Alleviation of the effect of environmental stresses using co-inoculation of mungbean by Bradyrhizobium and rhizobacteria containing stress-induced ACC deaminase enzyme. J Soil Sci Plant Nutr 59(4):559–571

Tiwari PK, Srivastava AK, Singh R, Srivastava AK (2025) Microbial alliance of *Paenibacillus* sp. SPR11 and *Bradyrhizobium yuanmingense* PR3 enhances nitrogen fixation, yield, and salinity tolerance in black gram under saline, nutrient-depleted soils. Nitrogen 6(3):66. https://doi.org/10.3390/nitrogen6030066

Tiwari S, Singh P, Tiwari R, Meena KK, Yandigeri M, Singh DP, Arora DK (2011) Salt-tolerant rhizobacteria-mediated induced tolerance in wheat (*Triticum aestivum*) and chemical diversity in rhizosphere enhance plant growth. Biol Fertil Soils 47(8):907–916

Tripathi M, Kumar S, Makarana G, Goel R (2023) Metal-tolerant bioinoculant *Pseudomonas putida* KNP9 mediated enhancement of soybean growth under heavy metal stress suitable for biofuel production at the metal-contaminated site. Energies 16(11):4508. https://doi.org/10.3390/en16114508

Tuheteru FD, Wu QS (2017) Arbuscular mycorrhizal fungi and tolerance of waterlogging stress in plants. In: Wu QS (ed) Arbuscular mycorrhizas and stress tolerance of plants. Springer, Singapore, pp 43–66

Ukkola AM, Thomas S, Vogel E, Bende-Michl U, Siems S, Matic V, Sharples W (2024) Future changes in seasonal drought in Australia. https://doi.org/10.31223/X56110 Accessed 27 Apr 2026

UNEP (2022) Global environmental outlook. https://www.unep.org/geo/. Accessed 27 Apr 2026

Upadhyay SK, Singh DP (2015) Effect of salt-tolerant plant growth-promoting rhizobacteria on wheat plants and soil health in a saline environment. Plant Biol (Stuttg) 17:288–293. https://doi.org/10.1111/plb.12173

Vafadar F, Amooaghaie R, Otroshy M (2014) Effects of plant-growth-promoting rhizobacteria and arbuscular mycorrhizal fungus on plant growth, stevioside, NPK, and chlorophyll content of *Stevia rebaudiana*. J Plant Interact 9:128–136. https://doi.org/10.1080/17429145.2013.822931

Van Zelm E, Zhang Y, Testerink C (2020) Salt tolerance mechanisms of plants. Annu Rev Plant Biol 71(1):403–433. https://doi.org/10.1146/annurev-arplant-050718-100059

Vanlerberghe GC, Robson CA, Yip JY (2002) Induction of mitochondrial alternative oxidase in response to a cell signal pathway down-regulating the cytochrome pathway prevents programmed cell death. Plant Physiol 129(4):1829–1842. https://doi.org/10.1104/pp.003616

Vardharajula S, Zulfikar Ali S, Grover M, Reddy G, Bandi V (2011) Drought-tolerant plant growth promoting Bacillus spp.: effect on growth, osmolytes, and antioxidant status of maize under drought stress. J Plant Interact 6(1):1–14

Velásquez AC, Castroverde CDM, He SY (2018) Plant and pathogen warfare under changing climate conditions. Curr Biol 28(10):619–634. https://doi.org/10.1016/j.cub.2018.03.050

Verma P, Bhattacharya A, Bharti C, Arora NK (2024a) Diazotrophs for enhancement of productivity of major cereal crops under abiotic stresses. In: Arora NK, Bouizgarne B (eds) Microbial biotechnology for sustainable agriculture, vol 2. Springer Nature, Singapore, pp 131–158

Verma P, Bhattacharya A, Bharti C et al (2024b) Antifungal metabolites produced by *Pseudomonas hunanensis* SPT26 effective in biocontrol of fusarium wilt of *Lycopersicum esculentum* under saline conditions. World J Microbiol Biotechnol 40:305. https://doi.org/10.1007/s11274-024-04112-2

Verma S, Verma R, Verma P, Bharti C, Arora NK (2021) Salt tolerant endophytic and diazotrophic strain of *Proteus mirabilis* Pd25 and its effect on the growth of wheat under saline conditions. Nat Volatiles Essent Oils 8(5):13172–13183

Vick SH, Fabian BK, Dawson CJ, Foster C, Asher A, Hassan KA et al (2021) Delving into defence: identifying the *Pseudomonas protegens* Pf-5 gene suite involved in defence against secreted products of fungal, oomycete and bacterial rhizosphere competitors. Microb Genom 7(11):000671. https://doi.org/10.1099/mgen.0.000671

Wang X, Tan NWK, Chung FY, Yamaguchi N, Gan ES, Ito T (2023) Transcriptional regulators of plant adaptation to heat stress. Int J Mol Sci 24(17):13297. https://doi.org/10.3390/ijms241713297

Wang J, Sun S, Fang W, Fu X, Cao F, Liu S (2025) Gamma-aminobutyric acid: a novel biomolecule to improve plant resistance and fruit quality. Plants 14(14):2162. https://doi.org/10.3390/plants14142162

Waqas M, Van Der Straeten D, Geilfus CM (2023) Plants' cry for help through acoustic signals. Trends Plant Sci 28(9):984–986. https://doi.org/10.1016/j.tplants.2023.06.001

Whipps JM (2001) Microbial interactions and biocontrol in the rhizosphere. J Exp Bot 52:487–511

WHO (2023) Air quality, energy and health. https://www.who.int/teams/environment-climate-change-and-health/air-quality-energy-and-health. Accessed 27 Apr 2026

Williams AP, Cook BI, Smerdon JE (2022) Rapid intensification of the emerging southwestern North American megadrought in 2020–2021. Nat Clim Chang 12(3):232–234. https://doi.org/10.1038/s41558-022-01290-z

Wu MW, Zhao H, Zhang JD, Guo L, Liu CM (2020) RADICLELESS 1 (RL1)-mediated nad4 intron 1 splicing is crucial for embryo and endosperm development in rice (*Oryza sativa* L.). Biochem Biophys Res Commun 523(1):220–225. https://doi.org/10.1016/j.bbrc.2019.12.060

Wu TY, Wang YH, Wu F, Wu XQ (2023) Dual inoculation with rhizosphere-promoting bacterium *Bacillus cereus* and beneficial fungus *Peniophora cinerea* improves salt stress tolerance and productivity in willow. Microbiol Res 268:127280. https://doi.org/10.1016/j.micres.2022.127280

WWF (2024) Living planet report. https://www.wwf.org.uk/sites/default/files/2024-10/living-planet-report-2024.pdf. Accessed 27 Apr 2026

Yuan YH, Li J, Ma HC, Yang QH, Liu CJ, Feng BL (2021) Salt-tolerant broomcorn millet (*Panicum miliaceum* L.) resists salt stress via modulation of cell wall biosynthesis and Na+ balance. Land Degrad Dev 32(1):518–532. https://doi.org/10.1002/ldr.3722

Zandalinas SI, Fritschi FB, Mittler R (2021) Global warming, climate change, and environmental pollution: recipe for a multifactorial stress combination disaster. Trends Plant Sci 26(6):588–599. https://doi.org/10.1016/j.tplants.2021.02.011

Zavala JA, Nabity PD, DeLucia EH (2017) An emerging understanding of mechanisms governing insect herbivory under elevated CO_2. Annu Rev Entomol 62:25–43. https://doi.org/10.1146/annurev-ento-031616-035444

Zhang S, Li X, Sun Z, Shao S, Hu L, Ye M, Zhou Y (2015) Antagonism between phytohormone signalling underlies the variation in disease susceptibility of tomato plants under elevated CO2. J Exp Bot 66(7):1951–1963. https://doi.org/10.1093/jxb/eru487

Zhang S, Zhang H, You X, Yang Z, Wang Y (2022) Potassium transporter HcKUP12 from *Halostachys caspica* improved transgenic *Arabidopsis* salt tolerance through maintaining potassium homeostasis. Trees 36(2):737–747. https://doi.org/10.1007/s00468-021-02248-2

Zhang S, Li F, Chang L, Zhang Q, Zhang Z, Wu T et al (2025) Broad-spectrum applications of plant growth-promoting rhizobacteria (PGPR) across diverse crops and intricate planting systems. Microbiol Spectrum 13(3):e01879-24. https://doi.org/10.1128/spectrum.01879-24

Zhao S, Zhang Q, Liu M, Zhou H, Ma C, Wang P (2021) Regulation of plant responses to salt stress. Int J Mol Sci 22(9):4609. https://doi.org/10.3390/ijms22094609

Zhao W, Wu Z, Amde M, Zhu G, Wei Y, Zhou P et al (2023) Nano enabled enhancement of plant tolerance to heat and drought stress on molecular response. J Agric Food Chem 71(51):20405–20418. https://doi.org/10.1021/acs.jafc.3c06180

Zhao L, Li T, Zhang Y, Hu B, Chen S (2024) A novel phenazine-producing *Pseudomonas chlororaphis* strain suppresses *Sclerotinia sclerotiorum* through antibiosis and competition. Plant Dis. https://doi.org/10.1094/PDIS-10-23-2136-RE

Zhong S, Wang W, Tang W, Zhou X, Bu T, Tang Z, Li Q (2025) *Serendipita indica*-dominated synthetic microbial consortia enhanced tartary buckwheat growth and improved its tolerance to drought stress. Front Microbiol 16:1562341. https://doi.org/10.3389/fmicb.2025.1562341

Zhou Y, Lv W, Zhu M, Wang C, Yu L, Bao Y et al (2025) Genome-wide analysis of the Hsp70 gene family in rice reveals that OsHsp70-9 plays a significant role in heat stress response. BMC Plant Biol. https://doi.org/10.1186/s12870-025-07975-9

Ziska LH (2022) Climate change and the language of plant competition: invasive species, CO_2, and the role of trait plasticity. J Exp Bot 73(15):5153–5157. https://doi.org/10.1093/jxb/erac243

Chapter 7
Advancing Plant Microbiome Engineering: Recent Innovations in Bioinoculant-Based and Host-Mediated Strategies

Inga Tamošiūnė, Elena Andriūnaitė, Muhammad Fahad Hakim, Odeta Buzaitė, and Danas Baniulis

Abstract The plant microbiome plays a pivotal role in maintaining plant and soil health and is considered the cornerstone of sustainable agriculture. Plant-associated microbial communities display considerable structural and functional plasticity in response to environmental conditions and to agronomic practices. This dynamic nature is the basis for rhizoengineering approaches aimed at enhancing plant health, stress tolerance, nutrient uptake efficiency, and pathogen biocontrol. The application of bioinoculants containing beneficial microorganisms is regarded as a promising strategy for establishing resilient microbiomes that enhance crop productivity while reducing the dependence on chemical inputs. Although inconsistent efficacy across diverse environmental conditions has long been a significant barrier to the widespread adoption of bioinoculants, rapid advances in plant microbiome research are opening new avenues for innovative bioinoculant design and rhizoengineering strategies. This review explores modern approaches to plant microbiome optimization, with particular emphasis on bioinoculant design and host-mediated strategies that enhance plant nutrient uptake, stress tolerance, and biocontrol of pathogens. It presents examples of improving single-strain bioinoculants using genetic engineering, precise gene editing, and adaptive acclimation. This chapter covers the principles for the rational design of multistrain synthetic microbial consortia and discusses natural community transplantation and host-mediated microbiome engineering approaches. The rapid progress in the strategies outlined in this review holds strong potential for practical application in sustainable agriculture.

I. Tamošiūnė (✉) · E. Andriūnaitė · M. F. Hakim · D. Baniulis
Institute of Horticulture, Lithuanian Research Centre for Agriculture and Forestry, Babtai, Kaunas Region, Lithuania
e-mail: inga.tamosiune@lammc.lt

O. Buzaitė
Department of Biochemistry, Vytautas Magnus University, Akademija, Kaunas Region, Lithuania

S. Mehnaz, I. Shahid (eds.), *Frontiers in Rhizoengineering*, Microorganisms for Sustainability 65, https://doi.org/10.1007/978-981-92-3665-7_7

Keywords Biocontrol · Genetic engineering · Host-mediated microbiome engineering · Microbial consortia · Plant growth-promoting traits · Sustainable agriculture

7.1 Introduction

Plant-associated microbial communities, collectively known as the plant microbiome, play a crucial role in regulating the growth and health of the host plant (Falkowski et al. 2008; Hacquard et al. 2015). Plants actively shape the microbiome structure through root exudates, such as sugars, amino acids, and flavonoids, which recruit beneficial microbes (Sasse et al. 2018). In return, microbes enhance nutrient acquisition, suppress pathogens, stimulate growth, and improve host plant stress tolerance (Berendsen et al. 2012; Glick 2012; Miliute et al. 2015). These plant–microbe interactions not only enhance plant fitness but also improve overall soil health, making the microbiome a vital component of sustainable ecosystems that contribute to sustainable agriculture within the broader One Health framework (Banerjee and van der Heijden 2022; Yusuf et al. 2025; Trivedi et al. 2022).

Soil microbiomes are increasingly recognized as a cornerstone for long-term solutions to the problems faced by agriculture in the twenty-first century, including global food security, sustainable agriculture, and resilience against climate change and soil degradation (Bender et al. 2016; Singh et al. 2020). Harnessing beneficial microbes, such as plant growth-promoting (PGP) bacteria and mycorrhizal fungi, could help to reduce the environmental impacts of conventional agricultural practices caused by synthetic fertilizers and pesticides, mitigating issues like eutrophication, greenhouse gas emissions, and biodiversity loss (Backer et al. 2018; Busby et al. 2017; Trivedi et al. 2017). In 2021, the International Network on Soil Biodiversity was established by the Global Soil Partnership, hosted by the Food and Agriculture Organization of the United Nations (FAO), to advance the conservation and sustainable use of soil biodiversity (FAO 2021). The FAO microbiome working group published a landmark report that addressed different microbiome ecosystems and their relatively underexplored potential to alleviate global problems, including transforming agrifood systems toward sustainability (Kendzior et al. 2022). This initiative promotes the use of microorganisms in agriculture, focusing on the sustainable management of soil biodiversity.

The plant microbiome is structurally and functionally dynamic and is largely shaped by environmental factors. Diverse rhizoengineering techniques are employed to shape plant–microbiome and intermicrobial interactions, delivering beneficial outcomes such as enhanced crop productivity, improved nutrient cycling, and better soil health (Compant et al. 2025; Dessaux et al. 2016). The functional activity of beneficial microorganisms colonizing the rhizosphere, phyllosphere, and endosphere can be promoted by agronomic practices such as crop rotation, reduced tillage, and the application of organic amendments (Chen et al. 2020; Hartmann et al. 2015; Venter et al. 2016). An alternative, more targeted, and controlled strategy for reshaping the rhizosphere microbiome involves the addition of microbial inoculants. The application of bioinoculant formulations containing live beneficial microorganisms to enhance nutrient cycling and availability, promote plant growth and abiotic/biotic stress resilience, and suppress pathogens has been extensively researched for decades (Bashan et al. 2014; French et al. 2021; Malusá et al. 2012; Santos et al. 2019). The development of microbial bioinoculants involves isolating beneficial microorganisms from their natural environments (e.g., endosphere, rhizosphere, or soil), cultivating them under controlled conditions, and applying them to plants or soil using techniques such as seed coating, foliar spraying, or direct soil incorporation (French et al. 2021). This approach resulted in the development of a variety of commercial products, most commonly including rhizobia (Kebede 2021; Mendoza et al. 2021), free-living nitrogen fixers (Veresoglou and Menexes 2010), PGP rhizobacteria (Backer et al. 2018), arbuscular mycorrhizae (Basiru et al. 2021; Ghorui et al. 2024), and entomopathogens (Arthurs and Dara 2019).

Despite substantial research and development efforts, the widespread adoption of bioinoculants in agriculture has been hindered by their inconsistent efficacy across diverse environmental conditions (Samonty et al. 2025; Singh et al. 2020). However, alongside ongoing efforts to improve inoculant efficiency through optimized formulation, delivery, and strain selection, rapid progress in plant microbiome research is paving the way for innovative rhizoengineering. Rapid developments in high-throughput sequencing technologies have shed new light on the remarkable complexity of soil- and plant-associated microbial communities and their vast functional diversity within microbial genomes (Banerjee and van der Heijden 2022; Fierer 2017). The use of multiomics (genomics, metagenomics, and transcriptomics) has deepened the molecular understanding of plant–microbe interactions, helping to identify genes for key PGP traits involved in nutrient cycling and hormone production (Kimotho and Maina 2024). Advances in computational tools, such as machine learning and constraint-based modeling, have revolutionized the ability to model intricate microbial interactions and metabolic networks (Blonde et al. 2025; Tarzi et al. 2024). The advent of culturomics has greatly facilitated the isolation and functional screening of beneficial microbes that are difficult to culture (Clagnan et al. 2024). Overall, these advances provide essential tools for developing novel microbiome engineering approaches and their effective translation into practical agricultural applications in the future.

In this chapter, we examine recent developments in novel microbial bioinoculant design strategies, ranging from genetically engineered, gene-edited, or adaptively acclimated single-strain inoculants to multistrain synthetic microbial consortia (SynComs) that promote plant growth and pathogen biocontrol. Furthermore, we discuss the potential applications of natural community (NatComs) transplantation and host-mediated microbiome engineering (HMME) to enhance abiotic stress tolerance and disease resistance in plants.

7.2 Emerging Strategies for Bioinoculant Design

PGP inoculants are a practical and environmentally friendly option for sustainable farming. However, their success depends on proper delivery and application, as well as their ability to survive within the plant and in rhizosphere soil. Inoculants developed using conventional approaches often fail to establish and sustain associations with plants under field conditions, resulting in unsatisfactory outcomes (Nadeem et al. 2014; Zimmer et al. 2016). For conventional bioinoculants, selection strategies involve screening large isolate collections for characteristic traits, such as nitrogen fixation, phosphate solubilization, or indole-3-acetic acid (IAA) production (Mondal et al. 2020). Based on in vitro assay results, strains showing the strongest expression of these traits are selected for greenhouse or field trials (Compant et al. 2019). Selecting PGP solely based on in vitro metabolic phenotypes neglects critical colonization determinants essential for establishing durable plant–microbe associations (Backer et al. 2018; Liu et al. 2019). With the exception of a few extensively studied systems, such as rhizobium–legume and plant–mycorrhizal fungi interactions, there is still no clear or consistent link between PGP traits observed in vitro and their contribution to stable plant–microbe associations or their effect on plant growth (De Souza et al. 2019; Finkel et al. 2017). Thus, PGP trait expression under optimal laboratory conditions often fails to reflect real-world situations, and strong PGP performance in laboratory assays does not necessarily translate into field success (Koziol et al. 2025; Li et al. 2023). Inoculant populations often decline after application, limiting their ability to enhance plant health. For example, the addition of the inoculant *Bacillus* spp. to the soil resulted in a rapid decline of up to 1000-fold within 3 days, and in some treatments, the inoculant fell below the detection limit after several weeks post-inoculation (Mawarda et al. 2020). Similarly, in a field study tracking rhizobial strains used to inoculate legumes, the inoculant strains often failed to persist from one cropping cycle to the next, being replaced by native rhizobia or losing their symbiotic genes to local populations (Mendoza et al. 2021). Even when the inoculum was still detectable, microbial strains performed less effectively under field conditions than in controlled laboratory settings (Bashan et al. 2014; Martinez Viveros et al. 2010).

Negative outcomes under field conditions often stem from the failure of sterile pot experiments under laboratory conditions to predict rhizosphere competence, which is the capacity of the inoculant to colonize, persist, and functionally express beneficial traits in the presence of variable edaphic factors and indigenous microbial communities (Amaya-Gómez et al. 2020; Santoyo et al. 2021; Shi et al. 2022). Soil conditions can compromise the effectiveness of inoculants, as factors such as pH, moisture, salinity, temperature, and texture directly affect microbial activity. Research has shown that drought decreases motility and organic acid diffusion, salinity causes osmotic stress, and extreme temperatures reduce survival rates, especially for bacteria that do not form spores (Lopes et al. 2021; Pang et al. 2024). Heavy metal contamination further reduces effectiveness, as toxic elements such as cadmium and lead disrupt microbial metabolism and enzyme function (Das et al. 2025).

A key challenge in introducing new bacteria into the soil is competition with native microbiota, which often prevents long-term colonization, leaving only transient effects (Čaušević et al. 2024; Neuhoff et al. 2024). The success of establishing a new species in an uncolonized environment depends on niche availability, competition with existing microbes, and interactions with native species (Čaušević et al. 2024; Papin et al. 2024). For instance, Thompson et al. (1990) showed that *Pseudomonas fluorescens* and *Azospirillum brasilense* maintained higher and more stable populations in heat-sterilized soil than in untreated soil, indicating that biotic competition limits their persistence. Similarly, Postma et al. (1990) reported that bacterial populations in sterilized soil were stabilized regardless of inoculum size, whereas they declined in nonsterile soil due to competition. Recent research by Wertz et al. (2007) revealed that sterilized soils initially act as a "biotic desert," allowing rapid establishment of inoculated strains before re-colonization by native microbes suppresses them. Mawarda et al. (2022) confirmed that *Bacillus* sp. inoculants exhibited higher survival and detectability in sterilized soils than in natural soils, again indicating that resident microbial communities are the dominant barrier. Even high-dose applications rarely cause lasting shifts in the resident microbiome (Liu et al. 2022). Repeated applications may increase the probability of establishment; however, they still depend heavily on local soil and community conditions (Klimasmith et al. 2024).

To overcome the limitations of rhizoengineering using single-strain bioinoculants, new approaches have been explored to enhance PGP traits and improve their persistence and performance in the field. Over the past decade, the rapid expansion of our understanding of the complexity of plant and soil microbiomes and their role in plant fitness and soil health has been the most transformative factor in advancing the development of new techniques and strategies required to address rhizoengineering problems. Advances in high-throughput sequencing technologies, standardized methodologies, and large-scale international collaborations have dramatically increased the volume and resolution of plant and soil microbiome data. There is a large number of studies characterizing plant-associated microbiomes, among which the most notable examples present large-scale or worldwide studies that have profiled microbial communities associated with plants or roots across diverse soil types and multiple plant phyla (Yeoh et al. 2017; Ochoa Hueso et al. 2024) or analyzed large datasets of soil metagenomes (Graham et al. 2024; Shaffer et al. 2022; Skidmore et al. 2022; Tedersoo et al. 2021). The wealth of available metagenomic and genomic data has significantly advanced the field of microbial ecology, and the accumulated insights are increasingly being applied to address practical challenges in the use of bioinoculants in agriculture. Several approaches offer complementary routes for developing effective and resilient microbial bioinoculants and community-based or host-mediated microbiome optimization techniques (Fig. 7.1).

The selection and improvement of PGP traits remain central to bioinoculant studies. Genetic engineering is increasingly used to optimize microbial metabolism and refine PGP functions (Ke et al. 2021). This is greatly benefited by the growing

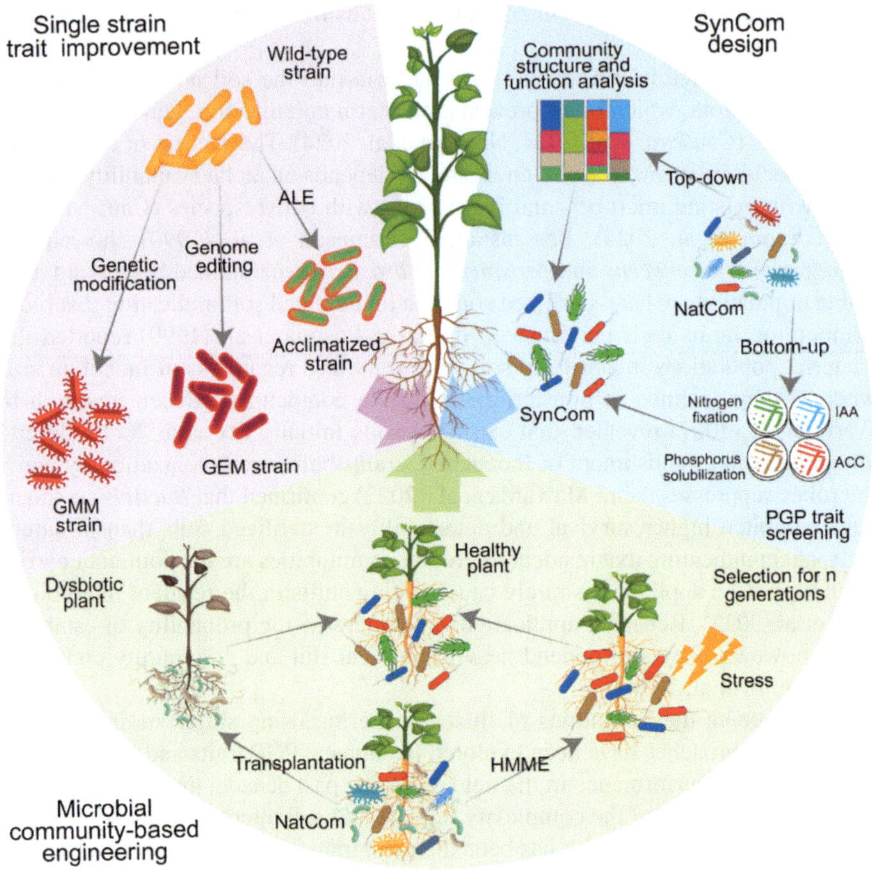

Fig. 7.1 Emerging strategies for microbial bioinoculant- or community-based rhizoengineering

availability of microbial genomic and metagenomic data. The introduction or modification of genes or metabolic pathways using these methods is primarily applied to single-strain bioinoculants to avoid excessive complexity in design, performance evaluation, and regulation. Four functional trait groups of PGP traits were pivotal across both single-strain inoculants and microbial consortia. These include (i) nutrient mobilization, including biological nitrogen fixation, siderophore production, and the solubilization or mineralization of phosphorus, potassium, and micronutrients (Devi et al. 2022c; Kaur et al. 2022); (ii) phytohormone and signaling traits, such as the production of auxins, gibberellins, cytokinins, and 1-aminocyclopropane-1-carboxylic acid (ACC) deaminase (Almazrouei et al. 2025; Bal et al. 2013; Gupta and Pandey 2019; Gupta et al. 2023); (iii) biocontrol-related traits, including the synthesis of lipopeptide antibiotics, volatile organic compounds, hydrogen cyanide, and lytic enzymes (e.g., chitinases, glucanases, and proteases), as well as competitive iron sequestration (Jaroszuk-Ściseł et al. 2019; Naz et al.

2022); and (iv) colonization and persistence traits, which enable effective root association through chemotaxis toward root exudates, motility, biofilm formation, exopolysaccharide (EPS) production, oxidative stress tolerance, and in some taxa, endospore formation (Wang et al. 2024b).

Although genetically modified (GMM) or genome-edited (GEM) microorganisms hold great potential for advancing PGP traits, current regulatory restrictions continue to limit their use in many regions (Kebede and Mengstie 2025; Rozas et al. 2022). An alternative approach involves selecting isolates that are naturally acclimated or adapted to harsh environmental conditions (Andrade Linares et al. 2016; Berninger et al. 2018; Margarita et al. 2023).

While improving single-strain bioinoculants remains an important strategy, considerable research focus has shifted toward multistrain and microbial community-based rhizoengineering strategies, including the design of SynComs, direct manipulation of the indigenous rhizosphere microbiome, and selective enrichment of PGP bacteria (Geerdes et al. 2025; Tsegaye et al. 2022). The bottom-up approach is the predominant paradigm for the rational design of small, defined SynComs (typically comprising two to dozens of compatible strains) (Xu et al. 2025). This reductionist strategy begins with the characterization of individual microbial strains, which are then deliberately combined into functional communities. In contrast, top-down strategies often employ methods that selectively enrich or manipulate complex native rhizosphere communities without isolating individual strains. The top-down approach to SynCom design starts with a complex, naturally occurring microbial community, which is then progressively simplified to identify and isolate key functional members. A top-down approach is used to retain the overall beneficial functions and ecological interactions of NatComs in simplified, manageable consortia that can be used as bioinoculants.

Owing to the limited understanding of intricate plant–microbiome interactions and complex intermicrobial networks, progress toward the practical implementation of SynComs-based bioinoculants remains slow (Copeland et al. 2025; Gupta et al. 2021; Li et al. 2024). Thus, alternative rhizoengineering strategies that leverage and enhance naturally occurring plant-associated and soil microbial communities have been developed. NatCom transplantation relies on the application of intact, unaltered microbial communities collected directly from the environment (Garrido Sanz and Keel 2025). Host-mediated microbiome engineering (HMME) uses the plant itself to select and enrich beneficial microbiomes across successive generations, allowing beneficial communities to emerge through plant-driven selection (Durán et al. 2021).

7.3 Bioinoculant Trait Improvement Using Genetic or Acclimation Techniques

7.3.1 Acclimatized or Adaptively Evolved Microorganisms

Microorganisms naturally evolve to suit their environments, and these changes can significantly improve plant health, productivity, and stress resilience (Lau et al. 2012). A broad group of acclimatized or adapted microorganisms has evolved to survive and function under conditions generally regarded as extreme, and these microorganisms inherently exhibit high stress tolerance. The stress tolerance of microorganisms can also be enhanced through adaptation- or selection-based methods (Andrade Linares et al. 2016; Berninger et al. 2018; Margarita et al. 2023). The most common method for improving the efficiency of stress-reducing PGP inoculants is to select adapted or acclimated strains that confer stress resistance in environments where the corresponding stressor is typically present. For example, *Pseudomonas* spp. strains isolated from Antarctic soils can grow at temperatures ranging from 4 to 30 °C. These strains exhibited plant growth-promoting (PGP)-like solubilization of inorganic phosphates, produced IAA, siderophores, and hydrogen cyanide, and inhibited the growth of certain pathogenic fungi. In pot experiments conducted in a temperature-controlled growth chamber (~14 °C), these isolates significantly promoted root and shoot growth in wheat (*Triticum aestivum*) (Yarzábal et al. 2018). A similar result was observed with *Pseudomonas* spp. isolated from Venezuelan tropical glaciers, where selected isolates significantly elongated wheat roots and shoots grown at 15 °C and increased resistance to pathogenic fungi such as *Pythium ultimum* (Rondón et al. 2019). Further evidence was provided by studies on cold-active Antarctic bacterial strains (i.e., *Pseudomonas* sp., *Psychrobacter* sp., and *Bacillus* sp.), which enhanced the growth of alfalfa (*Medicago sativa*), increasing shoot length and fresh biomass by up to 26% and 46%, respectively, while their metabolites enhanced soil iron bioavailability by up to 40% (Styczynski et al. 2022).

This concept has also been applied to other abiotic stressors, such as drought and salinity. *Bacillus* spp. isolated from the hyperarid Atacama Desert and used as inoculants demonstrated increased lettuce (*Lactuca sativa*) growth and increased biomass production under drought conditions compared to noninoculated plants. These changes are also associated with several beneficial plant traits, including improved water status, increased chlorophyll content, and increased photosynthetic activity (Santander et al. 2024). Likewise, *B. flexus*, isolated from salinity-affected soils in India, effectively improved sesame plant (*Sesamum indicum*) growth under salt stress, increasing chlorophyll and carotenoid content compared to the control, enhancing photosynthesis, and increasing the availability of essential nutrients such as nitrogen, calcium, and iron in the soil (Sridhar et al. 2025).

However, the application of microorganisms isolated from extreme environments is limited because there is often no symbiotic relationship between the introduced microorganisms and native plants or microbiota, thereby reducing their ability to colonize new host. As an alternative, it has been proposed to use local strains exposed to stressors under controlled conditions, so that following treatment, these strains could potentially withstand harmful stressors upon reintroduction into their original environment (Andrade Linares et al. 2016). There are several ways to enhance bacterial performance under stress. One method is stress priming (also called preconditioning, acclimation, acquired stress resistance, adaptive response, or cross-protection), in which bacteria are exposed to mild stress that activates stress genes, and the resulting stress response is then used to inoculate plants. For example, in the *Kosakonia radicincitans* strain, preconditioning through osmoadaptation (treatment with 1%, 4%, and 4% sodium chloride + hydroxyectoine) increased phosphate solubilization and improved radish (*Raphanus sativus*) growth (Cruz Barrera et al. 2019). Inoculation with osmoadapted cells also enhanced the root-colonizing ability of *K. radicincitans*. Sun et al. (2025) showed that *Glutamicibacter endophyticus* grown under salt-stress conditions simulating coastal saline-alkali soil could secrete proteases, siderophores, and IAA. As a result of this exposure, under 4% salt stress, wheat seed inoculation significantly increased seed germination rates from 37.5% to 95%, increased maize (*Zea mays*) seedling dry weight by 42%, and notably improved maize root development. Although microbial priming shows promise, it remains underexplored and is uncommon in agricultural biotechnology.

A more common method is adaptive laboratory evolution (ALE), in which bacteria are cultivated under stressful conditions for many generations. During ALE, bacteria acquire natural mutations, and natural selection favors the most stress-resistant mutants, allowing them to become dominant and produce a permanently tolerant strain that can be stored, revived, and reused (Dragosits and Mattanovich 2013; Wang et al. 2022). Sarapat et al. (2020) cultivated *Bradyrhizobium* sp. in minimal medium with gradually increasing ACC concentrations. After the strain successfully adapted to 3.0 mM ACC, mutations at the ACC-binding site were observed, leading to increased ACC deaminase activity in the ACC-treated group. This new strain has been shown to enhance nodulation early in mung bean (*Vigna radiata*) growth and promote nitrogen fixation under normal conditions. Under water-deficit conditions, inoculation with the modified strain enhanced plant growth, nodule formation, and dry weight (Sarapat et al. 2020).

Further studies have demonstrated that ALE can enhance microbial traits associated with plant colonization, thereby improving the effectiveness of any bioinoculant. Pomerleau et al. (2024) demonstrated that *Bacillus subtilis* repeatedly grew with Arabidopsis (*Arabidopsis thaliana*) seedlings in liquid medium for 21 cycles, resulting in nonsynonymous mutations in the *ywcC* or *sinR* genes, which encode regulators that repress biofilm formation. Further experiments revealed a two-fold increase in root colonization in tomato (*Solanum lycopersicum*) seedlings. Interestingly, ALE application during the cocultivation of different bacterial isolates improved strain compatibility and cooperative performance. For example, Pomerleau et al. (2024) demonstrated that cocultivation of *Bacillus* sp. and *Pseudomonas* spp. strains, known for their PGP traits, negatively affected root colonization by *Bacillus* sp. However, ALE experiments with *B. subtilis* cocultured with *P. fluorescens* on tomato plant roots resulted in stronger colonization of tomato and Arabidopsis roots than the ancestral *B. subtilis* strain in the presence of a few other

Pseudomonas spp. strains (Pomerleau et al. 2024). Similarly, repeated passaging of *Pseudomonas bijieensis* through eight cycles of inoculation onto sterile wheat roots resulted in key mutations in the flagellar regulator *fleN* (Li et al. 2025). These mutations increased bacterial flagellar biosynthesis and motility, improving root colonization and competitive fitness in wheat.

7.3.2 Genetically Modified and Genome-Edited Microorganisms

Identifying microorganisms that enhance soil health and promote plant growth involves examining genes associated with plant–microbe interactions, nutrient absorption, stress tolerance, and pathogen protection. However, simply identifying these microbes is insufficient, as some beneficial bacteria harbor genetic controls that suppress certain traits unless environmental conditions support their expression (Kebede and Mengstie 2025). To address this problem, researchers have focused on identifying the bacterial genes and regulatory mechanisms underlying PGP traits, providing opportunities to enhance their expression. These strategies, commonly employed in metabolic pathway engineering, can also be used to introduce or combine multiple PGP traits within a single bacterial strain. Such improvements can be achieved through deliberate genetic manipulation, either via traditional genetic modification or precise genome editing. These methods vary in their techniques, regulatory status, technological development, and applications, representing a range of genetic engineering approaches (Kebede and Mengstie 2025; Rozas et al. 2022).

Genetically engineered organisms can be categorized into two groups based on the genetic modifications in microorganisms: those that are modified (trait transfer) and those that are edited (precision editing). Traditional genetic modification methods, also known as transgenesis, involve the insertion, overexpression, or horizontal transfer of foreign genes into microbial strains. This approach can broadly improve beneficial plant traits but often faces strict regulations, limiting its use in the field. Genome editing methods, such as CRISPR-Cas9 and base editing, focus on more precise alterations of native microbial genes without necessarily adding foreign DNA. However, they encounter challenges like delivering edits into nonmodel microbes, off-target effects, and achieving field readiness (Ahmad et al. 2023; Miklau et al. 2024; Permyakova and Deineko 2024; Rozas et al. 2022).

7.3.2.1 Genetically Modified Microorganisms

GMMs are produced using recombinant DNA technology by introducing foreign genes or synthetic constructs into their genomes. This targeted engineering enables GMMs to improve plant growth, enhance nutrient uptake, and increase stress tolerance, often exhibiting better PGP traits than wild-type strains (Table 7.1). As low

Table 7.1 Genetically modified microorganisms used as PGP inoculants or biofertilizers

Engineered trait/modification	Microorganism/host crop	Observed effect	References
Nitrogen fixation			
Addition of chimeric pr-c *NifHcDK* operon (nitrogenase structural genes)	*Rhizobium etli* Common bean (*Phaseolus vulgaris*)	Enhanced symbiosis Higher nitrogenase activity Increased plant biomass Greater nitrogen content Higher seed yield Increased seed nitrogen	Peralta et al. (2004)
Addition of *NiFe* hydrogenase (Hup) minitransposon (nitrogenase-associated H_2 uptake)	*Rhizobium leguminosarum* *Mesorhizobium loti* Vetch (*Vicia sativa*) Lotus (*Lotus corniculatus*)	No hydrogen evolution in engineered strain Higher nitrogen fixation Increased shoot biomass Greater nitrogen accumulation	Sotelo et al. (2021)
Addition of heterologous *nif* island (nitrogen fixation gene cluster)	*Pseudomonas protegens* *Arabidopsis thaliana* Alfalfa (*Medicago sativa*) Fescue (*Schenodorus arundinaceus*) Maize (*Zea mays*)	Higher soil ammonium levels Improved plant productivity under nitrogen deficiency	Setten et al. (2013)
Phosphate solubilization			
Refactored phytase genes *HAPs, BPPs, CPs* (phosphorus-mobilizing enzymes)	*Pseudomonas putida* *Pseudomonas simiae* *Ralstonia* sp. Arabidopsis (*Arabidopsis thaliana*)	Increased fresh and dry weight Larger rosette size Improved phosphate availability	Sharma et al. (2013)
Addition of *pqq* and *gad* gene clusters (phosphorus solubilization and acid tolerance pathways)	*Herbaspirillum seropedicae* Rice (*Oryza sativa*)	High secretion of PQQ, gluconic, and 2-ketogluconic acids Enhanced solubilization of phosphorus and potassium Improved overall plant growth Increased tolerance to cadmium	Wagh et al. (2016)
Addition of PQQ biosynthetic genes (cofactor for glucose dehydrogenase -mediated phosphorus solubilization)	*B. subtilis* Arabidopsis (*Arabidopsis thaliana*)	Improved plant growth	Fang et al. (2025)

(continued)

Table 7.1 (continued)

Engineered trait/modification	Microorganism/host crop	Observed effect	References
Phytohormone biosynthesis			
Expression of the *ipt* gene (cytokinin biosynthetic)	*Sinorhizobium meliloti* Alfalfa (*Medicago sativa*)	Enhanced drought tolerance Higher cytokinin levels Greater chlorophyll content Higher leaf water potential Increased plant vitality	Xu et al. (2012)
Addition of *acdR* and *acdS* genes (ACC deaminase activity)	*Bradyrhizobium* sp. Mung bean (*Vigna radiata*)	Enhanced early-stage nodulation Maintained nitrogen fixation under water deficit Increased biomass after rehydration	Sarapat et al. (2020)
Addition of *acdS* gene (ACC deaminase activity)	*Enterobacter* sp. *Kosakonia* sp. Rice (*Oryza sativa*)	Higher ACC deaminase activity Reduced ethylene levels in rice sprouts Improved saline stress tolerance Higher germination rate Longer roots and shoots	Liu et al. (2017)
Biocontrol			
Addition of multiple copies of the *myc* gene cluster (mycosubtilin biosynthesis)	*Bacillus subtilis* Tomato (*Lycopersicon esculentum*)	Higher germination rate Improved seedling growth Antifungal activity against *Pythium aphanidermatum*	Leclère et al. (2005)
Addition of *phzA–G* operon (phenazine antibiotic biosynthesis)	*Pseudomonas fluorescens* Wheat (*Triticum aestivum*)	Expanded antibiotic profile Strong root colonization Antifungal activity against *Rhizoctonia* and *Pythium*	Huang et al. (2004)
Addition of *chiA* gene (chitinase activity)	*Pseudomonas fluorescens* Common bean (*Phaseolus vulgaris*)	Effective against *Rhizoctonia solani*	Downing and Thomson (2000)
Colonization improvement			
Addition of extra *envZ*-like gene copies (signaling regulation)	*Mesorhizobium mediterraneum* Chickpea (*Cicer arietinum*)	Higher shoot dry weight Enhanced symbiosis	Da Silva et al. (2025)
Overexpression of *exoY* gene (succinoglycan biosynthesis)	*Sinorhizobium meliloti* Alfalfa (*Medicago sativa*) *Medicago truncatula*	Enhanced symbiosis	Jones (2012)

(continued)

Table 7.1 (continued)

Engineered trait/modification	Microorganism/host crop	Observed effect	References
Response to the environment			
Overexpression of *ReOtsA* gene (trehalose biosynthesis)	*Rhizobium etli* Common bean (*Phaseolus vulgaris*)	Improved nodulation Increased biomass and grain yield Higher nitrogenase activity Increased drought recovery and tolerance to osmotic stress	Suárez et al. (2008)
Overexpression of *otsA* gene (trehalose biosynthesis)	*Mesorhizobium* sp. Peanut (*Arachis hypogaea*)	Improved nodulation Higher nitrogenase activity Increased drought recovery	Liu et al. (2025a)

nitrogen availability is a global challenge for plant growth and development, efforts in genetic engineering have increasingly targeted the *nif* gene system, which controls biological nitrogen fixation and offers a sustainable alternative to synthetic fertilizers. To improve nitrogen fixation efficiency, Peralta et al. (2004) created a chimeric, complete nitrogenase nifHDK operon regulated by a strong *nifHc* promoter and expressed it in symbiosis with common bean (*Phaseolus vulgaris*). In greenhouse experiments, the GMM strain showed approximately 58% higher nitrogenase activity than the wild-type strain. It also led to an approximately 58% higher plant weight, a 32% higher nitrogen content in plants, and notable increases in seed yield (~36%), nitrogen content (~25%), and nitrogen yield (~72%) (Peralta et al. 2004). Nitrogen fixation efficiency can be further improved by recycling hydrogen produced by nitrogenase through the introduction of uptake hydrogenase (*hup*) genes via TnHB 100 (hydrogen recycling) into hydrogenase-positive *Rhizobium leguminosarum* and *Mesorhizobium loti* strains (Sotelo et al. 2021). Inoculation of vetch (*Vicia sativa*) and lotus (*Lotus corniculatus*) plants with these GMM strains resulted in a notable increase in nitrogen incorporation into the host legumes. The level of improvement in symbiotic performance depends on both the recipient strain and the legume host. These findings suggest that hydrogen recycling enhances symbiotic nitrogen fixation in forage plants. Plants inoculated with *hup*-containing strains formed nodules with virtually no hydrogen production, whereas the wild type produced high levels of hydrogen across all hosts. The GMM strain also significantly improved plant performance by increasing nitrogen fixation, shoot dry weight (by approximately 22–24%), and total nitrogen accumulation compared to the wild type (Sotelo et al. 2021). Genetic modifications can also introduce *nif* genes into bacterial strains that previously lacked this function. Setten et al. (2013) demonstrated that the beneficial rhizobacterium *Pseudomonas protegens*, which was genetically modified to fix nitrogen by introducing nitrogenase genes via the X940 cosmid, increased soil ammonium levels and enhanced growth in Arabidopsis, alfalfa, fescue (*Schenodorus arundinaceus*), and maize under nitrogen-deficient conditions in a greenhouse.

Phosphorus availability also limits plant growth, leading researchers to focus on phosphate-solubilizing microbes that convert insoluble phosphorus into forms accessible to plants. Shulse et al. (2019) demonstrated this by engineering three root-colonizing Proteobacteria: *Pseudomonas putida*, *P. simiae*, and *Ralstonia* sp. They refactored (codon-optimized) and synthesized various microbial phytase genes (*HAPs*, *BPPs*, and *CPs*). These strains secrete active phytases, which are absent in their wild-type hosts, allowing them to hydrolyze phytate in vitro into inorganic phosphate available to plants. A selected subset of GMM strains was used to inoculate Arabidopsis, resulting in significant increases in plant fresh and dry weights and rosette size by enhancing phosphate availability (Shulse et al. 2019). To improve inorganic phosphorus solubilization, Wagh et al. (2016) added *pqq* and *gad* gene clusters to *Herbaspirillum seropedicae*. This strain produces a pyrroloquinoline quinone (PQQ)-dependent glucose dehydrogenase, which is key to mineral phosphate solubilization, but cannot synthesize PQQ itself. The modified strain exhibited high PQQ secretion and produced gluconic and 2-ketogluconic acids, facilitating the solubilization of mineral phosphate and potassium. When used to inoculate rice plants (*Oryza sativa*), these transformants promoted overall growth, increased nitrogen and phosphorus levels, and conferred tolerance to cadmium. Similarly, heterologous PQQ synthesis genes have been introduced into *Bacillus subtilis*, resulting in enhanced bacterial and plant growth (Fang et al. 2025). In Arabidopsis, inoculation with the PQQ transgenic strain resulted in increased leaf width, leaf number, siliques, and primary stem length compared to the nontransgenic *B. subtilis* strain.

In addition to nutrient limitations, deficiencies in endogenous hormones also restrict plant growth, thereby driving increased interest in phytohormone-producing microorganisms. Xu et al. (2012) developed a GMM strain of *Sinorhizobium meliloti* that introduced the *ipt* gene, encoding isopentenyl transferase, a key enzyme in cytokinin biosynthesis, under the control of a stress-inducible promoter. This modified strain improved alfalfa (*Medicago sativa*) drought tolerance by producing higher cytokinin levels, better physiological responses to stress, higher leaf water potential, greater chlorophyll content, and overall plant vitality compared with plants inoculated with the wild-type strain. Sarapat et al. (2020) demonstrated how *Bradyrhizobium* sp. was modified to increase ACC deaminase activity, which inhibits ethylene biosynthesis and promotes plant growth by degrading ethylene precursors. The modified strain, with the addition of *acdR* and *acdS* genes responsible for synthesizing ACC deaminase, enhanced nodulation during the early stage of mung bean growth, maintained nitrogen fixation under water deficit conditions, and increased plant biomass after rehydration. Similarly, the addition of the *acdS* deaminase gene to endophytic *Enterobacter* sp. and *Kosakonia* sp. cells increased deaminase activity in rice sprouts, thereby reducing ethylene levels (Liu et al. 2017). These changes were linked to increased tolerance to saline stress, improved germination rates in rice seeds, and longer sprout root and shoot lengths. Occasionally, GMMs can provide PGP benefits similar to those achieved by directly genetically modifying the host plant. For example, Ravanbakhsh et al. (2021) showed that Arabidopsis could increase shoot micronutrient levels by either overexpressing the ethylene biosynthesis gene ETO1 or adding the bacterial *acdS* gene to the rhizosphere microbiome. Both approaches reduce ethylene signaling and improve nutrient uptake, demonstrating the significant potential of engineered PGP bacteria to enable the acquisition of traits typically achieved through plant transgenesis.

Because environmental conditions greatly affect plant performance, research has focused on identifying microbial genes involved in responses to environmental changes. In this context, one potential target is the biosynthesis of trehalose, a disaccharide widely found in bacteria, fungi, and plants that helps protect cellular structures and preserve cell viability under stress (Raza et al. 2024). Changes in trehalose-6-phosphate synthase caused by the overexpression of the *ReOtsA* gene in the symbiotic bacterium *Rhizobium etli* improved drought tolerance in common bean (*Phaseolus vulgaris*) (Suárez et al. 2008). The modified *R. etli* produced more nodules, had increased nitrogenase activity, and showed higher biomass, with grain yield nearly doubling. Similarly, in *Mesorhizobium* sp., overexpression of the *otsA* genes improved both stress tolerance and nitrogen fixing capacity under salinity stress, resulting in more nodules and greater nitrogenase activity in peanut (*Arachis hypogaea*) (Liu et al. 2025a).

In addition to growth-limiting abiotic factors, biotic stress imposed by plant pathogens significantly reduces crop productivity, suggesting that biocontrol-related genes can be targeted for GMM engineering to suppress pathogens through antibiotic production, competition, and induced systemic resistance. Leclère et al. (2005) investigated the overproduction of mycosubtilin, a powerful antifungal lipopeptide, by *Bacillus subtilis* through the insertion of multiple copies of the regulatory region of the *myc* gene cluster. The addition of these modified strains to tomato seedlings increased germination rates, seedling growth, and biomass while effectively inhibiting *Pythium aphanidermatum* infection. In some cases, biocontrol-initiating genes can be added to endophytic bacteria that previously showed no such effect. Downing and Thomson (2000) demonstrated this by adding the gene coding for chitinase *chiA* to an endophytic strain of *Pseudomonas fluorescens*. This strain, which did not display antifungal activity in its wild form, became an effective biocontrol agent against *Rhizoctonia solani* after this modification. Similarly, Huang et al. (2004) enhanced biocontrol potential by introducing a seven-gene operon (*phzA–G*), responsible for phenazine-1-carboxylic acid synthesis, into *Pseudomonas fluorescens*, resulting in antibiotic production while maintaining a strong root colonization. In wheat plants infected with *Pythium irregulare*, inoculation with a modified *Pseudomonas fluorescens* strain increased plant height across all tested inoculum levels, whereas the wild-type strain was effective only at the highest dose (Huang et al. 2004).

For functional traits to translate into consistent plant benefits when applied as bioinoculants, successful root colonization is essential for long-lasting plant–microbe interactions. Accordingly, traits associated with microbial attachment and persistence in the rhizosphere have been considered targets for genetic modification. Among these is the two-component signal transduction system EnvZ/OmpR, which regulates processes such as virulence and flagella production, both of which are important for successful symbiosis. Da Silva et al. (2025) investigated this by overexpressing an *envZ*-like gene from *Mesorhizobium mediterraneum* in *Mesorhizobium* spp. strains. Chickpea (*Cicer arietinum*) inoculation with these strains resulted in significantly higher shoot dry weight and greater symbiotic effectiveness compared to the control strains. The study also demonstrated that overexpression of the e*nvZ*-like protein can increase the number of infection threads and promote earlier nodulation during infection (Da Silva et al. 2025). Effective symbiosis also depends on symbiotic EPS, which enables the initiation and maintenance of infections. Overexpression of *exoY*, which influences succinoglycan production in *Sinorhizobium meliloti*, enhances symbiosis with the host plant (*Medicago truncatula*) (Jones 2012).

7.3.2.2 Genome-Edited Microorganisms

GEMs are modified using precise tools, with changes made through targeted chromosomal editing that does not introduce foreign DNA. Depending on the method used, they may be regulated differently from traditional genetically modified organisms (Arroyo Olarte et al. 2021). However, because genetic editing is a relatively new technology, most studies seeking to enhance the PGP traits of microorganisms still concentrate on characterizing edited gene activity in vitro rather than assessing their actual impact on plant growth (Dudeja et al. 2025). Consequently, many recent publications report GEMs as PGP candidates but evaluate only molecular or metabolic changes resulting from specific genetic edits, without demonstrating their effects on plant performance. For example, Serantes et al. (2024) applied a CRISPR/Cas9 genome-editing system to two widely used crop inoculants, *Bradyrhizobium japonicum*, to improve nitrogen fixation efficiency. They targeted edits in the genes *napA* (nitrate reductase) and *glxA* (catalase), which reduced nitrate reductase activity and increased catalase activity, respectively. Mi et al. (2020) used a CRISPR interference system in *Klebsiella pneumoniae* to enhance PQQ production, an important cofactor for phosphate-solubilizing enzymes, by suppressing several sugar-transport genes and redirecting carbon toward PQQ biosynthesis. Guo et al. (2025) used the CRISPR-Cas9 system to modify *Bacillus velezensis* by replacing the native bacilysin promoter (*slrR* gene) with the P43 promoter and integrating an additional copy of the *bacG* gene, resulting in overexpression of the *bac* gene cluster. This genetic manipulation generated a strain with increased antimicrobial metabolite production and enhanced antibacterial activity against *Escherichia coli* and *Lactobacillus diolivorans*.

Few studies have provided evidence of the effects of GEM strains on plant growth. Wen et al. (2021) edited the diazotroph *Klebsiella variicola* using a plasmid-based guided microbial remodeling and allelic-exchange platform. In this process, the regulatory gene *nifL* was replaced with a strong native *infC* promoter, removing the NifL inhibitor and increasing *nifA* expression, thereby enabling nitrogenase activity even in nitrogen-rich conditions. This edit resulted in a 122-fold increase in nitrogen fixation activity per cell and an approximately 23-fold higher ammonium production per cell under nitrogen-rich conditions. Inoculation with this edited strain improved the colonization of maize (*Zea mays*) roots and increased nitrogenase activity. Martinez Feria et al. (2024) expanded this study by engineering two diazotrophic species, *Kosakonia sacchari* and *Klebsiella variicola*, using a suicide plasmid. They deleted *nifL* and replaced it with a strong native *nifA* promoter in both strains. Additionally, they modified the *glnD* gene by completely deleting it in *K. sacchari* and deleting the uridylyl transferase domain in *K. variicola*. In *K. sacchari*, the adenylyl removal domain of the *glnE* gene was also removed, which kept glutamine synthetase inactive and promoted ammonium excretion. These genetic changes enabled the bacteria to fix nitrogen even in nitrogen-rich conditions, which usually inhibit this process, and to release ammonium into the rhizosphere, providing fixed nitrogen to plants in multisite field trials. These gene edits improved maize plant fresh weight and leaf chlorophyll content, increasing early season aboveground nitrogen by approximately 14% and lowering the optimal nitrogen fertilizer requirement by 10–27 kg/ha (Martinez Feria et al. 2024).

7.4 Plant Growth Stimulation by Multistrain Bioinoculants

In the natural environment, 99% of microorganisms exist as microbial consortia (Dong et al. 2017). Within these consortia, the interactions between PGP bacteria and plants synergistically enhance plant health and microbiome functions (Santos and Olivares 2021). Building on this principle, the rational design of microbial inoculants increasingly relies on the selection of strains that can function compatibly within a consortium. Compatibility is typically assessed using in vitro coculture assays and rhizosphere colonization tests to ensure that the strains do not exhibit antagonistic effects when grown together (Castanheira et al. 2017; Jain et al. 2012; Liu et al. 2018; Santiago et al. 2017). Such compatibility ensures microorganism growth and promotes effective cocolonization, which is critical for sustained plant–microbe interactions and health.

Several studies have demonstrated the potential of bacterial consortia to improve plant growth under greenhouse and field conditions. For example, Wang et al. (2024a) demonstrated that two five-strain consortia and one two-strain consortium selected from 23 isolates achieved superior cocolonization and plant growth promotion compared with single-strain inoculants. Jha and Saraf (2012) found that the inoculation of a multispecies consortium composed of *Micrococcus* sp., *Acinetobacter calcoaceticus*, *Brevibacillus brevis*, and *Bacillus licheniformis* with the ability to produce IAA, ACC deaminase, and siderophores and to solubilize inorganic phosphorus improved the yield of physic nut (*Jatropha curcas*). Devi et al. (2022c) evaluated the effects of *Pseudomonas gessardii* and *Erwinia rhapontici* on amaranth (*Amaranthus hypochondriacus*) under greenhouse conditions using field-collected soil. Inoculation with these strains resulted in significant improvements in plant growth and physiological parameters compared with the untreated control. Similarly, in pot experiments, a bacterial consortium comprising *Enterobacter ludwigii*, *Micrococcus indicus*, and *Pseudomonas gessardii* improved growth and enhanced key physiological indicators, including chlorophyll content, proline accumulation, and antioxidant enzyme activity, in eggplant (*Solanum melongena*) (Kaur et al. 2024). In a polyhouse experiment, Desai et al. (2020) showed that the addition of *Bacillus sonorensis* and *Funneliformis mosseae* to the planting medium in propagation trays significantly improved growth compared to uninoculated seedlings of tomato (*Solanum lycopersicum*) and capsicum (*Capsicum annuum*). In a field experiment, inoculation of potato plants with a combination of *Microbacterium laevaniformans* or *Pantoea agglomerans* strains and *Pseudomonas putida* resulted in increased plant growth, as evidenced by higher root and aerial dry weights, nitrogen content, and final yield (Malboobi et al. 2009). Kumar et al. (2021) developed indigenous PGP consortia composed of a tetra-strain combination (*Bacillus megaterium, Arthrobacter chlorophenolicus*, *Enterobacter* sp., and *Pseudomonas aeruginosa*) that significantly enhanced wheat growth, yield, and nutrient acquisition under both greenhouse and field conditions. Under pot conditions, the consortium significantly increased plant height, grain yield by 79.8%, straw yield by 78.6%, and test weight by 17.5, 79.8, 78.6, and 26.7%, respectively, whereas under field conditions, increases of 29.4, 27.5, 29.5, and 17.6% were observed for these parameters. In another study, Molina-Romero et al. (2017) showed that a consortium of desiccation-tolerant bacteria (*Pseudomonas putida*, *Acinetobacter* sp., *Sphingomonas* sp., and *Azospirillum brasilense*) adhered to seeds and colonized the rhizosphere of maize plants. This experiment showed that different bacterial consortia can coexist with plants without antagonistic effects and increase maize height, diameter, and shoot–root dry weight compared with the uninoculated control.

7.4.1 Plant Growth Stimulation Under Stress Conditions

Stress conditions can impair plant morphology, physiology, biochemistry, and gene regulation, as well as soil microbial and physicochemical properties, resulting in considerable yield loss (Goswami and Suresh 2020). Under stress conditions, plant stress manifests as reduced photosynthetic efficiency, oxidative damage through reactive oxygen species (ROS) accumulation, impaired nutrient uptake, ion toxicity, membrane instability, hormonal imbalances, and ultimately stunted growth and diminished yields (Anjum et al. 2011; Kumar and Verma 2018). Microbial consortia composed of selected PGP microorganisms have emerged as a sustainable biotechnological solution for alleviating multifaceted stresses, including drought, salinity, and environmental contaminants (Meena et al. 2023; Sahoo et al. 2025).

7.4.1.1 Drought Stress

Under drought stress, inoculation with microbial consortia can assist plants in regulating water uptake and photosynthetic performance and in producing stress-related metabolites to mitigate drought-induced damage (Saleem et al. 2021). For example, tomato plants grown under drought stress in pots inoculated with a hexa-strain PGP consortium (*Bacillus megaterium*, *Pseudomonas fluorescens*, *P. aeruginosa*, *P. putida*, *Paenibacillus polymyxa*, and *Trichoderma harzianum*) exhibited lower cellular damage, improved growth, and higher yield under drought stress than the uninoculated control (Krishna et al. 2022). These beneficial effects are frequently associated with the activation of antioxidant defense mechanisms, such as changes in catalase, superoxide dismutase, and ascorbate peroxidase activities. Tyagi et al. (2023) demonstrated higher activity of these antioxidant enzymes under drought stress, followed by inoculation with multistrain consortia (*Rhizophagus intraradices*, *Azotobacter chroococcum*, and *Serendipita indica*). This inoculation also had a positive effect on maize growth, as shown by increases in root and shoot lengths, plant height, and chlorophyll content (Tyagi et al. 2023).

Drought stress often elevates ethylene levels in plants, restricting root and shoot growth (Chandwani and Amaresan 2022). Microbial ACC deaminase mitigates this effect by degrading excess ACC, an ethylene precursor, thereby restoring hormonal balance and promoting plant growth (Goswami and Suresh 2020; Naing et al. 2021). For example, Saikia et al. (2018) observed that in black gram (*Vigna mungo*) and pea (*Pisum sativum*) under drought conditions, inoculation with a consortium of three ACC-deaminase-producing rhizobacteria (*Ochrobactrum pseudogrignonense*, *Pseudomonas* sp., and *Bacillus subtilis*) significantly improved seed germination, root and shoot length, and dry weight. A similar effect was also reported by Rehman et al. (2022), where moringa (*Moringa oleifera*) grown under water deficit was inoculated with ACC-deaminase-producing endophytic fungal consortium (*Trichoderma gamsii* and *Fusarium proliferatum*), resulting in enhanced root and shoot growth as well as higher fresh and dry biomass. Thus, ACC-deaminase-producing microbial consortia play a critical role in mitigating ethylene-induced growth inhibition under drought stress.

Plant adaptation to drought is commonly assessed by changes in plant biomass (Eziz et al. 2017; Yan et al. 2023; Tufail et al. 2022). Supporting this, a meta-analysis revealed that PGP consortia exert greater effects on root and shoot length and biomass than single-strain inoculants under drought conditions (Zhao et al. 2023b). In a recent study, eight bacterial consortia constructed from 25 bacterial isolates with more than three functions or strong growth-promoting ability significantly improved the root dry weight of Mongolian milkvetch (*Astragalus propinquus*) by 78.8% under drought stress (Lin et al. 2022). The functional advantage of multistrain consortia over single inoculants has been further validated. For instance, Benmrid et al. (2025) evaluated three drought-tolerant bacterial consortia, each comprising functionally diverse PGP strains. These consortia were designed to enhance stability and synergistic interactions among individuals with varying PGP traits. All designed consortia emerged as highly effective under low phosphorus availability and drought stress, while the consortium composed of *Rhizobium laguerreae*, *Psychrobacillus psychrodurans*, *Bacillus tequilensis*, and *Arthrobacter oryzae* strains demonstrated the most significant increase, as observed by higher shoot dry weight in faba bean (*Vicia faba*) (by 50%) and wheat (by 41%). Similarly, Notununu et al. (2024) demonstrated that a four-member consortium (*Bacillus cereus*, *B. pseudomycoides*, *Lelliottia amnigena*, *Leclercia adecarboxylata*) significantly improved maize tolerance to combined drought and heat stress in greenhouse trials, with the consortium increasing root dry weight by 33% and shoot dry weight by 48% compared to the control while simultaneously enhancing leaf relative water content by 32%, a critical indicator of cellular hydration status and drought adaptation.

7.4.1.2 Salinity Stress

Salinity stress imposes multifaceted constraints on plant productivity by inducing osmotic stress, which restricts water availability, triggers ionic toxicity through excessive ion accumulation, and disrupts nutrient homeostasis, resulting in impaired root architecture, reduced photosynthetic efficiency, and stunted growth (Balasubramaniam et al. 2023; Hasanuzzaman and Fujita 2022). These effects can be mitigated by microbial consortia. Some PGP microorganisms are halotolerant and employ multifaceted strategies to alleviate salinity stress, thereby enabling plants to maintain metabolic function, structural integrity, and productivity, even under severe saline conditions.

One characteristic of halotolerant rhizobacteria is their capacity to produce diverse extracellular hydrolytic enzymes that facilitate the decomposition of organic matter, promote nutrient cycling, and support plant growth under saline conditions. For example, Damodaran et al. (2023) engineered halotolerant rhizobacterial consortia comprising *Lysinibacillus fusiformis*, *L. sphaericus*, and *Bacillus licheniformis*, which produced extracellular enzymes such as amylase, protease, cellulase, and lipase. This consortium enhanced tomato growth and yield in sodic soils by increasing IAA production, ACC deaminase, and antioxidative enzyme activity and efficiently excluding Na^+ ions from the rhizosphere (Damodaran et al. 2023). Similarly, Wael et al. (2024) isolated five halotolerant plant-associated bacterial strains (*Bacillus amyloliquefaciens*, *Providencia rettgeri*, *B. licheniformis*, *Salinicoccus sesuvii*, and *Paenalcaligenes suwonensis*) from three halophytic plant species. In a pot experiment with faba bean (*Vicia faba*), these strains demonstrated enhanced growth and stress tolerance compared with uninoculated controls. A study by Kapadia et al. (2021) demonstrated that tomato growth under salinity stress may be improved by using a tetra-inoculant (*Bacillus* sp., *Delftia* sp., *Enterobacter* sp., *Achromobacter* sp.). Inoculation with this consortium resulted in more secondary roots, leading to increased mineral uptake and transport throughout the plant, thereby increasing biomass and growth (Kapadia et al. 2021). Cortes-Tolalpa et al. (2018) developed a halotolerant lignocellulose-degrading microbial consortium in the laboratory through enrichment of the salt marsh microbiome. The dominant bacteria in the consortium were *Joostella marina*, *Flavobacterium beibuense*,

Algoriphagus ratkowskyi, *Pseudomonas putida*, and *Halomonas meridiana*. These bacterial consortia demonstrated superior performance on highly recalcitrant materials, achieving higher cellulose and lignin degradation (Cortes-Tolalpa et al. 2018).

Bacterial consortia can also help plants modulate ionic homeostasis under salinity stress by producing EPS. Thakur and Yadav (2024) demonstrated that a biofilm-forming, EPS-producing, halotolerant bacterial consortium significantly enhanced wheat salt tolerance under salt stress by increasing shoot length, chlorophyll content, fresh weight, and membrane stability while reducing electrolyte leakage. Moreover, electron microscopy revealed that EPS directly binds to sodium and chloride ions, thereby sequestering these elements in the rhizosphere and preventing their uptake into plant tissues.

7.4.1.3 Heavy Metal Stress

Heavy metal contamination imposes severe physiological constraints on crops, primarily by inducing oxidative stress through ROS overproduction, disrupting electron transport chains, inhibiting key metabolic enzymes, and displacing essential cofactors like zinc and iron in metalloproteins (Chen et al. 2023; Devi et al. 2022b; Kaushal and Pati 2025; Mohamed et al. 2025). Consequently, heavy metal stress disrupts overall plant physiology by impairing photosynthesis and nutrient uptake, restricting root and shoot growth, and ultimately reducing plant productivity (Nagajyoti et al. 2010). Although plants employ endogenous defense mechanisms, such as phytochelatin synthesis and the activation of antioxidant enzymes, these are often insufficient under high contamination levels (Michalak 2006; Sytar et al. 2013). Microbial consortia work simultaneously to immobilize metals, detoxify the rhizosphere, and promote plant growth via diverse synergistic mechanisms.

Microbial consortia enhance the plant antioxidant defense system by upregulating genes encoding superoxide dismutase, catalase, and peroxidase. For example, in a controlled experiment, Pandey et al. (2023) investigated the effect of a bacterial consortium of *Bacillus nealsonii*, *Pseudomonas nitritireducens*, *Exiguobacterium aestuarii*, *Bacillus tequilensis*, and *Microbacterium paraoxydans* on arsenic-stressed rice plants. Inoculation with the consortium significantly reduced oxidative stress by upregulating the genes encoding antioxidant enzymes. Similarly, under cadmium stress, Seang-On et al. (2023) evaluated a biofertilizer containing a cadmium-resistant microbial consortium on rice growth in a pot experiment. The biofertilizer significantly mitigated cadmium toxicity by modulating photosynthetic pigments, enhancing antioxidative activity, and reducing oxidative lipid damage. These effects resulted in increased shoot and root biomasses and reduced grain cadmium content compared to the control (Seang-On et al. 2023).

Microbial consortia solubilize essential nutrients while simultaneously immobilizing heavy metals through two distinct mechanisms: (i) chelation through the secretion of organic acids (e.g., gluconic, succinic, and citric acids) that bind metals and reduce their bioavailability and (ii) microbially mediated precipitation of insoluble metal salts. The combined metabolic activity of a consortium produces a broader spectrum of organic acids, thereby enhancing its chelation capacity for diverse heavy metals at multicontaminated sites, which is a common agricultural problem. For example, in a pot experiment with turnip (*Brassica rapa*), Wang et al. (2020) demonstrated that a consortium of *Enterobacter* sp. and *Comamonas* sp. immobilized the cadmium through synergistic biofilm formation, hydrogen sulfide-mediated cadmium sulfide precipitation, and succinic acid-driven adsorption, significantly reducing plant cadmium uptake. Furthermore, a recent study on microbially induced calcium carbonate precipitation by ureolytic consortia confirmed the stabilization of metals such as cadmium (99% removal) and nickel (77% removal) through coprecipitation within calcite crystals, highlighting the enhanced remediation potential of multispecies inoculants over single strains in multimetal-contaminated agricultural environments (Omoregie et al. 2025).

7.4.2 *Biocontrol Applications*

7.4.2.1 Modes of Action of Microbial Biocontrol

Direct or indirect pathogen suppression mechanisms are employed in the design of microbial biocontrol inoculants. Beneficial microorganisms can directly antagonize pathogens by secreting antimicrobial compounds and enzymes or by outcompeting them for space and nutrients. Direct antibiosis includes the synthesis of antimicrobial compounds, such as lipopeptides (surfactins, fengycins, and iturins), lytic enzymes (proteases, cellulases, chitinases, chitosanases, and beta-1,3-glucanase), 2,4-diacetilfosfoglucerate, pyrrolnitrin, phenazine, hydrogen cyanide, ammonia, and other metabolites and enzymes (Ayaz et al. 2023; Nguyen et al. 2025). These antimicrobial compounds exert their toxic effects on pathogens through the inhibition of key physiological processes (growth and spore germination), blocking essential metabolic pathways, inactivation of virulence factors, induction of oxidative stress, or disruption of cellular structure (Ayaz et al. 2023; Al Raish et al. 2025; Yu et al. 2022; Köhl et al. 2019; Nguyen et al. 2025).

Furthermore, microbial biocontrol agents can directly suppress plant pathogens through niche exclusion and competition for resources. Accordingly, the presence and propagation of nonpathogenic microorganisms on the surface or in the rhizosphere can benefit plants by providing a barrier against potential pathogens. However, most biocontrol agents rapidly colonize surfaces, grow quickly, and deploy specialized traits to dominate the available resources. Competition for nitrogen, oxygen, and micronutrients limits pathogen proliferation. Siderophore production is a common pathogen growth-suppressing mechanism employed in biocontrol (Backer et al. 2018; Köhl et al. 2019; Miethke and Marahiel 2007). In iron-limited environments, high-affinity siderophores produced by biocontrol microorganisms chelate iron, thereby making it unavailable to pathogens with lower-affinity siderophore systems. In addition, beneficial microorganisms thrive on plant-secreted root exudates or wound leachates, starving pathogens of energy for germination and growth. Root exudates play an important role in recruiting rhizosphere microorganisms (Park et al. 2023). Some microorganisms can affect plant physiology by altering root exudate composition, thereby affecting the composition and functional balance of the rhizosphere microbiota and exerting complex, multidimensional effects on the plant (Matilla et al. 2010).

Alternatively, biocontrol microbial agents can indirectly improve plant disease resistance by stimulating overall plant growth and vigor. Many PGP bacteria promote root development by the production of phytohormones. These microorganisms primarily synthesize auxins, such as IAA and other indole-related compounds, which stimulate root elongation, increase lateral root formation, and improve the overall root architecture (Etesami and Glick 2024). Additionally, many PGP bacteria produce gibberellins and cytokinins, which further support root and shoot growth by promoting cell division, differentiation, and enhanced root hair development (Backer et al. 2018). A robust root system enables better nutrient and water uptake, leading to improved development of structural barriers against infection (such as thicker cell walls and higher accumulation of defensive compounds) and the building of energy reserves required to support the plant's response to infection and stress. Furthermore, the nutrient acquisition facilitation properties of beneficial microorganisms, such as phosphate solubilization, nitrogen fixation, and siderophore production, contribute to optimal plant nutrition and health (Backer et al. 2018; Glick 2012). Another indirect biocontrol effect is the priming of the plant immune system through induced systemic resistance (Dimkić et al. 2022; Yu et al. 2022). Mediated by salicylic acid, jasmonic acid, or ethylene signaling pathways, induced systemic resistance leads to broad-spectrum protection against diverse pathogens. Biocontrol microorganisms can also disrupt quorum-sensing communication among pathogens through various mechanisms, including the degradation of quorum-sensing signal molecules (Fan et al. 2020; Zhu et al. 2023).

7.4.2.2 Composition, Compatibility, and Stress Tolerance of Biocontrol Consortia

A consortium of diverse biocontrol microorganisms is a promising tool for effective pathogen suppression, offering advantages over single-strain formulations. Combining multiple strains enables a broader spectrum of direct and indirect biocontrol mechanisms, and consortia can be more effective in specific niches and nutrient condition competition (Minchev et al. 2021). Furthermore, a multispecies consortium can target multiple pathogen groups and diseases (Niu et al. 2020). Different biocontrol strains can have distinct environmental preferences, occupy different niches in the plant environment, and utilize different carbon sources, thereby covering a wide range of habitats and reducing opportunities for pathogens to invade, thereby increasing robustness across variable field conditions (Ntakirutimana et al. 2024). Additionally, an important benefit of the consortium is its positive effect on the host plant, as multiple strains deliver both biocontrol and plant growth promotion (e.g., improved germination, biomass, and yield) through the combined PGP traits (Liu et al. 2018).

A combination of multiple species or strains provides resilience to the biocontrol system, as consortia can retain functionality even when environmental factors disable certain members. Thus, there are greater opportunities to maintain a stable biocontrol effect across environmental conditions shaped by plant genetics, soil type, agricultural methods, and climatic challenges during the vegetative period of the plant. Even when a single strain outperforms biocontrol, the presence of several interdependent partners can enhance overall process stability. For example, the removal of *Enterobacter cloacae* from a 7-member consortium disrupted the balance of the consortium (Niu et al. 2017). Furthermore, the induction of secondary metabolism in biocontrol strains can be triggered by competitive conditions within the consortia (Maciag et al. 2023). Bacterial and fungal consortia can aid each other during colonization; for instance, fungal partners can aid bacterial biofilm formation, and bacteria can promote root colonization by fungi (Bo et al. 2022). Synergistic interactions among bacterial taxa have also been reported previously. Notably, three bacterial rhizosphere isolates of *Xanthomonas*, *Microbacterium*, and *Stenotrophomonas* sp. showed synergistic biofilm formation in response to downy mildew (*Hyaloperonospora arabidopsidis*) infection in *Arabidopsis* roots (Berendsen et al. 2018). Additionally, certain microorganisms can facilitate the spread of other species by transporting bacteria or fungal spores, further promoting the establishment of the community (Niu et al. 2020). Despite these cooperative interactions, microorganisms within a consortium often exhibit distinct growth dynamics, community establishment, and bioactive compound synthesis; therefore, their individual contributions to biocontrol may differ. In some cases, the reverse phenomenon is possible, as confirmed by the case described by Kaki et al. (2026), in which an individual strain exhibited higher biocontrol efficiency than a broader multistrain consortium, but its metabolic and biochemical potential (including the synthesis of bioactive compounds) was not realized under consortium conditions.

It is important to assess the compatibility of biocontrol agents to avoid antagonistic interactions. Consortia members are chosen primarily for their antagonistic properties, including parasitism, predation, antibiosis, competition, and the production of lytic enzymes and antibiotic compounds. However, these properties, along with the disruption of quorum-sensing communication, can suppress other biocontrol agents or reduce the likelihood of successful colonization. Therefore, compatibility analysis is usually performed using pairwise tests on media. Pairwise tests can also reveal synergistic stimulation between strains (Berendsen et al. 2018). Although such tests help eliminate cases of direct antagonism between species, they only partially reflect the actual interactions among the members of the consortium. For example, the interaction between *Streptomyces microflavus* and *Trichoderma harzianum* was context-dependent: they showed antagonism on agar plates, *T. harzianum* suppressed the growth of *S. microflavus* in liquid cocultures, and in semicontrolled conditions with sterilized soil, the growth of *T. harzianum* was negatively affected (Prigigallo et al. 2023). Together, they showed a synergistic biocontrol effect on tomato seedlings, inducing stronger resistance responses than those induced by single inoculants. In vitro antagonistic cultures can occupy different niches; therefore, they do not necessarily compete directly under natural conditions. However, if two species cooperate through metabolite exchange, adding metabolites can shift their relationship from cooperation to competition (Lindemann et al. 2016). One can expect antagonistic or cooperative relationships to be influenced by environmental factors, such as the abundance of specific nutrients or the presence of other species involved in the metabolism network.

One of the most common approaches for biocontrol consortia involves combining fungal and bacterial strains, typically pairing *Trichoderma* spp. or arbuscular mycorrhizal fungi (*Rhizophagus* and *Glomus* spp.) with bacteria of *the Bacillus* or *Pseudomonas* genera, chosen for their proven compatibility and complementary modes of action (Joshi et al. 2025; Niu et al. 2020; Poveda and Eugui 2022). *Trichoderma* acts as a mycoparasite and a strong trigger of induced systemic resistance; therefore, it is used as a key component of microbiological inoculants in agriculture (Kredics et al. 2024). Several *Trichoderma* products are commercially available (Maciag et al. 2023; Menéndez and Paço 2020). Furthermore, mixed consortia, including bacteria, fungi, and nematodes, have been shown to increase insect mortality (Spescha et al. 2023). Consortia with fewer members are easier to manage, more applicable, and can be equally effective, yet greater diversity may buffer against the unpredictable conditions encountered under natural environments.

7.4.2.3 Use of Microbial Consortia as Biocontrol Agents

In agricultural ecosystems, plant health is shaped by complex interactions between host plants and pathogens and the resident microbiome. The use of microbial consortia as bioinocula leverages this complexity by assembling multiple beneficial microbes with distinct ecological roles to enhance disease suppression and resilience to pathogen invasion.

Microbial consortia have emerged as an effective strategy for controlling bacterial plant diseases, often simultaneously promoting plant growth and activating host defense response. Bioinoculant consortia have consistently exhibited disease-suppressive effects across crops. For example, Wang et al. (2019) showed that a rhizosphere-derived consortium consisting of *Bacillus subtilis* and *Trichoderma harzianum* significantly reduced potato common scab caused by *Streptomyces* spp. while simultaneously increasing potato yield. More recent studies have extended these findings to bacterial leaf blight in rice, caused by *Xanthomonas oryzae* pv. *oryzae*. Bora et al. (2025) demonstrated that a consortium of *Trichoderma asperellum* and *Pseudomonas fluorescens* effectively reduced disease severity while simultaneously improving rice biometric traits and modulating host defense responses. Similarly, Sai Teja et al. (2025) evaluated a more complex consortium comprising *Trichoderma harzianum*, *T. asperellum*, and *Pseudomonas fluorescens* under field conditions in Basmati rice. This treatment reduced bacterial leaf blight severity, accompanied by increased accumulation of defense-related molecules, including superoxide dismutase and hydrogen peroxide, indicating that the microbial consortia activated plant immune responses. The applicability of consortium-based biocontrol was also demonstrated in common bean, as Sallam and Aldayel (2025) reported that a synergistic combination of *Rahnella aquatilis* and *Trichoderma orientale* effectively controlled bacterial blight caused by *Xanthomonas axonopodis* pv. *phaseoli*. In addition to reducing disease incidence, this consortium significantly improved seed germination rates, further highlighting the multifunctional benefits of microbial consortia for bacterial disease management.

Microbial consortia have been particularly effective in managing fungal and oomycete plant pathogens, which are often persistent in the soil, exhibit broad host ranges, and are difficult to control using single biocontrol agents. Field-based evidence for the effectiveness of microbial combinations was provided by Lucas et al. (2009), who demonstrated that a combination of *Pseudomonas fluorescens* and *Chryseobacterium balustinum* significantly reduced rice blast caused by *Pyricularia oryzae*. Additional examples include endophytic bacterial consortia mixed from two *Bacillus* spp., *B. subtilis*, *Pseudomonas fluorescens*, and two *Serratia marcescens* strains, which were effective in suppressing purple blotch caused by *Alternaria porri* and improving the growth and yield of shallots (*Allium ascalonicum*) (Resti et al. 2021).

Subsequent studies have predominantly focused on major soil-borne fungal pathogens, particularly species of *Fusarium*, *Rhizoctonia*, and *Sclerotium*, which are frequently targeted using multistrain consortia because of their ecological persistence and broad host range (Niu et al. 2020). In tomato, Devi et al. (2022a) reported that arbuscular mycorrhizal fungi (*Funneliformis mosseae* and *Glomus fasciculatum*) combined with *Bacillus velezensis* and *Bacillus* spp. significantly reduced wilt caused by *Fusarium oxysporum* f. sp. *lycopersici* while simultaneously enhancing plant growth and chlorophyll content under both pot and field conditions. In cluster bean (*Cyamopsis tetragonoloba*), Singh et al. (2024) reported that a consortium comprising *Trichoderma afroharzianum*, *Pseudomonas fluorescens*, *Bacillus licheniformis*, and *B. subtilis* reduced root rot caused by *Rhizoctonia bataticola*. This effect coincided with the increased accumulation of phenolics, flavonoids, and antioxidant enzymes, ultimately resulting in increased cluster bean growth, biomass, pod number, and seed yield. Recently, Tian et al. (2024) demonstrated that a synthetic bacterial consortium effectively suppressed tree peony (*Paeonia suffruticosa*) root rot caused by *Fusarium solani* and *F. oxysporum*, accompanied by enhanced plant growth, as indicated by increased plant height and root length. Complementing these findings, Narine et al. (2025) demonstrated that indigenous *Bacillus* and *Pseudomonas* isolates were effective against tomato rot and wilt caused by *Sclerotium rolfsii* and *F. oxysporum* f. sp. *lycopersici*. The consortium not only suppressed *Sclerotium* and *Fusarium* rot and wilt but also significantly enhanced plant growth, reinforcing the dual role of bacterial consortia in disease control and growth promotion.

Further experimental evidence for the benefits of microbial consortia in brinjal (*Solanum melongena*) cultivation was provided by Upamanya et al. (2020), who evaluated integrated pest and disease management modules under field conditions. The most effective treatment comprised a consortium of the fungi *Beauveria bassiana*, *Metarhizium anisopliae, Trichoderma harzianum*, and a biofertilizer mixture containing *Rhizobium* sp., *Azotobacter* spp., and *Bacillus* sp. Application of this consortium significantly reduced fungal diseases, including Phomopsis fruit rot (*Phomopsis vexans*), Cercospora leaf spot (*Cercospora melongenae*), and early blight (*Alternaria solani*), while also improving plant growth parameters, yield, soil nutrient availability, and microbial biomass carbon. Minchev et al. (2021) reported that consortia including *Bacillus* spp., *Pseudomonas* spp., *Trichoderma* spp., and the arbuscular mycorrhizal fungus *Rhizophagus irregularis*, applied via seed inoculation, suppressed the root pathogen *Fusarium oxysporum* and induced resistance against the foliar pathogen *Botrytis cinerea* in tomato. The broad applicability of consortium-based approaches was also demonstrated by Kampa et al. (2025), who showed that a *Trichoderma asperellum*–*Bacillus subtilis* consortium effectively controlled chickpea wilt (*F. oxysporum* f. sp. *ciceris*) and groundnut (*Arachis hypogaea*) collar rot caused by *Aspergillus niger*.

Similar to fungal pathosystems, microbial consortia have been successfully applied against oomycete pathogens, which are phylogenetically distinct from true fungi and are often difficult to control using single biocontrol agents. Phytophthora blight of pepper, caused by *Phytophthora capsici,* has served as a prominent model. Kim et al. (2008) demonstrated that a consortium of chitinolytic bacteria (*Serratia plymuthica*, *Chromobacterium* sp., and *Lysobacter enzymogenes*) significantly reduced disease severity under field and greenhouse cultivation in chilli pepper (*Capsicum annuum*). Consistent with these findings, Zhang et al. (2019) reported that a consortium of PGP bacteria (e.g., *Bacillus cereus*, *B. subtilis*, and *Serratia* sp.) suppressed sweet pepper (*Capsicum annuum*) phytophthora blight by restructuring the rhizosphere microbiota and improving soil nutrient availability. Consortia have also proven to be effective against *Pythium* spp.-associated diseases. Muthukumar et al. (2011) showed that a combined application of *Trichoderma viride* and *Pseudomonas fluorescens* significantly reduced chilli pepper damping-off caused by *Pythium aphanidermatum* while enhancing defense-related enzyme activities and phenolic compound accumulation. Recently, Kaki et al. (2026) demonstrated that a diverse bacterial consortium effectively promoted corn (*Zea mays*) growth and suppressed root rot caused by *Pythium torulosum*, further supporting the role of functionally complementary consortia as a possible biocontrol solution.

Microbial biocontrol consortia have been increasingly explored for the management of plant-parasitic nematodes and insect pests, as they allow the combination of microorganisms with complementary ecological functions and modes of action. By integrating multiple biocontrol mechanisms, these consortia often provide more consistent and robust pest suppression than single-strain applications. Several studies have demonstrated the effectiveness of microbial consortia against nematodes. For example, Bo et al. (2022) showed that the nematophagous fungus *Pochonia chlamydosporia*, when combined with *Bacillus nematocida*, significantly reduced populations of the root-knot nematode *Meloidogyne incognita* under field conditions in tomatoes. The addition of *B. nematocida* enhanced *P. chlamydosporia* rhizosphere colonization and altered root-associated volatile profiles, including hexanal, furfural, benzaldehyde, and 2-nonanone, which are implicated in nematode suppression (Bo et al. 2022). Similarly, Asyiah et al. (2020) reported that a consortium of *Bacillus* and *Pseudomonas* species effectively suppressed the lesion nematode *Pratylenchus coffeae* while promoting the growth of coffee (*Coffea canephora*). Nematode suppression was positively correlated with increased plant height, number of leaves, and leaf area.

Beyond nematodes, microbial consortia provide an opportunity to control other taxonomically distinct soil-borne pathogens. Zhang et al. (2022) demonstrated that an intergenus bacterial consortium composed of *Bacillus cereus*, *Lysobacter antibioticus*, and *Lysobacter capsici* effectively suppressed clubroot disease of cabbage (*Brassica rapa*), caused by the soil-borne protist *Plasmodiophora brassicae*. The consortium produces a broad spectrum of secondary metabolites that significantly reduce disease incidence. In addition, the treatment altered the rhizosphere bacterial community composition and reduced soil acidity, thereby improving plant yield and disease suppression.

Microbial consortia have also been successfully applied to insect pest management across diverse crop systems. For example, a consortium comprising *Pseudomonas chlororaphis*, *Steinernema feltiae* with *Xenorhabdus bovienii*, and *Metarhizium brunneum* increased insect mortality and reduced time to kill in both foliar-feeding (*Pieris brassicae*) and root-feeding (*Diabrotica balteata*) pests in maize, demonstrating effective control across distinct insect feeding guilds (Spescha et al. 2023). Similarly, in tomatoes, both *Beauveria bassiana* and *Metarhizium anisopliae* colonized plant tissues and suppressed larval development and weight of the fall armyworm (*Spodoptera frugiperda*) under greenhouse conditions (Mwamburi 2021). In maize, the addition of endophytic *B. bassiana* and *M. anisopliae* led to high larval mortality in *S. frugiperda*, with both fungi detected in roots, stems, and leaves, resulting in >75–100% mortality in early instars (Ramos et al. 2020). Additionally, greenhouse and field studies combining *B. bassiana*, *M. anisopliae*, and *Bacillus thuringiensis* against the tomato leaf miner (*Tuta absoluta*) showed significant reductions in leaf and fruit damage and improved marketable yield (Aynalem et al. 2022). Similarly, in corn, inoculation with PGP strains (*Bacillus pumilus*, *B. subtilis*, and *Fictibacillus solisalsi*) reduced the performance and attractiveness of the common cutworm (*Spodoptera litura*), with larvae exhibiting lower body weights and relative growth rates. Consortium-treated plants also exhibited less leaf damage and received fewer eggs from ovipositing females than untreated controls (Cortez et al. 2024).

The functional versatility of microbial consortia extends to the simultaneous management of insect pests and other plant pathogens. Senthilraja et al. (2013) showed that a consortium of the entomopathogenic fungus *Beauveria bassiana* and the PGP rhizobacterium *Pseudomonas fluorescens* simultaneously reduced damage by collar rot caused by *Sclerotium rolfsii* and decreased infestation by the leafminer *Aproaerema modicella* in peanut (Senthilraja et al. 2013).

7.4.3 *Metagenome-Guided Design of Bioinoculants*

7.4.3.1 Metagenome-Guided Design of Microbial Consortia

Microbial SynComs composed of an arbitrary combination of microbial strains based on single PGP activities often fail to enhance bioinoculant performance, as these consortia fail to establish ecologically relevant cross-feeding and antagonistic interactions between the inoculated or indigenous soil microbiota. Thus, integrating soil microbiome data into the design of microbial inoculants represents a transformative approach to enhancing their efficacy. The application of high-throughput sequencing and computational modeling reduces the need for trial-and-error iterations in SynCom design, helping to identify beneficial microbial taxa and functional genes that promote nutrient cycling, stress resilience, and disease suppression in crops. This approach allows (i) the incorporation of keystone taxa and syntrophic partners identified in situ, (ii) the prioritization of functionally validated or predicted PGP genes, and (iii) the identification of traits that, in addition to specific beneficial functions, could also address host colonization efficacy or antagonistic and cross-feeding intermicrobial interactions. Thus, metagenome-guided rational design enables researchers to formulate minimal reproducible consortia with enhanced host colonization and stability, which are adapted to specific soil and crop environments (De Souza et al. 2020). Currently, two strategies are used for the metagenome-guided rational design of SynComs. Taxonomy- or PGP function-based approaches leverage the strengths of available metagenomic data generated using different DNA sequencing methods while addressing their current limitations in taxonomic resolution, functional annotation, or ecological context.

Taxonomy-based strategies analyze microbiome composition across diverse natural environments, including the identification of a core microbiome or keystone taxa. Taxonomic assessment of microbiomes can be readily implemented using widely available amplicon-based (primarily 16S rRNA gene or internal transcribed spacer region) short-read metataxonomic data. This method remains the most accessible and cost-effective for large-scale taxonomic profiling of the microbiome. Taxonomic resolution, at least at the genus level, is sufficient for SynCom design (Delgado Baquerizo 2022). However, short-read amplicon data provide only limited information on the functional attributes of the microbiome. Although broad functional differences across diverse environments can be identified from such data, their sensitivity in detecting subtle changes or specific gene functions that could be useful for SynComs design is limited (Liu et al. 2021; Matchado et al. 2024).

The implementation of function-based metagenome-guided rational design of SynComs requires data from shotgun metagenomic sequencing, which can be used in combination with deep amplicon sequencing of functional gene markers (*nifH*, *phoD*, *acdS*, etc.) to identify taxa carrying key PGP traits (e.g., nitrogen fixation, phosphate solubilization, ACC deaminase activity, siderophore production, or pathogen antagonism) and other metabolic capabilities essential for the design of microbial consortia, such as host colonization or intermicrobial interactions (Toju et al. 2018; Vorholt et al. 2017). Metagenomic data can be further supplemented with metatranscriptomics or metaproteomics to compare functional profiles of healthy, high-yielding plants or distinct soil environments. In addition, the number of microbial genomes available in public databases is rapidly expanding, providing an important foundation for identifying microbes with the desired traits (Cole et al. 2017; De Souza et al. 2019; Levy et al. 2017). This allows for the selection of strains that reconstitute the desired functions and ecological interactions in the designed SynComs.

Computational techniques can be used to process large datasets and bridge insights from controlled laboratory experiments with the outcomes expected in dynamic natural ecosystems. Sophisticated metagenome analysis methods and metabolic modeling techniques are used to identify keystone functional taxa and genes required to design robust and effective SynComs (De Souza et al. 2020; Du et al. 2023). Given the vast scale of modern omics datasets, there is growing interest in the application of machine learning and artificial intelligence methods for identifying promising microbial candidates from large-scale metagenomic or genomic data (De Souza et al. 2020; Zhao et al. 2023a). In addition, complex mathematical modeling is essential for predicting emergent ecosystem functions, as these cannot be deduced by linearly summing the individual contributions of each community member (van den Berg et al. 2022). Several studies have demonstrated the potential of such computational methods for selecting microbial candidates for SynCom design or predicting their application outcomes.

A three-step approach to identify candidate taxa for SynComs with PGP potential was developed by Delgado Baquerizo (2022). Initial soil microbiome complexity was reduced by selecting common, relatively ubiquitous taxa across the studied ecosystems. The selected dominant taxa were further classified into eight ecological clusters using correlation networks to predict the taxa that co-occur under similar environmental conditions and potentially coexist in microbial consortia. Finally, the potential consortia were associated with plant productivity using environmental regressions. The complexity of the microbial community was reduced to 0.4% of all investigated taxa using this method. In another study, Goncalves et al. (2023) analyzed 270 metagenome-assembled genomes from nutrient-impoverished soils using multigenome metabolic modeling to reconstruct genome-scale metabolic networks, scoring interspecies complementarity and competition. Key species required to assemble minimal SynComs of five to ten strains were identified and are expected to enhance crop yields through nutrient solubilization, plant hormone modulation, PGP traits, and improved plant–microbe interactions. Although the studies described above have not tested the performance of the selected taxa under experimental or field conditions, several other studies have shown that inoculation with metagenome-guided SynComs significantly enhances nutrient uptake and confers improved stress resilience in host plants.

Chang et al. (2017) conducted a metagenome-wide association study using bulk soil samples from soybean (*Glycine max*) fields and comparing high- and low-productivity plots. Using shotgun metagenomic sequencing, they identified microbiome compositional differences linked to yield variation (explaining ~20–30% of the variance). Random forest machine learning ranked the taxa and functional genes associated with nitrogen fixation and utilization. Herrera Paredes et al. (2018) developed a machine-learning framework to rationally design SynComs and predict plant responses to phosphate starvation in Arabidopsis. The authors constructed a collection of 55 highly abundant bacterial strains isolated from the Arabidopsis rhizosphere and assembled eight partially overlapping SynComs, each comprising 8 to 18 members. These SynComs were inoculated into plants grown under low- and high-phosphate conditions, and shoot phosphate accumulation was quantified as the key host phenotype. To infer the causal relationships between microbiota composition and host phosphate status, the team used two linear approaches and an artificial neural network. The latter method accurately predicted the shoot phosphate content based on microbial community membership. Specific bacterial taxa, such as *Variovorax* sp. or bacteria of the Comamonadaceae family, were identified as enhancers of phosphate accumulation.

Improved plant resilience to abiotic stress and disease has been demonstrated by applying metagenome-guided SynComs. Yamazaki et al. (2025) modeled the performance of more than 800 SynComs using an Elastic Net regression model trained on plant biomass data. The selected model-informed SynComs improved tomato heat stress tolerance under controlled laboratory conditions and enhanced plant growth during field trials. Liu et al. (2025b) analyzed amplicon and metagenomic data from poplar rhizosphere and bulk soil microbiomes under normal conditions and under three stresses (drought, salt, and anthracnose disease). A set of biomarker taxa was identified using Random Forest models and co-occurrence network analysis. Six SynComs incorporating stress-specific microbes, constructed from isolated bacterial strains, were evaluated in inoculation experiments, demonstrating that these communities significantly enhanced poplar fresh weight under stress and upregulated key stress-related genes in poplars. In another study, specific strains were prioritized for the design of protective SynComs from metagenomic data using ML algorithms (Carlstrom et al. 2019; Emmenegger et al. 2023). Random Forest and gradient boosting methods were used to assess microbiome composition data and predict the minimal consortia that could stably colonize *A. thaliana* and protect against bacterial pathogens.

Ruan et al. (2024b) used the metagenome-guided SynComs approach to facilitate the degradation of pesticides by the wheat-associated microbial community. The combined overlapping results of network (within-module (Zi) and among-module (Pi) connectivity) analysis, linear discriminant analysis effect size (LEfSe) biomarker discovery, and Random Forest machine learning were used as keystones for the simplification and optimization of key microbial species in the soil. The results were further refined using genome-scale metabolic models to design a targeted synthetic microbiome that promotes wheat seedling growth. The designed SynComs comprised seven bacterial strains that assisted indigenous bacteria in degrading bromoxynil herbicides. In another study, Ruan et al. (2024a) developed a computational microbiome modeling framework for SynCom design by mimicking the natural metabolic networks of microbial communities to optimize bromoxynil biodegradation performance. Eighteen keystone taxa screened from microbiomes developed in herbicide-contaminated soils were used to model the performance of simplified microbiomes with different combinations of these taxa. The simulation predicted the optimized combination of the four strains, which were subsequently used for experimental testing.

7.4.3.2 Integration of Microbiome Data into the Application of Bioinoculants

Rather than focusing on the design of new inoculants, another research direction emphasizes enhancing bioinoculant efficacy using microbiome-informed strategies. To address the long-standing issue of inconsistent field performance of bioinoculants, detailed soil and plant microbiome profiles could be used (i) to predict inoculant success, (ii) to adjust field management based on microbiome composition, or (iii) to monitor the persistence and impact of inoculants in real time. Microbiome profiling of soils and crops provides valuable insights for selecting bioinoculants with the highest likelihood of success under specific field conditions, thereby improving crop yields, ensuring more consistent performance, and reducing environmental impact.

It has been suggested that metagenome data could provide detailed information on the existing beneficial microbes and their functional traits in soil, which could guide the selection of inoculum strains better matching crop genotype, soil properties, and existing microbial communities for enhanced efficacy (Ptaszek et al. 2023; Saad et al. 2020; Wang et al. 2024c). Large-scale microbiome datasets help identify core beneficial microbes in specific crops and environments (Ling et al. 2022; Masenya et al. 2024; Toju et al. 2018). Bioinoculants containing core taxa are more likely to integrate successfully with native communities, thereby increasing the likelihood of successful colonization and plant benefits. In addition, Lutz et al. (2023) reported that the composition of the soil microbiome, particularly the abundance of pathogenic fungi, can be used to predict plant growth responses to arbuscular mycorrhizal fungal inoculation. This allows for accurate forecasting of inoculation success at the onset of the growing season.

The effects of agricultural practices on the structure and functioning of the rhizosphere microbiome have been researched extensively (Nadarajah and Abdul Rahman 2023; Nwachukwu and Babalola 2022; Schmidt et al. 2019; Town et al. 2022). Thus, microbiome data could guide the selection of agricultural practices (e.g., crop rotation, intercropping, and tillage intensity) that favor the establishment and persistence of microbial inoculants under various soil conditions. For example, Paudel et al. (2023) described the organic and conventional cropping practice-specific effects of rhizobial inoculation on rhizosphere microbial diversity.

Monitoring shifts in microbial communities within agroecosystems could ensure the sustainable and targeted application of inoculants without disrupting ecosystem stability (Wang et al. 2024c). The timing, dosage, and environmental conditions of inoculum application can be optimized based on post-inoculation microbial community dynamics. A metadata review by Mawarda et al. (2020) highlighted that microbial inoculants can induce transient or more persistent changes in indigenous microbial communities. In some cases, introducing pure bacterial strains or SynComs has been shown to increase the complexity of network interactions in the plant rhizosphere, thereby encouraging beneficial plant–microbe interactions and improving nutrient uptake (Da Cunha et al. 2024; Dong et al. 2024). However, studies examining the long-term effects of microbial inoculants often reveal limited resilience in native communities, which frequently fail to return to their initial state following inoculation (Mawarda et al. 2020). Thus, post-inoculation monitoring of soil microbial community resistance could be a helpful indicator to guide bioinoculant use or cropping practices to maintain soil health. However, microbiome analysis currently requires considerable expertise and specialized resources, which restricts its application to research settings. Therefore, accessible, farmer-friendly diagnostic tools that translate complex microbiome data into practical insights are needed.

7.5 Community-Based Modern Rhizoengineering Strategies

7.5.1 Transplantation of NatComs

Soil microbiota are key determinants of plant productivity, nutrition, and resistance to biotic and abiotic stresses. In recent decades, plant research on microbiota interactions has shown that the rhizosphere functions as a complex, coevolved system in which multiple microbial networks support nutrient transformation, phytohormone balance, and pathogen defense (Fitzpatrick et al. 2020; Hassani et al. 2018). However, commercial biopreparations (usually composed of single microorganisms or SynComs) often lose their effectiveness in the field because of competition, environmental stochasticity, and incompatibility with existing soil communities. This discrepancy is called the "ecological complexity gap," where simplified inoculants fail to perform the functions that millions of microorganisms support in natural ecosystems.

The transplantation of NatComs is an emerging and rapidly growing field of rhizoengineering, offering an alternative to traditional biopreparations and SynComs. Unlike SynComs, NatComs are ecologically balanced, naturally coevolved microbial communities capable of providing stable functional effects on plants and soil under field conditions (Fitzpatrick et al. 2020; Toju et al. 2018).

7.5.1.1 Core Principles of NatComs

NatComs transplantation is distinguished by the fact that its functioning does not depend on individual species but on the structure of the entire network. NatComs support a wide range of metabolic interactions, including cross-feeding, niche sharing, competition for substrates, and mutualistic interactions (Hacquard and Martin 2024). These communities have four main characteristics that make them practical tools for rhizoengineering.

NatComs transplantation is characterized by high ecological stability because its structure is supported by functional redundancy and functional groups. This means that several species perform the same functions; therefore, the community maintains its function even after the loss of some taxa. More diverse rhizosphere communities are more resilient to environmental fluctuations and better support nutrient transformation and biocontrol functions (Fitzpatrick et al. 2020; Hartman and Tringe 2019; Hassani et al. 2018). Furthermore, NatComs transplantation is characterized by metabolic activity, as it performs many ecologically essential processes, including phosphorus solubilization, nitrogen transformation, siderophore production, phytohormone synthesis, ACC-deaminase activity, and pathogen antagonism. This complex metabolic network arises from microbial interactions (cross-feeding and complementarity), ensuring that NatComs transplantation functionality is sustained even under stress (Bziuk et al. 2022; Toju et al. 2018). Many NatComs have evolved through coevolution with specific plant species. This results in better rhizosphere colonization, more sensitive responses to root exudates, and higher mutualistic efficiency of the microorganisms. Native plant microbiomes are often more functionally diverse than those of cultivated plant species; therefore, their translocation can enhance productivity and stress resistance (Barnes et al. 2024; Toju et al. 2018). NatComs transplantation can stably integrate into existing soil communities while maintaining their functions. This is important under field conditions, where the transplanted microbiome constantly interacts with the native microbiome. Jiang et al. (2022) showed that NatComs transplantation from disease-suppressing soils can integrate and provide protection even in a competitive rhizosphere.

7.5.1.2 Experimental Evidence: Effects of NatComs Transplants on Plants

NatComs transplantation can significantly enhance plant health and their response to environmental stressors and has great potential for application in sustainable agriculture. They can contribute to pathogen suppression, yield enhancement, improved nutrient availability, enhanced drought and salinity tolerance, and the restoration of soil functions in regenerative systems. These properties make NatComs a promising candidate for rhizoengineering.

Experimental studies suggest that NatComs can be applied in three main areas: (i) plant protection against pathogens and pests, (ii) enhancement of plant physiology and stress tolerance, and (iii) restoration of rhizosphere functions through modulation of community structure. In the context of plant protection, barley rhizosphere transplants from healthy plants have been shown to reduce powdery mildew caused by *Blumeria graminis* (Bziuk et al. 2022), and tomato rhizobiome transfer from bacterial wilt (*Ralstonia solanacearum*)-resistant lines has been shown to increase resistance in susceptible plants (Choi et al. 2020). Together, these studies show that

NatComs transplantation can not only inhibit pathogens but also restore plant immunity at the functional level, thereby fundamentally distinguishing it from single inoculants.

The second application area is the modulation of plant abiotic stress tolerance, where NatComs from Mediterranean plants altered the root architecture, metabolome, and water-use efficiency of sage (*Salvia officinalis*), revealing the role of microbiota in drought response (Abou Jaoudé et al. 2025). The third direction suggests that the functional activity of NatComs depends directly on community composition and diversity. In dilution experiments conducted in the wheat rhizosphere, more diverse communities were more effective at suppressing the fungus *Bipolaris sorokiniana*. The effect of the biocontrol agent *Pseudomonas inefficax* depends on the level of NatComs diversity at which it acts. Some effects are manifested only in the presence of specific community structures (Nishisaka et al. 2025). Finally, the importance of NatComs transplantation in natural plant defense is highlighted using the wild tomato (*Solanum pimpinellifolium*) model. This study shows that when the natural rhizosphere microbiota is removed (by soil sterilization), plants lose their entire resistance to the invasive insect *Prodiplosis longifila*. This suggests that the defense phenotype is not solely a genetic trait of the plant but arises from the interaction between the plant and its NatComs, which likely modulate the plant's metabolic and immune pathways, leading to resistance to pests (Sarango Flores et al. 2025).

7.5.2 Host-Mediated Microbiome Engineering

Traditionally, rhizoengineering strategies have primarily focused on directly modifying soil microbiomes through inoculation with beneficial microbes or soil amendments. However, contemporary rhizoengineering strategies increasingly rely on HMME (Thakur et al. 2023), in which the plant phenotype is used as a selection marker to shape the microbiome over several generations, thereby improving plant growth, stress resistance, and other desirable traits (Jochum et al. 2019). This approach exploits the natural ability of plants to shape and select beneficial microbial communities. While rhizoengineering strategies aim to modify the plant root microbiome to improve plant growth and stress resistance, HMME distinguishes itself by targeting the entire microbial consortium to enhance the plant phenotype rather than individual microorganisms or their functions (Afridi et al. 2022).

Traditional rhizoengineering strategies are usually based on two approaches: the first aims to adjust the structure or function of the microbiota through external interventions (e.g., soil management, nutrient availability, and inoculants), and in the second, synthetic microbial consortia with specific, predefined functions are created and introduced (Lyu et al. 2024). HMME is a host-centric strategy based on phenotypic selection, in which the plant "chooses" the microbiome that best supports its growth under specific conditions (Mueller and Sachs 2015). This approach allows the microbiome to evolve to adapt to the host, as selection occurs over several generations of plants and is based on changes in the plant stress response rather than on predetermined microbial properties (Jochum et al. 2019). Thus, HMME relies on natural plant–microbiome coevolution, providing greater ecological compatibility and potentially more stable results in the long term (Lebeis 2014).

HMME is a relatively recent technique that has emerged from broader microbiome research, particularly since the early 2000s, when high-throughput sequencing revealed the critical roles of individual microbes in complex ecosystems, as outlined in the holobiont concept, and their functional potential for host health (Foo et al. 2017; Kaul et al. 2021). HMME is a significant biological strategy because it leverages the powerful beneficial interactions between hosts and their microbial communities to improve host health (immunity and disease prevention), enhance agricultural sustainability (plant stress tolerance, productivity, and reduced chemical inputs), and drive biotechnology (developing new biofertilizers, biopesticides, and personalized treatments) (Durán et al. 2021).

7.5.2.1 Core Principles of HMME

The theoretical foundation of the HMME strategy is grounded in the assumption that plants are not passive hosts of microbiota; rather, they actively shape the composition of their microbiomes through immune, physiological, and biochemical mechanisms as a selective force to recruit protective microbes while managing potentially harmful ones (Mueller and Sachs 2015; Zhang et al. 2023). First, the plant immune system plays an important role in determining which microorganisms can colonize the rhizosphere and endosphere. Plants recognize microbes through pattern recognition receptors that identify microbe-associated molecular patterns, thereby initiating pattern-triggered immunity (Bulgarelli et al. 2013; Lebeis 2014). This mechanism rejects harmful microbes and preserves beneficial or mutualistic microbes, collectively referred to as the plant's second genome, which is crucial for plant health (Berendsen et al. 2012). Second, plant root exudates, a diverse array of metabolites (sugars, amino acids, organic acids, phenolic compounds, and phytohormones), act as chemosignals and carbon sources, shaping the rhizosphere microbial community according to the plant's physiological state and environmental conditions (Chaparro et al. 2014; Solomon et al. 2024). Third, plant hormonal regulation (particularly abscisic acid, ethylene, jasmonates, and salicylic acid) coordinates both the immune response and the release of root exudates, thus ensuring a dynamic balance between defense and symbiosis in the microbiome (Berendsen et al. 2012). Some plant species also exhibit behavioral traits, such as changes in root architecture or specific growth direction, that allow them to optimize conditions for colonization by beneficial microbes to enhance plant growth. The interaction of these mechanisms forms the basis for the HMME strategy: by leveraging the host's natural ability to shape the microbiome selectively, it enables the targeted enhancement of desired plant–microbe relationships through artificial, yet biologically based, selection (Mueller and Sachs 2015).

One of the fundamental principles of HMME is the application of artificial selection to microbiome communities using the host phenotype as a selection indicator (Jochum et al. 2019). In traditional selection, the target organism or its genotype is the organism itself, whereas in HMME, the microbiome is selected based on its effect on the host phenotype (Mueller and Sachs 2015; Swenson et al. 2000). This indicates that a plant's response to environmental stress or growth parameters (e.g., biomass accumulation, root length, and water stress tolerance) is a direct indicator of the quality and effectiveness of its associated microbiome, which provides beneficial functions, such as nutrient uptake and stress protection (Khan 2023). This strategy involves a cyclical process: (i) first-generation plants are grown in an initial source of microbiota, (ii) individuals exhibiting desired phenotypes are selected, (iii) the microbiome is harvested from their rhizosphere, and (iv) it is used to inoculate the next generation of plants, thus transmitting the selectively shaped microbial community to the next generation. After several cycles of selection, the microbiome acquires properties that enhance the plant's response to the selected factor (Jochum et al. 2019; Mueller and Sachs 2015; Styer et al. 2024). This concept is based on the ecosystem-based artificial selection model, in which the unit of selection is not a single microbe but the entire microbial community, whose functional interactions determine the phenotypic response of the plant (Blouin et al. 2015; Swenson et al. 2000). Thus, HMME enables the logic of artificial selection to be applied not for the genetic enhancement of the host but for the ecologically guided evolution of the microbiome.

In recent years, HMME studies have shifted from taxonomy-based to trait-based and quantitative microbiome assessments. Traditional methods based on 16S rRNA gene sequencing provide information about the composition of the microbiota but do not reveal its functional role. Microbiome-associated phenotypic profiling is increasingly used to elucidate how the microbiome influences host phenotypes (Escalas et al. 2022; Mueller and Sachs 2015; Oyserman et al. 2018). This methodological shift enables the assessment of the influence of functional genes, metabolic pathways, and microbiome traits on host response. These quantitative approaches rely on metagenomics, metatranscriptomics, and metabolomics data, which are integrated with plant phenotypic data to identify causal relationships between microbiome functions and plant traits (Fierer 2017; Widder et al. 2016). This shift to trait-based assessment enables the modeling of the functional properties of the microbiome and prediction of their impact on the host, thereby guiding the microbiome engineering process more precisely. In the context of HMME, this means that microbiome selection can be based not only on the observed phenotype but also on the well-characterized microbial traits that determine the phenotype.

7.5.2.2 Methodologies and Experimental Strategies of HMME

HMME evolved from earlier concepts of artificial ecosystem selection, which focused on applying selection principles to an entire microbial community to repeatedly alter its functional properties (Swenson et al. 2000). This approach provided a methodological basis for later work in which the host plant itself became the agent of microbiome selection, acting as a phenotypic indicator of the functional effects of the microbiota. One of the first clearly formulated applications of HMME was performed in Arabidopsis, in which plant phenotypes (aboveground biomass and flowering time) were used as selection criteria for the microbiome across multiple selection cycles (Panke-Buisse et al. 2015). These experiments demonstrated that repeated transfer of the microbiome from the best-performing plants can generate a stable microbiota that transmits improved growth traits to the next generation of plants (Panke-Buisse et al. 2015). Moreover, these results provide the first empirical evidence that a plant-driven selection process can direct the microbiome toward beneficial functional states.

Subsequent studies have extended the HMME approach to agricultural systems and stress-related contexts. For example, in HMME studies of wheat, selection for delayed onset of water stress symptoms resulted in microbiomes that extended seedling drought tolerance and increased root biomass, demonstrating that the plant can act as a "filter" that shapes the microbiome with beneficial functions (Jochum et al. 2019; Mueller et al. 2021). Another successful application has been described in tomato (*Solanum lycopersicum*), in which Antarctic soil microbiota was used as the initial inoculant, and selection was performed over ten generations of host selection. The resulting microbiomes significantly increased tomato tolerance to water deficit, and analysis revealed an increased abundance of *Bacillus* spp. and archaea *Candidatus Nitrosocosmicus*, possibly related to improved tolerance (Rodríguez et al. 2023). Similar results were obtained in rice (*Oryza sativa*), where iterative selection of plants with the lowest drought stress resulted in simplified, more stable microbiomes, improved plant growth, and drought tolerance (Styer et al. 2024). Bacterial taxa identified in these studies, such as *Pseudomona*s and *Sphingomonas*, were strongly associated with improved host phenotype, and metagenomic analysis revealed functions related to glycerol-3-phosphate and iron transport, which are important processes for plant–microbe interactions during drought. These examples demonstrate that the HMME strategy can be applied to various plant species and environmental contexts, enabling the identification of functional components of the microbiota and the creation of new, simplified microbiome consortia that reliably transmit beneficial traits to the host across multiple generations.

One of the most crucial methodological aspects of HMME is the control of the accuracy of the experimental results (Jochum et al. 2019; Rodríguez et al. 2023; Styer et al. 2024). Plant phenotypes depend not only on the microbiota but also on soil structure, nutrients, moisture, and climate; therefore, it is necessary to include appropriate control groups. The most common ones are (i) autoclaved inocula, which allow for the separation of biological effects from nonbiological factors; (ii) comparisons between sterile soil and nonautoclaved substrate; (iii) randomized block designs to reduce spatial variability; and (iv) repeated inoculation tests (e.g., dilution series or "common garden" experiments) to test the effect of the microbiome (Lau et al. 2012; Panke-Buisse et al. 2015). The most significant challenges arise from the complexity of the microbiome, numerous unknown interspecific interactions, and environmental effects; therefore, a systematic approach is necessary for interpreting the data.

Modern HMME methodology relies heavily on "omics" technologies, which allow the tracking of structural and functional changes between the host, microbiome, and environment during selection cycles (Jain et al. 2024; Thakur et al. 2023). Metagenomic sequencing data reveal taxonomic diversity and functional genes, whereas metatranscriptomics and metabolomics enable the identification of active metabolic pathways associated with plant phenotypes. Quantitative trait-based approaches allow the transition from simple taxonomy to the analysis of functional traits in the microbiome, linking these traits to host phenotypic indicators (Louca et al. 2018). The integration of these data with network and machine learning analyses provides a basis for targeted microbiome engineering, enabling the identification of key microorganisms or functional groups that increase plant resistance and productivity (Escalas et al. 2022; Widder et al. 2016).

7.5.2.3 Applications of HMME

HMME is currently used primarily in experimental models and controlled environments; however, its potential may extend to applied agronomy and ecosystem management. To date, most research has focused on enhancing plant resistance to abiotic stresses, particularly water stress, salinity, and temperature. In studies with wheat, rice, and tomato, HMME selection cycles have yielded microbiomes that enhance plant desiccation tolerance and biomass accumulation (Mueller et al. 2021; Styer et al. 2024; Taiwo et al. 2025). These results suggest that HMME can be used to develop a new generation of bioinoculants, in which selected community-adapted microbial consortia replace single microorganisms.

HMME is considered a promising tool for plant biocontrol, as repeated selection can develop disease-suppressing microbiota and promote plant health without chemical inputs (Afridi et al. 2022; Noman et al. 2021; Orozco-Mosqueda et al. 2018). This has been demonstrated, for example, by multigenerational HMME in tomatoes, which reduced the incidence *of Alternaria solani*-induced blight and improved plant productivity (Taiwo et al. 2025), as well as by proof-of-concept studies showing that the host can selectively select for protective microbes (Berendsen et al. 2012; Durán et al. 2021). However, this strategy can also have unintended consequences, as host exposure to soil sometimes promotes the accumulation of pathogenic organisms (Bever et al. 2012; Luo et al. 2019), and inappropriate microbiome selection can increase the abundance of opportunistic taxa (Mueller and Sachs 2015). Therefore, the application of HMME for biocontrol requires strict experimental controls and microbiome monitoring to prevent pathogen proliferation.

Furthermore, the application of HMME is increasingly associated with sustainable agriculture and crop rotation concepts, particularly in the context of catch crops and soil health. Review papers highlight that HMME can be integrated into practices such as catch cropping, reduced tillage, and diversified crop rotations to purposefully shape microbiomes that support soil biological activity, nutrient cycling, and yield stability (Afridi et al. 2022; Kaul et al. 2021; Thakur et al. 2023; Trivedi et al. 2021).

To date, HMME has rarely been applied at industrial or field scales, as it requires a long, multigenerational cycle, ongoing environmental management, and integrated omics analyses. However, semiautomated selection models are being developed, in which plant phenotypes are assessed with the help of artificial intelligence (e.g., growth rate and photosynthetic activity), and microbiome data are analyzed in real time (Ma et al. 2025; Maraveas 2024). Such methods may enable the use of HMME not only in the laboratory but also in precision agriculture systems, where microbiome selection can occur directly in the field.

7.6 Conclusion

Amidst the growing demands for food security, soil health restoration, and reduced chemical inputs, rhizoengineering is considered a key pillar of sustainable agriculture. Innovative developments in microbial inoculant formulations are largely driven by rapid advances in the vastly expanding field of plant microbiome research. Combining accumulated microbial genomic and metagenomic data with genetic engineering or genome-editing techniques offers promising prospects for enhancing PGP traits. Mainly used for single-strain applications, these methods ensure precise targeting and improved performance of specific genes and metabolic pathways important for plant nutrient acquisition, stress tolerance, or biocontrol of phytopathogens. Biosafety concerns and regulatory restrictions often limit the practical application of GMM in agriculture; thus, adaptive acclimation may be a viable alternative for microbial strain improvement. These include the selection of naturally tolerant strains from extreme environments or the use of controlled techniques such as stress priming and ALE. Furthermore, advances in the understanding of soil microbial ecology and plant–microbe interactions have largely shifted the focus from single-strain bioinoculants to the development of multistrain SynComs. To improve field stability and performance, rationally designed approaches that integrate metagenomics, machine learning, and metabolic modeling have been employed. This allows the selection of key microbial taxa and improves cross-feeding and synergistic interactions within the microbial consortium. As natural counterparts to SynComs, NatCom transplants and HMME offer promising alternatives with substantial potential for agricultural applications, particularly in conferring stress tolerance and reducing disease incidence. Together, these modern approaches present a viable solution for microbiome optimization, opening new avenues for improved productivity and resilience of agricultural systems in the face of global challenges.

Glossary

Biofertilizers are a type of bioinoculants containing living microorganisms that, when applied to seeds, plant surfaces, or soil, colonize the rhizosphere or the interior of the plant and promote growth by increasing the supply or availability of primary nutrients to the host plant (Vessey 2003).

Bioinoculants are biological preparations containing live beneficial microorganisms that are applied to seeds, plants, or the environment to perform specific functions, such as biocontrol or plant growth promotion (O'Callaghan et al. 2022).

Bottom-Up Bioinoculant Design is a reductionist approach for creating SynComs by selecting and combining individual microbial strains with characterized PGP traits into consortia to achieve the desired plant-beneficial functions (Xu et al. 2025).

The Core Microbiome is a set of microbial taxa that are consistently present and characteristic of a particular host or environment (Neu et al. 2021).

Host-Mediated Microbiome Engineering (HMME) is a microbiome engineering strategy that shapes plant-associated microbial communities by leveraging host traits that evolved to influence microbiomes and indirectly select for beneficial microbes based on the host plant phenotype (French et al. 2021; Mueller and Sachs 2015).

Keystone Taxa are microbial species or other groups (taxa) that have considerable influence on the structure, stability, and function of a microbial community, regardless of their overall abundance (Banerjee et al. 2018).

Microbial Biocontrol Agents are a type of bioinoculant used to suppress or eliminate plant pathogens, pests, or weeds by competing for nutrients, producing antibiotics, parasitizing pathogens, inducing plant defenses, or directly killing harmful organisms (Chaudhary et al. 2024; Eilenberg et al. 2001).

The Microbiome is a characteristic microbial community occupying a reasonably well-defined habitat with distinct physicochemical properties, including microbes, their genomes, and the theatre of activity (interactions and environmental conditions) (Berg et al. 2020).

Microbiome Engineering is the process of enhancing the performance of microbial communities through targeted manipulation of NatComs composition or the rational design and construction of synthetic consortia (Hu et al. 2022; Lawson et al. 2019).

Natural Microbial Community (NatComs) Transplantation refers to the transfer of a complex, coadapted soil microbial community from a natural source (such as healthy or suppressive soil/rhizosphere) to a recipient environment (e.g., degraded soil or plant rhizosphere) to restore or enhance microbial diversity, functionality, and associated benefits, such as plant growth promotion or disease suppression (Chen et al. 2025).

Plant Growth-Promoting (PGP) Traits are microbial functions and mechanisms that directly or indirectly enhance plant growth, development, nutrient uptake, and stress tolerance, including nutrient mobilization, production of phytohormones, alleviation of stress tolerance, and induction of systemic resistance or antagonistic activity against pathogens (Glick 1995).

Rhizoengineering involves the manipulation of plants and their root-associated microorganisms to improve plant health and productivity. This can involve engineering plants (e.g., modifying root exudates to create a "biased rhizosphere" favoring beneficial microbes), microbes (e.g., introducing or enhancing PGP traits), and soil amendments. (Dessaux et al. 2016; Ryan et al. 2009).

Synthetic Microbial Consortia (SynComs) are artificially constructed microbiomes or microbial communities assembled using a bottom-up combination of selected strains isolated from different sources that act in a synergistic and/or complementary manner (Vorholt et al. 2017).

Top-Down Bioinoculant Design is an approach for constructing SynComs by selectively refining existing complex microbial NatComs to replicate the diversity and beneficial functionality of a specific targeted microbiome (Xu et al. 2025).

References

Abou Jaoudé R, Luziatelli F, Ficca AG, Ruzzi M (2025) Soil microbiome transplantation to enhance the drought response of Salvia officinalis L. Front Microbiol 16:1553922. https://doi.org/10.3389/fmicb.2025.1553922

Afridi MS, Javed MA, Ali S, De Medeiros FHV, Ali B, Salam A, Sumaira, et al. (2022) New opportunities in plant microbiome engineering for increasing agricultural sustainability under stressful conditions. Front Plant Sci 13:899464. https://doi.org/10.3389/fpls.2022.899464

Ahmad A, Jamil A, Munawar N (2023) GMOs or non-GMOs? The CRISPR conundrum. Front Plant Sci 14:1232938. https://doi.org/10.3389/fpls.2023.1232938

Al Raish SM, Sourani OM, Abu-Elsaoud AM (2025) Plant growth-promoting microorganisms as biocontrol agents: mechanisms, challenges, and future prospects. Appl Microbiol 5(2):44. https://doi.org/10.3390/applmicrobiol5020044

Almazrouei LM, Purayil GP, Alnuaimi LS, Barhumi AI, Elameen AJ, Ibrahim SM, El-Tarabily KA (2025) Rhizosphere-competent actinobacterial consortium alleviates salinity stress in tomato through plant growth-promoting traits and ACC deaminase activity. BMC Plant Biol 25(1):1443. https://doi.org/10.1186/s12870-025-07446-1

Amaya-Gómez CV, Porcel M, Mesa-Garriga L, Gómez-Álvarez MI (2020) A framework for the selection of plant growth-promoting rhizobacteria based on bacterial competence mechanisms. Appl Environ Microbiol 86(14):e00760–e00720. https://doi.org/10.1128/AEM.00760-20

Andrade Linares DR, Lehmann A, Rillig MC (2016) Microbial stress priming: a meta-analysis. Environ Microbiome 18(4):1277–1288. https://doi.org/10.1111/1462-2920.13223

Anjum SA, Xie X, Wang LC, Saleem MF, Man C, Lei W (2011) Morphological, physiological and biochemical responses of plants to drought stress. Afr J Agric Res 6(9):2026–2032. https://doi.org/10.5897/AJAR10.027

Arroyo Olarte RD, Bravo Rodríguez R, Morales-Ríos E (2021) Genome editing in bacteria: CRISPR-Cas and beyond. Microorganisms 9(4):844. https://doi.org/10.3390/microorganisms9040844

Arthurs S, Dara SK (2019) Microbial biopesticides for invertebrate pests and their markets in the United States. J Invertebr Pathol 165:13–21. https://doi.org/10.1016/j.jip.2018.01.008

Asyiah IN, Mudakir I, Hoesain M, Pradana AP, Djunaidy A, Sari RF (2020) Consortium of endophytic bacteria and rhizobacteria effectively suppresses the population of *Pratylenchus coffeae* and promotes the growth of Robusta coffee. Biodiversitas 21(10). https://doi.org/10.13057/biodiv/d211032

Ayaz M, Li C-H, Ali Q, Zhao W, Chi Y-K, Shafiq M, Ali F et al (2023) Bacterial and fungal biocontrol agents for plant disease protection: journey from lab to field, current status, challenges, and global perspectives. Molecules 28(18):6735. https://doi.org/10.3390/molecules28186735

Aynalem B, Muleta D, Jida M, Shemekite F, Aseffa F (2022) Biocontrol competence of *Beauveria bassiana*, *Metarhizium anisopliae* and *Bacillus thuringiensis* against tomato leaf miner, *Tuta absoluta* Meyrick 1917 under greenhouse and field conditions. Heliyon 8(6):e09694. https://doi.org/10.1016/j.heliyon.2022.e09694

Backer R, Rokem JS, Ilangumaran G, Lamont J, Praslickova D, Ricci E, Subramanian S et al. (2018) Plant growth-promoting rhizobacteria: context, mechanisms of action, and roadmap to commercialization of biostimulants for sustainable agriculture. Front Plant Sci 9:1473. https://doi.org/10.3389/fpls.2018.01473

Bal HB, Das S, Dangar TK, Adhya TK (2013) ACC deaminase and IAA producing growth promoting bacteria from the rhizosphere soil of tropical rice plants. J Basic Microbiol 53(12):972–984. https://doi.org/10.1002/jobm.201200445

Balasubramaniam T, Shen G, Esmaeili N, Zhang H (2023) Plants' response mechanisms to salinity stress. Plants 12(12):2253. https://doi.org/10.3390/plants12122253

Banerjee S, van der Heijden MGA (2022) Soil microbiomes and one health. Nat Rev Microbiol 21(1):6–20. https://doi.org/10.1038/s41579-022-00779-w

Banerjee S, Schlaeppi K, van der Heijden MGA (2018) Keystone taxa as drivers of microbiome structure and functioning. Nat Rev Microbiol 16(9):567–576. https://doi.org/10.1038/s41579-018-0024-1

Barnes CJ, Bahram M, Nicolaisen M, Gilbert MTP, Vestergård M (2024) Microbiome selection and evolution within wild and domesticated plants. Trends Microbiol 33:447–458. https://doi.org/10.1016/j.tim.2024.11.011

Bashan Y, de Bashan LE, Prabhu SR, Hernandez JP (2014) Advances in plant growth-promoting bacterial inoculant technology: formulations and practical perspectives (1998-2013). Plant Soil 378(1–2):1–33. https://doi.org/10.1007/s11104-013-1956-x

Basiru S, Mwanza HP, Hijri M (2021) Analysis of arbuscular mycorrhizal fungal inoculant benchmarks. Microorganisms 9(1):81. https://doi.org/10.3390/microorganisms9010081

Bender SF, Wagg C, van der Heijden MGA (2016) An underground revolution: biodiversity and soil ecological engineering for agricultural sustainability. Trends Ecol Evol 31(6):440–452. https://doi.org/10.1016/j.tree.2016.02.016

Benmrid B, Idbella M, Bonanomi G, Khourchi S, Gherardelli M, Bargaz A, Cherki G (2025) Drought-tolerant rhizobacterial consortia with diverse plant growth promoting traits enhance wheat and faba bean growth under water and low-P availability promising multi-traits. BMC Microbiol 25(1):771. https://doi.org/10.1186/s12866-025-04492-5

Berendsen RL, Pieterse CMJ, Bakker PAHM (2012) The rhizosphere microbiome and plant health. Trends Plant Sci 17:478–486. https://doi.org/10.1016/j.tplants.2012.04.001

Berendsen RL, Vismans G, Yu K, Song Y, de Jonge R, Burgman WP, Burmølle M et al (2018) Disease-induced assemblage of a plant-beneficial bacterial consortium. ISME J 12(6):1496–1507. https://doi.org/10.1038/s41396-018-0093-1

Berg G, Rybakova D, Fischer D, Cernava T, Verges MC, Charles T, Chen X et al (2020) Microbiome definition re-visited: old concepts and new challenges. Microbiome 8(1):103. https://doi.org/10.1186/s40168-020-00875-0

Berninger T, González López Ó, Bejarano A, Preininger C, Sessitsch A (2018) Maintenance and assessment of cell viability in formulation of non-sporulating bacterial inoculants. Microb Biotechnol 11(2):277–301. https://doi.org/10.1111/1751-7915.12880

Bever JD, Platt TG, Morton ER (2012) Microbial population and community dynamics on plant roots and their feedbacks on plant communities. Ann Rev Microbiol 66:265–283. https://doi.org/10.1146/annurev-micro-092611-150107

Blonde C, Caddeo A, Nasser W, Reverchon S, Peyraud R, Haichar FE (2025) New insights in metabolism modelling to decipher plant-microbe interactions. New Phytol 246(4):1485–1493. https://doi.org/10.1111/nph.70063

Blouin M, Karimi B, Mathieu J, Lerch TZ (2015) Levels and limits in artificial selection of communities. Ecol Lett 18(10):1040–1048. https://doi.org/10.1111/ele.12486

Bo T, Kong C, Zou S, Mo M, Liu Y (2022) *Bacillus nematocida* B16 enhanced the rhizosphere colonization of *Pochonia chlamydosporia* ZK7 and controlled the efficacy of the root-knot nematode *Meloidogyne incognita*. Microorganisms 10(2):218. https://doi.org/10.3390/microorganisms10020218

Bora P, Chetia SK, Sharma A, Ahmed SS, Sharma P, Bhattacharyya A, Borgohain R et al (2025) Microbial biocontrol agents engineer plant biometrics and host response against *Xanthomonas oryzae* pv. *oryzae* in rice. Microbiol Res 16(7):151. https://doi.org/10.3390/microbiolres16070151

Bulgarelli D, Schlaeppi K, Spaepen S, Van Themaat EVL, Schulze-Lefert P (2013) Structure and functions of the bacterial microbiota of plants. Annu Rev Plant Biol 64:807–838. https://doi.org/10.1146/annurev-arplant-050312-120106

Busby PE, Soman C, Wagner MR, Friesen ML, Kremer J, Bennett A, Morsy M et al. (2017) Research priorities for harnessing plant microbiomes in sustainable agriculture. PLoS Biol 15(3):e2001793. https://doi.org/10.1371/journal.pbio.2001793

Bziuk N, Maccario L, Sørensen SJ, Schikora A, Smalla K (2022) Barley rhizosphere microbiome transplantation – a strategy to decrease susceptibility of barley grown in soils with low microbial diversity to powdery mildew. Front Microbiol 13:830905. https://doi.org/10.3389/fmicb.2022.830905

Carlstrom CI, Field CM, Bortfeld-Miller M, Muller B, Sunagawa S, Vorholt JA (2019) Synthetic microbiota reveal priority effects and keystone strains in the *Arabidopsis phyllosphere*. Nat Ecol Evol 3(10):1445–1454. https://doi.org/10.1038/s41559-019-0994-z

Castanheira NL, Dourado AC, Pais I, Semedo J, Scotti-Campos P, Borges N, Carvalho G et al (2017) Colonization and beneficial effects on annual ryegrass by mixed inoculation with plant growth promoting bacteria. Microbiol Res 198:47–55. https://doi.org/10.1016/j.micres.2017.01.009

Čaušević S, Dubey M, Morales M, Salazar G, Sentchilo V, Carraro N, Ruscheweyh H-J et al (2024) Niche availability and competitive loss by facilitation control proliferation of bacterial strains intended for soil microbiome interventions. Nat Commun 15(1):2557. https://doi.org/10.1038/s41467-024-46933-1

Chandwani S, Amaresan N (2022) Role of ACC deaminase producing bacteria for abiotic stress management and sustainable agriculture production. Environ Sci Pollut Res 29(16):22843–22859. https://doi.org/10.1007/s11356-022-18745-7

Chang HX, Haudenshield JS, Bowen CR, Hartman GL (2017) Metagenome-wide association study and machine learning prediction of bulk soil microbiome and crop productivity. Front Microbiol 8:519. https://doi.org/10.3389/fmicb.2017.00519

Chaparro JM, Badri DV, Vivanco JM (2014) Rhizosphere microbiome assemblage is affected by plant development. ISME J 8(4):790–803. https://doi.org/10.1038/ismej.2013.196

Chaudhary R, Nawaz A, Khattak Z, Butt MA, Fouillaud M, Dufossé L, Munir M et al. (2024) Microbial bio-control agents: a comprehensive analysis on sustainable pest management in agriculture. J Agric Food Res 18:101421. https://doi.org/10.1016/j.jafr.2024.101421

Chen HH, Dai ZM, Veach AM, Zheng JQ, Xu JM, Schadt CW (2020) Global meta-analyses show that conservation tillage practices promote soil fungal and bacterial biomass. Agric Ecosyst Environ 293:106841. https://doi.org/10.1016/j.agee.2020.106841

Chen J, Calderone LA, Pan L, Quist T, Pandelia ME (2023) The Fe and Zn cofactor dilemma. Biochim Biophys Acta Proteins Proteom 1871(5):140931. https://doi.org/10.1016/j.bbapap.2023.140931

Chen W, Zhu B, Wei CQ, He YF, Fan FY, Liu XY, Yang JY et al (2025) Soil transplantation: complex interplay between soil microbes and the local environment. New Phytol 247(6):2903–2913. https://doi.org/10.1111/nph.70393

Choi K, Choi J, Lee PA, Roy N, Khan R, Lee HJ, Weon HY et al. (2020) Alteration of bacterial wilt resistance in tomato plant by microbiota transplant. Front Plant Sci 11:1186. https://doi.org/10.3389/fpls.2020.01186

Clagnan E, Costanzo M, Visca A, Di Gregorio L, Tabacchioni S, Colantoni E, Sevi F et al. (2024) Culturomics- and metagenomics-based insights into the soil microbiome preservation and application for sustainable agriculture. Front Microbiol 15:1473666. https://doi.org/10.3389/fmicb.2024.1473666

Cole BJ, Feltcher ME, Waters RJ, Wetmore KM, Mucyn TS, Ryan EM, Wang G et al (2017) Genome-wide identification of bacterial plant colonization genes. PLoS Biol 15(9):e2002860. https://doi.org/10.1371/journal.pbio.2002860

Compant S, Samad A, Faist H, Sessitsch A (2019) A review on the plant microbiome: ecology, functions, and emerging trends in microbial application. J Adv Res 19:29–37. https://doi.org/10.1016/j.jare.2019.03.004

Compant S, Cassan F, Kostic T, Johnson L, Brader G, Trognitz F, Sessitsch A (2025) Harnessing the plant microbiome for sustainable crop production. Nat Rev Microbiol 23(1):9–23. https://doi.org/10.1038/s41579-024-01079-1

Copeland C, Schulze-Lefert P, Ma KW (2025) Potential and challenges for application of microbiomes in agriculture. Plant Cell 37(8):koaf185. https://doi.org/10.1093/plcell/koaf185

Cortes-Tolalpa L, Norder J, van Elsas JD, Falcao Salles J (2018) Halotolerant microbial consortia able to degrade highly recalcitrant plant biomass substrate. Appl Microbiol Biotechnol 102(6):2913–2927. https://doi.org/10.1007/s00253-017-8714-6

Cortez AO Jr, Yoshinaga N, Mori N, Hwang S-Y (2024) Plant growth-promoting rhizobacteria modulate induced corn defense against *Spodoptera litura* (Lepidoptera: Noctuidae). Biosci Biotechnol Biochem 88(8):872–884. https://doi.org/10.1093/bbb/zbae073

Cruz Barrera M, Jakobs-Schoenwandt D, Gómez MI, Becker M, Patel AV, Ruppel S (2019) Salt stress and hydroxyectoine enhance phosphate solubilisation and plant colonisation capacity of *Kosakonia radicincitans*. J Adv Res 19:91–97. https://doi.org/10.1016/j.jare.2019.03.012

Da Cunha IDAM, Da Silva AVOR, Boleta EHM, Pellegrinetti TA, Zagatto LFG, Zagatto SDS, de Chaves MG et al. (2024) The interplay between the inoculation of plant growth-promoting rhizobacteria and the rhizosphere microbiome and their impact on plant phenotype. Microbiol Res 283:127706. https://doi.org/10.1016/j.micres.2024.127706

Da Silva JR, Menéndez E, Oliveira S, Alexandre A (2025) The overexpression of an *EnvZ*-like protein improves the symbiotic performance of *Mesorhizobia*. Agronomy 15(5):1235. https://doi.org/10.3390/agronomy15051235

Damodaran T, Jha SK, Kumari S, Gupta G, Mishra VK, Sharma PC, Jat HS (2023) Development of halotolerant microbial consortia for salt stress mitigation and sustainable tomato production in sodic soils: an enzyme mechanism approach. Sustainability 15(6):5186. https://doi.org/10.3390/su15065186

Das K, Masud MAA, Sarker A, Arafa RA, Patel M (2025) Unveiling the sustainable and biological remediation of heavy metals contaminations in soils and water ecosystems through potential microbes – a review. Sustainability 17(16):7357. https://doi.org/10.3390/su17167357

De Souza RSC, Armanhi JSL, Damasceno NB, Imperial J, Arruda P (2019) Genome sequences of a plant beneficial synthetic bacterial community reveal genetic features for successful plant colonization. Front Microbiol 10:1779. https://doi.org/10.3389/fmicb.2019.01779

De Souza RSC, Armanhi JSL, Arruda P (2020) From microbiome to traits: designing synthetic microbial communities for improved crop resiliency. Front Plant Sci 11:1179. https://doi.org/10.3389/fpls.2020.01179

Delgado Baquerizo M (2022) Simplifying the complexity of the soil microbiome to guide the development of next-generation SynComs. J Sustain Agric Environ 1(1):9–15. https://doi.org/10.1002/sae2.12012

Desai S, Bagyaraj DJ, Ashwin R (2020) Inoculation with microbial consortium promotes growth of tomato and capsicum seedlings raised in pro trays. Proc Natl Acad Sci India Sect B Biol Sci 90(1):21–28. https://doi.org/10.1007/s40011-019-01078-w

Dessaux Y, Grandclément C, Faure D (2016) Engineering the rhizosphere. Trends Plant Sci 21(3):266–278. https://doi.org/10.1016/j.tplants.2016.01.002

Devi NO, Tombisana Devi RK, Debbarma M, Hajong M, Thokchom S (2022a) Effect of endophytic *Bacillus* and arbuscular mycorrhiza fungi (AMF) against *Fusarium* wilt of tomato caused by *Fusarium oxysporum* f. sp. *lycopersici*. Egypt J Biol Pest Control 32(1):1. https://doi.org/10.1186/s41938-021-00499-y

Devi R, Kaur T, Kour D, Hricovec M, Mohan R, Yadav N, Yadav AN (2022b) Microbes-mediated alleviation of heavy metal stress in crops: current research and future challenges. J App Biol Biotechnol 10(2), 25–37. https://doi.org/10.7324/JABB.2022.10s203

Devi R, Kaur T, Kour D, Yadav AN (2022c) Microbial consortium of mineral solubilizing and nitrogen fixing bacteria for plant growth promotion of amaranth (*Amaranthus hypochondrius* L.). Biocatal Agric Biotechnol 43:102404. https://doi.org/10.1016/j.bcab.2022.102404

Dimkić I, Janakiev T, Petrović M, Degrassi G, Fira D (2022) Plant-associated *Bacillus* and *Pseudomonas* antimicrobial activities in plant disease suppression via biological control mechanisms – a review. Physiol Mol Plant Pathol 117:101754. https://doi.org/10.1016/j.pmpp.2021.101754

Dong W, Liu K, Wang F, Xin F, Zhang W, Zhang M, Jiang M (2017) The metabolic pathway of metamifop degradation by consortium ME-1 and its bacterial community structure. Biodegradation 28(2):181–194. https://doi.org/10.1007/s10532-017-9787-8

Dong SJ, Wei YY, Yu Q, Gao YF, Chen HS, Zhou KY, Cheng MD et al. (2024) Inoculating functional bacteria improved the humification process by regulating microbial networks and key genera in straw composting by adding different nitrogen sources. Bioresour Technol 393:130022. https://doi.org/10.1016/j.biortech.2023.130022

Downing KJ, Thomson JA (2000) Introduction of the *Serratia marcescens chiA* gene into an endophytic *Pseudomonas fluorescens* for the biocontrol of phytopathogenic fungi. Can J Microbiol 46(4):363–369. https://doi.org/10.1139/w99-147

Dragosits M, Mattanovich D (2013) Adaptive laboratory evolution – principles and applications for biotechnology. Microb Cell Factories 12(1):64. https://doi.org/10.1186/1475-2859-12-64

Du H, Li M, Liu Y (2023) Towards applications of genome-scale metabolic model-based approaches in designing synthetic microbial communities. Quant Biol 11(1):15–30. https://doi.org/10.15302/J-QB-022-0313

Dudeja C, Mishra A, Ali A, Singh PP, Jaiswal AK (2025) Microbial genome editing with CRISPR–Cas9: recent advances and emerging applications across sectors. Fermentation 11(7):410. https://doi.org/10.3390/fermentation11070410

Durán P, Tortella G, Sadowsky MJ, Viscardi S, Barra PJ, Mora ML (2021) Engineering multigenerational host-modulated microbiota against soilborne pathogens in response to global climate change. Biology 10(9):865. https://doi.org/10.3390/biology10090865

Eilenberg J, Hajek A, Lomer C (2001) Suggestions for unifying the terminology in biological control. BioControl 46(4):387–400. https://doi.org/10.1023/A:1014193329979

Emmenegger B, Massoni J, Pestalozzi CM, Bortfeld-Miller M, Maier BA, Vorholt JA (2023) Identifying microbiota community patterns important for plant protection using synthetic communities and machine learning. Nat Commun 14(1):7983. https://doi.org/10.1038/s41467-023-43793-z

Escalas A, Paula FS, Guilhaumon F, Yuan M, Yang Y, Wu L, Liu F et al (2022) Macroecological distributions of gene variants highlight the functional organization of soil microbial systems. ISME J 16(3):726–737. https://doi.org/10.1038/s41396-021-01120-8

Etesami H, Glick BR (2024) Bacterial indole-3-acetic acid: a key regulator for plant growth, plant-microbe interactions, and agricultural adaptive resilience. Microbiol Res 281:127602. https://doi.org/10.1016/j.micres.2024.127602

Eziz A, Yan Z, Tian D, Han W, Tang Z, Fang J (2017) Drought effect on plant biomass allocation: a meta-analysis. Ecol Evol 7(24):11002–11010. https://doi.org/10.1002/ece3.3630

Falkowski PG, Fenchel T, Delong EF (2008) The microbial engines that drive Earth's biogeochemical cycles. Science 320(5879):1034–1039. https://doi.org/10.1126/science.1153213

Fan X, Ye T, Li Q, Bhatt P, Zhang L, Chen S (2020) Potential of a quorum quenching bacteria isolate Ochrobactrum intermedium D-2 against soft rot pathogen Pectobacterium carotovorum subsp. carotovorum. Front Microbiol 11:898. https://doi.org/10.3389/fmicb.2020.00898

Fang T, Lo S-C, Yu Y-N, Sou N-L, Hung S-HW, Peng J-H, Chiang E-PI et al (2025) Development of an engineered bacterial endophyte: promoting plant growth through pyrroloquinoline quinone (PQQ) synthesis. Microorganisms 13(2):293. https://doi.org/10.3390/microorganisms13020293

FAO (2021) Keep soil alive, protect soil biodiversity – Global Symposium on Soil Biodiversity 19–22 April 2021: Proceedings. FAO, 927 p.

Fierer N (2017) Embracing the unknown: disentangling the complexities of the soil microbiome. Nat Rev Microbiol 15:579–590. https://doi.org/10.1038/nrmicro.2017.87

Finkel OM, Castrillo G, Herrera Paredes S, Salas González I, Dangl JL (2017) Understanding and exploiting plant beneficial microbes. Curr Opin Plant Biol 38:155–163. https://doi.org/10.1016/j.pbi.2017.04.018

Fitzpatrick CR, Salas-González I, Conway JM, Finkel OM, Gilbert S, Russ D, Teixeira PJPL et al (2020) The plant microbiome: from ecology to reductionism and beyond. Ann Rev Microbiol 74:81–100. https://doi.org/10.1146/annurev-micro-022620-014327

Foo JL, Ling H, Lee YS, Chang MW (2017) Microbiome engineering: current applications and its future. Biotechnol J 12:1600099. https://doi.org/10.1002/biot.201600099

French E, Kaplan I, Lyer-Pascuzzi A, Nakatsu CH, Enders L (2021) Emerging strategies for precision microbiome management in diverse agroecosystems. Nat Plants 7(3):256–267. https://doi.org/10.1038/s41477-020-00830-9

Garrido Sanz D, Keel C (2025) Seed-borne bacteria drive wheat rhizosphere microbiome assembly via niche partitioning and facilitation. Nat Microbiol 10(5):1130–1144. https://doi.org/10.1038/s41564-025-01973-1

Geerdes N, de Lorimier P, Howe A, Shade A (2025) Synthetic microbial communities: bridging research and application in second-generation bioenergy feedstock microbiomes. Plant Soil 517:1533–1561. https://doi.org/10.1007/s11104-025-07937-y

Ghorui M, Chowdhury S, Balu P, Burla S (2024) Arbuscular mycorrhizal inoculants and its regulatory landscape. Heliyon 10(9):e30359. https://doi.org/10.1016/j.heliyon.2024.e30359

Glick BR (1995) The enhancement of plant-growth by free-living bacteria. Can J Microbiol 41(2):109–117. https://doi.org/10.1139/m95-015

Glick BR (2012) Plant growth-promoting bacteria: mechanisms and applications. Scientifica 2012:963401. https://doi.org/10.6064/2012/963401

Goncalves OS, Creevey CJ, Santana MF (2023) Designing a synthetic microbial community through genome metabolic modeling to enhance plant-microbe interaction. Environ Microbiome 18(1):81. https://doi.org/10.1186/s40793-023-00536-3

Goswami M, Suresh DEKA (2020) Plant growth-promoting rhizobacteria – alleviators of abiotic stresses in soil: a review. Pedosphere 30(1):40–61. https://doi.org/10.1016/S1002-0160(19)60839-8

Graham EB, Garayburu-Caruso VA, Wu R, Zheng J, McClure R, Jones GD (2024) Genomic fingerprints of the world's soil ecosystems. mSystems 9(6):e0111223. https://doi.org/10.1128/msystems.01112-23

Guo N, Wang S, Whitfield CT, Batchelor WD, Wang Y, Blersch D, Higgins BT et al (2025) High-efficiency CRISPR-Cas9 genome editing unveils biofilm insights and enhances antimicrobial activity in *Bacillus velezensis* FZB42. Biotechnol Bioeng 122(4):983–994. https://doi.org/10.1002/bit.28933

Gupta S, Pandey S (2019) ACC deaminase producing bacteria with multifarious plant growth promoting traits alleviates salinity stress in French bean (*Phaseolus vulgaris*) plants. Front Microbiol 10:1506. https://doi.org/10.3389/fmicb.2019.01506

Gupta R, Anand G, Gaur R, Yadav D (2021) Plant-microbiome interactions for sustainable agriculture: a review. Physiol Mol Biol Plants 27(1):165–179. https://doi.org/10.1007/s12298-021-00927-1

Gupta S, Pandey S, Kotra V, Kumar A (2023) Assessing the role of ACC deaminase-producing bacteria in alleviating salinity stress and enhancing zinc uptake in plants by altering the root architecture of French bean (*Phaseolus vulgaris*) plants. Planta 258(1):3. https://doi.org/10.1007/s00425-023-04159-3

Hacquard S, Martin FM (2024) The chemical language of plant-microbe-microbe associations: an introduction to a virtual issue. New Phytol 244(3):739–742. https://doi.org/10.1111/nph.20124

Hacquard S, Garrido-Oter R, Gonzalez A, Spaepen S, Ackermann G, Lebeis S, McHardy AC et al (2015) Microbiota and host nutrition across plant and animal kingdoms. Cell Host Microbe 17(5):603–616. https://doi.org/10.1016/j.chom.2015.04.009

Hartman K, Tringe SG (2019) Interactions between plants and soil shaping the root microbiome under abiotic stress. Biochem J 476(19):2705–2724. https://doi.org/10.1042/BCJ20180615

Hartmann M, Frey B, Mayer J, Mäder P, Widmer F (2015) Distinct soil microbial diversity under long-term organic and conventional farming. ISME J 9(5):1177–1194. https://doi.org/10.1038/ismej.2014.210

Hasanuzzaman M, Fujita M (2022) Plant responses and tolerance to salt stress: physiological and molecular interventions. Int J Mol Sci 23(9):4810. https://doi.org/10.3390/ijms23094810

Hassani MA, Durán P, Hacquard S (2018) Microbial interactions within the plant holobiont. Microbiome 6:58–58. https://doi.org/10.1186/s40168-018-0445-0

Herrera Paredes S, Gao T, Law TF, Finkel OM, Mucyn T, Teixeira P, Salas Gonzalez I et al (2018) Design of synthetic bacterial communities for predictable plant phenotypes. PLoS Biol 16(2):e2003962. https://doi.org/10.1371/journal.pbio.2003962

Hu H, Wang M, Huang Y, Xu Z, Xu P, Nie Y, Tang H (2022) Guided by the principles of microbiome engineering: accomplishments and perspectives for environmental use. mLife 1(4):382–398. https://doi.org/10.1002/mlf2.12043

Huang Z, Bonsall RF, Mavrodi DV, Weller DM, Thomashow LS (2004) Transformation of *Pseudomonas fluorescens* with genes for biosynthesis of phenazine-1-carboxylic acid improves biocontrol of rhizoctonia root rot and in situ antibiotic production. FEMS Microbiol Ecol 49(2):243–251. https://doi.org/10.1016/j.femsec.2004.03.010

Jain A, Singh S, Kumar Sarma B, Bahadur Singh H (2012) Microbial consortium–mediated reprogramming of defence network in pea to enhance tolerance against *Sclerotinia sclerotiorum*. J Appl Microbiol 112(3):537–550. https://doi.org/10.1111/j.1365-2672.2011.05220.x

Jain A, Sarsaiya S, Singh R, Gong Q, Wu Q, Shi J (2024) Omics approaches in understanding the benefits of plant-microbe interactions. Front Microbiol 15:1391059. https://doi.org/10.3389/fmicb.2024.1391059

Jaroszuk-Ściseł J, Tyśkiewicz R, Nowak A, Ozimek E, Majewska M, Hanaka A, Janusz G (2019) Phytohormones (auxin, gibberellin) and ACC deaminase in vitro synthesized by the mycoparasitic *Trichoderma* DEMTkZ3A0 strain and changes in the level of auxin and plant resistance markers in wheat seedlings inoculated with this strain conidia. Int J Mol Sci 20(19):4923. https://doi.org/10.3390/ijms20194923

Jha CK, Saraf M (2012) Evaluation of multispecies plant-growth-promoting consortia for the growth promotion of *Jatropha curcas* L. J Plant Growth Regul 31(4):588–598. https://doi.org/10.1007/s00344-012-9269-5

Jiang G, Zhang Y, Gan G, Li W, Wan W, Jiang Y, Yang T et al. (2022) Exploring rhizo-microbiome transplants as a tool for protective plant-microbiome manipulation. ISME Commun 2(1):10. https://doi.org/10.1038/s43705-022-00094-8

Jochum MD, McWilliams KL, Pierson EA, Jo YK (2019) Host-mediated microbiome engineering (HMME) of drought tolerance in the wheat rhizosphere. PLoS One 14(12):e0225933. https://doi.org/10.1371/journal.pone.0225933

Jones KM (2012) Increased production of the exopolysaccharide succinoglycan enhances *Sinorhizobium meliloti* 1021 symbiosis with the host plant *Medicago truncatula*. J Bacteriol 194(16):4322–4331. https://doi.org/10.1128/jb.00751-12

Joshi B, Chattopadhyay A, Patel M, Singh N, Purohit J (2025) Microbial consortium-based biocontrol technologies for pulse crops. J Food Legumes 38(2):179–186. https://doi.org/10.53550/jfl.v38.i2.258

Kaki P, Schlatter D, Khokhani D (2026) Refining the root-associated microbial consortia for enhanced biocontrol of the root-rot pathogen of corn. Front Microbiol 17:1714069. https://doi.org/10.3389/fmicb.2026.1714069

Kampa R, Vittal R, Basha SA, Pushpavalli SNCVL, Bokka V (2025) Plant growth-promoting microbial consortia for effective biocontrol of groundnut collar rot and chickpea wilt. Plant Pathol 74:1121–1141. https://doi.org/10.1111/ppa.14078

Kapadia C, Sayyed RZ, El Enshasy HA, Vaidya H, Sharma D, Patel N, Zuan ATK (2021) Halotolerant microbial consortia for sustainable mitigation of salinity stress, growth promotion, and mineral uptake in tomato plants and soil nutrient enrichment. Sustainability 13(15):8369. https://doi.org/10.3390/su13158369

Kaul S, Choudhary M, Gupta S, Dhar MK (2021) Engineering host microbiome for crop improvement and sustainable agriculture. Front Microbiol 12:635917. https://doi.org/10.3389/fmicb.2021.635917

Kaur T, Devi R, Kumar S, Sheikh I, Kour D, Yadav AN (2022) Microbial consortium with nitrogen fixing and mineral solubilizing attributes for growth of barley (Hordeum vulgare L.). Heliyon 8(4):e09326. https://doi.org/10.1016/j.heliyon.2022.e09326

Kaur T, Devi R, Negi R, Kumar S, Singh S, Rustagi S, Yadav AN (2024) Microbial consortium with multifunctional attributes for the plant growth of eggplant (*Solanum melongena* L.). Folia Microbiol 69(6):1255–1266. https://doi.org/10.1007/s12223-024-01168-x

Kaushal P, Pati AM (2025) Unleashing rhizobacteria for sustainable soil remediation: PGPR roles in heavy metal tolerance, detoxification, and plant productivity. Front Microbiol 16:1662000. https://doi.org/10.3389/fmicb.2025.1662000

Ke J, Wang B, Yoshikuni Y (2021) Microbiome engineering: synthetic biology of plant-associated microbiomes in sustainable agriculture. Trends Biotechnol 39(3):244–261. https://doi.org/10.1016/j.tibtech.2020.07.008

Kebede E (2021) Competency of rhizobial inoculation in sustainable agricultural production and biocontrol of plant diseases. Front Sustain Food Syst 5:728014. https://doi.org/10.3389/fsufs.2021.728014

Kebede MT, Mengstie GY (2025) Optimization the efficacy of plant growth-promoting rhizobacteria via genetic engineering. Discov Plants 2(1):154. https://doi.org/10.1007/s44372-025-00240-1

Kendzior J, Warren Raffa D, Bogdanski A (2022) The soil microbiome: a game changer for food and agriculture – executive summary for policymakers and researchers. FAO, 32 p. https://doi.org/10.4060/cc0717en

Khan AL (2023) The phytomicrobiome: solving plant stress tolerance under climate change. Front Plant Sci 14:1219366. https://doi.org/10.3389/fpls.2023.1219366

Kim YC, Jung H, Kim KY, Park SK (2008) An effective biocontrol bioformulation against Phytophthora blight of pepper using growth mixtures of combined chitinolytic bacteria under different field conditions. Eur J Plant Pathol 120(4):373–382. https://doi.org/10.1007/s10658-007-9227-4

Kimotho RN, Maina S (2024) Unraveling plant-microbe interactions: can integrated omics approaches offer concrete answers? J Exp Bot 75(5):1289–1313. https://doi.org/10.1093/jxb/erad448

Klimasmith IM, Wang B, Yu S, Yoshikuni Y, Kent AD (2024) Translating macroecological models to predict microbial establishment probability in an agricultural inoculant introduction. Front Microbiomes 3:1452476. https://doi.org/10.3389/frmbi.2024.1452476

Köhl J, Kolnaar R, Ravensberg WJ (2019) Mode of action of microbial biological control agents against plant diseases: relevance beyond efficacy. Front Plant Sci 10:845. https://doi.org/10.3389/fpls.2019.00845

Koziol L, McKenna TP, Bever JD (2025) Meta-analysis reveals globally sourced commercial mycorrhizal inoculants fall short. New Phytol 246(3):821–827. https://doi.org/10.1111/nph.20278

Kredics L, Buchner R, Balazs D, Allaga H, Kedves O, Racic G, Varga A et al (2024) Recent advances in the use of Trichoderma-containing multicomponent microbial inoculants for pathogen control and plant growth promotion. World J Microbiol Biotechnol 40(5):162. https://doi.org/10.1007/s11274-024-03965-5

Krishna R, Jaiswal DK, Ansari WA, Singh S, Soumia PS, Singh AK, Verma JP (2022) Potential microbial consortium mitigates drought stress in tomato (*Solanum lycopersicum* L.) plant by up-regulating stress-responsive genes and improving fruit yield and soil properties. J Soil Sci Plant Nutr 22(4):4598–4615. https://doi.org/10.1007/s42729-022-00929-2

Kumar A, Verma JP (2018) Does plant – microbe interaction confer stress tolerance in plants: a review? Microbiol Res 207:41–52. https://doi.org/10.1016/j.micres.2017.11.004

Kumar A, Maurya BR, Raghuwanshi R (2021) The microbial consortium of indigenous rhizobacteria improving plant health, yield and nutrient content in wheat (*Triticum aestivum*). J Plant Nutr 44(13):1942–1956. https://doi.org/10.1080/01904167.2021.1884706

Lau JA, Lennon JT, Kellogg WK, Karl DM (2012) Rapid responses of soil microorganisms improve plant fitness in novel environments. Proc Natl Acad Sci USA 109(35), 14058–14062. https://doi.org/10.5061/dryad.qc537

Lawson CE, Harcombe WR, Hatzenpichler R, Lindemann SR, Löffler FE, O'Malley MA, Martín HG et al (2019) Common principles and best practices for engineering microbiomes. Nat Rev Microbiol 17(12):725–741. https://doi.org/10.1038/s41579-019-0255-9

Lebeis SL (2014) The potential for give and take in plant-microbiome relationships. Front Plant Sci 5:287. https://doi.org/10.3389/fpls.2014.00287

Leclère V, Béchet M, Adam A, Guez J-S, Wathelet B, Ongena M, Thonart P et al (2005) Mycosubtilin overproduction by *Bacillus subtilis* BBG100 enhances the organism's antagonistic and biocontrol activities. Appl Environ Microbiol 71(8):4577–4584. https://doi.org/10.1128/AEM.71.8.4577-4584.2005

Levy A, Salas Gonzalez I, Mittelviefhaus M, Clingenpeel S, Herrera Paredes S, Miao J, Wang K et al (2017) Genomic features of bacterial adaptation to plants. Nat Genet 50(1):138–150. https://doi.org/10.1038/s41588-017-0012-9

Li C, Jia Z, Ma S, Liu X, Zhang J, Müller C (2023) Plant and native microorganisms amplify the positive effects of microbial inoculant. Microorganisms 11(3):570. https://doi.org/10.3390/microorganisms11030570

Li X, Zheng X, Yadav N, Saha S, Salama ES, Li X, Wang L et al (2024) Rational management of the plant microbiome for the second green revolution. Plant Commun 5(4):100812. https://doi.org/10.1016/j.xplc.2024.100812

Li J, Zhang Y, Jiang W, Zhang LQ (2025) Experimental evolution of plant rhizobacteria reveals emerging adaptive mutations. MBio 16(8):e0102325. https://doi.org/10.1128/mbio.01023-25

Lin Y, Zhang H, Li P, Jin J, Li Z (2022) The bacterial consortia promote plant growth and secondary metabolite accumulation in *Astragalus mongholicus* under drought stress. BMC Plant Biol 22(1):475. https://doi.org/10.1186/s12870-022-03859-4

Lindemann SR, Bernstein HC, Song H-S, Fredrickson JK, Fields MW, Shou W, Johnson DR et al (2016) Engineering microbial consortia for controllable outputs. ISME J 10(9):2077–2084. https://doi.org/10.1038/ismej.2016.26

Ling N, Wang TT, Kuzyakov Y (2022) Rhizosphere bacteriome structure and functions. Nat Commun 13(1):836. https://doi.org/10.1038/s41467-022-28448-9

Liu Y, Cao L, Tan H, Zhang R (2017) Surface display of ACC deaminase on endophytic *Enterobacteriaceae* strains to increase saline resistance of host rice sprouts by regulating plant ethylene synthesis. Microb Cell Factories 16(1):214. https://doi.org/10.1186/s12934-017-0831-5

Liu K, McInroy JA, Hu CH, Kloepper JW (2018) Mixtures of plant-growth-promoting rhizobacteria enhance biological control of multiple plant diseases and plant-growth promotion in the presence of pathogens. Plant Dis 102(1):67–72. https://doi.org/10.1094/PDIS-04-17-0478-RE
Liu F, Hewezi T, Lebeis SL, Pantalone V, Grewal PS, Staton ME (2019) Soil indigenous microbiome and plant genotypes cooperatively modify soybean rhizosphere microbiome assembly. BMC Microbiol 19(1):201. https://doi.org/10.1186/s12866-019-1572-x
Liu YX, Qin Y, Chen T, Lu M, Qian X, Guo X, Bai Y (2021) A practical guide to amplicon and metagenomic analysis of microbiome data. Protein Cell 12(5):315–330. https://doi.org/10.1007/s13238-020-00724-8
Liu X, Le Roux X, Salles JF (2022) The legacy of microbial inoculants in agroecosystems and potential for tackling climate change challenges. iScience 25(3):103821. https://doi.org/10.1016/j.isci.2022.103821
Liu J, Wang D, Tong R, Ye S, Zhao Y, Wu J, Gan Y (2025a) Engineered rhizobia with trehalose-producing genes enhance peanut growth under salinity stress. Agronomy 15(4):974. https://doi.org/10.3390/agronomy15040974
Liu SJ, Wu JD, Cheng Z, Wang HF, Jin ZL, Zhang X, Zhang DQ et al. (2025b) Microbe-mediated stress resistance in plants: the roles played by core and stress-specific microbiota. Microbiome 13(1):111. https://doi.org/10.1186/s40168-025-02103-z
Lopes MJD, Dias MB, Gurgel ESC (2021) Successful plant growth-promoting microbes: inoculation methods and abiotic factors. Front Sustain Food Syst 5:606454. https://doi.org/10.3389/fsufs.2021.606454
Louca S, Polz MF, Mazel F, Albright MBN, Huber JA, O'Connor MI, Ackermann M et al (2018) Function and functional redundancy in microbial systems. Nat Ecol Evol 2:936–943. https://doi.org/10.1038/s41559-018-0519-1
Lucas JA, Ramos Solano B, Montes F, Ojeda J, Megias M, Gutierrez Mañero FJ (2009) Use of two PGPR strains in the integrated management of blast disease in rice (*Oryza sativa*) in southern Spain. Field Crop Res 114(3):404–410. https://doi.org/10.1016/j.fcr.2009.09.013
Luo L, Guo C, Wang L, Zhang J, Deng L, Luo K, Huang H et al. (2019) Negative plant-soil feedback driven by re-assemblage of the rhizosphere microbiome with the growth of Panax notoginseng. Front Microbiol 10:1597. https://doi.org/10.3389/fmicb.2019.01597
Lutz S, Bodenhausen N, Hess J, Valzano-Held A, Waelchli J, Deslandes-Herold G, Schlaeppi K et al (2023) Soil microbiome indicators can predict crop growth response to large-scale inoculation with arbuscular mycorrhizal fungi. Nat Microbiol 8(12):2277–2289. https://doi.org/10.1038/s41564-023-01520-w
Lyu X, Nuhu M, Candry P, Wolfanger J, Betenbaugh M, Saldivar A, Zuniga C et al. (2024) Top-down and bottom-up microbiome engineering approaches to enable biomanufacturing from waste biomass. J Ind Microbiol Biotechnol 51:kuae025. https://doi.org/10.1093/jimb/kuae025
Ma J, Cheng Z, Cao Y (2025) Artificial intelligence-assisted breeding for plant disease resistance. Int J Mol Sci 26:5324. https://doi.org/10.3390/ijms26115324
Maciag T, Kozieł E, Rusin P, Otulak-Kozieł K, Jafra S, Czajkowski R (2023) Microbial consortia for plant protection against diseases: more than the sum of its parts. Int J Mol Sci 24(15):12227. https://doi.org/10.3390/ijms241512227
Malboobi MA, Behbahani M, Madani H, Owlia P, Deljou A, Yakhchali B, Hassanabadi H (2009) Performance evaluation of potent phosphate solubilizing bacteria in potato rhizosphere. World J Microb Biot 25(8):1479–1484. https://doi.org/10.1007/s11274-009-0038-y
Malusá E, Sas-Paszt L, Ciesielska J (2012) Technologies for beneficial microorganisms inocula used as biofertilizers. Sci World J 2012:491206. https://doi.org/10.1100/2012/491206
Maraveas C (2024) Image analysis artificial intelligence technologies for plant phenotyping: current state of the art. AgriEngineering 6:3375–3407. https://doi.org/10.3390/agriengineering6030193
Margarita S, Sviatoslav M, Yelena O, Makhpal Y (2023) Can salt-adapted microorganisms alleviate salt stress in plants and enhance their non-specific resilience? Front Agron 5:1287108. https://doi.org/10.3389/fagro.2023.1287108

Martinez Feria R, Simmonds MB, Ozaydin B, Lewis S, Schwartz A, Pluchino A, McKellar M et al (2024) Genetic remodeling of soil diazotrophs enables partial replacement of synthetic nitrogen fertilizer with biological nitrogen fixation in maize. Sci Rep 14(1):27754. https://doi.org/10.1038/s41598-024-78243-3

Martinez Viveros O, Jorquera MA, Crowley DE, Gajardo G, Mora ML (2010) Mechanisms and practical considerations involved in plant growth promotion by rhizobacteria. J Soil Sci Plant Nutr 10:293–319. https://doi.org/10.4067/S0718-95162010000100006

Masenya K, Manganyi MC, Dikobe TB (2024) Exploring cereal metagenomics: unravelling microbial communities for improved food security. Microorganisms 12(3):510. https://doi.org/10.3390/microorganisms12030510

Matchado MS, Ruhlemann M, Reitmeier S, Kacprowski T, Frost F, Haller D, Baumbach J et al. (2024) On the limits of 16S rRNA gene-based metagenome prediction and functional profiling. Microb Genom 10(2):001203. https://doi.org/10.1099/mgen.0.001203

Matilla MA, Ramos JL, Bakker PAHM, Doornbos R, Badri DV, Vivanco JM, Ramos-González MI (2010) *Pseudomonas putida* KT2440 causes induced systemic resistance and changes in Arabidopsis root exudation. Environ Microbiol Rep 2(3):381–388. https://doi.org/10.1111/j.1758-2229.2009.00091.x

Mawarda PC, Le Roux X, Van Elsas JD, Salles JF (2020) Deliberate introduction of invisible invaders: a critical appraisal of the impact of microbial inoculants on soil microbial communities. Soil Biol Biochem 148:107874. https://doi.org/10.1016/j.soilbio.2020.107874

Mawarda PC, Mallon CA, Le Roux X, van Elsas JD, Salles JF (2022) Interactions between bacterial inoculants and native soil bacterial community: the case of spore-forming Bacillus spp. FEMS Microbiol Ecol 98(12), 1–11. https://doi.org/10.1093/femsec/fiac127

Meena M, Yadav G, Sonigra P, Nagda A, Mehta T, Swapnil P, Zehra A (2023) Advantageous features of plant growth-promoting microorganisms to improve plant growth in difficult conditions. In: Swapnil P et al (eds) Plant-microbe interaction-recent advances in molecular and biochemical approaches. Academic Press, pp 279–296. https://doi.org/10.1016/B978-0-323-91876-3.00019-1

Mendoza SM, Andersen SU, Poole PS, Sanchez-Canizares C (2021) Competition, nodule occupancy, and persistence of inoculant strains: key factors in the legume symbioses. Front Plant Sci 12:690567. https://doi.org/10.3389/fpls.2021.690567

Menéndez E, Paço A (2020) Is the application of plant probiotic bacterial consortia always beneficial for plants? Exploring synergies between rhizobial and non-rhizobial bacteria and their effects on agro-economically valuable crops. Life 10(3):24. https://doi.org/10.3390/life10030024

Mi Z, Sun Z, Huang Z, Zhao P, Li Q, Tian P (2020) Engineering CRISPR interference system to enhance the production of pyrroloquinoline quinone in *Klebsiella pneumonia*. Lett Appl Microbiol 71(3):242–250. https://doi.org/10.1111/lam.13311

Michalak A (2006) Phenolic compounds and their antioxidant activity in plants growing under heavy metal stress. Pol J Environ Stud 15(4), 523–530.

Miethke M, Marahiel MA (2007) Siderophore-based iron acquisition and pathogen control. Microbiol Mol Biol R 71(3):413. https://doi.org/10.1128/Mmbr.00012-07

Miklau M, Burn S-J, Eckerstorfer M, Dolezel M, Greiter A, Heissenberger A, Hörtenhuber S et al. (2024) Horizon scanning of potential environmental applications of terrestrial animals, fish, algae and microorganisms produced by genetic modification, including the use of new genomic techniques. Front Genome Ed 6:1376927. https://doi.org/10.3389/fgeed.2024.1376927

Miliute I, Buzaite O, Baniulis D, Stanys V (2015) Bacterial endophytes in agricultural crops and their role in stress tolerance: a review. Zemdirbyste 102(4):465–478. https://doi.org/10.13080/z-a.2015.102.060

Minchev Z, Kostenko O, Soler R, Pozo MJ (2021) Microbial consortia for effective biocontrol of root and foliar diseases in tomato. Front Plant Sci 12:756368. https://doi.org/10.3389/fpls.2021.756368

Mohamed HI, Ullah I, Toor MD, Tanveer NA, Din MMU, Basit A, Rehman MU (2025) Heavy metals toxicity in plants: understanding mechanisms and developing coping strategies for remediation: a review. Bioresour Bioprocess 12(1):95. https://doi.org/10.1186/s40643-025-00930-4

Molina-Romero D, Baez A, Quintero-Hernández V, Castañeda-Lucio M, Fuentes-Ramírez LE, Bustillos-Cristales MDR, Muñoz-Rojas J (2017) Compatible bacterial mixture, tolerant to desiccation, improves maize plant growth. PLoS One 12(11):e0187913. https://doi.org/10.1371/journal.pone.0187913

Mondal S, Halder SK, Yadav AN, Mondal KC (2020) Microbial consortium with multifunctional plant growth-promoting attributes: future perspective in agriculture. In: Yadav A, Rastegari A, Yadav N, Kour D (eds) Advances in plant microbiome and sustainable agriculture: functional annotation and future challenges. Springer, pp 219–258. https://doi.org/10.1007/978-981-15-3204-7_10

Mueller UG, Sachs JL (2015) Engineering microbiomes to improve plant and animal health. Trends Microbiol 23:606–617. https://doi.org/10.1016/j.tim.2015.07.009

Mueller UG, Juenger TE, Kardish MR, Carlson AL, Burns KM, Edwards JA, Smith CC et al. (2021) Artificial selection on microbiomes to breed microbiomes that confer salt tolerance to plants. mSystems 6:e01125–21. https://doi.org/10.1128/msystems.01125-21

Muthukumar A, Eswaran A, Sangeetha G (2011) Induction of systemic resistance by mixtures of fungal and endophytic bacterial isolates against *Pythium aphanidermatum*. Acta Physiol Plant 33(5):1933–1944. https://doi.org/10.1007/s11738-011-0742-8

Mwamburi LA (2021) Endophytic fungi, *Beauveria bassiana* and *Metarhizium anisopliae*, confer control of the fall armyworm, *Spodoptera frugiperda* (J. E. Smith) (Lepidoptera: Noctuidae), in two tomato varieties. Egypt J Biol Pest Control 31(1):7. https://doi.org/10.1186/s41938-020-00357-3

Nadarajah K, Abdul Rahman NSN (2023) The microbial connection to sustainable agriculture. Plants 12(12):2307. https://doi.org/10.3390/plants12122307

Nadeem SM, Ahmad M, Zahir ZA, Javaid A, Ashraf M (2014) The role of mycorrhizae and plant growth promoting rhizobacteria (PGPR) in improving crop productivity under stressful environments. Biotechnol Adv 32(2):429–448. https://doi.org/10.1016/j.biotechadv.2013.12.005

Nagajyoti PC, Lee KD, Sreekanth TVM (2010) Heavy metals, occurrence and toxicity for plants: a review. Environ Chem Lett 8(3):199–216. https://doi.org/10.1007/s10311-010-0297-8

Naing AH, Maung TT, Kim CK (2021) The ACC deaminase-producing plant growth-promoting bacteria: influences of bacterial strains and ACC deaminase activities in plant tolerance to abiotic stress. Physiol Plant 173(4):1992–2012. https://doi.org/10.1111/ppl.13545

Narine S, Jayaraj P, Ali O, Ramsubhag A, Jayaraman J (2025) Indigenous bacterial consortia for growth promotion and biocontrol of tomato root pathogens in tropical agroecosystems. Biocontrol Sci Tech 35(7):719–739. https://doi.org/10.1080/09583157.2025.2503776

Naz R, Khushhal S, Asif T, Mubeen S, Saranraj P, Sayyed RZ (2022) Inhibition of bacterial and fungal phytopathogens through volatile organic compounds produced by *Pseudomonas* sp. In: Sayyed RZ, Uarrota VG (eds) Secondary metabolites and volatiles of PGPR in plant-growth promotion. Springer, pp 95–118. https://doi.org/10.1007/978-3-031-07559-9_6

Neu AT, Allen EE, Roy K (2021) Defining and quantifying the core microbiome: challenges and prospects. Proc Natl Acad Sci USA 118(51):e2104429118. https://doi.org/10.1073/pnas.2104429118

Neuhoff D, Neumann G, Weinmann M (2024) Testing plant growth promoting microorganisms in the field – a proposal for standards. Front Plant Sci 14:1324665. https://doi.org/10.3389/fpls.2023.1324665

Nguyen HT, Pham TT, Nguyen PT, Dinh NCG, Le MT, Nguyen TD, Nguyen TT et al (2025) Microbial biocontrol in agriculture: from mechanistic understanding to field application. Discov Plants 2(1):334. https://doi.org/10.1007/s44372-025-00421-y

Nishisaka CS, Quevedo HD, Ventura JP, Andreote FD, Mauchline TH, Mendes R (2025) Soil microbial diversity: a key factor in pathogen suppression and inoculant performance. Geoderma 460:117444. https://doi.org/10.1016/j.geoderma.2025.117444

Niu B, Paulson JN, Zheng X, Kolter R (2017) Simplified and representative bacterial community of maize roots. Proc Natl Acad Sci 114(12):E2450–E2459. https://doi.org/10.1073/pnas.1616148114

Niu B, Wang W, Yuan Z, Sederoff RR, Sederoff H, Chiang VL, Borriss R (2020) Microbial interactions within multiple-strain biological control agents impact soil-borne plant disease. Front Microbiol 11:585404. https://doi.org/10.3389/fmicb.2020.585404

Noman M, Ahmed T, Ijaz U, Shahid M, Azizullah LD, Manzoor I et al. (2021) Plant–microbiome crosstalk: dawning from composition and assembly of microbial community to improvement of disease resilience in plants. Int J Mol Sci 22(13):6852. https://doi.org/10.3390/ijms22136852

Notununu I, Moleleki L, Roopnarain A, Adeleke R (2024) Enhancing maize drought and heat tolerance: single vs combined plant growth promoting rhizobacterial inoculation. Front Plant Sci 15:1480718. https://doi.org/10.3389/fpls.2024.1480718

Ntakirutimana R, Ishimwe J, Ukobukeye J (2024) Multiple-strain biological control agents and their impact on soil borne plant diseases. Int J Res Agron 7(4):577–583. https://doi.org/10.33545/2618060X.2024.v7.i4h.607

Nwachukwu BC, Babalola OO (2022) Metagenomics: a tool for exploring key microbiome with the potentials for improving sustainable agriculture. Front Sustain Food Syst 6:886987. https://doi.org/10.3389/fsufs.2022.886987

O'Callaghan M, Ballard RA, Wright D (2022) Soil microbial inoculants for sustainable agriculture: limitations and opportunities. Soil Use Manag 38(3):1340–1369. https://doi.org/10.1111/sum.12811

Ochoa Hueso R, Eldridge DJ, Berdugo M, Trivedi P, Sokoya B, Cano Diaz C, Abades S et al. (2024) Unearthing the soil-borne microbiome of land plants. Glob Chang Biol 30(5):e17295. https://doi.org/10.1111/gcb.17295

Omoregie AI, Kan FK, Basri HF, Silini MOE, Rajasekar A (2025) Enhanced MICP for soil improvement and heavy metal remediation: insights from landfill leachate-derived ureolytic bacterial consortium. Microorganisms 13(1):174. https://doi.org/10.3390/microorganisms13010174

Orozco-Mosqueda MC, Rocha-Granados MC, Glick BR, Santoyo G (2018) Microbiome engineering to improve biocontrol and plant growth-promoting mechanisms. Microbiol Res 208:25–31. https://doi.org/10.1016/j.micres.2018.01.005

Oyserman BO, Medema MH, Raaijmakers JM (2018) Road MAPs to engineer host microbiomes. Curr Opin Microbiol 43:46–54. https://doi.org/10.1016/j.mib.2017.11.023

Pandey N, Xalxo R, Chandra J, Keshavkant S (2023) Bacterial consortia mediated induction of systemic tolerance to arsenic toxicity via expression of stress responsive antioxidant genes in *Oryza sativa* L. Biocatal Agric Biotechnol 47:102565. https://doi.org/10.1016/j.bcab.2022.102565

Pang F, Li Q, Solanki MK, Wang Z, Xing YX, Dong DF (2024) Soil phosphorus transformation and plant uptake driven by phosphate-solubilizing microorganisms. Front Microbiol 15:1383813. https://doi.org/10.3389/fmicb.2024.1383813

Panke-Buisse K, Poole AC, Goodrich JK, Ley RE, Kao-Kniffin J (2015) Selection on soil microbiomes reveals reproducible impacts on plant function. ISME J 9:980–989. https://doi.org/10.1038/ismej.2014.196

Papin M, Philippot L, Breuil MC, Bru D, Dreux-Zigha A, Mounier A, Le Roux X et al (2024) Survival of a microbial inoculant in soil after recurrent inoculations. Sci Rep 14(1):4177. https://doi.org/10.1038/s41598-024-54069-x

Park I, Seo YS, Mannaa M (2023) Recruitment of the rhizo-microbiome army: assembly determinants and engineering of the rhizosphere microbiome as a key to unlocking plant potential. Front Microbiol 14:1163832. https://doi.org/10.3389/fmicb.2023.1163832

Paudel D, Wang LP, Poudel R, Acharya JP, Victores S, de Souza CHL, Rios E et al. (2023) Elucidating the effects of organic vs. conventional cropping practice and rhizobia inoculation on rhizosphere microbial diversity and yield of peanut. Environ Microbiome 18(1):60. https://doi.org/10.1186/s40793-023-00517-6

Peralta H, Mora Y, Salazar E, Encarnación S, Palacios R, Mora J (2004) Engineering the *nifH* promoter region and abolishing poly-beta-hydroxybutyrate accumulation in *Rhizobium etli* enhance nitrogen fixation in symbiosis with *Phaseolus vulgaris*. Appl Environ Microbiol 70(6):3272–3281. https://doi.org/10.1128/aem.70.6.3272-3281.2004

Permyakova NV, Deineko EV (2024) Crop improvement: comparison of transgenesis and gene editing. Horticulturae 10(1):57. https://doi.org/10.3390/horticulturae10010057

Pomerleau M, Charron Lamoureux V, Léonard L, Grenier F, Rodrigue S, Beauregard PB (2024) Adaptive laboratory evolution reveals regulators involved in repressing biofilm development as key players in *Bacillus subtilis* root colonization. mSystems 9(2):e00843–e00823. https://doi.org/10.1128/msystems.00843-23

Postma J, Hok-A-Hin CH, Voshaar JHO (1990) Influence of the inoculum density on the growth and survival of rhizobium leguminosarum biovar trifolii introduced into sterile and non-sterile loamy sand and silt loam. FEMS Microbiol Ecol 6(1):49–57. https://doi.org/10.1111/j.1574-6968.1990.tb03924.x

Poveda J, Eugui D (2022) Combined use of *Trichoderma* and beneficial bacteria (mainly *Bacillus* and *Pseudomonas*): Development of microbial synergistic bio-inoculants in sustainable agriculture. Biol Control 176:105100. https://doi.org/10.1016/j.biocontrol.2022.105100

Prigigallo MI, Staropoli A, Vinale F, Bubici G (2023) Interactions between plant-beneficial microorganisms in a consortium: *Streptomyces microflavus* and *Trichoderma harzianum*. Microb Biotechnol 16(12):2292–2312. https://doi.org/10.1111/1751-7915.14311

Ptaszek M, Canfora L, Pugliese M, Pinzari F, Gilardi G, Trzcinski P, Malusa E (2023) Microbial-based products to control soil-borne pathogens: methods to improve efficacy and to assess impacts on microbiome. Microorganisms 11(1):224. https://doi.org/10.3390/microorganisms11010224

Ramos Y, Taibo AD, Jiménez JA, Portal O (2020) Endophytic establishment of *Beauveria bassiana* and *Metarhizium anisopliae* in maize plants and its effect against *Spodoptera frugiperda* (J. E. Smith) (Lepidoptera: Noctuidae) larvae. Egypt J Biol Pest Control 30(1):20. https://doi.org/10.1186/s41938-020-00223-2

Ravanbakhsh M, Kowalchuk GA, Jousset A (2021) Targeted plant hologenome editing for plant trait enhancement. New Phytol 229(2):1067–1077. https://doi.org/10.1111/nph.16867

Raza A, Bhardwaj S, Atikur Rahman M, García-Caparrós P, Habib M, Saeed F, Charagh S et al (2024) Trehalose: a sugar molecule involved in temperature stress management in plants. Crop J 12(1):1–16. https://doi.org/10.1016/j.cj.2023.09.010

Rehman B, Javed J, Rauf M, Khan SA, Arif M, Hamayun M, Lee IJ (2022) ACC deaminase-producing endophytic fungal consortia promotes drought stress tolerance in *M. oleifera* by mitigating ethylene and H_2O_2. Front Plant Sci 13:967672. https://doi.org/10.3389/fpls.2022.967672

Resti Z, Warnita, Liswarni Y (2021) Endophytic bacterial consortia as biocontrol of purple blotch and plant growth promoters of shallots. IOP Conf Ser Earth Environ Sci 741(1):012009. https://doi.org/10.1088/1755-1315/741/1/012009

Rodríguez R, Barra PJ, Larama G, Carrion VJ, de la Luz Mora M, Hale L, Durán P (2023) Microbiome engineering optimized by Antarctic microbiota to support a plant host under water deficit. Front Plant Sci 14:1241612. https://doi.org/10.3389/fpls.2023.1241612

Rondón JJ, Ball MM, Castro LT, Yarzábal LA (2019) Eurypsychrophilic *Pseudomonas* spp. isolated from Venezuelan tropical glaciers as promoters of wheat growth and biocontrol agents of plant pathogens at low temperatures. Environ Sustain 2(3):265–275. https://doi.org/10.1007/s42398-019-00072-2

Rozas P, Kessi Pérez EI, Martínez C (2022) Genetically modified organisms: adapting regulatory frameworks for evolving genome editing technologies. Biol Res 55(1):31. https://doi.org/10.1186/s40659-022-00399-x

Ruan ZP, Chen K, Cao WM, Meng L, Yang BG, Xu MJ, Xing YW et al. (2024a) Engineering natural microbiomes toward enhanced bioremediation by microbiome modeling. Nat Commun 15(1):176047. https://doi.org/10.1038/s41467-024-49098-z

Ruan ZP, Xu MJ, Xing YW, Yang KQ, Xu XH, Jiang JD, Qiu RL (2024b) Enhanced growth of wheat in contaminated fields via synthetic microbiome as revealed by genome-scale metabolic modeling. Sci Total Environ 953:176047. https://doi.org/10.1016/j.scitotenv.2024.176047

Ryan PR, Dessaux Y, Thomashow LS, Weller DM (2009) Rhizosphere engineering and management for sustainable agriculture. Plant Soil 321(1–2):363–383. https://doi.org/10.1007/s11104-009-0001-6

Saad MM, Eida AA, Hirt H (2020) Tailoring plant-associated microbial inoculants in agriculture: a roadmap for successful application. J Exp Bot 71(13):3878–3901. https://doi.org/10.1093/jxb/eraa111

Sahoo A, Yadav G, Mehta T, Meena M, Swapnil P (2025) Omics-driven insights into plant growth-promoting microorganisms for sustainable agriculture. Discov Sustain 6(1):659. https://doi.org/10.1007/s43621-025-01582-2

Sai Teja B, Gupta V, Gupta SK, Sharma A, Amin Z, Ahmad Mohiddin F, Ashraf S et al. (2025) Microbial consortia as a biocontrol strategy for bacterial blight in basmati rice. Rev Argent Microbiol 58(2), 218–226. https://doi.org/10.1016/j.ram.2025.08.003

Saikia J, Sarma RK, Dhandia R, Yadav A, Bharali R, Gupta VK, Saikia R (2018) Alleviation of drought stress in pulse crops with ACC deaminase producing rhizobacteria isolated from acidic soil of Northeast India. Sci Rep 8(1):3560. https://doi.org/10.1038/s41598-018-21921-w

Saleem M, Nawaz F, Hussain MB, Ikram RM (2021) Comparative effects of individual and consortia plant growth promoting bacteria on physiological and enzymatic mechanisms to confer drought tolerance in maize (*Zea mays* L.). J Soil Sci Plant Nutr 21(4):3461–3476. https://doi.org/10.1007/s42729-021-00620-y

Sallam NMA, Aldayel MF (2025) Synergistic effects of *Rahnella aquatilis* and *Trichoderma orientale* in biocontrol of common bacterial blight in bean. Egypt J Biol Pest Control 35(1):9. https://doi.org/10.1186/s41938-025-00847-2

Samonty I, Hasan MZ, Hossain MR, Mahamud AGMSU (2025) The PGPB paradox: a critical review of field performance and practical constraints in agriculture. World J Microb Biot 41(9):321. https://doi.org/10.1007/s11274-025-04552-y

Santander C, González F, Pérez U, Ruiz A, Aroca R, Santos C, Cornejo P et al (2024) Enhancing water status and nutrient uptake in drought-stressed lettuce plants (*Lactuca sativa* L.) via inoculation with different *Bacillus* spp. isolated from the Atacama desert. Plants 13(2):158. https://doi.org/10.3390/plants13020158

Santiago CD, Yagi S, Ijima M, Nashimoto T, Sawada M, Ikeda S, Ohwada T (2017) Bacterial compatibility in combined inoculations enhances the growth of potato seedlings. Microbes Environ 32(1):14–23. https://doi.org/10.1264/jsme2.ME16127

Santos LF, Olivares FL (2021) Plant microbiome structure and benefits for sustainable agriculture. Curr Plant Biol 26:100198. https://doi.org/10.1016/j.cpb.2021.100198

Santos MS, Nogueira MA, Hungria M (2019) Microbial inoculants: reviewing the past, discussing the present and previewing an outstanding future for the use of beneficial bacteria in agriculture. AMB Express 9(1):205. https://doi.org/10.1186/s13568-019-0932-0

Santoyo G, Urtis-Flores CA, Loeza-Lara PD, Orozco-Mosqueda MDC, Glick BR (2021) Rhizosphere colonization determinants by plant growth-promoting rhizobacteria (PGPR). Biology 10(6):475. https://doi.org/10.3390/biology10060475

Sarango Flores S, Cordovez V, Oyserman BO, Arias Giraldo LM, Stopnisek N, Raaijmakers JM, van 't Hof P (2025) Microbiome-mediated resistance of wild tomato to the invasive insect Prodiplosis longifila. Environ Microbiol Rep 17(5):e70190. https://doi.org/10.1111/1758-2229.70190

Sarapat S, Songwattana P, Longtonglang A, Umnajkitikorn K, Girdthai T, Tittabutr P, Boonkerd N et al. (2020) Effects of increased 1-aminocyclopropane-1-carboxylate (ACC) deaminase activity in Bradyrhizobium sp. SUTN9-2 on mung bean symbiosis under water deficit conditions. Microbes Environ 35(3):ME20024. https://doi.org/10.1264/jsme2.ME20024

Sasse J, Martinoia E, Northen T (2018) Feed your friends: do plant exudates shape the root microbiome? Trends Plant Sci 23(1):25–41. https://doi.org/10.1016/j.tplants.2017.09.003

Schmidt JE, Kent AD, Brisson VL, Gaudin ACM (2019) Agricultural management and plant selection interactively affect rhizosphere microbial community structure and nitrogen cycling. Microbiome 7(1):146. https://doi.org/10.1186/s40168-019-0756-9

Seang-On L, Meeinkuirt W, Koedrith P (2023) Alleviation of cadmium toxicity in Thai rice cultivar (PSL2) using biofertilizer containing indigenous cadmium-resistant microbial consortia. Plants 12(20):3651. https://doi.org/10.3390/plants12203651

Senthilraja G, Anand T, Kennedy JS, Raguchander T, Samiyappan R (2013) Plant growth promoting rhizobacteria (PGPR) and entomopathogenic fungus bioformulation enhance the expression of defense enzymes and pathogenesis-related proteins in groundnut plants against leafminer insect and collar rot pathogen. Physiol Mol Biol Plants 82:10–19. https://doi.org/10.1016/j.pmpp.2012.12.002

Serantes L, Stritzler M, Brambilla S, Soto G, Ayub N (2024) Genome editing of soybean inoculant using CRISPR/Cas9 system: enhancing agricultural sustainability. Plant Cell Tissue Organ Cult 157(2):35. https://doi.org/10.1007/s11240-024-02764-y

Setten L, Soto G, Mozzicafreddo M, Fox AR, Lisi C, Cuccioloni M, Angeletti M et al (2013) Engineering *pseudomonas protegens* Pf-5 for nitrogen fixation and its application to improve plant growth under nitrogen-deficient conditions. PLoS One 8(5):e63666. https://doi.org/10.1371/journal.pone.0063666

Shaffer JP, Nothias LF, Thompson LR, Sanders JG, Salido RA, Couvillion SP, Brejnrod AD et al (2022) Standardized multi-omics of Earth's microbiomes reveals microbial and metabolite diversity. Nat Microbiol 7(12):2128-2150. https://doi.org/10.1038/s41564-022-01266-x

Sharma SB, Sayyed RZ, Trivedi MH, Gobi TA (2013) Phosphate solubilizing microbes: sustainable approach for managing phosphorus deficiency in agricultural soils. SpringerPlus 2:587. https://doi.org/10.1186/2193-1801-2-587

Shi P, Zhang J, Li X, Zhou L, Luo H, Wang L, Wei G (2022) Multiple metabolic phenotypes as screening criteria are correlated with the plant growth-promoting ability of rhizobacterial isolates. Front Microbiol 12:747982. https://doi.org/10.3389/fmicb.2021.747982

Shulse CN, Chovatia M, Agosto C, Wang G, Hamilton M, Deutsch S, Yoshikuni Y et al. (2019) Engineered root bacteria release plant-available phosphate from phytate. Appl Environ Microbiol 85(18), e01210–19. https://doi.org/10.1128/aem.01210-19

Singh BK, Trivedi P, Egidi E, Macdonald CA, Delgado-Baquerizo M (2020) Crop microbiome and sustainable agriculture. Nat Rev Microbiol 18(11):601–602. https://doi.org/10.1038/s41579-020-00446-y

Singh D, Geat N, Jadon KS, Verma A, Sharma R, Rajput LS, Mahla HR et al (2024) Isolation and characterization of biocontrol microbes for development of effective microbial consortia for managing *Rhizoctonia bataticola* root rot of cluster bean under hot arid climatic conditions. Microorganisms 12(11):2331. https://doi.org/10.3390/microorganisms12112331

Skidmore AK, Siegenthaler A, Wang TJ, Darvishzadeh R, Zhu X, Chariton A, de Groot GA (2022) Mapping the relative abundance of soil microbiome biodiversity from eDNA and remote sensing. Sci Remote Sensing 6:100065. https://doi.org/10.1016/j.srs.2022.100065

Solomon W, Janda T, Molnár Z (2024) Unveiling the significance of rhizosphere: implications for plant growth, stress response, and sustainable agriculture. Plant Physiol Biochem 206:108290. https://doi.org/10.1016/j.plaphy.2023.108290

Sotelo M, Ureta AC, Muñoz S, Sanjuán J, Monza J, Palacios J (2021) Introduction of H_2-uptake hydrogenase genes into rhizobial strains improves symbiotic nitrogen fixation in Vicia sativa and Lotus corniculatus forage legumes. Front Agron 3:661534. https://doi.org/10.3389/fagro.2021.661534

Spescha A, Zwyssig M, Hess Hermida M, Moix A, Bruno P, Enkerli J, Campos-Herrera R et al (2023) When competitors join forces: consortia of entomopathogenic microorganisms increase killing speed and mortality in leaf- and root-feeding insect hosts. Microb Ecol 86(3):1947–1960. https://doi.org/10.1007/s00248-023-02191-0

Sridhar D, Alheswairini SS, Barasarathi J, Enshasy HAE, Lalitha S, Mir SH, Nithyapriya S et al. (2025) Halophilic rhizobacteria promote growth, physiology and salinity tolerance in Sesamum

indicum L. grown under salt stress. Front Microbiol 16:1590854. https://doi.org/10.3389/fmicb.2025.1590854
Styczynski M, Biegniewski G, Decewicz P, Rewerski B, Debiec Andrzejewska K, Dziewit L (2022) Application of psychrotolerant Antarctic bacteria and their metabolites as efficient plant growth promoting agents. Front Bioeng Biotechnol 10:772891. https://doi.org/10.3389/fbioe.2022.772891
Styer A, Pettinga D, Caddell D, Coleman-Derr D (2024) Improving rice drought tolerance through host-mediated microbiome selection. eLife 13:RP97015. https://doi.org/10.7554/eLife.97015.1
Suárez R, Wong A, Ramírez M, Barraza A, Orozco MC, Cevallos MA, Lara M et al (2008) Improvement of drought tolerance and grain yield in common bean by overexpressing trehalose-6-phosphate synthase in rhizobia. Mol Plant-Microbe Interact 21(7):958–966. https://doi.org/10.1094/mpmi-21-7-0958
Sun L, Sun S, Liu T, Lei X, Liu R, Zhang J, Dai S et al (2025) Association analysis of the genomic and functional characteristics of halotolerant *Glutamicibacter endophyticus* J2-5-19 from the rhizosphere of *Suaeda salsa*. Microorganisms 13(1):208. https://doi.org/10.3390/microorganisms13010208
Swenson W, Wilson DS, Elias R (2000) Artificial ecosystem selection. Proc Natl Acad Sci USA 97(16):9110–9114. https://doi.org/10.1073/pnas.150237597
Sytar O, Kumar A, Latowski D, Kuczynska P, Strzałka K, Prasad MNV (2013) Heavy metal-induced oxidative damage, defense reactions, and detoxification mechanisms in plants. Acta Physiol Plant 35(4):985–999. https://doi.org/10.1007/s11738-012-1169-6
Taiwo MO, Abiola-Kuforiji OT, Ojesola CO (2025) Treatment effect of host-mediated rhizospheric microbiota on disease expression and yield of tomato under the influence of *Alternaria solani* stress. Biocontrol Sci Tech 35(6):653–666. https://doi.org/10.1080/09583157.2025.2498158
Tarzi C, Zampieri G, Sullivan N, Angione C (2024) Emerging methods for genome-scale metabolic modeling of microbial communities. Trends Endocrin Met 35(6):533–548. https://doi.org/10.1016/j.tem.2024.02.018
Tedersoo L, Mikryukov V, Anslan S, Bahram M, Khalid AN, Corrales A, Agan A et al (2021) The global soil mycobiome consortium dataset for boosting fungal diversity research. Fungal Divers 111(1):573–588. https://doi.org/10.1007/s13225-021-00493-7
Thakur R, Yadav S (2024) Biofilm forming, exopolysaccharide producing and halotolerant, bacterial consortium mitigates salinity stress in *Triticum aestivum*. Int J Biol Macromol 262:130049. https://doi.org/10.1016/j.ijbiomac.2024.130049
Thakur N, Nigam M, Mann NA, Gupta S, Hussain CM, Shukla SK, Shah AA et al. (2023) Host-mediated gene engineering and microbiome-based technology optimization for sustainable agriculture and environment. Funct Integr Genom 23:57. https://doi.org/10.1007/s10142-023-00982-9
Thompson IP, Cook KA, Lethbridge G, Burns RG (1990) Survival of two ecologically distinct bacteria (*Flavobacterium* and *Arthrobacter*) in unplanted and rhizosphere soil: laboratory studies. Soil Biol Biochem 22(8):1029–1037. https://doi.org/10.1016/0038-0717(90)90028-X
Tian J, Zhang M, Huang Y, Luo X, Shao F, Lei W, Xing W et al (2024) Biocontrol potential of an artificial synthetic bacterial consortium against peony root rot disease. Biol Control 195:105563. https://doi.org/10.1016/j.biocontrol.2024.105563
Toju H, Peay KG, Yamamichi M, Narisawa K, Hiruma K, Naito K, Fukuda S et al (2018) Core microbiomes for sustainable agroecosystems. Nat Plants 4:247–257. https://doi.org/10.1038/s41477-018-0139-4
Town JR, Gregorich EG, Drury CF, Lemke R, Phillips LA, Helgason BL (2022) Diverse crop rotations influence the bacterial and fungal communities in root, rhizosphere and soil and impact soil microbial processes. Appl Soil Ecol 169:104241. https://doi.org/10.1016/j.apsoil.2021.104241
Trivedi P, Schenk PM, Wallenstein MD, Singh BK (2017) Tiny microbes, big yields: enhancing food crop production with biological solutions. Microb Biotechnol 10(5):999–1003. https://doi.org/10.1111/1751-7915.12804

Trivedi P, Mattupalli C, Eversole K, Leach JE (2021) Enabling sustainable agriculture through understanding and enhancement of microbiomes. New Phytol 230:2129–2147. https://doi.org/10.1111/nph.17319

Trivedi P, Batista BD, Bazany KE, Singh BK (2022) Plant-microbiome interactions under a changing world: responses, consequences and perspectives. New Phytol 234(6):1951–1959. https://doi.org/10.1111/nph.18016

Tsegaye Z, Alemu T, Desta FA, Assefa F (2022) Plant growth-promoting rhizobacterial inoculation to improve growth, yield, and grain nutrient uptake of teff varieties. Front Microbiol 13:896770. https://doi.org/10.3389/fmicb.2022.896770

Tufail MA, Ayyub M, Irfan M, Shakoor A, Chibani CM, Schmitz RA (2022) Endophytic bacteria perform better than endophytic fungi in improving plant growth under drought stress: a meta-comparison spanning 12 years (2010–2021). Physiol Plant 174(6):e13806. https://doi.org/10.1111/ppl.13806

Tyagi J, Mishra A, Kumari S, Singh S, Agarwal H, Pudake RN, Joshi NC (2023) Deploying a microbial consortium of *Serendipita indica*, *Rhizophagus intraradices*, and *Azotobacter chroococcum* to boost drought tolerance in maize. Environ Exp Bot 206:105142. https://doi.org/10.1016/j.envexpbot.2022.105142

Upamanya GK, Bhattacharyya A, Dutta P (2020) Consortia of entomo-pathogenic fungi and bio-control agents improve the agro-ecological conditions for brinjal cultivation of Assam. 3 Biotech 10(10):450. https://doi.org/10.1007/s13205-020-02439-3

van den Berg NI, Machado D, Santos S, Rocha I, Chacon J, Harcombe W, Mitri S et al (2022) Ecological modelling approaches for predicting emergent properties in microbial communities. Nat Ecol Evol 6(7):855–865. https://doi.org/10.1038/s41559-022-01746-7

Venter ZS, Jacobs K, Hawkins HJ (2016) The impact of crop rotation on soil microbial diversity: a meta-analysis. Pedobiologia 59(4):215–223. https://doi.org/10.1016/j.pedobi.2016.04.001

Veresoglou SD, Menexes G (2010) Impact of inoculation with *Azospirillum* spp. on growth properties and seed yield of wheat: a meta-analysis of studies in the ISI web of science from 1981 to 2008. Plant Soil 337(1–2):469–480. https://doi.org/10.1007/s11104-010-0543-7

Vessey JK (2003) Plant growth promoting rhizobacteria as biofertilizers. Plant Soil 255(2):571–586. https://doi.org/10.1023/A:1026037216893

Vorholt JA, Vogel C, Carlstrom CI, Muller DB (2017) Establishing causality: opportunities of synthetic communities for plant microbiome research. Cell Host Microbe 22(2):142–155. https://doi.org/10.1016/j.chom.2017.07.004

Wael D, El-Amier Y, Saber WI, Elsayed A (2024) Plant-associated halotolerant bacteria improving growth of *Vicia faba* L. Mariout-2 under salinity conditions. Sci Rep 14(1):16737. https://doi.org/10.1038/s41598-024-66504-0

Wagh J, Chanchal K, Sonal S, Praveena B, Archana G, Kumar GN (2016) Inoculation of genetically modified endophytic *Herbaspirillum seropedicae* Z67 endowed with gluconic and 2-ketogluconic acid secretion, confers beneficial effects on rice (*Oriza sativa*) plants. Plant Soil 409(1):51–64. https://doi.org/10.1007/s11104-016-2937-7

Wang Z, Li Y, Zhuang L, Yu Y, Liu J, Zhang L, Gao Z et al (2019) A rhizosphere-derived consortium of *Bacillus subtilis* and *Trichoderma harzianum* suppresses common scab of potato and increases yield. Comput Struct Biotechnol J 17:645–653. https://doi.org/10.1016/j.csbj.2019.05.003

Wang X, Hu K, Xu Q, Lu L, Liao S, Wang G (2020) Immobilization of Cd using mixed *Enterobacter* and *Comamonas* bacterial reagents in pot experiments with *Brassica rapa* L. Environ Sci Technol 54(24):15731–15741. https://doi.org/10.1021/acs.est.0c03114

Wang G, Li Q, Zhang Z, Yin X, Wang B, Yang X (2022) Recent progress in adaptive laboratory evolution of industrial microorganisms. J Ind Microbiol Biotechnol 50(1):kuac023. https://doi.org/10.1093/jimb/kuac023

Wang Y, Dall'Agnol RF, Bertani I, Bez C, Venturi V (2024a) Identification of synthetic consortia from a set of plant-beneficial bacteria. Microb Biotechnol 17(2):e14330. https://doi.org/10.1111/1751-7915.14330

Wang X, Chi Y, Song S (2024b) Important soil microbiota's effects on plants and soils: a comprehensive 30-year systematic literature review. Front Microbiol 15:1347745. https://doi.org/10.3389/fmicb.2024.1347745

Wang Z, Fu X, Kuramae EE (2024c) Insight into farming native microbiome by bioinoculant in soil-plant system. Microbiol Res 285:127776. https://doi.org/10.1016/j.micres.2024.127776

Wen A, Havens KL, Bloch SE, Shah N, Higgins DA, Davis-Richardson AG, Sharon J et al (2021) Enabling biological nitrogen fixation for cereal crops in fertilized fields. ACS Synth Biol 10(12):3264–3277. https://doi.org/10.1021/acssynbio.1c00049

Wertz S, Czarnes S, Bartoli F, Renault P, Commeaux C, Guillaumaud N, Clays-Josserand A (2007) Early-stage bacterial colonization between a sterilized remoulded soil clod and natural soil aggregates of the same soil. Soil Biol Biochem 39(12):3127–3137. https://doi.org/10.1016/j.soilbio.2007.07.005

Widder S, Allen RJ, Pfeiffer T, Curtis TP, Wiuf C, Sloan WT, Cordero OX et al (2016) Challenges in microbial ecology: building predictive understanding of community function and dynamics. ISME J 10:2557–2568. https://doi.org/10.1038/ismej.2016.45

Xu J, Li XL, Luo L (2012) Effects of engineered *Sinorhizobium meliloti* on cytokinin synthesis and tolerance of alfalfa to extreme drought stress. Appl Environ Microbiol 78(22):8056–8061. https://doi.org/10.1128/aem.01276-12

Xu X, Dinesen C, Pioppi A, Kovács ÁT, Lozano-Andrade CN (2025) Composing a microbial symphony: synthetic communities for promoting plant growth. Trends Microbiol 33(7):738–751. https://doi.org/10.1016/j.tim.2025.01.006

Yamazaki S, Nakayasu M, Kanai K, Mizuno R, Kaida R, Masuda S, Shibata A et al (2025) From model to field: predictive design of microbial communities for resilient crop performance. BioRxiv 2025(2010):2021.683643. https://doi.org/10.1101/2025.10.21.683643

Yan S, Weng B, Jing L, Bi W (2023) Effects of drought stress on water content and biomass distribution in summer maize (*Zea mays* L.). Front Plant Sci 14:1118131. https://doi.org/10.3389/fpls.2023.1118131

Yarzábal LA, Monserrate L, Buela L, Chica E (2018) Antarctic *Pseudomonas* spp. promote wheat germination and growth at low temperatures. Polar Biol 41(11):2343–2354. https://doi.org/10.1007/s00300-018-2374-6

Yeoh YK, Dennis PG, Paungfoo-Lonhienne C, Weber L, Brackin R, Ragan MA, Schmidt S et al. (2017) Evolutionary conservation of a core root microbiome across plant phyla along a tropical soil chronosequence. Nat Commun 8:215. https://doi.org/10.1038/s41467-017-00262-8

Yu Y, Gui Y, Li Z, Jiang C, Guo J, Niu D (2022) Induced systemic resistance for improving plant immunity by beneficial microbes. Plants 11(3):386. https://doi.org/10.3390/plants11030386

Yusuf A, Li M, Zhang SY, Odedishemi-Ajibade F, Luo RF, Wu YX, Zhang TT et al (2025) Harnessing plant-microbe interactions: strategies for enhancing resilience and nutrient acquisition for sustainable agriculture. Front Plant Sci 16:1503730. https://doi.org/10.3389/fpls.2025.1503730

Zhang LN, Wang DC, Hu Q, Dai XQ, Xie YS, Li Q, Liu HM et al. (2019) Consortium of plant growth-promoting rhizobacteria strains suppresses sweet pepper disease by altering the rhizosphere microbiota. Front Microbiol 10:1668. https://doi.org/10.3389/fmicb.2019.01668

Zhang J, Ahmed W, Dai Z, Zhou X, He Z, Wei L, Ji G (2022) Microbial consortia: an engineering tool to suppress clubroot of chinese cabbage by changing the rhizosphere bacterial community composition. Biology 11(6):918. https://doi.org/10.3390/biology11060918

Zhang J, Liu W, Bu J, Lin Y, Bai Y (2023) Host genetics regulate the plant microbiome. Curr Opin Microbiol 72:102268. https://doi.org/10.1016/j.mib.2023.102268

Zhao L, Walkowiak S, Fernando WGD (2023a) Artificial intelligence: a promising tool in exploring the phytomicrobiome in managing disease and promoting plant health. Plants 12(9):1852. https://doi.org/10.3390/plants12091852

Zhao X, Yuan X, Xing Y, Dao J, Zhao D, Li Y, Wang Z (2023b) A meta-analysis on morphological, physiological and biochemical responses of plants with PGPR inoculation under drought stress. Plant Cell Environ 46(1):199–214. https://doi.org/10.1111/pce.14466

Zhu X, Chen WJ, Bhatt K, Zhou Z, Huang Y, Zhang LH, Chen S et al. (2023) Innovative microbial disease biocontrol strategies mediated by quorum quenching and their multifaceted applications: a review. Front Plant Sci 13:1063393. https://doi.org/10.3389/fpls.2022.1063393

Zimmer S, Messmer M, Haase T, Piepho H-P, Mindermann A, Schulz H, Habekuß A et al (2016) Effects of soybean variety and *Bradyrhizobium* strains on yield, protein content and biological nitrogen fixation under cool growing conditions in Germany. Eur J Agron 72:38–46. https://doi.org/10.1016/j.eja.2015.09.008

Part IV
Physiological and Anatomical Impacts of Drought Stress

Chapter 8
Mitigation of Drought Stress in Crop Plants Through Rhizoengineering

Kausar Hussain Shah and Tehreem Ghafoor

Abstract Drought is one of the biggest threats to global food security, and it affects the survival, physiological function, and potential yield of a large number of plants. Adaptive and sustainable methods for improving crop resilience are becoming increasingly necessary as climate change increases the intensity of water scarcity. Despite their value, traditional breeding and genetic engineering techniques are costly time-consuming, and have legal and environmental limitations. The rhizosphere engineering, also known as rhizoengineering, has emerged as a promising area of agricultural biotechnology that involves precise alteration of the rhizosphere, which is a small area of soil that plant roots and their microbiota share with each other. It is possible to reduce the detrimental effects of drought stress on plants by improving the interactions between roots, soil, and microbial communities. This can be achieved through biological, chemical, and physical methods. Beneficial microbes, such as arbuscular mycorrhizal fungi (AMF), plant growth-promoting rhizobacteria (PGPR), and endophytes, are important for enhancing plant resistance to drought. They do this by helping plants take up more water, controlling stress-responsive genes, maintaining hormonal balance, and boosting antioxidant defenses. This chapter highlights various ways in which rhizoengineering can help plants to withstand the drought stress. These include the use of microbial inoculants, synthetic microbial consortia, soil amendments such as biochar and compost, and molecular tools to change the profiles of root exudates. There are case studies of different types of crops, such as wheat, maize, and oats, that show how the applications work. The chapter additionally discusses the physiological and molecular processes that help engineered rhizospheres survive drought, such as osmotic adjustment and phytohormone modulation. Despite its potential, the widespread adoption of rhizoengineering faces challenges related to formulation consistency, environmental variability, and farmer awareness issues. The chapter concludes by discussing future directions, including the integration of rhizosphere microbiome engineering

K. H. Shah (✉) · T. Ghafoor
Institute of Botany, Bahauddin Zakariya University Multan, Multan, Pakistan
e-mail: kausar.shah@bzu.edu.pk

S. Mehnaz, I. Shahid (eds.), *Frontiers in Rhizoengineering*, Microorganisms for Sustainability 65, https://doi.org/10.1007/978-981-92-3665-7_8

with precision agriculture and climate-resilient microbiome discovery, paving the way for biologically intelligent and climate-resilient farming systems.

Keywords Rhizosphere · Rhizoengineering · Rhizobacteria · Drought stress mitigation · Biocontrol Agents · PGPR · Sustainable Agriculture

8.1 Introduction

Stress is defined as any unfavorable circumstance that interferes with or inhibits a plant's metabolism, growth, development, and yield. In the coming decades, plants will likely face more severe issues owing to the combined effects of drought, salinity, and temperature stress. Drought is a noteworthy abiotic stress that significantly limits agricultural outcomes and food production globally (Begna 2020). Food security is becoming a major global concern owing to drought stress, which is a major limitation on agricultural production. Drastic and rapid changes in global climate conditions have worsened the situation. Drought stress induces changes in plant morphology, physiology, and biochemistry (Iqbal et al. 2020).

The significant physiological changes caused by dehydration slow down or even stop growth, putting yield stability at risk. Drought-related field yield losses vary by crop type and usually fall between 30% and 90%. The susceptibility of the yield to drought differs depending on the type of agricultural goods harvested, such as tap roots, fruits, seeds, shoots, or leaves (Dietz et al. 2021). In several regions of the world, droughts are expected to increase in frequency in the future. Unfortunately, many parts of the world may experience increased droughts as a result of global climate change, which may have a significant effect on crops. Therefore, there is an urgent need to produce crop varieties that can adapt to these conditions and maintain high levels of output (Begna 2022).

Increasing crop output to meet consumer demand and feed the world's rising population necessitates the extensive use of pesticides and chemical fertilizers, which are frequently misused in soil (Naqqash et al. 2022). Transgenic plant production will enable the production of more palatable food and feeding of the expanding population. Contaminated soils have also been bioremediated by transgenic plants. Genes encoding bacterial enzymes in transgenic plants have been employed to purify soils from organic pollutants, such as selenium, mercury, and polychlorinated biphenyls (PCBs) (Meagher 2000). The use of transgenic crops has long been controversial, raising questions about the potential health effects of genetically modified materials. For instance, transgenic crops have been linked to allergies in certain individuals, although it is unclear whether this reaction is caused by transgenic crops (Ferber 1999). The question of whether transgenic crops harm the environment has become a major problem. One such example is pollen from transgenic corn, which can destroy the larvae of monarch butterflies. To produce transgenic crops that are safe for the environment and nonhazardous, their apparent benefits and drawbacks must be balanced (Rani and Usha 2013). In comparison to artificial agricultural methods, natural alternatives are more economical and environmentally friendly. Utilizing natural processes, such as microbial interactions, biofertilizers, and biocontrol agents, can help in reducing reliance on chemical inputs, limiting the ecological effects, and in improving crop and soil health in the long run. The use of transgenic plants has made significant progress in crop improvement; however, incorporating other strategies, such as inoculating crops with beneficial microbial strains and enhancing root system architecture, offers a viable and sustainable way to increase crop yield, nutrient uptake, and adaptability to environmental stresses (Mehmood et al. 2018). Drought causes water shortage, crop failure, hunger, poverty, and environmental degradation, all of which make it more difficult to achieve the UN Sustainable developmental Goals (SDGs) (Sena et al. 2016).

The rhizosphere is known to be highly dynamic, enriched with bacteria from local soil, and responsive to specific alterations associated with various plant growth phases. Furthermore, its colonization is significantly regulated and controlled by the composition of host-specific exudates, which modify the physicochemical characteristics of the soil and, hence, the microbiome (Michl et al. 2023). Rhizosphere engineering is an innovative method for enhancing soil and crop productivity through increased input usage efficiency. It uses the biophysical characteristics of the soil to affect plant-microbiome-soil interactions. Rhizosphere engineering is a branch of science and engineering that focuses on manipulating the rhizosphere, the area of soil occupied and impacted by roots, to improve plant growth, soil health, and environmental sustainability (Velmourougane and Blaise 2021). Biocontrol products based on plant growth-promoting rhizobacteria have recently emerged as an alternative to chemical pesticides. *Bacillus* spp. are among the most promising biocontrol agents (Berg et al. 2021). Competent bacteria in the rhizosphere improve root colonization ability, create intricate and dynamic interactions between and within species, and catalyze chemical changes (Ishaq et al. 2023). PGPR can uphold plant health, promote nutrient uptake, ease mineral transport, and combat phytopathogens (Bhat et al. 2023). Rhizobia improve plant development by fixing nitrogen. Several researchers have observed that inoculating seeds with *Rhizobium* dramatically boosts legume crop growth and yield (Raza et al. 2004). PGPR can aid in enhancing plant growth, nutrition, root growth pattern, plant competitiveness, and responses to external stresses. They can also suppress soil-borne plant diseases by creating growth-promoting compounds and inducing plant resistance (Egamberdiyeva and Höflich 2004).

8.2 Drought Stress Physiology in Plants

8.2.1 Plant Responses to Water Deficit

Water is essential in all physiological processes in plants, accounting for 80–95% of nonwoody plant biomass. Water deficit is caused not only by a shortage of water but also by environmental stresses such as extreme temperature or elevated salt levels resulting in physiological drought (Hirt and Shinozaki 2003). The amount, rate, and duration of water loss from a stressful situation all affect plant reactions to water deficit. The generation of active oxygen species (AOS) is a natural byproduct of oxygen-dependent life. In response, plants increase their production of the stress hormone abscisic acid (ABA), which tells guard cells to shut their stomata. CO_2 cannot enter the leaf effectively when stomata are closed, limiting the quantity available for photosynthesis. Consequently, declining the growth, biomass, and carbon absorption in plants. Drought causes an imbalance in the internal environment of a plant, namely, altering its water status and photosynthesis rate. Oxidative stress, which occurs when the production of reactive oxygen species (ROS) surpasses the plant's antioxidant defenses, one of the main effects of this imbalance (Navari-Izzo and Rascio 1999).

Drought-induced hormonal imbalance is a complex physiological response that aids plants in adjusting to water restriction. A number of important plant hormones are perturbed in their signaling and balance by drought stress. Research on the effects of Salicylic Acid (SA) on biotic and abiotic stress has been extensive, but little is known about its effects on drought stress. Nevertheless, several studies have indicated that it may play a role in drought stress by altering the transcriptional regulation of genes linked to drought and stomatal aperture. Drought tolerance and sensitivity are affected by the quantity of SA used (Signorelli et al. 2019). In addition, drought stress responses and ABA production are regulated by auxin homeostasis (Farhangi-Abriz and Torabian 2018). Gibberellin (GA) insufficiency encourages the distribution of reserves to the roots. Primary metabolism is significantly altered by impaired GA production, primarily due to the severe consequences of drought. Gibberellin deficiency alleviates drought stress by maintaining water content. Symptoms of gibberellin deficiency resemble those of drought stress (Salazar-Cerezo et al. 2018). Drought is becoming more severe and frequent as a result of global warming, potentially reducing food output worldwide. Simultaneously, a continually rising human population, which could reach 9 billion by 2050, necessitates an increase in food production and supplies. This problem will worsen in the future as desertification increases, and the current level of annual loss of arable land may quadruple by the end of the century due to global warming (Naveed et al. 2014).

8.2.2 Biomarkers of Drought Tolerance

Drought tolerance biomarkers are quantifiable indicators of a plant's capacity to endure or thrive in water-deficient environments. These biomarkers, which may be morphological, physiological, biochemical, or molecular characteristics, are frequently employed in biotechnology and plant breeding to identify genotypes that are resistant to drought. Relative water content indicates the water status of plants. Reports state that plants with higher amounts of relative water can withstand drought, which results from either decreased tissue cell wall flexibility or increased osmotic control (Keyvan 2010). A decline in chlorophyll concentration may be regarded as a nonstomatal limiting factor under drought stress situations, as it has been demonstrated to be an index for assessing the source. In plants under drought stress, electrolyte leakage is a common biochemical indicator of membrane breakdown. Water stress accelerates the degradation of proteins, membrane lipids, photosynthetic pigments, and cell membranes and boosts the production of ROS. Osmolytes accumulation is a crucial adaptation mechanism for plants to withstand drought and other abiotic challenges. Osmolytes accumulation (OA) in plant cells diminishes the osmotic potential of cells, thereby maintaining water absorption and cell turgor pressure. This may help maintain physiological functions, such as stomatal opening, photosynthesis, and growth. To maintain a stable water potential gradient and to allow the plant to continue absorbing water from the soil, OA lowers the plant water potential when the soil dries (Serraj and Sinclair 2002).

8.3 The Rhizosphere: A Dynamic Interface

8.3.1 Rhizosphere Composition

In 1904, German botanist Lorenz Hiltner coined the term "rhizosphere," which is partially derived from the Greek word "rhiza," meaning root, to describe the interaction between plants and roots. The rhizosphere extends 15 mm from the root surface into the surrounding soil despite the lack of a clear geographical border. The endorhizosphere consists of parts of the endodermis and cortex, where cations and microorganisms can live in the "free space" (apoplastic space) between cells. The rhizoplane is the middle zone that contains mucilage and the root epidermis. The outermost zone, known as the ectorhizosphere, extends from the rhizoplane to the bulk soil (McNear Jr 2013).

Exudates from the roots of plants, such as amino acids, sugars, and flavonoids, are a complex mixture of inorganic and organic substances that are released into the rhizosphere. Plant root exudates play an essential role in promoting plant growth and interacting with the biological and physicochemical elements of the rhizosphere. Under stressful conditions, root exudation helps plants use resources more efficiently and facilitates communication between soil bacteria and plants to reduce stress (Chai and Schachtman 2022). The metabolites secreted by roots into the rhizosphere play several roles. To adapt and ensure their survival in harsh environments, plants can alter the soil quality by changing the composition of their root exudates (Vives-Peris et al. 2020).

Plants harbor a fascinating variety of microorganisms, called the microbiome, within and around their roots. Oomycetes, bacteria, fungi, and archaea constitute the majority of the microbiome and can be either harmful or advantageous to the health and fitness of plants (Pascale et al. 2020). Inoculating plant rhizospheres with root-associative plant growth-promoting rhizobacteria (PGPR) is an environmentally friendly method that reduces fertilizer rates by enhancing nutrient-use efficiency. PGPR increase nutrient usage efficiency in plants, resulting in optimal growth and output with reduced fertilizer inputs (Ejaz et al. 2020). PGPR are a broad category of beneficial soil bacteria that live in plant roots and improve plant development and resistance to stress, including drought. Rhizobacteria that promote plant development are a sustainable substitute for pesticides in crop protection and agricultural output (Santoyo et al. 2021). They can also be further separated into two groups, iPGPR and ePGPR, according to their habitat. While ePGPR are free-living bacteria that are frequently found outside plants and aid in plant growth, iPGPR are symbiotic bacteria that live inside plant cells and create nodules (Asghari et al. 2020). PGPB secrete siderophores and lytic enzymes that inhibit the growth of phytopathogens in the rhizosphere. Furthermore, some PGPB release antibiotics that are especially important for rhizosphere and rhizoplane colonization by pathogens. Some well-known examples include hydrogen cyanide, 2,4-diacetylphloroglucinol (DAPG), oomycin A, pyoluteorin, phenazine, pyrrolnitrin, tropolone, thiotropocin, cyclic rhamnolipids, lipopeptides, Oligomycin A, Kanosamine, Zwittermicin A, and Xanthobaccin (Compant et al. 2010). Extensive research has shown that rhizobacteria secrete a variety of secondary metabolites (SMs). Rhizobacteria and their SMs produced by them can operate on pathogens by denying them nutrients (competition), lysing cells, and/or preventing certain activities associated with pathogen growth (antibiosis) and hence act as biocontrol agents (Compant et al. 2013).

Soil type can be a major predictor of microbial communities because soils have a variety of features, such as nutrient and water content, cation exchange capability, and texture and structure. Soil texture and structure can impact the size and distribution of particles and pore spaces, influencing the movement of water and nutrients and consequently leading to changes in the soil microbial community. Particle size also has a major impact on microbial community structure, with smaller silt and clay particles exhibiting higher microbial diversity than coarse sand fractions (Saad et al. 2020).

The most common relationship between microorganisms and higher plants is mycorrhiza. Mycorrhizal relationships can be mutualistic, neutral, or parasitic for plants. Endomycorrhiza, which forms structures called arbuscules for nutrient exchange and vesicles for storage, and ectomycorrhiza, which forms a sheath around plant roots and penetrates between root cells, are the two main mycorrhizal groups based on differences in morphology and physiology (Marschner 2012). The effects of roots and arbuscular-mycorrhizal (AM) fungi on the composition of soil bacterial colonies and the impact of AM fungus-rhizobacterium interactions on plant and soil development are not well understood. Mycorrhizal fungi do not pose a significant threat to other microbes for nutritional resources during the initial phase of propagule germination and growth because of their limited or nonexistent saprophytic ability. They are, on the other hand, susceptible to both helpful responses to certain soil microbes and microbial antagonistic interactions during these stages (Andrade et al. 1997).

Leguminous plants, such as beans, peas, lentils, and clover, have symbiotic interactions with a group of soil bacteria called rhizobia. These bacteria are well known for their capacity to fix atmospheric nitrogen (N_2) into ammonia (NH_3), a chemical form that plants may use. Over the past 20 years, the taxonomy of rhizobia-bacteria that can nodulate leguminous plants has undergone significant modification. The initial genus of alpha-Proteobacteria, *Rhizobium*, has now split into multiple genera (Willems 2006). In general, legume nodule growth can be categorized as determinate or indeterminate. Indeterminate nodules maintain meristematic tissue, whereas determinate nodules exhibit transient meristems. The type of nodule is determined by the host plant; therefore, it is uncommon for legume species to generate both determinate and indeterminate nodules (Sprent et al. 2013). The production of a variety of chemicals, primarily flavonoids, by legumes triggers the synthesis of NodD protein in rhizobia, initiating the nodulation process in practically all legumes investigated (Downie 2014). Different legumes produce a variety of chemicals. Rhizobia release signal molecules known as "nod factors," which the plant recognizes as the source of nodule organogenesis. NodD protein encourages the transcription of additional genes associated with the nodulation procedure, such as those required to produce Nod factors (Oldroyd and Downie 2008).

8.3.2 Root-Microbiome Interactions Under Stress

Throughout their lives, plants encounter various abiotic and biotic stresses. The microbial communities associated with flora may shift due to alterations in transcriptomics and metabolomics, which may also affect the root and leaf exudates. New data show that modifications, particularly the increase in commensal microbial abundance after stress, can help plants survive and leave a legacy that improves the fitness of their progeny (Liu et al. 2020). For instance, it has been shown that a variety of bacteria can alter the amount of ethylene in crops by the release of 1-aminocyclopropane-1-carboxylate deaminase. The primary precursor of ethylene generated by plants is ACC. By cleaving ACC and lowering the amount of ethylene that plants can biosynthesize, bacteria that produce ACC deaminase reduce plant damage under stress (Glick 2005). The impact of drought as a water stressor influencing soil and root-associated microbial communities has been the subject of far more research than that of flooding. However, it has been demonstrated that AMF root colonization levels typically decrease during drought. AMF can increase plant drought fitness through improved nutrition or more direct impacts on stomatal conductance and increased water-use efficiency; however, the outcomes are frequently species specific (Augé 2001). The drought-responsive bacterial populations were characterized. During drought stress, there were more Actinobacteria in the root biological communities of rice (*Oryza sativa*) and sorghum (*Sorghum bicolor*). The study revealed that when 30 plant species were subjected to drought stress, *Streptomyces* species dominated the root-endosphere population. Indeed, as *the abundance of Streptomyces* species increased, plant species became more resilient to drought stress (Hilker et al. 2016).

8.4 Rhizoengineering Strategies for Drought Tolerance

8.4.1 Microbial Inoculants

A growing number of inoculants are currently being marketed for a variety of crops, and their application is thought to be highly appealing because it would significantly minimize the use of chemical pesticides and fertilizers.

8.4.1.1 Plant Growth-Promoting Rhizobacteria (PGPR)

Rhizosphere-dwelling bacteria improve plant health, which in turn promotes plant growth. The use of plant growth-promoting rhizobacteria (PGPR) can reduce fertilizer costs and environmental hazards and they can suppress diseases. PGPR are well known for their involvement in plant growth promotion, specifically biological nitrogen fixation, phytohormone synthesis, and biocontrol (Mehnaz 2011). In plants, an increase in ethylene levels is directly proportional to the content of 1-am inocyclopropane-1-carboxylic acid (ACC) in plant tissues. ACC deaminase-containing PGPR may operate as a sink for ACC, ensuring that the quantity of ethylene does not rise above the point at which root growth is hindered. As a result, PGPR in growing or stressed plants could lower ethylene levels, increase seedling life, and help plants generate longer roots. Furthermore, PGPR can produce phytohormones that accelerate cell growth and division and help plants tolerate environmental challenges (Fahad et al. 2015). PGPR are assemblies of beneficial soil bacteria that not only improve plant development but also enhance drought resistance. Plants grow their root systems when the rhizosphere changes, which improves nutrition and water intake. It also detoxifies free radicals and reduces the harmful effects of reactive oxygen species (ROS) (Ilyas et al. 2020). *Burkholderia*, *Rhizobium*, *Bradyrhizobium*, *Serratia*, *Frankia*, *Thiobacillus*, *Bacillus*, *Pseudomonas*, *Arthobacter*, *Azotobacter*, *Azospirillum*, and *Acinetobacter* comprise the majority of the credible groups of PGPR (Goswami et al. 2016). Over the last decade, the genus *Bacillus* has been taxonomically split into numerous distinct genera based on detailed investigations of 16S rRNA gene sequences and other chemotaxonomic data (Ahmed et al. 2007).

8.4.1.1.1 Direct Mechanisms of PGPR

Understanding the mechanics of plant growth promotion (PGP) is essential for manipulating rhizospheric flora to maximize activities that greatly increase plant yields. PGP mechanisms have historically been categorized into direct and indirect mechanisms. Although it is not always easy to distinguish between the two, direct processes occur within the plant and directly regulate its metabolism, whereas indirect mechanisms typically occur outside the plant (Siddikee et al. 2010).

Biological Nitrogen Fixation Bacterial strains that exhibit nitrogen-fixing characteristics are classified into two categories. The first category includes symbiotic bacteria connected to roots and legumes, including *Rhizobium* strains. These bacterial populations are specialized and infect the roots to form nodules. Another group of bacteria is the so-called free-living nitrogen fixers, which are not restricted to plants. In agricultural trials, the most commonly utilized species are *Azotobacter* and *Azospirillum*. Many strains of these species have become more significant because, in addition to fixing nitrogen, they enhance plant growth by generating phytohormones, such as gibberellic acid, cytokinins, and IAA (Oberson et al. 2013).

Solubilization of Phosphate After nitrogen, phosphorus is the nutrient that most limits plant growth. Despite having extremely large deposits, phosphorus is not in a form that plants can absorb. Only soluble forms of phosphate, mono- and dibasic phosphate, can be absorbed by plants (Jha and Saraf 2015). Tricalcium phosphate, rock phosphate, aluminum phosphate, and other complex-structured phosphates are dissolved by microorganisms, which mineralize organic phosphorus in the soil and convert it to an inorganic form, which helps plants access phosphates. These bacteria that solubilize phosphate employ several methods to dissolve the insoluble forms of phosphate. Microorganisms secrete organic acids as a result of sugar metabolism, which is the main mechanism of phosphate solubilization. Rhizosphere-dwelling organisms utilize sugars in root exudates and break them down to create organic acids (Goswami et al. 2014).

Phytohormone Production Well-known phytohormone classes include ethylene, abscisic acid, gibberellins, cytokinins, and auxins. Soil microbes, especially rhizosphere bacteria, have the capacity to produce such hormones. Any phytohormone in the rhizosphere, created by the resident microbial flora or supplemented externally, will cause a plant to react. These phytohormones can mediate plant cell expansion, division, and extension in both symbiotic and nonsymbiotic roots (Glick 2014).

8.4.1.1.2 Indirect Mechanics of PGPR

Siderophores Production Siderophores are small molecules with functional groups that can reversibly bind iron and typically weigh less than 1 kDa. The group distances of catechol and hydroximates, the most prevalent functional groups, are ideal for binding iron. The concentration of siderophores in soil is roughly 10–30 M. Bacteria that produce siderophores typically belong to the genus *Pseudomonas*, of which *Pseudomonas fluorescens* and *Pseudomonas aeruginosa* are the most extensively researched species. These bacteria release siderophores of pyochelin and pyoverdine types (Haas and Défago 2005).

Antibiotic Production by PGPR As an alternative to conventional pesticides, antimicrobial organisms have been proposed to combat plant diseases in agricultural crops. Pathogenic microorganisms that produce antibiotics are actively suppressed by PGPR, such as *Bacillus* and *Pseudomonas* species. By secreting extracellular metabolites at low concentrations, these bacterial antagonists reduce plant diseases. Numerous antibacterial and antifungal antibiotics are produced by *Bacillus* species. Some of these substances, such as lipopeptides, rhizocticins, bacillaene, difficidin, mycobacillin, bacilysin, and chlorotetain from the families of iturin, surfactin, and fengycin, are produced by polyketide synthases and/or nonribosomal peptide synthetases (NRPSs). Other substances, such as sublancin, TasA, subtilin, and subtilosin A, are commonly used and originate from ribosomes (Leclère et al. 2005). In brief, PGPR form a capsular envelope to protect itself and raises the pH to avoid soil acidity. This occurs directly or indirectly by changing root exudates through beneficial microorganisms such as arbuscular mycorrhizal fungi (AMF). Additionally, PGPR can break down toxins generated by soil-dwelling bacteria and plant roots (Dutta and Podile 2010).

8.4.1.2 Mycorrhizal Fungi

The mutually beneficial partnership between fungi and plant roots was referred to as "mycorrhiza" by Albert Bernhard Frank. In literal terms, mycorrhiza means "fungus root." Mutualism between the roots of higher plants and specific fungi produces mycorrhizae. Fossil mycorrhizae discovered in carbonaceous deposits provide evidence that mycorrhizal fungi co-evolved with plants for approximately 400 million years to become a component of the root system. The roots of most terrestrial plants create symbiotic partnerships with these soil-dwelling fungi, which are found all over the planet. Plants without mycorrhizal root systems are uncommon in natural environments. Thus, mycorrhizal associations can be considered a widespread or almost ubiquitous phenomenon in the plant kingdom (Bagyaraj 2011).

8.4.1.2.1 Types of Mycorrhizae

Different types of mycorrhizal connections exist. Ectomycorrhiza, ericoid mycorrhiza, orchidaceous mycorrhiza, and arbuscular mycorrhiza are the most important types of mycorrhizae. The most prevalent ectomycorrhiza is present in tree species found in temperate forests belonging to the *Salicaceae*, *Betulaceae*, *Fagaceae*, *Tiliaceae*, and *Pinaceae* families and in certain *Juglandaceae*, *Myrtaceae*, *Leguminaceae*, and *Rosaceae* members. Plants of the *Epacridaceae*, *Empterraceae*, and *Ericaceae*, often called heath plants, such as rhododendrons, blueberries, cranberries, and azaleas, are typically colonized by ericoid mycorrhizal fungi. These plants are found in the temperate regions of the world. The apothecium-forming *ascomycete Hymenoscyphus ericae*, formerly known as *Pezizella ericeae*, establishes mycorrhizal associations. The family Orchidaceae includes orchid mycorrhizae. The species in this family are close to 30,000. Orchid seeds contain very little fat or carbohydrate reserves. The structure known as the "protocorm" is exposed when seeds ingest water, swell somewhat, and rupture their seed coat during germination. To develop into a plant, the mycorrhizal fungus must infect the protocorms. Protocorms can wait up to six months for mycorrhizal fungi to infect them. If the fungus does not infect it, it dies. Therefore, mycorrhizal fungi are essential for orchids (Krishnamurthy and Senthilkumar 2005).

The majority of tropical tree species, most shrubs, agricultural crops, and certain temperate tree species have arbuscular mycorrhizal connections. Glomeromycetous nonseptate fungi produce arbuscular mycorrhizae (AM). In the phylum Glomeromycota, there are members of 3 classes (Paraglomeromycetes, Glomeromycetes, and Archaeosporomycetes), 5 orders (Paraglomerales,

Archaeosporales, Gigasporales, Diversisporales, and Glomerales), 14 families, and 26. *Entrophospora*, *Glomus*, *Gigaspora*, *Scutellospora*, and *Acaulospora* are genera of AM fungi that are frequently seen. These fungi have not been cultivated in nutritional media and are obligatory symbionts. Although there is growing evidence that some endophytes may create preferred associations with specific host plants, some endophytes are not host specific. Numerous studies have shown that plants react to inoculation with effective strains of arbuscular mycorrhizae, even in nonsterile soils (Bagyaraj 2014).

8.4.2 Synthetic and Bioinspired Consortia

Multistrain microbial formulations are products that contain two or more distinct microbial strains that are intended to cooperate to produce increased biological activity. They are extensively used in agriculture, wastewater treatment, animal and human health, and environmental cleanup. Although microbial inoculation has been used for centuries to improve crops, single-strain microbial inoculants are not very adaptable to changing environmental conditions, leading to variable outcomes when applied in the field. Co-inoculation and multistrain inoculation, also known as microbial consortia, are mixed inoculants of various microorganisms that can increase the adaptability of bacterial inoculants, according to simultaneous plant experiments (Khan et al. 2017).

Because the microbes added to multistrain inoculants typically require an extensive range of temperature, pH, and moisture for activities such as root invasion, resistance against plant pathogens, metabolic activities, stabilization of nitrogen, phosphorous solubilization, and hormone and antibiotic production, these inoculants can function better in a variety of soil types (Upadhyay et al. 2012). Owing to the high competence and wide range of actions of microbial strains when utilized in consortia without the need for genetic modification, multistrain inoculation is a more dependable technique than other methods for boosting agricultural production and promoting healthy plant growth (Ma 2019).

8.4.3 Organic Amendments and Soil Conditioners

Organic additions improve the rhizosphere zone by promoting root-microbe interactions, boosting microbial activity, improving nutrient availability, and strengthening soil structure. Biochar is becoming increasingly popular as a sustainable addition to contemporary farming, providing several environmental advantages, such as nutrient cycling, carbon sequestration, and greenhouse gas emission reduction (Khare et al. 2021). Organic elements used to make organic fertilizers come from both plant and animal sources. Compost, animal dung, and decomposed plant waste are common forms of organic manure. These materials are rich in micronutrients (iron, zinc, and manganese) and macronutrients (nitrogen, phosphorus, and potassium). They enhance their influence on nutrient cycling and plant health by incorporating helpful microbes, such as mycorrhizal fungi, nitrogen-fixing bacteria, and phosphate-solubilizing bacteria, in addition to organic materials (Zhao et al. 2024).

The cation exchange capacity (CEC), a crucial soil attribute, measures the soil's capacity to store and exchange cations or positively charged ions, including ammonium (NH_4), potassium (K), magnesium (Mg_2), and calcium (Ca_2). Enhancing CEC in the rhizosphere is mostly dependent on organic fertilizers, especially those that are high in organic matter (Ross and Skyllberg 2023).

Soil conditioning can be defined as any procedure that improves the capacity of a soil to raise crop yields or its performance for any purpose. Any product used for soil conditioning is referred to as a soil conditioner. Soil conditioning is necessary to alleviate water repellency, decrease soil crusting and compaction, and enhance soil air-water interaction, drainage, and aggregation. Many types of natural deposits, including gypsum, lime, organic compounds, living plants, microorganisms, and numerous industrial waste products, are considered soil conditioners. Soil conditioners are beneficial because they improve ecosystem function and the soil's ability to support crops (Shinde et al. 2019).

Soil conditioners can be categorized based on two criteria: (1) the origin of the material and (2) their composition. Regarding the materials' origin, they might be either natural or synthetic. The materials used to prepare soil conditioners can be classified as organic or inorganic. In the presence of oxygen, microbes break down organic waste and form compost. There are several advantages to using compost as a soil conditioner, such as increased microbial activity and organic C content, a source of nutrients for plants, such as phosphorus, potassium, nitrogen, and magnesium, and strengthened root growth (Donn et al. 2014). Compost can affect soil microflora by preventing a variety of soil-borne diseases, including those caused by *Phytophthora*, *Fusarium*, and *Pythium* species. Therefore, generally speaking, compost preserves and enhances the stability of agricultural soil (Zhou et al. 2016).

8.4.4 Genetic Rhizosphere Engineering

Genetic rhizosphere engineering is the process of genetically modifying plants to change their root exudates to improve symbiotic effectiveness or root architecture and draw in beneficial bacteria. There is a growing global presence of genetically modified (GM) crops in agriculture worldwide. Internal mechanisms in plants are pleiotropic effects of genetic alterations caused by internal plant processes. Rhizosphere-mediated mechanisms alter the biotic or abiotic interactions of plants, which subsequently affect the plants themselves. Internal impacts on plants may arise if the altered gene affects root functions without specifically expressing itself in the root tissue. For instance, aerial portions may express a gene that is added to improve resistance to soil diseases. These impacts may be detrimental, such as when the gene product disrupts the main metabolism in the aerial portion, or beneficial, such as when the expression in aerial portions also confers aversion to aboveground diseases. If the gene is only expressed in the roots, but the product created by the altered gene is transferred from the roots into the aerial portions, this can also have internal effects on the plant (Fig. 8.1, pathway B). Similar to pathway a, these impacts can be favorable or unfavorable (Kabouw et al. 2012).

Rice is the most significant crop in the world, as it feeds almost 2 billion people. Genetic manipulation that causes plants to produce more or deeper developing roots is one method for improving drought tolerance in rice. However, shoot growth may be hampered by a greater focus on roots (Price et al. 2002). Utilizing a constitutive or root-specific promoter, rice's *OsNAC10* gene was overexpressed, a gene implicated in auxin signaling, leading to enhanced drought tolerance in laboratory conditions. Plants produced using a root-specific promoter exhibited a greater root

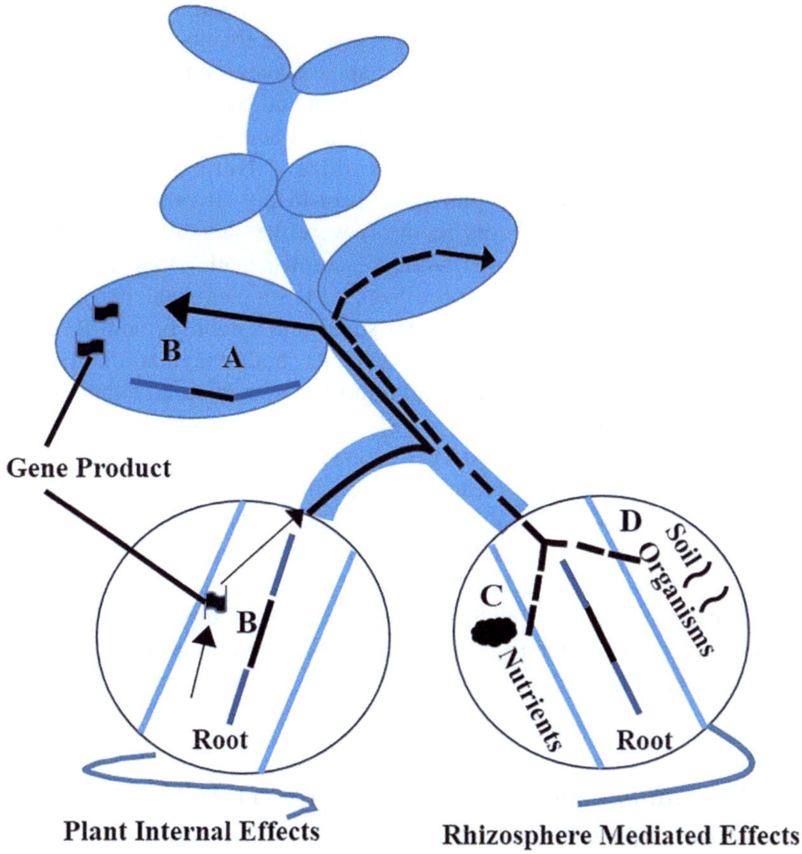

Fig. 8.1 Genetic alterations in roots drive shoot functional responses either directlyor in indirect way. (**A**) Direct way of altered gene expression in the shoot (**B**) Indirect way of altered gene expression in the shoot through transport of gene products or signals from root (**C**) Through the variation in soil nutrients, their availability and mobility (**D**) Through altered interactions with beneficial or harmful soil organisms. Plant-internal effects are indicated by solid lines; rhizosphere-mediated effects are indicated by dotted lines

diameter than those transformed with a constitutive promoter (Jeong et al. 2010) under laboratory conditions (Redillas et al. 2011) and increased yield in the field under drought (Jeong et al. 2010). Regardless of the mechanism, changes linked to root-specific gene expression had a net positive impact on yield during drought conditions. This effect appeared to be reversed if the gene was constitutively expressed.

8.5 Case Studies and Applications

8.5.1 Rhizobia-Enhanced Drought Resilience in Wheat Under Supplementary Irrigation

In the 2017–2018 and 2018–2019 crop years, a study (Najafi Vafa et al. 2022) was conducted in Iran. The primary factors were irrigation treatments (no irrigation or dry land, two irrigations in the booting and spike stages, and one irrigation in the booting stage) and two cultivars of wheat (Sirvan and Sardari) as secondary variables. The eight levels of biological fertilizer treatment are as follows: Seaweed extract + Nitrozist and Phosphozist, Mycorrhiza + Nitrozist and Phosphozist, Mycorrhiza + Nitrozist and Phosphozist, Mycorrhiza + Seaweed extract, and no application of biological fertilizers (control) were the sub-sub-factors taken into consideration. The study found that while a lack of irrigation raised malondialdehyde (MDA) levels, the use of biofertilizer with supplemental irrigation levels applied once or twice improved leaf water status and soluble protein levels in the plants. The quantity and efficacy of antioxidants, both enzymatic and nonenzymatic, were elevated in both years by biofertilizers, particularly their mixtures. Additionally, it eliminated active oxygen species and improved the suppression of harmful radicals by 2–2-diphenyl picryl hydrazyl.

The findings showed that wheat yield was higher with biological fertilizers (Nitrozist and phosphozist + seaweed extract + mycorrhiza) than with the control fertilizer. Additionally, in the first and second years, Mycorrhiza + Seaweed Extract + Nitrozist and Phosphozist increased the yield by 8.04% and 6.96%, respectively. Therefore, in dry and semiarid environments, the negative impacts of drought stress can be lessened by using the right mix of microbes, helpful biological chemicals, and additional irrigation.

8.5.2 Mitigating Drought Stress in Maize via ACC Deaminase-Producing PGPR

A glass jar study was conducted to assess the effectiveness of inoculating maize seedlings with drought-resistant, ACC deaminase-producing PGPR under axenic conditions (Danish et al. 2020). In maize seedlings, several ACC deaminase-containing PGPR significantly increased root fresh and dry weights, shoot and root lengths, and shoot fresh and dry weights at various concentrations (i.e., 0, 10, and 20%) of polyethylene glycol (PEG)-induced drought stress.

8.5.3 Impact of Rhizoengineering on Microbial Communities and Soil Dynamics in Postharvest Oat Fields

This study investigated the effects of certain PGPR, including *Klebsiella aerogenes* BHUSR3, *Serratia marcescens* BHUSR2, and *Enterobacter cloacae* BHUSR1, on microbial communities and soil nutrient mobilization in rhizospheric soils. The selected PGPR were used to bioprime *Avena sativa* seeds prior to planting. The fields were subjected to two water stress levels, with one plot receiving irrigation every 7 days and the other receiving irrigation every 10 days. The rhizosphere microbial flora, pH, EC, moisture, and nutrients were analyzed in fully grown crops. The water status of the soil and accessible nutrients, except for nitrate, showed notable improvements in the oat rhizosphere. A reduction in fungal count was noted by PGPR, although bacterial counts considerably improved even in regions exposed to drought. The most significant contribution to plant growth from the chosen PGPR was *E. cloacae* BHUSR1, which mobilized soil nutrients and retained moisture in the rhizosphere under drought conditions (Gupta et al. 2025).

8.6 Field Implementation and Limitations

8.6.1 Field Implementation

The purposeful modification of the plant rhizosphere through the use of beneficial microorganisms (such as endophytes, PGPR, mycorrhizal fungi, and engineered microbial consortia) has demonstrated encouraging outcomes under controlled conditions. However, careful preparation and flexible approaches are required for the field-scale implementation of these techniques. Some important implementation factors are as follows:

- **Selection of Host-Specific Microbial Strains**

  Finding and utilizing microbial strains that are appropriate for the crop species and local soil conditions are essential for success. Regional agroclimatic zones must be considered when designing field trials (Provorov and Vorobyov 2009).

- **Formulation and Delivery Systems**

  Inoculants need to be created in stable forms with sufficient shelf lives, such as liquid suspensions, granules, or powders. Various crops benefit from different application methods, including root dipping, soil drenching, and seed coating. Microbial inoculants are administered in various ways depending on the pathogen's mode of infection and survival. The main methods for applying these bioagents in the field are seed treatment, soil implementation, root dipping, and foliar application. They cannot be used as spore suspensions (Jambhulkar et al. 2016).

- **Integration with Agronomic Practices**

Rhizoengineering is most effective when combined with conventional drought reduction techniques, such as crop rotation, mulching, and deficit irrigation. One of the most important strategies for tackling the problems caused by climate change and its effects on agriculture is climate-smart agriculture (CSA). It seeks to enhance farming methods to reduce greenhouse gas emissions, increase output, and adjust to a changing climate. The FAO estimates that to meet demand, global crop and livestock production must rise by 60% from 2006 levels by 2050, highlighting the growing need for CSA as climate change threatens food and nutrition security (Chandel et al. 2024).

- **Farmer Training and Adoption**

Demonstration plots and field-level training boost farmers' confidence and encourage the wider adoption of rhizoengineered solutions.

8.6.2 Limitations

Despite its immense potential, rhizoengineering for drought stress mitigation has certain biological and practical limitations such as:

- **Soil and Environmental Variability:** Microbial inoculants may not function consistently depending on the type of soil, pH, and climate. The efficacy is also affected by the timing and severity of the drought.
- **Competition with Native Microbiota:** Because of competition with native microbial communities, introduced microbes frequently find it difficult to establish or endure in the rhizosphere.
- **Regulatory and Biosafety Issues:** The use of genetically modified microorganisms presents ecological and regulatory issues. Regional differences in approvals may postpone field use.
- **Scalability and Cost:** Producing, storing, and distributing efficient microbial inoculants on a large scale can be expensive, particularly in environments with limited resources.
- **Short-Term Efficacy:** Without consistent use, long-term resilience to recurrent drought cycles is unresolved, and certain microbial consortia provide temporary advantages.
- **Knowledge Gaps:** Little is known about how microbes, plants, and the environment interact in the field, particularly when stressors are simultaneously present at the same time (O'Callaghan et al. 2022).

8.7 Future Prospects

With developments in microbiology, synthetic biology, genomics, and sustainable agriculture, rhizoengineering as a method for mitigating drought stress in crop plants has a bright future. The scalability, effectiveness, and resilience of this strategy can be improved in several future directions.

8.7.1 System Biology and Synthetic Methods

- **Designer Microbiomes:** Resilience and performance can be enhanced by creating artificial microbial consortia suited to particular crops and drought conditions.
- **Gene Editing Tools** can be used to change the genomes of microorganisms to improve their ability to withstand drought, form biofilms, and colonize plant roots.

8.7.2 Precision Agriculture Integration

- The use of rhizoengineering in conjunction with smart irrigation, remote sensing, and soil moisture mapping will allow for the targeted application of microbial inoculants, thereby reducing waste and improving results.
- Using AI and machine learning to forecast interactions between plants, microbes, and the environment and optimize field-level implementation.

8.7.3 Crop Microbiome Breeding

To co-select crops with root exudate profiles that promote good microbial colonization under drought stress, future breeding programs may include microbiome selection traits.

8.7.4 Systems for Climate-Resilient Agriculture

- In climate-smart agriculture, rhizoengineering will be essential, particularly in arid and semiarid regions. It fits well with low-input systems and regenerative practices.
- In addition to reducing drought, improving soil structure and carbon sequestration via microbial activity may have other advantages.

8.7.5 Policy, Education, and Regulation

- To encourage farmers to embrace these sustainable solutions, governments are anticipated to implement incentives, such as tax breaks and subsidies. Customized education and training initiatives, in addition to policy support, are crucial to provide farmers, agronomists, and extension agents with the knowledge and abilities required for successful implementation.
- The creation of open, scientifically grounded regulatory frameworks will be essential as the prevalence of genetically modified or engineered microbial strains increases in the future. This will encourage innovation while guaranteeing the safe application of such technologies.
- To integrate rhizoengineering into future agricultural systems in a responsible and significant manner, these efforts must be coordinated across disciplines.

8.8 Conclusion

Enhancing drought tolerance in crops through rhizoengineering has revolutionary potential. This approach provides a low-cost, environmentally friendly route to climate-smart agriculture by utilizing the beneficial microbes and root-soil interaction. Interdisciplinary cooperation, farmer involvement, and supportive policy frameworks are necessary for its practical application. In summary, rhizoengineering has considerable potential for improving crop performance in drought-stressed environments. A path toward more robust and sustainable agricultural systems may be established by combining this strategy with conventional breeding and contemporary biotechnological tools.

References

Ahmed I, Yokota A, Yamazoe A, Fujiwara T (2007) Proposal of *Lysinibacillus boronitolerans* gen. nov. sp. nov. and transfer of *Bacillus fusiformis* to *Lysinibacillus fusiformis* comb. nov. and *Bacillus sphaericus* to *Lysinibacillus sphaericus* comb. nov. Int J Syst Evol Microbiol 57:1117–1125. https://doi.org/10.1099/ijs.0.64934-0

Andrade G, Mihara K, Linderman R, Bethlenfalvay GJ (1997) Bacteria from rhizosphere and hyphosphere soils of different arbuscular mycorrhizal fungi. Plant Soil 192:71–79. https://doi.org/10.1023/A:1004248623210

Asghari B, Khademian R, Sedaghati B (2020) Plant growth-promoting rhizobacteria (PGPR) confer drought resistance and stimulate the biosynthesis of secondary metabolites in pennyroyal (*Mentha pulegium* L.) under water shortage conditions. Sci Hortic 263:109132. https://doi.org/10.1016/j.scienta.2019.109132

Augé RM (2001) Water relations, drought, and vesicular-arbuscular mycorrhizal symbiosis. Mycorrhiza 11:3–42. https://doi.org/10.1007/s005720100097

Bagyaraj DJ (2011) Microbial biotechnology for sustainable agriculture, horticulture and forestry. New India Publishing Agency, New Delhi

Bagyaraj DJ (2014) Mycorrhizal fungi. Proc Indian Natl Sci Acad 80:415–428. https://doi.org/10.16943/ptinsa/2014/v80i2/55110
Begna T (2020) Effects of drought stress on crop production and productivity. Int J Res Stud Agric Sci 6:34–43. https://doi.org/10.20431/2454-6224.0607005
Begna T (2022) Impact of drought stress on crop production and management options. Int J Res Stud Agric Sci 8:1–13
Berg G et al (2021) Microbial bioprotectants for plant disease management. Burleigh Dodds Science Publishing, Cambridge
Bhat MA, Mishra AK, Jan S, Bhat MA, Kamal MA, Rahman S, Shah AA, Jan AT (2023) Plant growth promoting rhizobacteria in plant health: a perspective study of the underground interaction. Plants 12:629. https://doi.org/10.3390/plants12030629
Chai YN, Schachtman DP (2022) Root exudates impact plant performance under abiotic stress. Trends Plant Sci 27:80–91. https://doi.org/10.1016/j.tplants.2021.08.003
Chandel N, Kumar A, Kumar R (2024) Towards sustainable agriculture: integrating agronomic practices, environmental physiology and plant nutrition. Int J Plant Soil Sci 36:492–503
Compant S, Clément C, Sessitsch A (2010) Plant growth-promoting bacteria in the rhizo- and endosphere of plants: their role, colonization, mechanisms involved and prospects for utilization. Soil Biol Biochem 42:669–678. https://doi.org/10.1016/j.soilbio.2009.11.024
Compant S, Brader G, Muzammil S, Sessitsch A, Lebrihi A, Mathieu F (2013) Use of beneficial bacteria and their secondary metabolites to control grapevine pathogen diseases. BioControl 58:435–455. https://doi.org/10.1007/s10526-012-9479-6
Danish S, Zafar-Ul-Hye M, Hussain S, Riaz M, Qayyum MF (2020) Mitigation of drought stress in maize through inoculation with drought tolerant ACC deaminase containing PGPR under axenic conditions. Pak J Bot 52:49–60. https://doi.org/10.30848/PJB2020-1(20)
Dietz KJ, Zörb C, Geilfus CM (2021) Drought and crop yield. Plant Biol 23:881–893. https://doi.org/10.1111/plb.13302
Donn S, Wheatley RE, McKenzie BM, Loades KW, Hallett PD (2014) Improved soil fertility from compost amendment increases root growth and reinforcement of surface soil on slopes. Ecol Eng 71:458–465. https://doi.org/10.1016/j.ecoleng.2014.07.066
Downie JA (2014) Legume nodulation. Curr Biol 24:R184–R190. https://doi.org/10.1016/j.cub.2014.01.028
Dutta S, Podile AR (2010) Plant growth promoting rhizobacteria (PGPR): the bugs to debug the root zone. Crit Rev Microbiol 36:232–244. https://doi.org/10.3109/1040841X.2010.486896
Egamberdiyeva D, Höflich G (2004) Effect of plant growth-promoting bacteria on growth and nutrient uptake of cotton and pea in a semi-arid region of Uzbekistan. J Arid Environ 56:293–301. https://doi.org/10.1016/S0140-1963(03)00052-1
Ejaz S, Batool S, Anjum MA, Naz S, Qayyum MF, Naqqash T, Shah KH, Ali S (2020) Effects of inoculation of root-associative *Azospirillum* and *Agrobacterium* strains on growth, yield and quality of pea (*Pisum sativum* L.) grown under different nitrogen and phosphorus regimes. Sci Hortic 270:109401. https://doi.org/10.1016/j.scienta.2020.109401
Fahad S, Hussain S, Bano A, Saud S, Hassan S, Shan D, Khan FA et al (2015) Potential role of phytohormones and plant growth-promoting rhizobacteria in abiotic stresses: consequences for changing environment. Environ Sci Pollut Res 22:4907–4921. https://doi.org/10.1007/s11356-014-3754-2
Farhangi-Abriz S, Torabian S (2018) Biochar increased plant growth-promoting hormones and helped to alleviates salt stress in common bean seedlings. J Plant Growth Regul 37:591–601. https://doi.org/10.1007/s00344-017-9756-9
Ferber D (1999) Risks and benefits: GM crops in the cross hairs. Science 286:1662–1666. https://doi.org/10.1126/science.286.5445.1662
Glick BR (2005) Modulation of plant ethylene levels by the bacterial enzyme ACC deaminase. FEMS Microbiol Lett 251:1–7. https://doi.org/10.1016/j.femsle.2005.07.030
Glick BR (2014) Bacteria with ACC deaminase can promote plant growth and help to feed the world. Microbiol Res 169:30–39. https://doi.org/10.1016/j.micres.2013.09.002

Goswami D, Dhandhukia P, Patel P, Thakker JN (2014) Screening of PGPR from saline desert of Kutch: growth promotion in *Arachis hypogea* by *Bacillus licheniformis* A2. Microbiol Res 169:66–75. https://doi.org/10.1016/j.micres.2013.07.007

Goswami D, Thakker JN, Dhandhukia PC (2016) Portraying mechanics of plant growth promoting rhizobacteria (PGPR): a review. Cogent Food Agric 2:1127500. https://doi.org/10.1080/23311932.2016.1127500

Gupta S, Dube SP, Raghuwanshi R (2025) Rhizoengineering influences on soil dynamics and microbial communities in post-harvest oat fields. Soil Adv 3:100032. https://doi.org/10.1016/j.soilad.2025.100032

Haas D, Défago G (2005) Biological control of soil-borne pathogens by fluorescent pseudomonads. Nat Rev Microbiol 3:307–319. https://doi.org/10.1038/nrmicro1129

Hilker M, Schwachtje J, Baier M, Balazadeh S, Bäurle I, Geiselhardt S, Hincha DK et al (2016) Priming and memory of stress responses in organisms lacking a nervous system. Biol Rev 91:1118–1133. https://doi.org/10.1111/brv.12215

Hirt H, Shinozaki K (eds) (2003) Plant responses to abiotic stress. Springer, Berlin. https://doi.org/10.1007/b84369

Ilyas N, Mumtaz K, Akhtar N, Yasmin H, Sayyed R, Khan W, Enshasy HAE et al (2020) Exopolysaccharides producing bacteria for the amelioration of drought stress in wheat. Sustainability 12:8876. https://doi.org/10.3390/su12218876

Iqbal MS, Singh AK, Ansari MI (2020) Effect of drought stress on crop production. In: Kumar A, Singh AK, Upadhyay PK (eds) New frontiers in stress management for durable agriculture. Springer, Singapore, pp 35–47

Ishaq L, Simamora AV, Bako PO, Benggu YI, Airthur MM, Nguru ESO (2023) Abundance of arbuscular mycorrhizal fungi in the rhizosphere of healthy and declining citrus in East Nusa Tenggara, Indonesia. Asian J Agric Biol 11(1). https://doi.org/10.35495/ajab.2023.010

Jambhulkar PP, Sharma P, Yadav R (2016) Delivery systems for introduction of microbial inoculants in the field. In: Singh DP, Singh HB, Prabha R (eds) Microbial inoculants in sustainable agricultural productivity: Vol. 2: Functional applications. Springer, New Delhi, pp 199–218

Jeong JS, Kim YS, Baek KH, Jung H, Ha SH, Choi YD, Kim M et al (2010) Root-specific expression of OsNAC10 improves drought tolerance and grain yield in rice under field drought conditions. Plant Physiol 153:185–197. https://doi.org/10.1104/pp.110.154773

Jha CK, Saraf M (2015) Plant growth promoting rhizobacteria (PGPR). J Agric Res Dev 5:108–119

Kabouw P, van Dam NM, van der Putten WH, Biere A (2012) How genetic modification of roots affects rhizosphere processes and plant performance. J Exp Bot 63:3475–3483. https://doi.org/10.1093/jxb/ers097

Keyvan S (2010) The effects of drought stress on yield, relative water content, proline, soluble carbohydrates and chlorophyll of bread wheat cultivars. J Anim Plant Sci 8:1051–1060

Khan MY, Zahir ZA, Asghar HN, Waraich EA (2017) Preliminary investigations on selection of synergistic halotolerant plant growth promoting rhizobacteria for inducing salinity tolerance in wheat. Pak J Bot 49:1541–1551

Khare P, Deshmukh Y, Yadav V, Pandey V, Singh A, Verma K (2021) Biochar production: a sustainable solution for crop residue burning and related environmental issues. Environ Prog Sustain Energ 40:e13529. https://doi.org/10.1002/ep.13529

Krishnamurthy K, Senthilkumar S (2005) Mycorrhiza: role and applications. Allied Publishers Pvt Ltd, New Delhi

Leclère V, Béchet M, Adam A, Guez JS, Wathelet B, Ongena M, Thonart P et al (2005) Mycosubtilin overproduction by *Bacillus subtilis* BBG100 enhances the organism's antagonistic and biocontrol activities. Appl Environ Microbiol 71:4577–4584. https://doi.org/10.1128/AEM.71.8.4577-4584.2005

Liu H, Brettell LE, Qiu Z, Singh BK (2020) Microbiome-mediated stress resistance in plants. Trends Plant Sci 25:733–743. https://doi.org/10.1016/j.tplants.2020.03.014

Ma Y (2019) Seed coating with beneficial microorganisms for precision agriculture. Biotechnol Adv 37:107423. https://doi.org/10.1016/j.biotechadv.2019.107423

Marschner P (2012) Rhizosphere biology. In: Marschner's mineral nutrition of higher plants, 3rd edn. Elsevier, London, pp 369–388

McNear DH Jr (2013) The rhizosphere-roots, soil and everything in between. Nat Educ Knowl 4:1

Meagher RB (2000) Phytoremediation of toxic elemental and organic pollutants. Curr Opin Plant Biol 3:153–162. https://doi.org/10.1016/S1369-5266(99)00054-4

Mehmood U, Inam-ul-Haq M, Saeed M, Altaf A, Azam F, Hayat S (2018) A brief review on plant growth promoting rhizobacteria (PGPR): a key role in plant growth promotion. Plant Prot 2:77–82

Mehnaz S (2011) Plant growth-promoting bacteria associated with sugarcane. In: Maheshwari DK (ed) Bacteria in agrobiology: crop ecosystems. Springer, Berlin, pp 165–187

Michl K, Berg G, Cernava T (2023) The microbiome of cereal plants: the current state of knowledge and the potential for future applications. Environ Microbiome 18:28. https://doi.org/10.1186/s40793-023-00485-3

Najafi Vafa Z, Sohrabi Y, Mirzaghaderi G, Heidari G (2022) The effect of rhizobia in improving the protective mechanisms of wheat under drought and supplementary irrigation conditions. Front Sustain Food Syst 6:987654. https://doi.org/10.3389/fsufs.2022.987654

Naqqash T, Malik KA, Imran A, Hameed S, Shahid M, Hanif MK, Majeed A et al (2022) Inoculation with *Azospirillum* spp. acts as the liming source for improving growth and nitrogen use efficiency of potato. Front Plant Sci 13:929114. https://doi.org/10.3389/fpls.2022.929114

Navari-Izzo F, Rascio N (1999) Plant response to water-deficit conditions. In: Pessarakli M (ed) Handbook of plant and crop stress, 2nd edn. Marcel Dekker, New York, pp 231–270

Naveed M, Mitter B, Reichenauer TG, Wieczorek K, Sessitsch A (2014) Increased drought stress resilience of maize through endophytic colonization by *Burkholderia phytofirmans* PsJN and *Enterobacter* sp. FD17. Environ Exp Bot 97:30–39. https://doi.org/10.1016/j.envexpbot.2013.09.014

O'Callaghan M, Ballard RA, Wright D (2022) Soil microbial inoculants for sustainable agriculture: limitations and opportunities. Soil Use Manag 38:1340–1369. https://doi.org/10.1111/sum.12811

Oberson A, Frossard E, Bühlmann C, Mayer J, Mäder P, Lüscher A (2013) Nitrogen fixation and transfer in grass-clover leys under organic and conventional cropping systems. Plant Soil 371:237–255. https://doi.org/10.1007/s11104-013-1689-1

Oldroyd GE, Downie JA (2008) Coordinating nodule morphogenesis with rhizobial infection in legumes. Annu Rev Plant Biol 59:519–546. https://doi.org/10.1146/annurev.arplant.59.032607.092839

Pascale A, Proietti S, Pantelides IS, Stringlis IA (2020) Modulation of the root microbiome by plant molecules: the basis for targeted disease suppression and plant growth promotion. Front Plant Sci 10:1741. https://doi.org/10.3389/fpls.2019.01741

Price AH, Cairns JE, Horton P, Jones HG, Griffiths H (2002) Linking drought-resistance mechanisms to drought avoidance in upland rice using a QTL approach: progress and new opportunities to integrate stomatal and mesophyll responses. J Exp Bot 53:989–1004. https://doi.org/10.1093/jexbot/53.371.989

Provorov NA, Vorobyov NI (2009) Host plant as an organizer of microbial evolution in the beneficial symbioses. Phytochem Rev 8:519–534. https://doi.org/10.1007/s11101-008-9106-5

Rani SJ, Usha R (2013) Transgenic plants: types, benefits, public concerns and future. J Pharm Res 6:879–883

Raza W, Akhtar MJ, Arshad M, Yousaf S (2004) Growth, nodulation and yield of mungbean (*Vigna radiata* L.) as influenced by coinoculation with rhizobium and plant growth promoting rhizobacteria. Pak J Agric Sci 41:125

Redillas MC, Strasser RJ, Jeong JS, Kim YS, Kim JK (2011) The use of JIP test to evaluate drought-tolerance of transgenic rice overexpressing OsNAC10. Plant Biotechnol Rep 5:169–175. https://doi.org/10.1007/s11816-011-0173-1

Ross DS, Skyllberg U (2023) Cation exchange capacity and reactions. In: Goss MJ, Oliver M (eds) Encyclopedia of soils in the environment, 2nd edn. Elsevier, Amsterdam. https://doi.org/10.1016/B978-0-12-822974-3.00063-X

Saad MM, Eida AA, Hirt H (2020) Tailoring plant-associated microbial inoculants in agriculture: a roadmap for successful application. J Exp Bot 71:3878–3901. https://doi.org/10.1093/jxb/eraa232

Salazar-Cerezo S, Martínez-Montiel N, García-Sánchez J, Pérez-y-Terrón R, Martínez-Contreras RD (2018) Gibberellin biosynthesis and metabolism: a convergent route for plants, fungi and bacteria. Microbiol Res 208:85–98. https://doi.org/10.1016/j.micres.2018.01.010

Santoyo G, Urtis-Flores CA, Loeza-Lara PD, Orozco-Mosqueda MDC, Glick BR (2021) Rhizosphere colonization determinants by plant growth-promoting rhizobacteria (PGPR). Biology 10:475. https://doi.org/10.3390/biology10060475

Sena A, Freitas CM, Barcellos C, Ramalho W, Corvalan C (2016) Measuring the invisible: analysis of the sustainable development goals in relation to populations exposed to drought. Cienc Saude Coletiva 21:671–684. https://doi.org/10.1590/1413-81232015213.22412015

Serraj R, Sinclair TR (2002) Osmolyte accumulation: can it really help increase crop yield under drought conditions? Plant Cell Environ 25:333–341. https://doi.org/10.1046/j.1365-3040.2002.00778.x

Shinde R, Sarkar PK, Thombare N (2019) Soil conditioners. Agric Food e-Newsletter 1:1–5

Siddikee MA, Chauhan PS, Anandham R, Han GH, Sa T (2010) Isolation, characterization, and use for plant growth promotion under salt stress, of ACC deaminase-producing halotolerant bacteria derived from coastal soil. J Microbiol Biotechnol 20:1577–1584. https://doi.org/10.4014/jmb.1005.05016

Signorelli S, Tarkowski ŁP, Van den Ende W, Bassham DC (2019) Linking autophagy to abiotic and biotic stress responses. Trends Plant Sci 24:413–430. https://doi.org/10.1016/j.tplants.2019.02.009

Sprent J, Ardley J, James E (2013) From north to south: a latitudinal look at legume nodulation processes. S Afr J Bot 89:31–41. https://doi.org/10.1016/j.sajb.2013.06.011

Upadhyay SK, Singh JS, Saxena AK, Singh DP (2012) Impact of PGPR inoculation on growth and antioxidant status of wheat under saline conditions. Plant Biol 14:605–611. https://doi.org/10.1111/j.1438-8677.2011.00533.x

Velmourougane K, Blaise D (2021) Rhizoengineering: a strategy to enhance soil and crop productivity. In: Sharma SK, Singh UB, Sahu PK (eds) Omics science for rhizosphere biology. Springer, Singapore, pp 235–262

Vives-Peris V, De Ollas C, Gómez-Cadenas A, Pérez-Clemente RM (2020) Root exudates: from plant to rhizosphere and beyond. Plant Cell Rep 39:3–17. https://doi.org/10.1007/s00299-019-02447-5

Willems A (2006) The taxonomy of rhizobia: an overview. Plant Soil 287:3–14. https://doi.org/10.1007/s11104-006-9051-5

Zhao T, He A, Khan MN, Yin Q, Song S, Nie L (2024) Coupling of reduced inorganic fertilizer with plant-based organic fertilizer as a promising fertilizer management strategy for colored rice in tropical regions. J Integr Agric 23:93–107. https://doi.org/10.1016/j.jia.2023.06.026

Zhou H, Fang H, Mooney SJ, Peng X (2016) Effects of long-term inorganic and organic fertilizations on the soil micro and macro structures of rice paddies. Geoderma 266:66–74. https://doi.org/10.1016/j.geoderma.2015.12.001

Chapter 9
Microbial Inoculants for Climate-Smart, Drought-Resilient Agriculture: Mechanisms, Challenges, and Innovations

Silpi Sorongpong, Ankita Das, Nandita Das, Subhradeep Chakraborty, and Piyush Pandey

Abstract Climate change-induced drought poses a major threat to global food security by disrupting plant physiology, reducing yield, and degrading soil health. Bioinoculants containing plant growth-promoting microorganisms (PGPMs) offer a sustainable solution for enhancing drought resilience. These microbes alleviate water deficit stress by modulating phytohormones, accumulating osmolytes, activating antioxidants, synthesizing exopolysaccharides, and producing volatile organic compounds. Recent innovations, including synthetic microbial consortia (SynComs), genetic engineering, and nanotechnology-based delivery systems, have enhanced the efficacy and adaptability of bioinoculants under variable field conditions. Despite these advancements, challenges such as environmental variability, soil heterogeneity, temporal dynamics, and biosafety concerns hinder large-scale applications. Multiomics approaches (genomics, transcriptomics, proteomics, and metabolomics) are emerging as powerful tools for precise strain selection, formulation, and deployment. Integrating these tools with climate-smart agriculture can improve plant productivity, water-use efficiency, and soil carbon sequestration. Addressing technical, economic, and regulatory barriers is critical for translating laboratory innovations into consistent field performance and establishing bioinoculants as key components of resilient, future-ready agroecosystems.

Keywords Bioinoculants · Drought resilience · PGPMs · Climate-smart agriculture · Multiomics

---

S. Sorongpong · A. Das · N. Das · S. Chakraborty · P. Pandey (✉)
Soil and Environment Microbiology Laboratory, Department of Microbiology, Assam University, Silchar, Assam, India
e-mail: piyush.pandey@aus.ac.in

S. Mehnaz, I. Shahid (eds.), *Frontiers in Rhizoengineering*, Microorganisms for Sustainability 65, https://doi.org/10.1007/978-981-92-3665-7_9

9.1 Introduction

Global crop production is facing unprecedented challenges owing to the rising demand for food resulting from population growth and shifting dietary trends (Fukase and Martin 2020). According to the Food and Agriculture Organization of the United Nations (FAO) projections, agricultural output must increase by an estimated 60% by 2050 to meet this demand. However, climate change opposes the drive for higher food yields. The most recent assessment by the Intergovernmental Panel on Climate Change (IPCC) emphasizes how climate change increases food security risks, particularly for populations and places that are already at risk. According to various studies conducted by researchers, climate change is projected to result in a reduction in winter rainfall by approximately 15% by 2030 and up to 30% by 2070, along with a temperature increase of approximately 1.6 °C by 2050. The extent of the decline in plant productivity due to changing rainfall patterns also depends on soil fertility and is influenced by various management practices, including fertilization (Change 2022; Benmrid et al. 2023).

Drought, defined as prolonged subnormal rainfall that reduces soil moisture below crop requirements, is one of the most severe stressors affecting global agriculture and livelihoods. Reduced water availability disrupts plant growth and development, influencing processes at the molecular, physiological, biochemical, and ecological levels (Moore et al. 2023; Ortiz et al. 2015). Globally, drought has impacted nearly 64% of the land area and caused major socio-economic crises. The 2011–2012 Horn of Africa drought triggered famine, displacement, and economic loss in Ethiopia, Kenya, and Somalia, with over 250,000 deaths in Somalia and heavy livestock losses (Terefe 2012; Mahmud 2021). Similarly, Australia's Millennium Drought inflicted billions of agricultural losses and increased dependence on government aid (Van Dijk et al. 2013). Projections suggest that by 2050, droughts will threaten over half of the world's arable land (Vocciante et al. 2022). Yield losses may reach 60% in wheat, 40% in chickpeas, and 21% in maize (Khan et al. 2018).

Plants encounter diverse biotic and abiotic stresses that reduce crop productivity and soil health, particularly under climate change. These stresses cause osmotic imbalance, nutrient deficiency, and oxidative stress through excessive reactive oxygen species (ROS), which damage cellular macromolecules and may trigger programmed cell death (Kumar et al. 2022a, b; Sachdev et al. 2021). The stress hormone abscisic acid (ABA), which is vital for drought tolerance, can also be synthesized microbially. ABA regulates genes involved in dehydration tolerance and promotes stomatal closure (Zandalinas et al. 2016). Microbial inoculation mitigates the effects of drought by reducing ROS levels, enhancing antioxidant enzyme activity, and improving water-use efficiency through auxin-ABA crosstalk. Auxin signaling enhances water uptake and transport, whereas ABA minimizes transpiration, together maintaining plant tissue water content. Under severe drought conditions, inoculated plants display higher survival rates, whereas under mild drought conditions, they sustain growth and development (Bittencourt et al. 2023). For instance, inoculation of potatoes with *Bacillus subtilis* HAS31 lowered ROS and malondialdehyde (MDA) levels while boosting catalase, peroxidase, superoxide dismutase, and soluble sugar levels, thereby enhancing drought resilience (Batool et al. 2020).

Microorganisms secrete numerous phytohormones, exopolysaccharides (EPS), antioxidants, osmolytes, and other substances that play crucial roles in enhancing drought tolerance in situations where water is scarce. Indole-3-acetic acid (IAA), ABA, and cytokinin (CK) are crucial phytohormones that reduce drought stress. Extended periods of stress cause endogenous IAA levels to decrease, which limits plant growth. Auxins are essential for root architecture and growth. Microbes that produce auxin can enhance plant development during drought episodes by increasing endogenous IAA production (Goshwami and Deka 2020).

Drought stress disrupts plant metabolism, physiology, and morphology, thereby reducing agricultural productivity. Bioinoculants, which comprise beneficial microbes, offer a sustainable solution by enhancing growth and stress tolerance through the production of ACC deaminase, phytohormones, and volatile organic compounds (VOCs) that mitigate stress and activate defense responses (Kumar et al. 2022a, b). Jasmonic acid also improves drought resilience by regulating stomatal closure (Gowtham et al. 2022). Drought-adaptive microbiomes, including *Bacillus*, *Pseudomonas*, *Aspergillus*, and *Streptomyces*, support tolerance by improving nutrient uptake, producing growth hormones, lowering stress ethylene levels, and accumulating osmolytes. They further prevent oxidative damage under water-deficient conditions. When applied as biofertilizers or biopesticides, particularly in horticulture, these microbes enhance crop resilience, reduce dependence on agrochemicals, and promote sustainable agriculture. Their integration into farming practices represents a promising strategy for maintaining productivity and countering the adverse effects of drought stress under climate change (Kour et al. 2022).

Addressing climate-induced drought stress necessitates innovative strategies, such as the use of microbial inoculants, which enhance plant resilience under water-limited conditions. The development of site-specific multistrain formulations is crucial for ensuring consistent field efficacy. Advancing research on plant-microbe-environment interactions is key to optimizing microbial inoculants as sustainable tools for drought stress management in agriculture (Hart et al. 2018; Kavadia et al. 2020). The frequency, severity, and duration of drought can all impact the complex effects of drought on microorganisms (Kundel et al. 2020). Vurukonda et al. (2016) primarily emphasized individual mechanisms, such as ACC deaminase activity, phytohormone modulation, and EPS production. Research on microbial communities in the endosphere and rhizosphere highlights the lasting impact of microbiome composition on plant resilience (Singh et al. 2023). Microbial modulation of plant hormone levels and gene expression provides valuable insights into enhancing plant resilience under drought stress (Yasmin et al. 2022). Many studies have highlighted the vital role of microbial consortia in the endosphere and rhizosphere in coordinating plants' adaptive metabolic responses under the difficult conditions of drought stress. In addition to influencing the metabolic profile of plants, this symbiotic interaction increases plant resistance to stress (Nephali et al. 2021; Kaya 2024).

This chapter explores the crucial role of microbial inoculants in promoting climate-smart and drought-resilient agricultural practices. It begins by exploring the diverse mechanisms through which microbial inoculants confer drought resilience, including the modulation of phytohormones and the production of osmotolerant compounds, thereby enhancing plant tolerance to water scarcity. This chapter also

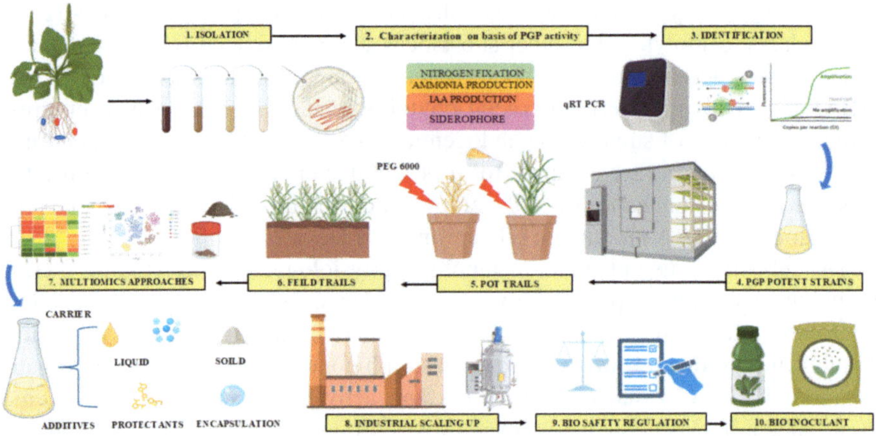

Fig. 9.1 Stepwise approach for developing plant growth-promoting microbial bioinoculants for drought resilience. The workflow outlines the steps from microbial isolation, PGP trait characterization, and molecular identification to the selection of potent strains, pot and field trials, and formulation. Multiomics tools (genomics, transcriptomics, proteomics, and metabolomics) are incorporated to unravel plant-microbe-environment interactions, optimize strain selection, and guide formulation design. This integrative approach ensures targeted, efficient, and scalable bioinoculant development, followed by industrial scale-up, biosafety compliance, and final product formulation

addresses the current challenges and limitations associated with the application of bioinoculants for drought resistance, providing a critical perspective on real-world barriers. Furthermore, innovative strategies such as bacterial genetic engineering, SynCom, and the integration of IoT-enabled microbial delivery and precision irrigation systems are discussed as potential pathways to optimize the efficacy of these biological solutions. By synthesizing recent advances and practical considerations, this chapter is highly relevant to researchers, practitioners, and policymakers seeking to promote sustainable agriculture in the face of the growing threat of climate-induced drought stress. A schematic workflow illustrating the development pipeline for drought-resilient bioinoculants is shown in Fig. 9.1.

9.2 Role of Microbial Inoculants in Climate-Smart Drought-Resilient Agriculture

Climate-smart agriculture (CSA) is a transformative approach to adapting and modifying farming methods in response to the dynamic challenges posed by climate change (Lipper et al. 2014). The primary goal of CSA is to guarantee food security and promote development, where productivity, adaptation, and mitigation arise as the three interlinked pillars vital for achieving this goal. Since 2000, there has been a notable 29% increase in both the frequency and duration of droughts, which has been attributed to climate change (Kamau et al. 2023).

Microbial inoculants are formulations of beneficial microorganisms, including bacteria (e.g., *Azospirillum*, *Bacillus*, *Pseudomonas*, *Rhizobium*, *Serratia*, *Stenotrophomonas*, and *Streptomyces*) and fungi (*Ampelomyces*, *Coniothyrium*, and *Trichoderma*), as well as arbuscular mycorrhizal fungi (AMF) (Berg 2009). Plant growth-promoting microorganisms (PGPMs), including PGPR and AMF, enhance soil health by improving nutrient availability, soil structure, and sustainable plant growth (Malgioglio et al. 2022). Beyond growth promotion, they protect against pathogens and pests, regulate phytohormone signaling, and strengthen stress resilience (Ansabayeva et al. 2025). When applied to seeds, foliage, or the rhizosphere, these inoculants function as biopesticides, biocontrol agents, or phytostimulators (Kumari et al. 2022). Microbial inoculants are increasingly recognized as sustainable, eco-friendly, and cost-effective strategies for mitigating climate change impacts, enhancing crop tolerance to abiotic stress, and improving productivity (Seenivasagan and Babalola 2021).

9.2.1 Mechanisms by Microbial Inoculants Confer Drought Resilience

The effectiveness of microbial inoculants against drought depends on their functional traits and soil-plant interaction. EPS-producing microbes enhance soil aggregation, water retention, and plant tolerance (Sandhya et al. 2015). Plant-associated inoculants that modulate physiology and development can improve agroecosystem performance and ecosystem services (Koskey et al. 2021). PGPR producing ACC deaminase and IAA enhance root development, enabling deeper soil exploration and improved water uptake under drought conditions (Etesami et al. 2014; Zahedi et al. 2025). Microbial inoculation also boosts biomass, increasing soil carbon inputs through litter and exudates. Combined with residue management, this practice promotes carbon sequestration and supports climate change mitigation (Nie et al. 2015). Moreover, PGPMs offer a sustainable and environmentally friendly alternative for reducing the impact of abiotic stress on crops. Their use supports overall ecosystem health by decreasing dependence on chemical fertilizers and pesticides, thus promoting biodiversity, including beneficial organisms such as pollinators (Di Benedetto et al. 2017; Hayat et al. 2012).

9.2.1.1 Phytohormone Modulation

Phytohormones regulate plant growth and stress responses (Fahad et al. 2015). Phytohormone-producing PGPR are vital for sustainable agriculture, as they secrete auxins, cytokinins (CK), abscisic acid (ABA), ethylene (ET), and gibberellins, which collectively enhance growth and drought tolerance. However, excess hormone levels may inhibit development (Song et al. 2025). Under drought conditions, IAA activates stress-responsive pathways (Zhang et al. 2020a, b) but can also trigger ACC synthase, elevating ethylene levels and impairing root growth (Vanderstraeten et al. 2019). PGPR producing ACC deaminase mitigates this by degrading ACC, lowering stress-induced ethylene, and promoting healthier root systems under drought stress (Ghosh et al. 2018).

The bacterial isolate, *Ochrobactrum pseudogrignonense,* produced 85 μg mL^{-1} of IAA, while *B. subtilis* and *Pseudomonas* sp. yielded 72 and 68 μg mL^{-1}, respectively, when applied to garden pea and black gram under −0.73 MPa osmotic stress conditions. These strains significantly enhanced root lengths (10.2 cm and 12.1 cm) (Saikia et al. 2018). Drought stress triggers the upregulation of the *YUCCA* (YUC) gene family, which encodes flavin-containing monooxygenase enzymes essential for auxin biosynthesis (Bai et al. 2013). This gene activation promotes apical dominance, stem elongation, and enhanced drought tolerance in plants (Cao et al. 2019).

Auxins contribute to drought resilience primarily by altering root system architecture, facilitating deeper soil exploration for water uptake (Fadiji et al. 2022). Additionally, several bacterial species expressing acdS, which encodes ACC deaminase, play a crucial role in mitigating stress (Shahid et al. 2023). These bacteria can stimulate the expression of drought-responsive genes, such as *Cadhn*, *VA*, *CaPR-10*, and *sHSP*, leading to improved drought tolerance (Hanaka et al. 2021). In one study, inoculated pepper plants exhibited an 80% survival rate 15 days after exposure to drought, whereas uninoculated plants failed to survive under the same conditions (Mendoza-Alatorre et al. 2024).

Inoculating *Mucuna* sp. with *Bacillus* and *Enterobacter* strains reduced drought stress, decreasing ACC accumulation in roots by 46% and 15% and in leaves by 41% and 21%, respectively, which lowered ethylene production by 45% and 65% and improved water stress tolerance (Saleem et al. 2018). Inoculation with *Azospirillum brasilense* enhanced drought tolerance by increasing endogenous phytohormones, particularly auxins, and raising yields by up to 19% (Bulegon et al. 2019). Similarly, three *Bacillus* strains mitigated drought stress in soybeans by modulating hormonal networks, emphasizing the critical role of microbially mediated auxin synthesis in maintaining plant performance under water-limited conditions (Dubey et al. 2021).

Emerging evidence highlights the metabolic versatility of *B. amyloliquefaciens* in auxin biosynthesis, including the production of IAA, indole-3-lactic acid (ILA), and indole-3-carboxylic acid (ICA) (Bisht et al. 2025). Inoculation with this strain has been shown to significantly enhance root system development and drought tolerance in wheat (Hakim et al. 2021). Furthermore, auxin-producing strains, such as *E. cloacae* and *Citrobacter* sp., have been reported to improve biomass accumulation in wheat under combined salinity and drought stress, underscoring the broad applicability of such microbes in promoting crop resilience under challenging agroecological conditions (Gontia-Mishra et al. 2017).

PGPR that produce ABA can modulate endogenous ABA levels in plants, which play a key role in regulating water use efficiency (Zhang et al. 2022). Fluctuations in ABA levels play a crucial role in the regulation of stomatal conductance, particularly under osmotic stress. Such conditions induce the upregulation of key ABA biosynthetic genes, including *NCED3*, *ZEP* (*LOS6/ABA1*), *MCSU* (*LOS5/ABA3*), and *AAO3,* leading to increased endogenous ABA accumulation and enhanced drought-responsive signaling (Xiong et al. 2002). Similarly, CK synthesized by *B. subtilis* stimulated drought tolerance in *Platycladus orientalis* (oriental thuja). Under drought conditions, non-inoculated plants experienced a 39.14% decline in shoot cytokinin content, whereas inoculation with CK-producing *B. subtilis* limited this reduction to 10.22%. Treated seedlings exhibited improved relative water content (RWC) and higher leaf water potentials than uninoculated controls. Additionally, both root and shoot dry weights increased by 13.99% and 19.23%, respectively. Enhanced root exudation of sugars, amino acids, and organic acids was also observed in this study. These findings suggest that cytokinin-producing PGPR can effectively alleviate drought stress and support the growth of *P. orientalis* seedlings (Liu et al. 2013).

9.2.1.2 Production of Osmotolerants

Under drought conditions, microbes play a pivotal role in helping plants maintain osmotic equilibrium by enhancing the uptake of essential inorganic ions, such as calcium (Ca), magnesium (Mg), and potassium (K) (Porcel et al. 2003). Additionally, AMF colonization contributes to the accumulation of compatible solutes, including polyamines, sugars, and proline (Rapparini and Peñuelas 2013; Zhang et al. 2021). Among these, proline functions as a key osmoprotectant, regulating the expression of stress-responsive genes and preserving the stability of cellular structures, such as lipid membranes and nucleic acids (Zhang et al. 2021). It also assists in osmotic regulation, reduces oxidative stress by scavenging ROS, and safeguards subcellular integrity (Ngumbi and Kloepper 2016).

In *Arabidopsis thaliana*, inoculation with *P. putida* results in increased proline accumulation, which is associated with enhanced water uptake, increased biomass production, and increased chlorophyll content (Ghosh et al. 2017). Two plant growth-promoting rhizobacterial strains, *Acinetobacter calcoaceticus* and *B. amyloliquefaciens*, demonstrated a strong capacity to synthesize key osmoprotectants under osmotic stress. The quantified levels included proline (2.21 and 1.75 μg mL^{-1}), salicylic acid (18.59 and 14.21 μg mL^{-1}), trehalose (28.35 and 22.74 μg mg^{-1} fresh weight), and glycine betaine (11.35 and 7.74 mg g^{-1}), respectively (Wang et al. 2019). The drought-mitigating potential of these strains was further validated by inoculating *Arachis hypogaea* L. under moderate (60% field capacity) and severe (40% field capacity) water deficit conditions, highlighting their efficacy in promoting stress tolerance in leguminous crops (Correa et al. 2024).

Microbial biostimulants significantly enhance osmotic adjustment in groundnuts under drought conditions, improving biomass, photosynthetic pigments, relative water content, and increasing proline and soluble carbohydrates compared to controls (Eswaran et al. 2024; Masoudi-Sadaghiani et al. 2011). Drought-tolerant *Bacillus* spp. enhance biomass and modulate stress responses by boosting water retention, proline, sugars, and amino acids while reducing antioxidant enzyme activity in crops such as *Helianthus annuus* and *Zea mays*. In *A. thaliana*, *A.*

baldaniorum Sp245 improved proline and water status under drought conditions (Vardharajula et al. 2011). Co-inoculation of AMF and PGPR also enhances proline accumulation, thereby strengthening drought resilience in diverse plant species (Vahdati and Lotfi 2013).

The application of plant PGPR inoculants, including the bacterial strains *P. fluorescens*, *Klebsiella variicola*, and *Roultella planticola*, showed enhanced development *of Z. mays* by elevating the levels of key osmoprotectants, including betaine, glycine, and choline (Gou et al. 2015). Similarly, the inoculation of bacterial sp. such as *Ochrobactrum* sp., *Enterobacter* sp., and *Microbacterium* sp. in drought-stressed sorghum plants led to marked improvements in plant growth and osmotic balance (Govindasamy et al. 2020). In another study, colonization by AMF facilitated the accumulation of sucrose, fructose, and glucose, which function as osmolytes and simultaneously enhance photosynthetic activity. This beneficial response has been observed in crops such as lettuce grown under varying water availability conditions, including two-thirds field capacity (FC), half FC, and cyclic drought stress (Baslam and Goicoechea 2012), as well as in *Z. mays* (Hu et al. 2020) and *Macadamia integrifolia* (Yooyongwech et al. 2013).

9.2.1.3 Production of Exopolysaccharides

EPS, high-molecular-weight polymers secreted by microbial cells, play a crucial role in maintaining cellular membrane stability and integrity, particularly under abiotic stress (Jha and Srivastava 2018; Naseem et al. 2018). These polymers exhibit an exceptional capacity for water retention, reportedly absorbing up to 70 times their weight in water (Flemming and Wingender 2010; Chandra et al. 2021). In drought-stressed *H. annuus*, the EPS-producing *P. putida* strain significantly enhanced biomass accumulation (64% increase in dry weight) and soil structural integrity as evidenced by a marked increase in water-stable soil aggregates (70.8% ± 0.8%) and EPS yield (63.30 ± 9.95 mg $plan^{-1}$). This strain efficiently colonizes both the rhizosphere and rhizoplane, thereby promoting rhizospheric cohesion (Saberi Riseh et al. 2021). Similarly, EPS-producing strains, such as *Proteus penneri* (Pp1), *P. aeruginosa* (Pa2), and *Alcaligenes faecalis*, improved maize leaf relative water content by 45% and increased foliar protein content to 1121.2 μg g^{-1} under drought conditions, suggesting mitigation of both osmotic and oxidative stress. Enhanced soil moisture retention of up to 68% has been observed in these inoculated soils (Naseem and Bano 2014). In *Setaria italica*, strains such as *P. fluorescens* and *E. hormaechei* produced substantial EPS (11.63 ± 0.51, 2.91 ± 0.19, and 5.44 ± 0.24 mg mg^{-1} protein, respectively), which supported improved plant performance under limited water availability. Root hair colonization by these strains was associated with greater soil-to-root adhesion, underscoring the mechanistic link between microbial EPS synthesis and enhanced drought resilience in host plants (Niu et al. 2018).

9.2.1.4 Production of Volatile Organic Compound (VOC)

Plant growth-promoting bacteria produce volatile organic compounds (VOCs) that enhance plant development and drought resistance (Sharifi and Ryu 2018). These VOCs stimulate root branching, photosynthesis, seedling emergence, and iron acquisition while improving yield and reducing disease incidence (Terra et al. 2018; Martins et al. 2022). *B. subtilis* volatiles, particularly 2,3-butanediol, markedly promoted growth. In *A. thaliana*, a 10 day exposure increased the leaf area fivefold, whereas the shoot biomass doubled in sweet basil. In *A. thaliana*, fresh and dry weights increased by 58% and 71%, respectively, after 2 weeks, with significant root and shoot biomass gains observed after 6 days (Wu et al. 2018).

In addition to *Bacillus* sp. and *P. chlororaphis* strains, which also synthesize 2,3-butanediol, a volatile compound known to induce systemic resistance in plants. This strain enhances drought tolerance by colonizing plant roots and reducing water loss through stomatal closure, a response mediated by 2,3-butanediol (Yi et al. 2016). Notably, bacterial strains lacking the ability to produce this compound failed to confer similar drought resilience. Similarly, *P. pseudoalcaligenes* produces VOCs such as dimethyl disulfide (DMDS), 2,3-butanediol, and 2-pentylfuran, which have been shown to induce systemic tolerance in maize during a 7 day drought period (Yasmin et al. 2021). Plants exposed to these volatiles exhibit improved stress resilience, characterized by reduced electrolyte leakage and MDA accumulation, along with elevated levels of proline, photosynthetic pigments, and stress-associated phytohormones (Negri et al. 2024).

Maize plants exposed to drought stress exhibited enhanced tolerance through increased antioxidant defense enzyme activity in both shoots and roots. Among the key volatile organic compounds, DMDS is a bioactive molecule produced by certain bacterial strains (Lopes et al. 2024). Inoculation of *A. thaliana* with *Burkholderia ambifaria* significantly promoted lateral root proliferation and overall biomass accumulation after 21 days of exposure, indicating a strong growth-promoting interaction. Among the mechanisms involved, DMDS, a volatile compound produced by this bacterium, plays a crucial role in activating induced systemic resistance (ISR), thereby enhancing the plant's capacity to withstand both abiotic and biotic stressors (Zhao et al. 2016). In parallel, *Bacillus* spp. strains have been shown to modulate plant developmental programs and root system architecture in *A. thaliana* through the emission of VOCs, suggesting that microbial volatiles are key regulators of plant morphogenesis and adaptive responses. Notably, the presence of DMDS is associated with a 66% reduction in disease severity, highlighting its dual function in promoting growth and enhancing resistance to biotic stress (Gutiérrez-Luna et al. 2010).

9.2.1.5 Production of Antioxidant Enzymes

Drought-induced stress triggers the overproduction of ROS, leading to perturbation of cellular redox homeostasis and resulting in oxidative damage to nucleic acids, membrane lipids, and structural or enzymatic proteins (Wu et al. 2015). This oxidative stress impairs normal physiological functions, thereby hindering overall plant growth and productivity (Laxa et al. 2019). To counteract these harmful effects, plants rely on an efficient antioxidant defense system that can scavenge ROS. The major forms of ROS generated under drought conditions include singlet oxygen ($^1O_2$), hydroxyl radicals (OH), hydrogen peroxide (H_2O_2), superoxide anions (O_2^-), and alkoxy radicals, all of which can inflict severe oxidative damage if not tightly regulated (Hasanuzzaman et al. 2020). In the drought-sensitive PRI-Red, CAT activity increased by 30% and 61% with the application of beneficial microorganisms compared to 28% and 56% without them. The activity of superoxide dismutase (SOD) increased significantly in plants inoculated with beneficial microorganisms, with increases of up to 24% and 63% observed under moderate and severe drought conditions, respectively. Similarly, the activity of peroxidase (POD) increased by 27% and 58% in plants treated with beneficial microorganisms, in contrast to the 22% and 51% observed in the uninoculated controls. The findings suggest that the application of beneficial microorganisms boosts antioxidant defense mechanisms, leading to enhanced drought resilience in both potato varieties (Batool et al. 2020). Murali et al. (2021) revealed that out of 96 rhizobacterial isolates from pearl millet, 28 exhibited ACC deaminase activity, with *B. amyloliquefaciens* displaying the most significant activity (2196.23 nmol α-ketobutyrate mg^{-1} protein h^{-1}). The chosen strains that produce ACC deaminase include *B. amyloliquefaciens*, *B. subtilis*, and *Stenotrophomonas maltophilia. These strains exhibited various* advantageous traits for plants, demonstrated effective root colonisation, and were found to be nonpathogenic (Murali et al. 2021).

Seed treatment with *B. amyloliquefaciens* significantly enhanced germination (91.75%) and seedling vigor (1213.73) under drought conditions. Inoculated plants exhibited improved growth traits, higher total chlorophyll content, and elevated RWC, indicative of enhanced drought tolerance (Lahlali et al. 2022). Furthermore, inoculation with *B. amyloliquefaciens* significantly increased proline accumulation, accompanied by enhanced activities of key antioxidant enzymes, such as ascorbate peroxidase (APX) and SOD, indicating an improved oxidative stress response under drought conditions. These biochemical changes were accompanied by a 2.3-fold reduction in MDA levels, reflecting lower lipid peroxidation and oxidative damage (Neeti et al. 2021). Gene expression analysis revealed a downregulation of drought-responsive (DREB-1E, by 3.3-fold) and ethylene-responsive (ERF-1B, by 1.8-fold) transcription factors, whereas the expression of antioxidant-related genes APX1 and SOD1 was upregulated by 2.2-fold and 2.9-fold, respectively. Murali et al. (2021) demonstrated that *B. amyloliquefaciens* enhances drought resilience in pearl millet by fortifying antioxidative defenses and modulating stress-related gene networks. Shahid et al. (2023) reported that inoculating *Variovorax paradoxus*, a beneficial microorganism, in drought-stressed pea plants significantly improved seed number, total yield, and nitrogen accumulation. Plants subjected to bacterial treatment exhibited a pronounced enhancement in enzymatic antioxidant defenses, as evidenced by the elevated activities of SOD, CAT, and POD. In a related investigation, Paul et al. (2023) reported that the seed priming of *Triticum aestivum* with a biofilm-forming strain of *B. pseudomycoides* significantly increased the activities of antioxidant enzymes, including CAT, SOD, POD, and ascorbate peroxidase (APX). At the molecular level, *B. pseudomycoides* was found to induce upregulation of the *Dreb1* gene, a key regulator of drought-responsive pathways, while concurrently suppressing the *MYB* gene encoding the transcription factor TaMpc1-D4, thereby enhancing plant resilience under water-deficit conditions. This suppression was mediated by bacterial VOCs. Among the 45 identified VOCs, 2,6-di-tert-butylcyclohexa-2,5-diene-1,4-dione and 3,5-di-tert-butylphenol were particularly effective in repressing Myb protein expression under drought stress, suggesting the potential of *B. pseudomycoides* as a promising microbial agent for sustainable agriculture under climate-induced stress. Numerous studies have demonstrated the potential of bioinoculants to mitigate drought stress in diverse crop species through various biochemical, physiological, and molecular mechanisms (Table 9.1). An illustration of the key mechanisms by which PGPMs enhance drought tolerance, including phytohormone modulation, stress-responsive gene activation, ACC deaminase activity, antioxidant induction, EPS and osmolyte production, and volatile emission, leading to improved root architecture, water uptake, and plant resilience, is shown in Fig. 9.2.

9.3 Challenges and Limitations in Bioinoculant Application for Drought Resistance

9.3.1 Environmental Variability Affecting Bioinoculant Performance

The dynamic and unpredictable nature of the environment poses a significant challenge to the development and application of climate-smart bioinoculants. Environmental variability, including fluctuations in temperature, precipitation patterns, soil moisture, pH, salinity, and nutrient availability, directly influences the establishment, survival, and functional efficacy of microbial inoculants in the rhizosphere. While laboratory and greenhouse trials often report consistent and promising results, the field performance of microbial strains under variable environmental conditions remains inconsistent and poorly predictable (Compant et al. 2010; Vurukonda et al. 2016).

9.3.1.1 Impact of Abiotic Stressors on Microbial Function

Abiotic stressors, such as drought, temperature extremes, salinity, heavy metals, and emerging contaminants such as antibiotics, have a profound impact on soil microbial communities, often altering their structure, diversity, and functional capabilities (Dos Santos et al. 2022). For instance, osmotic stress is caused by salt, drought results in water scarcity, and severe temperatures, whether hot or cold, can damage proteins and membranes of plant cells. In addition to having a

Table 9.1 Bioinoculant microbial strains and their mechanisms for enhancing drought tolerance in different crop species

Test crop	Bioinoculant (microbial strains)	Drought-related effects	Reference
T. aestivum L.	*A. brasilense*	Root system enhancement and drought resilience in wheat	Timmusk et al. (2014a)
	B. phytofirmans strain PsJN	ACC deaminase activity and siderophore production	Naveed et al. (2014)
	P. brasiccacearum subsp. *brasiccacearum*	ACC deaminase activity	Magnucka and Pietr (2015)
	Pantoea alhagi	Reduced breakdown of chlorophyll and increased soluble sugars	Chen et al. (2017)
	B. subtilis	LDR2 upregulated the expression of TaCTR1/TaDREB2 TFs under drought stress	
	B. thuringiensis, B. amyloliquefaciens, and *B. simplex*	Enhanced the production of indole-3-acetic acid under water stress. Upregulated stress-related gene seen in APX1, SAMS1, and HSP17.8	Raheem et al. (2018)
	V. paradoxus (sole inoculation), *O. anthropi* + *Pseudomonas* spp. (consortium)	Synthesis of siderophore, ACC deaminase activity, indole-3-acetic acid production, and phosphate solubilization	Chandra et al. (2019)
	P. lini and *S. marcescens*	The relative water content was boosted. Increased root, shoot, and plant height	Khan and Singh (2021)

Z. mays L.	*P. putida* GAP-P45	Enhances drought tolerance in maize via exopolysaccharide production	Sandhya et al. (2009)
	A. lipoferum	It elevated the production of abscisic acid and auxins. Increased levels of free amino acids and soluble carbohydrates	Cohen et al. (2009)
	P. penneri, P. aeruginosa, and *A. faecalis*	Improved plant biomass, leaf area, and growth parameters	Naseem and Bano (2014)
	B. thuringiensis	Enhanced nutritional content, water transport proteins, and decreased lipid oxidation in stressed plants	Ortiz et al. (2015)
	Azospirillum sp. and *Herbaspirillum seropedicae*	Increased chlorophyll contents, increased biomass output, and increased drought tolerance	Cura et al. (2017)
	P. penneri, and *A. faecalis*	Boosted plant biomass, leaf area, and the production of exopolysaccharides	Naseem et al. (2018)
	Azotobacter sp.	Increased siderophore synthesis and solubilization of phosphate and potassium	Shirinbayan et al. (2019)
	B. licheniformis	Showed higher root dry weight and more efficient utilization of water	Akhtar et al. (2020)
	Achromobacter xylosoxidans	Increased carotenoids, chlorophyll a, total chlorophyll, stomatal conductance, and photosynthetic rate	Danish et al. (2020)
	B. subtilis	Increased 1-aminocyclopropane-1-carboxylate synthase enzyme transcription	Sood et al. (2020)
	A. brasilense	Enhanced root development and reduced osmotic stress	Marques et al. (2021)
	B. licheniformis, B. thuringiensis, and *Paenibacillus favisporus*	The root adherent soil/root tissue ratio, relative water content, and leaf water potential were improved. Decreased leaf water loss, electrolyte leakage, and antioxidant enzyme activity	Etesami et al. (2023)
	Azospirillum sp.	Increased drought tolerance of *ZmVP14*, the gene for abscisic acid biosynthesis. Increased maize biomass	Gureeva and Gureev (2023)
	P. aeruginosa and B. cereus	Boosted the synthesis of exopolysaccharides (EPS). Improved water-holding capacity. Increased root and shoot growth and dry biomass weight	Naseem et al. (2024)

(continued)

Table 9.1 (continued)

Test crop	Bioinoculant (microbial strains)	Drought-related effects	Reference
Oryza sativa L.	*P. jessenii* and *Arthrobacter nitroguajacoli*	Increased proline accumulation	Gusain et al. (2015)
	P. flourescens	Activated genes involved in the abscisic acid-mediated signaling cascade, including bZIP1, AP2-EREBP, and Hsp20	Saakre et al. (2017)
	H. huttiense, *E. asburiae*, and *Staphylococcus* sp.	Produced indole-3-acetic acid	Andreozzi et al. (2019)
	Trichoderma sp. and *Pseudomonas* sp.	Enhanced the overexpression of the genes for dehydration (DREB), drought adaptation (DHN), and water permeability (OSPiP)	Singh et al. (2020)
	E. aerogenes	Enhanced health of the rhizosphere during mild drought by activating 1-aminocyclopropane-1-carboxylate deaminase	Zhang et al. (2020a, b)
Foxtail mellet (*S. Italica* L.)	*P. fluorescens*	ACC deaminase activity, ROS scavenging under drought	Niu et al. (2018)
Velvet bean (*Mucuna pruriens* L.) DC	*Bacillus* spp. *Enterobacter* spp.	Enhanced plant biomass and reduced ACC concentration in leaves and roots of inoculated plants	Saleem et al. (2018)

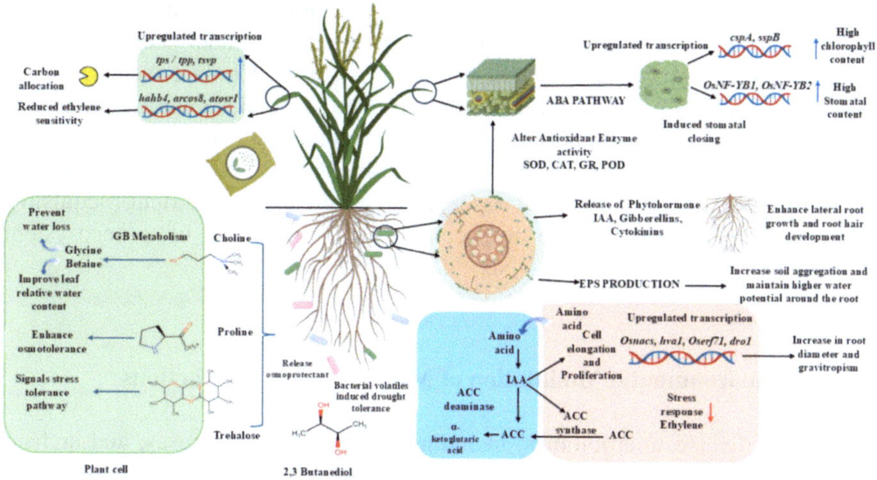

Fig. 9.2 Mechanisms by which PGPMs enhance drought tolerance in plants. PGPMs enhance drought tolerance by modulating phytohormones, upregulating stress-responsive genes, producing ACC deaminase, boosting antioxidant defenses, synthesizing EPS and osmoprotectants, and releasing beneficial volatiles, thereby improving root architecture, water retention, and overall plant resilience

direct effect on plant health, these pressures lower biodiversity and ecosystem services, which in turn affect agricultural sustainability and production (Ullah et al. 2025).

Drought stress alters the soil microenvironment by reshaping the microbial community structure and function. Reduced water availability restricts solute diffusion and microbial mobility, limiting access to nutrients and root exudates while intensifying osmotic stress, which induces dormancy or death in sensitive strains (Naylor and Coleman-Derr 2018). Plant growth-promoting rhizobacteria (PGPR) that produce IAA, ACC deaminase, or siderophores often downregulate these traits during desiccation. For instance, drought-induced oxidative stress disrupts bacterial IAA biosynthesis, thereby diminishing their plant growth-promoting potential (Timmusk et al. 2014b; Sorongpong et al. 2025). Thus, drought not only directly stresses plants but also constrains supportive microbial functions. Under drought conditions, reduced root exudation limits microbial access to carbon, shifting microbial populations and metabolism (Rolli et al. 2015). Biochemically, drought triggers osmolyte accumulation (proline, sugars, and betaine), which prevents dehydration and preserves cellular function. Furthermore, drought-induced ROS can oxidize proteins, lipids, and nucleic acids, thereby impairing cell integrity. To counteract this, plants employ antioxidative systems that scavenge ROS and maintain their homeostasis. Key enzymes include peroxidases, catalase, and SOD, which collectively regulate ROS levels and sustain cellular integrity during prolonged water stress (Impa et al. 2011). These mechanisms underpin plant resilience to drought conditions. Prolonged drought degrades soil quality by reducing soil organic carbon (SOC) and disrupting carbon dynamics, with serious consequences for agriculture and food security (Schimel 2018). The resilience of microbial communities under such pressures is vital for sustaining soil fertility and ecosystem stability. Climate factors, such as heat waves and elevated CO_2 levels, further reshape plant microbiomes, influencing their beneficial or pathogenic roles. Higher temperatures can increase microbial respiration, enhancing plant–microbe interactions but also elevating disease virulence or weakening symbiotic associations critical for nutrient acquisition (Muhammad et al. 2024). These shifts complicate the responses of agroecosystems to climate stress.

9.3.1.2 Environmental Modulation of Microbe-Plant Interactions

Plants host diverse microbiomes, including bacteria, fungi, viruses, and archaea, across the rhizosphere, endosphere, and phyllosphere, where they perform key functions that support plant health and fitness (Afridi et al. 2022). Environmental cues are integrated into plant biochemical and genetic pathways, which shape their interactions with microbes. Roots adjust exudation profiles under stress, selectively recruiting beneficial microbes such as ACC deaminase producers that regulate ethylene levels. This interaction is governed by a dynamic network of plant genetics, stress hormones, and exudates, enabling adaptive plant-microbe communication and resilience to environmental fluctuations (Wankhade et al. 2025). Such cooperation underpins plant survival under changing conditions.

Soil pH, fertility, and nutrient availability strongly influence plant-microbe interaction. Root exudates, comprising sugars, amino acids, and hormones, shift under nutrient limitations, recruiting beneficial microbes such as nitrogen fixers or phosphate solubilizers. Conversely, excessive synthetic fertilizer use can reduce plant reliance on mutualists, thereby diminishing their abundance (Pérez-Lorente et al. 2024). These relationships are highly context-dependent and are shaped by soil type, temperature, and nutrient dynamics (Fitzpatrick et al. 2018). In drought-prone soils, plants may modify exudates to recruit microbes that produce osmolytes or stress-responsive enzymes; however, inoculants may not always match these needs. Moreover, microbial communication via quorum sensing, which is crucial for colonization and biofilm formation, can be disrupted by pH or temperature extremes, reducing efficacy (Schikora et al. 2016). This mismatch between controlled studies and field realities underscores the importance of strain selection and formulation tailored to site-specific conditions. Maestro- Gaitán et al. (2023) examined how water stress and genotypes shape quinoa-associated bacterial communities. They found genotype-specific responses, with susceptible cultivars showing increased *Nocardiaceae* abundance under stress, whereas drought-resistant cultivars were dominated by *Pseudomonas*, *Burkholderia*, and *Sphingomonas* (Maestro- Gaitán et al. 2023; Pérez-Lorente et al. 2024).

9.3.1.3 Soil Heterogeneity and Microbial Survival

Soil heterogeneity, that is, variation in physical, chemical, and biological properties at multiple spatial and temporal scales, is a fundamental driver of microbial survival, community structure, and functional processes in terrestrial ecosystems (Nunan et al. 2020). Variations in organic matter content, bulk density, texture, and microbial competition from indigenous communities can hinder the survival and establishment of the introduced strains. Even within the same field, microhabitat variations can result in differential microbial persistence and colonization, contributing to inconsistent plant responses (Bashan et al. 2014). Moreover, environmental stress can cause selective pressure, favoring stress-tolerant native microbes over inoculated strains, thereby reducing the functional relevance of bioinoculants (Compant et al. 2021). Some studies have reported a rapid decline in inoculant populations after application owing to incompatibility with local edaphic conditions, highlighting the critical importance of adaptive strain selection and formulation (Sessitsch et al. 2019). Environmental heterogeneity disrupts direct competition, as many microorganisms exploit locally available substrates, thereby limiting competitive exclusion and supporting high biodiversity, even within small volumes (Nunan et al. 2020). Microbial taxa display a spectrum of r/K-selection strategies in response to resource unpredictability, with resource-rich microsites favoring fast-growing r-strategists and resource-poor patches favoring efficient, stress-tolerant K-strategists. Soil electrical conductivity, nutrient content, and organic matter dynamics further influence microbial survival by shifting community composition under wet-dry cycles, salinity changes, or management interventions (Zhang et al. 2023).

9.3.1.4 Temporal Dynamics and Resilience

Environmental conditions are not only variable across space but also fluctuate over time, particularly in drought-prone ecosystems. Seasonal shifts in rainfall and temperature can lead to fluctuations in soil microbial communities and their functional output. Inoculants that perform well during early plant development may not persist or remain active during critical drought periods later in the crop cycle (de Vries et al. 2020). Temporal resilience, the ability of a microbial inoculant to maintain functional stability across climatic fluctuations, is an emerging concept that is gaining importance in climate-smart agriculture. However, this requires screening and development of microbial strains capable of entering dormant states or producing protective extracellular substances (EPS) that enhance survival under fluctuating moisture regimes (Timmusk et al. 2011; Marasco et al. 2012).

According to Kashyap et al. (2017), temperature is a key factor that shapes the microbial community associated with plants, as well as their development and phenological characteristics. Microbial activity is significantly affected by temperature stress, as microbes require an ideal temperature range to support their growth, reproduction, and disease-causing capacity (Wang et al. 2022). Recent research has highlighted that soil microbial resilience is governed by both intrinsic factors, such as taxonomic and functional diversity, microbial life-history traits, and physiological tolerance, as well as extrinsic factors, including climate history, soil physicochemical properties, and vegetation interactions. For instance, drought conditions, a prevalent and intensifying stress globally, select for drought-tolerant microbial taxa (e.g., *Actinobacteria*) that exhibit high resistance, whereas fast-growing taxa (e.g., *Acidobacteria*) may show more rapid recovery post-disturbance, showcasing resilience strategies linked to microbial life-history strategies (Bardgett and Caruso 2020).

9.3.2 Scale-Up and Commercialization Issues

Scaling up PGPM formulations from lab to field remains challenging, requiring improved upstream productivity through advances in fermentation techniques and bioreactor models (Vassileva et al. 2021). Robust quality control, encompassing formulation stability, microbial viability, and cell density assessment, is critical for reliable PGPM applications (Neuhoff et al. 2024). In organic systems, PGPMs enhance soil aggregation and porosity via extracellular polysaccharide secretion, thereby improving water retention and reducing runoff and nitrate leaching. These processes strengthen nutrient acquisition and biocontrol efficacy, thereby sustaining crop productivity and promoting agroecosystem resilience (Parizad and Bera 2023). The effectiveness of bacterial inoculants depends on the application method, such as seed coating, soil application, or foliar spray, as they are sensitive to temperature, pH, and moisture. For example, soybean seeds with inoculants must be sown within 12 h in hot and dry conditions to prevent bacterial dehydration. Moreover, chemical fertilizers, pesticides, and herbicides can further reduce their viability and efficacy (dos Reis et al. 2024).

Logistics, financial resources, timetables, and end goals should be carefully considered when translating research from the laboratory scale to a manufacturing process (pilot) to large-scale commercial production (Crater and Lievense 2018). Although various methods exist for evaluating the difficulties in scaling up, one of the most effective approaches is to apply a process hazard analysis (PHA) method at each stage of the process (Ganeshan et al. 2021). Introducing microbial inoculants to the market must address the limited awareness of their benefits among growers. Many perceive biocontrol products as more expensive and less effective than chemicals, despite their strong market potential. The biofertilizer market is projected to reach USD 5.23 billion by 2029, with 9.85% annual growth (2024–2029), while the biopesticide market is expected to reach USD 11.41 billion by 2030, growing at 11.1% annually (2025–2030) (Díaz-Rodríguez et al. 2025.

9.3.3 Safety Concerns Using Bioinoculation in Sustainable Agriculture

Ensuring environmental compliance and safety is crucial for producing and using biofertilizers. Bioinoculants, including beneficial microbes such as actinomycetes, mycorrhizae, and rhizobacteria, are emerging as eco-friendly alternatives to chemical inputs in agriculture. They enhance nutrient uptake, induce stress tolerance, and support climate-resilient agriculture. However, their safety and efficacy, particularly under water-scarce conditions, require careful evaluation. One of the foremost safety concerns is the ecological unpredictability of the introduced microbial strains. Under drought stress, soil physicochemical conditions change drastically, which may affect the survival, colonization ability, and functional expression of inoculated microbes (Pérez-García et al. 2011). These altered dynamics may lead to unintended ecological shifts, such as microbial imbalance, pathogen resurgence, or horizontal gene transfer, particularly when genetically modified strains are used (Compant et al. 2021). Second, drought stress can intensify microbial competition for limited nutrients and root exudates, potentially leading to antagonistic interactions with the native microbiota. This could destabilize indigenous microbial communities and compromise soil biodiversity (Saad et al. 2020). Furthermore, some microbial strains may become opportunistic under extreme environmental stress, posing risks to plants or the environment, especially if their biosafety assessments are conducted only under optimal conditions.

Farmers should receive hands-on training through workshops, field demonstrations, and educational materials on the proper application, dosage, and timing of agricultural product use. Continuous support is vital during adoption. For producers, technical training in fermentation, quality control, and environmental management is crucial, along with guidance on optimizing production and marketing strategies. Such integrated training empowers both farmers and producers to enhance the effectiveness and sustainability of biofertilizers (Samantaray et al. 2024).

9.4 Innovation of Microbial Bioinoculants for Addressing Drought-Related Challenges

Innovation in microbial bioinoculants for drought stress now focuses on ecologically important strain selection with advanced delivery and desiccation-tolerant encapsulation to ensure their viability (Bittencourt et al. 2023; Chaudhary et al. 2025). These formulations enhance nutrient uptake, phytohormone production, antioxidant defense, and osmolyte accumulation, thereby supporting plant growth and productivity under drought conditions (Kumar et al. 2022a, b). Recent studies have shown notable yield gains and stress tolerance, positioning bioinoculants as sustainable alternatives to chemical fertilizers (Sati et al. 2023). Some recent innovative technologies are discussed below.

9.4.1 Next-Generation SynComs in Combating Drought Stress

Synthetic microbial consortia (SynComs) have emerged as a promising approach for mitigating drought stress in climate-smart agriculture, offering greater resilience, sustainability, and productivity than single-strain bioinoculants (Aleksieienko et al. 2025). By combining microbes with complementary traits, SynComs provide predictable drought tolerance through enhanced root growth, water and nutrient uptake, and antioxidant activities (Maheshwari et al. 2023; Tariq et al. 2025). For instance, SynComs comprising *Bacillus*, *Pseudomonas*, and *Azospirillum* improve plant survival by stimulating hormones such as abscisic acid and indole-3-acetic acid, which promotes deeper roots and elevates stress-protective compounds like proline and trehalose (Rigerte et al. 2025; Alzate Zuluaga et al. 2024). Researchers design SynComs using genomic, metabolic, and functional screening to identify strains with complementary traits and minimal negative interactions and assemble them in controlled proportions for reproducible performance under stress (Tariq et al. 2025; De Souza et al. 2020). Genomic data, gene markers, and expression profiles help in the selection of microbes with traits such as high colonization efficiency or stress tolerance, ensuring effective plant growth promotion. Furthermore, defined compositions enhance field reproducibility by minimizing variability due to soil or environmental factors (Zhang 2024). For example, a 7-strain SynComs enhanced the growth of *Astragalus mongholicus* by 50.10% in fresh weight, 55.29% in dry weight, and 76.40% in underground fresh weight in severe drought conditions (Li et al. 2025). Similarly, a 16-member synthetic microbial community called the "Drought-Tolerance Synthetic Microbial Community (DT-SynCom)," composed of Proteobacteria, Firmicutes, and Actinobacteria, decreased the number of wilting leaves and positively influenced barley growth under drought conditions (Rigerte et al. 2025). The proportions within SynComs are highly standardized, such that any alteration leads to different outcomes. For example, the removal of a single strain, *Enterobacter cloacae*, diminished the microbial consortium's activity, which correlated with a reduction in the severity of maize blight (Niu et al. 2018). Thus, the composition can be altered according to the application requirements. This makes SynComs a cornerstone of next-generation agricultural microbial solutions. Despite their benefits, SynComs face several key challenges. Many microbial species remain unculturable, limiting the construction of complete communities, and large-scale sequencing is costly (Price et al. 2023). Additionally, the complexity of soil microbiomes and limited understanding of microbe-environment interactions hinder the identification of optimal strains for biocontrol (Naylor et al. 2020). As a result, most SynComs are currently restricted to easily cultured bacteria with known plant growth-promoting traits.

9.4.2 Bacterial Genetic Engineering

Genetic engineering for drought mitigation has mainly focused on plants, but conventional breeding is slow, labor-intensive, and may result in the loss of desirable traits, often targeting a single crop at a time (Shinwari et al. 2020). Although transgenic approaches are faster, they still require long development timelines, face regulatory hurdles, and suffer from inconsistent consumer acceptance (Ullah et al. 2019). In contrast, genetically modified microbial bioinoculants offer a faster, adaptable, and potentially more acceptable strategy for enhancing crop drought tolerance under increasing climate stress (Mukherjee et al. 2018).

Genetically engineered bacteria can be tailored with drought tolerance traits by overexpressing genes for osmoprotectants such as proline (*p5cr, proA,* and *proB*), trehalose (*otsA and treS*), mannitol (*mtlD*), and glycine betaine (*betA* and *betB*), which help both microbes and plants maintain water balance (Singh et al. 2015; Sharma et al. 2020; Liu et al. 2022). Engineering can also enhance the production of phytohormones, including IAA, gibberellins, and cytokinins, to promote root growth and stress resilience (Fahad et al. 2015). For example, overexpression of the cytokinin biosynthesis gene *ipt* in *Sinorhizobium meliloti* significantly improved alfalfa drought tolerance (Xu et al. 2012), while metabolic engineering of *Paenibacillus polymyxa* to overproduce IAA via native (*patA, ilvB3, and fusE*) and heterologous pathways nearly doubled the IAA yield, enhancing plant growth (Sun et al. 2022). The study showed that overexpressing multiple synergistic genes enhances drought tolerance more effectively than single-gene approaches, with bottlenecks addressed through integrated metabolic engineering (Das et al. 2023). Similarly, the upregulation of exopolysaccharide (EPS) biosynthetic genes (*galU, pgmA*, and EPS operons) improves soil structure, water retention, and rhizosphere stability (Schmid et al. 2015; Sun and Zhang 2021). The expression of stress-responsive genes, such as DREB and cold shock proteins, further boosts plant-microbe survival under drought conditions (Agarwal et al. 2017). Genes related to nutrient solubilization, including *gcd* and *gdh* for mineral phosphate and *phoA, phoD, and appA* for organic phosphate, have been engineered in *Pseudomonas, Bacillus*, and *Rhizobium* to enhance phosphorus availability during stress (Pan and Cai 2023; Joshi et al. 2023; Kumar et al. 2025). Table 9.2 lists common bacterial genes that can be modified through genetic engineering to improve drought stress tolerance, along with the possible methods used for these modifications. Collectively, these strategies expand the toolkit for drought-adaptive biofertilizer development, offering a framework for sustainable agriculture under climate stress conditions.

9.4.3 Integration of Nanotechnology with Bioinoculants

Nanotechnology is transforming bioinoculant development for drought tolerance by improving microbial survival, delivery, and efficacy under stress. Incorporating nanoparticles (NPs), such as SiO_2, ZnO, and TiO_2, into bioinoculants (nanobioinoculants) enhances crop resilience (Chaudhary et al. 2025). For example, apatite

Table 9.2 Common gene targets and their functions that can be targeted during genetic engineering studies (Kumar and Sindhu 2024; Shaffique et al. 2022;Ullah et al. 2019;Shinwari et al. 2020)

Gene/operon	Function in drought stress mitigation	Precise engineering strategy
accd (ACC deaminase)	Lowers ethylene, delays senescence, promotes roots	Overexpress in rhizosphere-competent strains to enhance ACC degradation
eps genes (EPS synthesis)	Boost EPS, biofilm, and root-soil moisture	Upregulate biosynthetic genes to increase EPS yield
phaC (PHB synthase)	Improves osmotic stress tolerance via PHB storage	Overexpress with constitutive promoters for better stress fitness
ipdC (IAA biosynthesis)	Increases IAA, stimulates root elongation, water uptake	Multicopy insertion in root-colonizing strains to maximize IAA
katA, sodA, ahpC (antioxidants)	Scavenge ROS, protect microbes and plants	Use stress-inducible promoters for coordinated upregulation
pqq operon	Enhances phosphate solubilization and uptake	Transfer complete operon into PGPR suited for crop rhizospheres
otsA/*otsB*/*treS* (trehalose pathway)	Trehalose acts as a microbial and plant osmoprotectant	Overexpress trehalose genes for higher secretion
bet operon (glycine betaine)	Produces glycine betaine for osmoprotection	Introduce or upregulate in drought-tolerant PGPR
Volatile organic compound (VOC) genes (*aldH, alsS, alsD*)	Produce VOCs (e.g., acetoin, 2,3-butanediol) for plant defense	Engineer synthetic VOC clusters in root colonizers
nif/*fix* (nitrogen fixation)	Provides N during drought stress	Combine enhanced *nif*/*fix* with stress-tolerant strains

nanoparticles used as a phosphorus nanofertilizer increased soybean seed yield by 20% and growth by 30%, outperforming conventional fertilizers (Liu and Lal 2014). Nanofertilizers, nanoherbicides, and nanopesticides derived from beneficial microbes are also considered safe and sustainable alternatives (Adeleke et al. 2022). NPs enhance PGPR activity by boosting nitrogen fixation and secondary metabolite synthesis. Copper oxide NPs increased IAA production in *Pseudomonas chlororaphis*, whereas ZnO NPs stimulated siderophore synthesis (Dimkpa et al. 2012). *Nitrosomonas europaea* exposed to silver NPs showed enhanced nitrification via the upregulation of *amoA1* and *amoC2* genes (Yang et al. 2013), and ZnO-treated *B. subtilis* produced more antimicrobials and solubilized phosphate (Sabir et al. 2020). TiO_2 NPs also boosted nitrogenase activity in *Rhizobium* sp., improving legume symbiosis (Shahroz et al. 2024). In addition to microbes, nanobioinoculants modulate plant gene expression. In wheat, cerium-based formulations upregulated drought-responsive genes, including *MYB33, DREB2,* and *SnRK2.4*, thereby enhancing adaptive responses (Boora et al. 2024). Together, these findings demonstrate that NPs are compatible with beneficial microbes and expand the potential of bioinoculants for climate-smart and drought-resilient agriculture.

The encapsulation of beneficial microbes within NPs enhances their survival by protecting them from desiccation, UV radiation, and extreme temperatures while enabling their controlled release near plant roots to sustain rhizosphere activity under drought conditions (Karnwal et al. 2023). Mesoporous silica NPs (MSiNPs) gradually release microbes and nutrients in sync with crop needs (Sarkar et al. 2022), and chitosan-alginate nanocapsules with humic acid encapsulating *Pseudomonas fluorescens* and NPK provided sustained nutrient release for 30 days with superior plant performance (Hamed et al. 2024). NPs also boost antioxidant enzymes, such as catalase and ascorbate peroxidase, modulate phytohormones, such as abscisic acid, and enhance osmolyte accumulation. For example, ZnO NPs increased proline levels, reduced lipid peroxidation, and elevated antioxidant activity in *Zea mays* under drought conditions (Fatima et al. 2023). They further improve root architecture, soil moisture retention, microbial diversity, and nutrient solubilization (Sun et al. 2021; Gupta et al. 2024). Specific NPs offer distinct benefits such as silica for slow nutrient release and moisture retention, ZnO for micronutrient delivery and stress protection, TiO_2 for UV shielding and colonization, and chitosan for microbial encapsulation (Bueno et al. 2022; Dimkpa et al. 2019; Mohr et al. 2019; Isabel et al. 2024). Various methodologies for integrating nanotechnology with bioinoculants are listed in Table 9.3. The integration of nanobiotechnology with genome editing offers new avenues for climate-smart crops, although scaling and field testing remain essential. Thus, nanobioinoculants represent a promising and sustainable strategy for enhancing drought tolerance and global food security.

Table 9.3 Various methodologies for integrating nanotechnology with bioinoculants

Integration aspect	Description	Benefits	Example nanomaterial
Nanobiofertilizers	Nanotech-enhanced biofertilizers improve nutrient delivery under drought	Better nutrient uptake, growth, and tolerance	Silica NPs, carbon nanotubes
NP encapsulation of bioinoculants	NPs encapsulate PGPR (e.g., *Pseudomonas, Bacillus*) for stress survival	Higher viability, root colonization, and biomass	Alginate-silica NPs, carbon nanotubes
Synergistic effects of NPs and bioinoculants	NPs with bioinoculants boost plant stress resilience	Stronger antioxidants, photosynthesis, and nutrient uptake	Silica NPs, ZnO NPs
Nano-enabled soil and water management	Nanobubbles/particles improve soil oxygenation, water use	Better fertility, water efficiency, and yield	Micro-nanobubbles, magnetic NPs
Nanobiofertilizers to activate plant defenses	NPs enhance bioinoculant-induced resistance	Improved defense, reduced oxidative stress, and drought resilience	Cu NPs, silica NPs

9.4.4 IoT (Internet of Things)-Enabled Microbial Delivery and Precision Irrigation Systems

IoT-enabled microbial delivery and precision irrigation systems are at the forefront of innovative agricultural technologies aimed at combating drought stress, especially when integrated with bioinoculant approaches (Stočes et al. 2016). IoT devices monitor soil and crop conditions, enabling data-driven, real-time delivery of water and bioinoculants to the root zones. This automation minimizes manual effort while ensuring timely, controlled, and efficient microbial application, which is important for enhancing crop resilience during drought (Ali et al. 2024). Furthermore, precision irrigation, guided by IoT soil data, delivers water at optimal times and amounts, reducing wastage, avoiding under- or over-watering, and maintaining consistent moisture for roots and bioinoculants (Khanna and Kaur 2019). The precise co-delivery of water and microbes using smart irrigation has demonstrated increased yields (up to 23.9% over conventional methods in some crops), better nutrient uptake, and improved soil health, which are important considerations for drought-prone farming systems (Shravani et al. 2019). One such example is the Wi-Fi-enabled "iPOTs." This is an IoT-based automatic irrigation system that mimics drought by controlling bottom-up water supply and soil moisture at user-set levels while monitoring soil temperature, air temperature, humidity, and light intensity in each pot (Numajiri et al. 2021). The integration of AI enables the predictive scheduling of irrigation and bioinoculant delivery based on anticipated droughts, thereby reducing risks and enhancing crop resilience. This emerging field is gaining attention for improving drought mitigation.

9.5 Multiomics Approaches for Enhancing Crop Resilience in Drought

9.5.1 Genome-Wide Association Studies (GWAS)

Genome-wide association studies (GWAS) analyze genome-wide variations across natural populations to identify loci linked to complex traits. By associating single-nucleotide polymorphisms (SNPs) with agronomic traits, such as heat tolerance, salinity resistance, and drought endurance, GWAS have advanced crop stress tolerance research (Yasir et al. 2019; Devate et al. 2022; Thingujam et al. 2025). Genomic insights have deepened our understanding of plant physiology and gene function under drought conditions. Genomics-assisted breeding (GAB) approaches, including haplotype-based and marker-assisted breeding, are increasingly used in rice. These methods rely on DNA markers closely linked to desirable traits, thereby accelerating the development of climate-resilient cultivars (Sandhu et al. 2019). Multiomics approaches that integrate genomics, transcriptomics, proteomics, and metabolomics have advanced our understanding of sorghum stress tolerance mechanisms (Mwamahonje et al. 2021). Whole-genome sequencing and re-sequencing using NGS platforms, such as Illumina and Ion Torrent, have identified numerous QTLs and drought-related genes (Das et al. 2017). Owing to the high cost of genome sequencing, reduced-representation methods, such as genotyping-by-sequencing (GBS), are widely applied in GWAS, genomic selection, diversity analysis, genome profiling, marker discovery, and high-resolution QTL mapping (Scheben et al. 2017; Roy et al. 2025).

9.5.2 Transcriptomics: Gene Expression under Stress

9.5.2.1 RNA Sequencing (RNA-Seq) for Identifying Stress-Responsive Genes

RNA-Seq is a key tool in plant stress biology, enabling the comprehensive profiling of transcriptional changes under environmental stress. It identifies differentially expressed genes (DEGs) involved in stress perception, signaling, transcriptional regulation, and metabolic reprogramming (Kumar et al. 2022a, b; Meraj et al. 2020). In soybeans, drought and flooding altered bHLH, ERF, MYB, NAC, and WRKY transcription factors, whereas salt stress in *Arabidopsis* induced *LEA45* and *GSTF6*. In wheat, HSPs and *HsfA2* have been highlighted under heat stress. Time-course studies distinguish early and late responses (Chen et al. 2016; Seok et al. 2020; Thingujam et al. 2025), supporting molecular breeding and genetic engineering.

PGPMs support plants under abiotic stress, as shown by transcriptome analysis of *A. brasilense*, which revealed increased expression of osmoprotection and hormone production genes during drought conditions (Garciglia-Mercado et al. 2024). Transcriptomic studies have demonstrated that PGPMs modulate host gene expression, enhancing defense and stress responses. *P. fluorescens* primes plants by upregulating PR1, PDF1.2, and PR4, providing >50% protection against *Botrytis cinerea* and 60% protection against *Pst* DC3000 (Chen et al. 2024; Sahoo et al. 2025). Dual RNA-seq shows thousands of DEGs (1400–4484) during colonization, linked to defense, ethylene biosynthesis, ROS detoxification, and nutrient uptake (Ferrarezi et al. 2023; Nobori et al. 2022). In rice, *OsMIOX* overexpression enhanced drought tolerance, with 743 DEGs mainly associated with sugar metabolism and hormone signaling (Shi et al. 2020). Similarly, WR16 was identified as being highly drought tolerant among 133 rice accessions (Han et al. 2020). In sesame, 2030 drought-responsive genes showed contrasting expression patterns between tolerant and susceptible genotypes (You et al. 2019). Core drought genes were downregulated in photosynthesis, fatty acid and amino acid metabolism, DNA replication, and ribosome pathways but upregulated in ER function, galactose metabolism, and hormone signaling (Bogati and Walczak 2022).

9.5.3 Proteomics and Metabolomics

Proteomics provides direct insights into plant stress responses beyond transcriptome data, revealing changes in protein abundance, localization, and posttranslational modifications in response to heat, salinity, and drought (Kosová et al. 2011). Spatial proteomics aids in mapping complex protein-protein interactions, offering advantages over traditional methods like Cytotrap (Lundberg and Borner 2019; Mohan et al. 2023). Proteomic studies have shown PGPM-mediated modulation of host and microbial proteins. *P. polymyxa* NSY50 upregulated 56 cucumber proteins, mainly linked to defense (28.6%), protein metabolism (25%), and energy metabolism (19.6%) (Sahoo et al. 2025). Root-associated PGPMs, such as *B. subtilis*, enhance antibiotic production, biofilm formation, and stress-responsive proteins (SODs and peroxidases), aiding ROS detoxification and osmotic balance. The high ACC deaminase activity of *Pseudomonas* sp. Y1 (895 ± 35 nmol α-ketobutyrate/mg protein/hr) further promoted root growth and salt tolerance (Mukherjee 2022). Molecular chaperones also play key roles; Bip was overexpressed in transgenic soybean plants under drought conditions, as shown by proteomic and metabolomic analyses (Coutinho et al. 2021).

9.5.4 Epigenomics for Drought Tolerance

The epigenome encompasses biochemical alterations in nuclear polymers, proteins, and noncoding RNAs, with epigenomics studying the modifications that regulate gene activity. Epigenetics enables plants to cope with external stress through nonheritable mechanisms. Advances have clarified the signaling and metabolic pathways underlying stress responses (Ku et al. 2018; Kumar et al. 2019). Drought often induces DNA demethylation, which is tissue-specific, with up to 1% variation across cells, tissues, genotypes, and developmental stages (Zargar et al. 2022). In strawberries, time-course DNA methylation studies have revealed the regulation of ROS-related genes under drought conditions (Cao et al. 2019).

9.6 Future Perspectives

Soil microbial inoculants offer promising solutions for agricultural and soil management (Rafi et al. 2019; O'Callaghan et al. 2022). Enhancing their performance, persistence, and efficiency while minimizing risks is essential (Qiu et al. 2019). The rising demand for productivity, soil health, and sustainable farming underscores their importance. In barley, breeding drought-resistant genotypes could strengthen tolerance to water shortages. Advances in genetics, genomics, and phenotyping, combined with genome editing, genomic selection, and marker-assisted selection

(MAS), can accelerate the development of drought-tolerant varieties (Sallam et al. 2019; Ferioun et al. 2025). Advances in molecular insights into plant-microbe interactions offer new opportunities for breeding resilient crops (Nerva et al. 2022). Investigating rhizosphere interactions is key to understanding the plant "cry for help" strategy (Zancarini et al. 2021). Deciphering the signaling pathways and identifying functional genes will clarify the mechanisms driving plant development and stress responses (Depuydt and Vandepoele 2021).

In agro-business, microbial inoculants used as biofertilizers and biopesticides have gained global acceptance because of their eco-friendly nature and potential to replace chemical inputs. Despite their slower mode of action, bioinoculants are favoured for promoting sustainable agriculture, enhancing nutrient uptake, and improving soil health. Rhizospheric microbes contribute to plant growth, suppress pathogens, and facilitate the recycling of organic waste, especially when urban organic matter is integrated into agricultural soils. Effective application requires optimization of inoculum density and consistent performance evaluation to ensure efficacy under field conditions (Maitra et al. 2021). Recent advances in plant-microbe biotechnology have included the development of tailored microbial consortia, such as combinations of *Bacillus*, *Pseudomonas*, and *Rhizobium* species, which enhance drought and nutrient stress tolerance (Chieb and Gachomo 2023). The efficacy of microbial inoculants is inherently strain-specific, with functional heterogeneity evident even among conspecifics. The phylogenetic proximity to known PGPR does not guarantee similar performance. Optimal strain selection should integrate phenotypic traits with ecological adaptability, considering spatial and temporal abundance and tolerance to abiotic stresses in the target habitat. Locally adapted strains, such as desiccation-tolerant *R. leguminosarum* bv. *trifolii* from drought-prone soils demonstrated superior environmental persistence and functional stability under field conditions (O'Callaghan et al. 2022).

9.7 Conclusion

The integration of bioinoculants into climate-smart agricultural systems offers a sustainable pathway for mitigating drought-induced yield losses while improving soil health and ecosystem resilience. Microbial inoculants can significantly bolster plant adaptive responses to water scarcity through mechanisms such as phytohormone modulation, osmolyte accumulation, EPS production, VOC secretion, and enhanced antioxidant defense. However, their field efficacy remains constrained by environmental variability, soil heterogeneity, temporal dynamics, and challenges associated with scaling up. Innovations, including synthetic microbial consortia, genetic engineering, nanotechnology-enabled delivery systems, and IoT-based precision applications, hold promise for overcoming these bottlenecks. Furthermore, multiomics approaches offer valuable insights into plant-microbe-environment interactions, enabling targeted strain selection and the development of tailored formulations. For bioinoculants to become a reliable cornerstone of drought-resilient agriculture, future efforts must prioritize site-specific validation, robust formulation technologies, farmer-oriented capacity building, and regulatory frameworks that ensure safety and efficacy. By combining ecological principles with cutting-edge biotechnology, bioinoculants can be strategically deployed to enhance crop productivity, conserve water resources, and strengthen global food security against climate change.

Acknowledgments S.S. received research support from the NFST (Ministry of Tribal Affairs). A.D. received research support from the Department of Science and Technology (DST), Government of India (DST-INSPIRE fellowship ID: IF190463). N.D., S.C., and P.P. are grateful to the Department of Biotechnology, Ministry of Science and Technology, Government of India, for financial support.

References

Adeleke BS, Akinola SA, Adedayo AA, Glick BR, Babalola OO (2022) Synergistic relationship of endophyte-nanomaterials to alleviate abiotic stress in plants. Front Environ Sci 10:1015897

Afridi MS, Javed MA, Ali S, De Medeiros FHV, Ali B, Salam A, Santoyo G (2022) New opportunities in plant microbiome engineering for increasing agricultural sustainability under stressful conditions. Front Plant Sci 13:899464

Agarwal PK, Gupta K, Lopato S, Agarwal P (2017) Dehydration responsive element binding transcription factors and their applications for the engineering of stress tolerance. J Exp Bot 68:2135–2148

Akhtar SS, Amby DB, Hegelund JN, Fimognari L, Großkinsky DK, Westergaard JC, Roitsch T (2020) *Bacillus licheniformis* FMCH001 increases water use efficiency via growth stimulation in both normal and drought conditions. Front Plant Sci 11:297

Aleksieienko I, Fernandes Hertel M, Reilhan J, De Castro M, Légeret B, Caixeta Oliveira H, Reite IMSC (2025) Soil-gradient-derived bacterial synthetic communities enhance drought tolerance in *Quercus pubescens* and *Sorbus domestica* seedlings. Plants 14:1659

Ali T, Rehman SU, Ali S, Mahmood K, Obregon SA, Iglesias RC, Khurshaid T, Ashraf I (2024) Smart agriculture: utilizing machine learning and deep learning for drought stress identification in crops. Sci Rep 14:30062

Alzate Zuluaga MY, Fattorini R, Cesco S, Pii Y (2024) Plant-microbe interactions in the rhizosphere for smarter and more sustainable crop fertilization: the case of PGPR-based biofertilizers. Front Microbiol 15:1440978

Andreozzi A, Prieto P, Mercado-Blanco J, Monaco S, Zampieri E, Romano S, Bianco C (2019) Efficient colonization of the endophytes *Herbaspirillum huttiense* RCA24 and *Enterobacter cloacae* RCA25 influences the physiological parameters of *Oryza sativa* L. cv. Baldo rice. Environ Microbiol 21(9):3489–3504

Ansabayeva A, Makhambetov M, Rebouh NY, Abdelkader M, Saudy HS, Hassan KM, Ebrahim M (2025) Plant growth-promoting microbes for resilient farming systems: mitigating environmental stressors and boosting crops productivity—a review. Hortic 11(3):260

Bai B, Su YH, Yuan J, Zhang XS (2013) Induction of somatic embryos in *Arabidopsis* requires local YUCCA expression mediated by the down-regulation of ethylene biosynthesis. Mol Plant 6(4):1247–1260

Bardgett RD, Caruso T (2020) Soil microbial community responses to climate extremes: resistance, resilience and transitions to alternative states. Philos Trans R Soc B 375(1794):20190112

Bashan Y, de-Bashan LE, Prabhu SR, Hernandez JP (2014) Advances in plant growth-promoting bacterial inoculant technology: formulations and practical perspectives (1998–2013). Plant Soil 378(1–2):1–33. https://doi.org/10.1007/s11104-013-1956-x

Baslam M, Goicoechea (2012) Water deficit improved the capacity of arbuscular mycorrhizal fungi (AMF) for inducing the accumulation of antioxidant compounds in lettuce leaves. Mycorrhiza 22(5):347–359

Batool T, Ali S, Seleiman MF, Naveed NH, Ali A, Ahmed K, Mubushar M (2020) Plant growth promoting rhizobacteria alleviates drought stress in potato in response to suppressive oxidative stress and antioxidant enzymes activities. Sci Rep 10(1):16975

Benmrid B, Ghoulam C, Zeroual Y, Kouisni L, Bargaz A (2023) Bioinoculants as a means of increasing crop tolerance to drought and phosphorus deficiency in legume-cereal intercropping systems. Commun Biol 6(1):1016

Berg G (2009) Plant–microbe interactions promoting plant growth and health: perspectives for controlled use of microorganisms in agriculture. Appl Microbiol Biotechnol 84(1):11–18

Bisht N, Singh T, Ansari MM, Joshi H, Mishra SK, Chauhan PS (2025) Plant growth-promoting *Bacillus amyloliquefaciens* orchestrate homeostasis under nutrient deficiency exacerbated drought and salinity stress in *Oryza sativa* L. seedlings. Planta 261(1):8

Bittencourt PP, Alves AF, Ferreira MB, da Silva Irineu LES, Pinto VB, Olivare L (2023) Mechanisms and applications of bacterial inoculants in plant drought stress tolerance. Microorganisms 11:502

Bogati K, Walczak M (2022) The impact of drought stress on soil microbial community, enzyme activities and plants. Agronomy 12(1):189

Boora R, Rani N, Kumari S, Goel S, Arya A, Grewal S (2024) Exploring the role of green synthesized cerium nanoparticles in enhancing wheat's drought tolerance: a comprehensive study of biochemical parameters and gene expression. Cereal Res Commun 52:1487–1498

Bueno V, Wang P, Harrisson O, Bayen S, Ghoshal S (2022) Impacts of a porous hollow silica nanoparticle-encapsulated pesticide applied to soils on plant growth and soil microbial community. Environ Sci Nano 9:1476–1488

Bulegon LG, Guimarães VF, Battistus AG, Inagaki AM, Costa NVD (2019) Mitigation of drought stress effects on soybean gas exchanges induced by *Azospirillum brasilense* and plant regulators. Pesqui Agropecu Trop 49:e52807

Cao X, Yang H, Shang C, Ma S, Liu L, Cheng J (2019) The roles of auxin biosynthesis YUCCA gene family in plants. Int J Mol Sci 20(24):6343

Chandra D, Srivastava R, Gupta VV, Franc CM, Sharma AK (2019) Evaluation of ACC-deaminase-producing rhizobacteria to alleviate water-stress impacts in wheat (*Triticum aestivum* L.) plants. Can J Microbiol 65(5):387–403

Chandra KK, Bhardwaj AK, Singh A, Kumar R (2021) Impact of drought on tropical forests and plant mechanism to mitigate drought stress. In: Tropical dry forest: emerging features and ecological perspectives, pp 367–383

Change C (2022) Mitigating Climate Change. Working Group III contribution to the sixth assessment report of the Intergovernmental Panel on Climate Change

Chaudhary A, Agrawal V, Chaudhary P, Jaiswal DK, Aravindharajan S, Pathania D, Bhandari P, Chen S (2025) Nanobioinoculants and their applications in different sectors. Nanobioinoculants. Elsevier

Chen W, Yao Q, Patil GB, Agarwal G, Deshmukh RK, Lin L, Nguyen HT (2016) Identification and comparative analysis of differential gene expression in soybean leaf tissue under drought and flooding stress revealed by RNA-Seq. Front Plant Sci 7:1044

Chen C, Xin K, Liu H, Cheng J, Shen X, Wang Y, Zhang L (2017) *Pantoea alhagi*, a novel endophytic bacterium with ability to improve growth and drought tolerance in wheat. Sci Rep 7(1):41564

Chen S, Zhu B, Huang S, Hickey JW, Lin KZ, Snyder M, Ma Z (2024) Integration of spatial and single-cell data across modalities with weakly linked features. Nat Biotechnol 42(7):1096–1106

Chieb M, Gachomo EW (2023) The role of plant growth promoting rhizobacteria in plant drought stress responses. BMC Plant Biol 23(1):407

Cohen AC, Travaglia CN, Bottini R, Piccoli PN (2009) Participation of abscisic acid and gibberellins produced by endophytic *Azospirillum* in the alleviation of drought effects in maize. Botany 87(5):455–462

Compant S, Clément C, Sessitsch A (2010) Plant growth-promoting bacteria in the rhizo- and endosphere of plants: their role, colonization, mechanisms involved and prospects for utilization. Soil Biol Biochem 42(5):669–678. https://doi.org/10.1016/j.soilbio.2009.11.024

Compant S, Samad A, Faist HSA (2021) A review on the plant microbiome: ecology, functions, and emerging trends in microbial application. J Adv Res 19:29–37

Correa SS, Pacheco RS, Viana GC, Vidal MS, Xavier GR, de Araújo JLS (2024) Ability of nitrogen-fixing bacteria to alleviate drought stress in cowpea varies depending on the origin of the inoculated strain. Plant Soil 498(1):391–408

Coutinho FS, Rodrigues JM, Lima LL, Mesquit RO, Carpinetti PA, Machado JPB d ORHJ (2021) Remodeling of the cell wall as a drought-tolerance mechanism of a soybean genotype revealed by global gene expression analysis. aBIOTECH 2(1):14–31

Crater JS, Lievense JC (2018) Scale-up of industrial microbial processes. FEMS Microbiol Lett 365(13):fny138

Curá J, Franz D, Filosofía J, Balestrasse K, Burgueño L (2017) Inoculation with Azospirillum sp. and Herbaspirillum sp. bacteria increases the tolerance of maize to drought stress. Microorganisms 5(3):41. https://doi.org/10.3390/microorganisms5030041

Danish S, Zafar-ul-Hye M, Fahad S, Saud S, Brtnicky M, Hammerschmiedt T, Datta R (2020) Drought stress alleviation by ACC deaminase producing *Achromobacter xylosoxidans* and *Enterobacter cloacae*, with and without timber waste biochar in maize. Sustainability 12(15):6286

Das G, Patra JK, Baek KH (2017) Insight into MAS: a molecular tool for development of stress resistant and quality of rice through gene stacking. Front Plant Sci 8:985

Das N, Das A, Das S, Bhatawadekar V, Pandey P, Choure K, Damare S, Pandey P (2023) Petroleum hydrocarbon catabolic pathways as targets for metabolic engineering strategies for enhanced bioremediation of crude-oil-contaminated environments. Fermentation 9:196

De Souza RSC, Armanhi JSL, Arruda P (2020) From microbiome to traits: designing synthetic microbial communities for improved crop resiliency. Front Plant Sci 11:1179

de Vries FT, Griffiths RI, Bailey M, Woodward G (2020) Harnessing rhizosphere microbiomes for drought-resilient crop production. Science 368(6488):270–274. https://doi.org/10.1126/science.aaz5192

Depuydt T, Vandepoele K (2021) Multi-omics network-based functional annotation of unknown *Arabidopsis* genes. Plant J 108(4):1193–1212

Devate NB, Krishna H, Parmeshwarappa SKV, Manjunath KK, Chauhan D, Singh S, Singh PK (2022) Genome-wide association mapping for component traits of drought and heat tolerance in wheat. Front Plant Sci 13:943033

Di Benedetto NA, Corbo MR, Campaniello D, Cataldi MP, Bevilacqua A, Sinigaglia M, Flagella (2017) The role of plant growth promoting bacteria in improving nitrogen use efficiency for sustainable crop production: a focus on wheat. AIMS Microbiol 3(3):413

Díaz-Rodríguez AM, Parra Cota FI, Cira Chávez LA, García Ortega LF, Estrada Alvarado MI, Santoyo G, de Los Santos-Villalobos S (2025) Microbial inoculants in sustainable agriculture: advancements, challenges, and future directions. Plants 14(2):191

Dimkpa CO, Zeng J, McLean JE, Britt DW, Zhan J, Anderson AJ (2012) Production of indole-3-acetic acid via the indole-3-acetamide pathway in the plant-beneficial bacterium *Pseudomonas chlororaphis* O6 is inhibited by ZnO nanoparticles but enhanced by CuO nanoparticles. AEM 78:1404–1410

Dimkpa CO, Singh U, Bindraban PS, Elmer WH, Gardea-Torresdey JL, White JC (2019) Zinc oxide nanoparticles alleviate drought-induced alterations in sorghum performance, nutrient acquisition, and grain fortification. Sci Total Environ 688:926–934 dos Reis GA, Martínez-Burgos WJ, Pozzan R, Pastrana Puche Y, Ocán-Torres D, de Queiroz Fonseca Mota P, Soccol CR (2024) Comprehensive review of microbial inoculants: agricultural applications, technology trends in patents, and regulatory frameworks. Sustainability 16(19):8720

Dos Santos TB, Ribas AF, de Souza SGH, Budzinski IGF, Domingues DS (2022) Physiological responses to drought, salinity, and heat stress in plants: a review. Stress 2(1):113–135

Dubey A, Kumar A, Malla MA, Chowdhary K, Singh G, Gudasalamani Ravikanth H, Dames JF (2021) Approaches for the amelioration of adverse effects of drought stress on crop plants

Eswaran SUD, Sundaram L, Perveen K, Bukhari NA, Sayyed RZ (2024) Osmolyte-producing microbial biostimulants regulate the growth of *Arachis hypogaea* L. under drought stress. BMC Microbio 24(1):165

Etesami H, Hosseini HM, Alikhani HA, Mohammadi L (2014) Bacterial biosynthesis of 1-amin ocyclopropane-1-carboxylate (ACC) deaminase and indole-3-acetic acid (IAA) as endophytic preferential selection traits by rice plant seedlings. J Plant Growth Regul 33(3):654–670

Etesami H, Jeong BR, Glick BR (2023) Potential use of *Bacillus* spp. as an effective biostimulant against abiotic stresses in crops—a review. Curr res Biotechnol 5:100128

Fadiji AE, Santoyo G, Yadav AN, Babalola OO (2022) Efforts towards overcoming drought stress in crops: revisiting the mechanisms employed by plant growth-promoting bacteria. Front Microbiol 13:962427

Fahad S, Hussain S, Bano A, Saud S, Hassan S, Shan D, Huang J (2015) Potential role of phytohormones and plant growth-promoting rhizobacteria in abiotic stresses: consequences for changing environment. ESPR 22(7):4907–4921

Fatima A, Safdar N, Ain N-u, Yasmin A, Chaudhry G-e-S (2023) Abscisic acid-loaded ZnO nanoparticles as drought tolerance inducers in *Zea mays* L. with physiological and biochemical attributes. J Plant Growth Regul 42:7280–7293

Ferioun M, Zouitane I, Bouhraoua S, Elouattassi Y, Belahcen D, Errabbani A, El Ghachtouli N (2025) Applying microbial biostimulants and drought-tolerant genotypes to enhance barley growth and yield under drought stress. Front Plant Sci 15:1494987

Ferrarezi JA, Defant H, De Souza LF, Azevedo JL, Hungria M, Quecine MC (2023) Meta-omics integration approach reveals the effect of soil native microbiome diversity in the performance of inoculant *Azospirillum brasilense*. Front Plant Sci 14:1172839

Fitzpatrick CR, Copeland J, Wang PW, Guttman DS, Kotanen PM, Johnson MTJ (2018) Assembly and ecological function of the root microbiome across angiosperm plant species. PNAS 115(6):E1157–E1165. https://doi.org/10.1073/pnas.1717617115

Flemming HC, Wingender J (2010) The biofilm matrix. At. Rev Microbiol 8(9):623–633

Fukase E, Martin W (2020) Economic growth, convergence, and world food demand and supply. World Dev 132:104954

Goswami M, Suresh DEKA (2020) Plant growth-promoting rhizobacteria—alleviators of abiotic stresses in soil: a review. Pedosphere 30(1):40–61

Ganeshan S, Kim SH, Vujanovic V (2021) Scaling-up production of plant endophytes in bioreactors: concepts, challenges and perspectives. BIOB 8(1):63

Garciglia-Mercado C, Contreras CA, Choix FJ, de-Bashan LE, Gómez-Anduro GA, Palacios OA (2024) Metabolic and physiological adaptations of microalgal growth-promoting bacterium *Azospirillum brasilense* growing under biogas atmosphere: a microarray-based transcriptome analysis. Arch Microbiol 206(4):173

Ghosh D, Sen S, Mohapatra S (2017) Modulation of proline metabolic gene expression in *Arabidopsis thaliana* under water-stressed conditions by a drought-mitigating *Pseudomonas putida* strain. Ann Microbiol 67(10):655–668

Ghosh PK, De TK, Maiti TK (2018) Role of ACC deaminase as a stress ameliorating enzyme of plant growth-promoting rhizobacteria useful in stress agriculture: a review. In: Role of rhizospheric microbes in soil: volume 1: stress management and agricultural sustainability, pp 57–106

Gontia-Mishra I, Sapre S, Kachare S, Tiwari S (2017) Molecular diversity of 1-aminocyclopropane-1-carboxylate (ACC) deaminase producing PGPR from wheat (*Triticum aestivum* L.) rhizosphere. Plant Soil 414(1):213–227

Gou X, Deng Y, Gao L, Chen F, Cook E, Yang M, Zhang F (2015) Millennium tree-ring reconstruction of drought variability in the eastern Qilian Mountains, Northwest China. Clim Dyn 45(7):1761–1770

Govindasamy V, George P, Kumar M, Aher L, Raina SK, Rane J, Minhas S (2020) Multi-trait PGP rhizobacterial endophytes alleviate drought stress in a senescent genotype of sorghum [*Sorghum bicolor* (L.) Moench]. 3 Biotech 10(1):13

Gowtham HG, Singh SB, Shilpa N, Aiyaz M, Nataraj K, Udayashankar AC, Sayyed RZ (2022) Insight into recent progress and perspectives in improvement of antioxidant machinery upon PGPR augmentation in plants under drought stress: a review. Antioxidants 11(9):1763

Gupta S, Kumar D, Aziz A, AE AbdelRahman M, Mustafa A-rA, Scopa A, Radice RP, Drosos M, Moursy AR (2024) Nanoecology: exploring engineered nanoparticles' impact on soil ecosystem health and biodiversity. EJSS 64:1637–1655

Gureeva MV, Gureev AP (2023) Molecular mechanisms determining the role of bacteria from the genus *Azospirillum* in plant adaptation to damaging environmental factors. Int J Mol Sci 24(11):9122

Gusain YS, Singh US, Sharma AK (2015) Bacterial mediated amelioration of drought stress in drought tolerant and susceptible cultivars of rice (*Oryza sativa L.*). Afr J Biotechnol 14(9):764–773

Gutiérrez-Luna FM, López-Bucio J, Altamirano-Hernández J, Valencia-Cantero E, de la Cruz HR, Macías-Rodríguez L (2010) Plant growth-promoting rhizobacteria modulate root-system architecture in *Arabidopsis thaliana* through volatile organic compound emission. Symbiosis 51(1):75–83

Hakim S, Naqqash T, Nawaz MS, Laraib I, Siddique MJ, Zia R, Imran A (2021) Rhizosphere engineering with plant growth-promoting microorganisms for agriculture and ecological sustainability. Front Sustain Food Syst 5:617157

Hamed R, Jodeh S, Alkowni R (2024) Nano bio fertilizer capsules for sustainable agriculture. Sci Rep 14:13646

Han B, Ma X, Cui D, Geng L, Cao G, Zhang H, Han L (2020) Parallel reaction monitoring revealed tolerance to drought proteins in weedy rice (*Oryza sativa* f. *spontanea*). Sci Rep 10(1):12935

Hanaka A, Ozimek E, Reszczyńska E, Jaroszuk-Ściseł J, Stolarz M (2021) Plant tolerance to drought stress in the presence of supporting bacteria and fungi: an efficient strategy in horticulture. Horticulturae 7(10):390

Hart MM, Antunes PM, Chaudhary VB, Abbott LK (2018) Fungal inoculants in the field. Funct Ecol 32(1):126–135

Hasanuzzaman M, Bhuya MB, Zulfiqar F, Raza A, Mohsin SM, Mahmud JA, Fotopoulos V (2020) Reactive oxygen species and antioxidant defense in plants under abiotic stress: revisiting the crucial role of a universal defense regulator. Antioxidants 9(8):681

Hayat S, Hayat Q, Alyemeni MN, Wani AS, Pichtel J, Ahmad A (2012) Role of proline under changing environments: a review. PSB 7(11):1456–1466

Hu T, van Dijk AI, Renzullo LJ, Xu Z, He J, Tian S, Li H (2020) On agricultural drought monitoring in Australia using Himawari-8 geostationary thermal infrared observations. Int J Appl Earth Obs Geoinf 91:102153

Impa SM, Nadaradjan S, Jagadish SVK (2011) Drought stress induced reactive oxygen species and anti-oxidants in plants. In: Abiotic stress responses in plants: metabolism, productivity and sustainability. Springer, New York, pp 131–147

Isabel JB, Balamurugan A, Devi PR, Periyasamy S (2024) Chitosan-encapsulated microbial biofertilizer: a breakthrough for enhanced tomato crop productivity. Int J Biol Macromol 260:129462

Jha S, Srivastava R (2018) Impact of drought on vegetation carbon storage in arid and semi-arid regions. RSASE 11:22–29

Joshi S, Gangola S, Jaggi VSM (2023) Functional characterization and molecular fingerprinting of potential phosphate solubilizing bacterial candidates from Shisham rhizosphere. Sci Rep 13:7003

Kamau JN, Magutu J, Amadi H (2023) Climate smart agriculture policy interventions for inclusive and sustainable development. Pathways to African Feminism and Development: AWSC J 8(1):97–110

Karnwal A, Dohroo A, Malik T (2023) Unveiling the potential of bioinoculants and nanoparticles in sustainable agriculture for enhanced plant growth and food security. Biomed Res Int 2023:6911851

Kashyap AS, Pandey VK, Manzar N, Kannojia P, Singh UB, Sharma PK (2017) Role of plant growth-promoting rhizobacteria for improving crop productivity in sustainable agriculture. In: Singh DP, Singh HB, Prabha R (eds) Plant-microbe interactions in agro-ecological perspectives, Microbial interactions and agro-ecological impacts, vol 2. Springer, Singapore, pp 673–693

Kavadia A, Omirou M, Fasoula D, Ioannides IM (2020) The importance of microbial inoculants in a climate-changing agriculture in eastern Mediterranean region. Atmos 11(10):1136

Kaya C (2024) Microbial modulation of hormone signaling, proteomic dynamics, and metabolomics in plant drought adaptation. FES 13(1):e513

Khan A, Singh AV (2021) Multifarious effect of ACC deaminase and EPS producing *Pseudomonas* sp. and *Serratia marcescens* to augment drought stress tolerance and nutrient status of wheat. WJMB 37(12):198

Khan N, Bano A, Zandi P (2018) Effects of exogenously applied plant growth regulators in combination with PGPR on the physiology and root growth of chickpea (*Cicer arietinum*) and their role in drought tolerance. J Plant Interact 13(1):239–247

Khanna A, Kaur S (2019) Evolution of Internet of Things (IoT) and its significant impact in the field of precision agriculture. Comput Electr Agric 157:218–231

Koskey G, Mburu SW, Awino R, Njeru EM, Maingi JM (2021) Potential use of beneficial microorganisms for soil amelioration, phytopathogen biocontrol, and sustainable crop production in smallholder agroecosystems. Front Sustain Food Syst 5:606308

Kosová K, Vítámvás P, Prášil IT, Renaut J (2011) Plant proteome changes under abiotic stress—contribution of proteomics studies to understanding plant stress response. J Proteome 74(8):1301–1322

Kour D, Khan SS, Kaur T, Kour H, Singh G, Yadav A, Yadav AN (2022) Drought adaptive microbes as bioinoculants for the horticultural crops. Heliyon 8(5):e09493

Ku YS, Sintaha M, Cheung MY, Lam HM (2018) Plant hormone signaling crosstalks between biotic and abiotic stress responses. Int J Mol Sci 19(10):3206

Kumar S, Sindhu SS (2024) Drought stress mitigation through bioengineering of microbes and crop varieties for sustainable agriculture and food security. CRMICR 7:100285

Kumar M, Kesawat MS, Ali A, Lee SC, Gill SS, Kim HU (2019) Integration of abscisic acid signaling with other signaling pathways in plant stress responses and development. Plants 8(12):592

Kumar S, Seem K, Kumar S, Mohapatra T (2022a) RNA-seq analysis reveals the genes/pathways responsible for genetic plasticity of rice to varying environmental conditions on direct-sowing and transplanting. Sci Rep 12(1):2241

Kumar P, Singh S, Pranaw K, Kumar S, Singh B, Poria V (2022b) Bioinoculants as mitigators of multiple stresses: a ray of hope for agriculture in the darkness of climate change. Heliyon 8(11):e11269

Kumar S, Diksha Sindhu SS, Kumar R (2025) Harnessing phosphate-solubilizing microorganisms for mitigation of nutritional and environmental stresses, and sustainable crop production. Planta 261:95

Kumari M, Swarupa P, Kesari KK, Kumar A (2022) Microbial inoculants as plant biostimulants: a review on risk status. Life 13(1):12

Kundel D, Bodenhausen N, Jørgensen HB, Truu J, Birkhofer K, Hedlund K, Fliessbach A (2020) Effects of simulated drought on biological soil quality, microbial diversity and yields under long-term conventional and organic agriculture. FEMS Microbiol Ecol 96(12):fiaa205

Lahlali R, Ezrari S, Radouane N, Belabess Z, Jiang Y, Mokrini F, Peng G (2022) *Bacillus* spp.-mediated drought stress tolerance in plants: current and future prospects. In: *Bacilli* in agro-biotechnology: plant stress tolerance, bioremediation, and bioprospecting. Springer, Cham, pp 487–518

Laxa M, Liebthal M, Telman W, Chibani K, Dietz KJ (2019) The role of the plant antioxidant system in drought tolerance. Antioxidants 8(4):94

Li S, Li P, Ding H, Zhao S, Ai J, Wei G, Li Z (2025) Rhizosphere bacterial community confers drought tolerance to Astragalus mongholicus. Agric Water Manag 315:109543

Lipper L, Thornton P, Campbell BM, Baedeker T, Braimoh A, Bwalya M, Torquebiau EF (2014) Climate-smart agriculture for food security. Nat Clim Chang 4(12):1068–1072

Liu R, Lal R (2014) Synthetic apatite nanoparticles as a phosphorus fertilizer for soybean (*Glycine max*). Sci Rep 4:5686

Liu F, Xing S, Ma H, Du Z, Ma B (2013) Cytokinin-producing, plant growth-promoting rhizobacteria that confer resistance to drought stress in *Platycladus orientalis* container seedlings. Appl Microbiol Biotechnol 97(20):9155–9164

Liu M, Liu H, Mei F, Yang N, Zhao D, Ai G, Xiang H, Zheng Y (2022) Identification of the biosynthetic pathway of glycine betaine that is responsible for salinity tolerance in halophilic *Thioalkalivibrio versutus* D301. Front Microbiol 13:875843

Lopes T, Costa P, Cardoso P, Figueira E (2024) Bacterial volatile organic compounds as a strategy to increase drought tolerance in maize (*Zea mays* L.): influence on plant biochemistry. Plants 13(17):2456

Lundberg E, Borner GH (2019) Spatial proteomics: a powerful discovery tool for cell biology. Nat Rev Mol Cell Biol 20(5):285–302

Maestro- Gaitán I, Granado- Rodríguez S, Redondo- Nieto M, Battaglia A, Poza-Viejo L, Matías J (2023) Unveiling changes in rhizosphere-associated bacteria linked to the gen-otype and water stress in quinoa. Microb Biotechnol 16:2326

Magnucka EG, Pietr SJ (2015) Various effects of fluorescent bacteria of the genus *Pseudomonas* containing ACC deaminase on wheat seedling growth. Microbiol Res 181:112–119

Maheshwari DK, Das A, Dheeman S, Pandey P (2023) An overall insight into the attributes, interactions, and future applications of "microbial consortium" for plant growth promotion with contemporary approaches. In: Sustainable agrobiology: design and development of microbial consortia, pp 3–22

Mahmud NA (2021) The poverty in Somalia: how can economic reform policies and conflicts destroy a Country's way of life?

Maitra S, Brestic M, Bhadra P, Shankar T, Praharaj S, Palai JB, Hossain A (2021) Bioinoculants—natural biological resources for sustainable plant production. Microorganisms 10(1):51

Malgioglio G, Rizzo GF, Nigro S, Lefebvre du Prey V, Herforth-Rahmé J, Catara V, Branca F (2022) Plant-microbe interaction in sustainable agriculture: the factors that may influence the efficacy of PGPM application. Sustainability 14(4):2253

Marasco R, Rolli E, Vigani G, Borin S, Sorlini C, Ouzari H, Daffonchio D (2012) Are drought-resistance promoting bacteria cross-compatible with different plant models? Plant Signal Behav 7(8):951–954. https://doi.org/10.4161/psb.20862

Marques DM, Magalhães PC, Marriel IE, Gome CC, Silv ABD, Souza TCD (2021) Gas exchange, root morphology and nutrients in maize plants inoculated with *Azospirillum brasilense* cultivated under two water conditions. Braz Arch Biol Technol 64:e21190580

Martins J, Pétriacq P, Flandin A, Gomez-Cadenas A, Monteiro P, Pinto G, Canhoto J (2022) Genotype determines *Arbutus unedo* L. physiological and metabolomic responses to drought and recovery. Front Plant Sci 13:1011542

Masoudi-Sadaghiani F, Babak AM, Zardoshti MR, Hassan RSM, Tavakoli A (2011) Response of proline, soluble sugars, photosynthetic pigments and antioxidant enzymes in potato (*Solanum tuberosum* L.) to different irrigation regimes in greenhouse condition. J Crop Sci 5(1):55–60

Mendoza-Alatorre M, Infante-Ramírez R, González-Rangel MO, Nevárez-Moorillón GV, González-Horta MDC, Hernández-Huerta J, Delgado-Gardea MCE (2024) Enhancing drought

stress tolerance and growth promotion in chiltepin pepper (*Capsicum annuum* var. *glabriusculum*) through native *Bacillus* spp. Sci Rep 14(1):15383
Meraj TA, Fu J, Raza MA, Zhu C, Shen Q, Xu D, Wang Q (2020) Transcriptional factors regulate plant stress responses through mediating secondary metabolism. Genes 11(4):346
Mohan B, Thingujam D, Pajerowska-Mukhtar KM (2023) Cytotrap: an innovative approach for protein–protein interaction studies for cytoplasmic proteins. In: Protein-protein interactions: methods and protocols. Springer, New York, pp 9–22
Mohr L, Capelezzo A, Baretta C, Martins M, Fiori M, Mello J (2019) Titanium dioxide nanoparticles applied as ultraviolet radiation blocker in the polylactic acid bidegradable polymer. Polym Test 77:105867
Moore ER, Carter KR, Heneghan JP, Steadman CR, Nachtsheim AC, Anderson-Cook C, Albright MB (2023) Microbial drivers of plant performance during drought depend upon community composition and the greater soil environment. Microbiol Spectrum 11(2):e01476–e01422
Muhammad A, Kong X, Zheng S, Bai N, Li L, Khan MHU, Zhang Z (2024) Exploring plant-microbe interactions in adapting to abiotic stress under climate change: a review. Front Plant Sci 15:1482739
Mukherjee A (2022) What do we know from the transcriptomic studies investigating the interactions between plants and plant growth-promoting bacteria? Front Plant Sci 13:997308
Mukherjee S, Mishra A, Trenberth KE (2018) Climate change and drought: a perspective on drought indices. Curr Clim Chang Rep 4:145–163
Murali M, Singh SB, Gowtham HG, Shilpa N, Prasad M, Aiyaz M, Amruthesh KN (2021) Induction of drought tolerance in Pennisetum glaucum by ACC deaminase producing PGPR-*Bacillus amyloliquefaciens* through Antioxidant defense system. Microbiol Res 253:126891
Mwamahonje A, Eleblu JSY, Ofori K, Deshpande S, Feyissa T, Tongoona P (2021) Drought tolerance and application of marker-assisted selection in sorghum. Biology 10(12):1249
Naseem H, Bano A (2014) Role of plant growth-promoting rhizobacteria and their exopolysaccharide in drought tolerance of maize. Plant Interact 9(1):689–701
Naseem H, Ahsan M, Shahid MA, Khan N (2018) Exopolysaccharides producing rhizobacteria and their role in plant growth and drought tolerance. J Basic Microbiol 58(12):1009–1022
Naseem M, Chaudhry AN, Jilani G, Alam T, Naz F, Ullah R, Zaman S (2024) Exopolysaccharide-producing bacterial cultures of *Bacillus* cereus and *Pseudomonas aeruginosa* in soil augment water retention and maize growth. Heliyon 10(4)
Naveed M, Hussain MB, Zahir ZA, Mitter B, Sessitsch A (2014) Drought stress amelioration in wheat through inoculation with *Burkholderia phytofirmans* strain PsJN. PGR 73(2):121–131
Naylor D, Coleman-Derr D (2018) Drought stress and root-associated bacterial communities. Front Plant Sci 8:2223. https://doi.org/10.3389/fpls.2017.02223
Naylor D, Fansler S, Brislawn C, Nelson WC, Hofmockel KS, Jansson JK, McClure R (2020) Deconstructing the soil microbiome into reduced-complexity functional modules. MBio 11. https://doi.org/10.1128/mbio.01349-20
Neeti N, Murali CA, Chowdary VM, Rao NH, Kesarwani M (2021) Integrated meteorological drought monitoring framework using multi-sensor and multi-temporal earth observation datasets and machine learning algorithms: a case study of Central India. J Hydrol 601:126638
Negri S, Commisso M, Bisson L, Zorzi G, Pietrolucci F, Dusi V, Guzzo F (2024) Phylogenetic-guided bioprospection of the Italian flora: from the exploitation of bioactive phytochemicals to the study of chemo-evolutionary dynamics. In: Book of abstracts, p 109
Nephali L, Moodley V, Piater L, Steenkamp P, Buthelezi N, Dubery I, Tugizimana F (2021) A metabolomic landscape of maize plants treated with a microbial biostimulant under well-watered and drought conditions. Front Plant Sci 12:676632
Nerva L, Sandrini M, Moffa L, Velasco R, Balestrini R, Chitarra W (2022) Breeding toward improved ecological plant–microbiome interactions. Trends Plant Sci 27:1134–1143. https://doi.org/10.1016/j.tplants.2022.06.004
Neuhoff D, Neumann G, Weinmann M (2024) Testing plant growth promoting microorganisms in the field-a proposal for standards. Front Plant Sci 14:1324665

Ngumbi E, Kloepper J (2016) Bacterial-mediated drought tolerance: current and future prospects. Appl Soil Ecol 105:109–125
Nie M, Bell C, Wallenstein MD, Pendall E (2015) Increased plant productivity and decreased microbial respiratory C loss by plant growth-promoting rhizobacteria under elevated CO2. Sci Rep 5(1):9212
Niu X, Song L, Xiao Y, Ge W (2018) Drought-tolerant plant growth-promoting rhizobacteria associated with foxtail millet in a semi-arid agroecosystem and their potential in alleviating drought stress. Front Microbiol 8:2580
Nobori T, Cao Y, Entila F, Dahms E, Tsuda Y, Garrido-Oter R, Tsuda K (2022) Dissecting the cotranscriptome landscape of plants and their microbiota. EMBO Rep 23(12):e55380
Numajiri Y, Yoshino K, Teramoto S, Hayashi A, Nishijima R, Tanaka T, Hayashi T, Kawakatsu T, Tanabata T, Uga Y (2021) iPOTs: internet of things-based pot system controlling optional treatment of soil water condition for plant phenotyping under drought stress. Plant J 107:1569–1580
Nunan N, Schmidt H, Raynaud X (2020) The ecology of heterogeneity: soil bacterial communities and C dynamics. Philos Trans R Soc B 375(1798):20190249
O'Callaghan M, Ballard RA, Wrigh D (2022) Soil microbial inoculants for sustainable agriculture: limitations and opportunities. Soil Use Manag 38(3):1340–1369
Ortiz N, Armada E, Duque E, Roldán A, Azcón R (2015) Contribution of arbuscular mycorrhizal fungi and/or bacteria to enhancing plant drought tolerance under natural soil conditions: effectiveness of autochthonous or allochthonous strains. J Plant Physiol 174:87–96
Pan L, Cai B (2023) Phosphate-solubilizing bacteria: advances in their physiology, molecular mechanisms and microbial community effects. Microorganisms 11:2904
Parizad S, Bera S (2023) The effect of organic farming on water reusability, sustainable ecosystem, and food toxicity. Environ Sci Pollut Res Int 30(28):71665–71676
Paul A, Mondal S, Pal A, Biswas S, Chakraborty K, Mazumder A, Kundu R (2023) Seed priming with NaCl helps to improve tissue tolerance, potassium retention ability of plants, and protects the photosynthetic ability in two different legumes, chickpea and lentil, under salt stress. Planta 257(6):111
Pérez-García A, Romero D, De Vicente A (2011) Plant protection and growth stimulation by microorganisms: biotechnological applications of *Bacilli* in agriculture. Curr Opin Biotechnol 22(2):187–193
Pérez-Lorente AI, Romero D, Molina-Santiago C (2024) Unravelling the impact of environmental factors in shaping plant microbiomes. Microb Biotechnol 17(6):e14504
Porcel R, Barea JM, Ruiz-Lozano JM (2003) Antioxidant activities in mycorrhizal soybean plants under drought stress and their possible relationship to the process of nodule senescence. New Phytol 157(1):135–143
Price V, Ngwira LG, Lewis JM, Baker KS, Peacock SJ, Jauneikaite E, Feasey N (2023) A systematic review of economic evaluations of whole-genome sequencing for the surveillance of bacterial pathogens. Microb Genom 9:000947
Qiu Z, Egidi E, Liu H, Kaur S, Singh BK (2019) New frontiers in agriculture productivity: optimised microbial inoculants and in situ microbiome engineering. Biotechnol Adv 37(6):107371
Rafi MM, Krishnaveni MS, Charyulu PBBN (2019) Phosphate-solubilizing microorganisms and their emerging role in sustainable agriculture. In: Recent developments in applied microbiology and biochemistry, pp 223–233
Raheem A, Shaposhnikov A, Belimov AA, Dodd IC, Ali B (2018) Auxin production by rhizobacteria was associated with improved yield of wheat (*Triticum aestivum* L.) under drought stress. Arch Agron Soil Sci 64(4):574–587
Rapparini F, Peñuelas J (2013) Mycorrhizal fungi to alleviate drought stress on plant growth. In: Miransari M (ed) Use of microbes for the alleviation of soil stresses, vol 1. Springer, New York, pp 21–42
Rigerte L, Heintz-Buschart A, Reitz T, Tarkka MT (2025) Assembly and application of a synthetic bacterial community for enhancing barley tolerance to drought. Front Bacteriol 4:1572294

Rolli E, Marasco R, Vigani G, Ettoumi B, Mapelli F, Deangelis ML, Daffonchio D (2015) Improved plant resistance to drought is promoted by the root-associated microbiome as a water stress-dependent trait. Environ Microbiol 17(2):316–331. https://doi.org/10.1111/1462-2920.12439

Roy N, Debnath P, Gaur HS (2025) Adoption of multi-omics approaches to address drought stress tolerance in rice and mitigation strategies for sustainable production. Mol Biotechnol 68:1–13

Saad MM, Eida AA, Hirt H (2020) Tailoring plant-associated microbial inoculants in agriculture: a roadmap for successful application. J Exp Bot 71(13):3878–3901

Saakre M, Baburao TM, Salim AP, Ffancies RM, Achuthan VP, Thomas G, Sivarajan SR (2017) Identification and characterization of genes responsible for drought tolerance in rice mediated by *Pseudomonas fluorescens*. Rice Sci 24(5):291–298

Saberi Riseh R, Ebrahimi-Zarandi M, Gholizadeh Vazvani M, Skorik YA (2021) Reducing drought stress in plants by encapsulating plant growth-promoting bacteria with polysaccharides. Int J Mol Sci 22(23):12979

Sabir S, Zahoor MA, Waseem M, Siddique MH, Shafique M, Imran M, Hayat S, Malik IR, Muzammil S (2020) Biosynthesis of ZnO nanoparticles using *Bacillus subtilis*: characterization and nutritive significance for promoting plant growth in *Zea mays* L. Dose-Response 18:1559325820958911

Sachdev S, Ansari SA, Ansari MI, Fujita M, Hasanuzzaman M (2021) Abiotic stress and reactive oxygen species: generation, signaling, and defense mechanisms. Antioxidants 10(2):277

Sahoo A, Yadav G, Mehta T, Meena M, Swapnil P (2025) Omics-driven insights into plant growth-promoting microorganisms for sustainable agriculture. Disco Sustain 6(1):1–23

Saikia J, Sarma RK, Dhandia R, Yadav A, Bharali R, Gupta VK, Saikia R (2018) Alleviation of drought stress in pulse crops with ACC deaminase producing rhizobacteria isolated from acidic soil of Northeast India. Sci Rep 8(1):3560

Saleem AR, Brunetti C, Khalid A, Della Rocca G, Raio A, Emiliani G, Centritto M (2018) Drought response of *Mucuna pruriens* (L.) DC. inoculated with ACC deaminase and IAA producing rhizobacteria. PLoS One 13(2):e0191218

Sallam A, Alqudah AM, Dawood MF, Baenziger PS, Börner A (2019) Drought stress tolerance in wheat and barley: advances in physiology, breeding and genetics research. Int J Mol Sci 20(13):3137

Samantaray A, Chattaraj S, Mitra D, Ganguly A, Kumar R, Gaur A, Thatoi H (2024) Advances in microbial based bio-inoculum for amelioration of soil health and sustainable crop production. Curr Res Microb Sci 7:100251

Sandhu N, Dixit S, Swamy BPM, Raman A, Kumar S, Singh SP, Anandan A, Yadav S, Singh VK, Catolos M, Kumar A (2019) Marker-assisted breeding to develop multiple stress-tolerant varieties for food- and drought-prone areas. Rice 12:1–16. https://doi.org/10.1186/s12284-019-0270-4

Sandhya VZAS, SK. ZA, Grover M, Reddy G, Venkateswarlu BSSS (2009) Alleviation of drought stress effects in sunflower seedlings by the exopolysaccharides producing *Pseudomonas putida* strain GAP-P45. Biol Fertil Soils 46(1):17–26

Sandhya V, Reddy GP, Ali SZ, Kumar P (2015) Traditional knowledge and sustainable agriculture: the strategy to cope with climate change. In: Envisioning a future without food waste and food poverty: societal challenges. Wageningen Academic Publishers, pp 109–118

Sarkar MM, Mathur P, Mitsui T, Roy S (2022) A review on functionalized silica nanoparticle amendment on plant growth and development under stress. PGR 98:421–437

Sati D, Pande V, Pandey SC, Samant M (2023) Recent advances in PGPR and molecular mechanisms involved in drought stress resistance. J Soil Sci Plant Nutr 23:106–124

Scheben A, Batley J, Edwards D (2017) Genotyping-by-sequencing approaches to characterize crop genomes: choosing the right tool for the right application. Plant Biotechnol J 15(2):149–161. https://doi.org/10.1111/pbi.12645

Schikora A, Schenk ST, Hartmann A (2016) Beneficial effects of bacteria-plant communication based on quorum sensing molecules of the N-acyl homoserine lactone group. Plant Mol Biol 90:605–612. https://doi.org/10.1007/s11103-015-0413-8

Schimel JP (2018) Life in dry soils: effects of drought on soil microbial communities and processes. Annu Rev Ecol Evol Syst 49(1):409–432

Schmid J, Sieber V, Rehm B (2015) Bacterial exopolysaccharides: biosynthesis pathways and engineering strategies. Front Microbiol 6:496

Seenivasagan R, Babalola OO (2021) Utilization of microbial consortia as biofertilizers and biopesticides for the production of feasible agricultural product. Biology 10(11):1111

Seok HY, Nguyen LV, Nguyen DV, Lee SY, Moon YH (2020) Investigation of a novel salt stress-responsive pathway mediated by *Arabidopsis* DEAD-box RNA helicase gene AtRH17 using RNA-Seq analysis. Int J Mol Sci 5:1595

Sessitsch A, Pfaffenbichler N, Mitter B (2019) Microbiome applications from lab to field: facing complexity. Trends Plant Sci 24(3):194–198. https://doi.org/10.1016/j.tplants.2018.12.006

Shaffique S, Khan MA, Imran M, Kang SM, Park YS, Wani SH, Lee IJ (2022) Research progress in the field of microbial mitigation of drought stress in plants. Front Plant Sci 13:870626

Shahid M, Singh UB, Khan MS, Singh P, Kumar R, Singh RN, Singh HV (2023) Bacterial ACC deaminase: insights into enzymology, biochemistry, genetics, and potential role in amelioration of environmental stress in crop plants. Front Microbiol 14:1132770

Shahroz M, Ashiq M, Infal M, Riaz A, Chaudhry Z, Hassan MN, Rana BA, Arif A, Aslam R, Ulah Q (2024) Nanoparticles mediated modulation of nitrogen fixation in legumes: a biochemical perspective. Glob Acad J Agric Biosci 6:119

Sharifi R, Ryu CM (2018) Revisiting bacterial volatile-mediated plant growth promotion: lessons from the past and objectives for the future. Ann Bot 122(3):349–358

Sharma MP, Grover M, Chourasiya D, Bharti A, Agnihotri R, Maheshwari HS, Pareek A, Buyer JS, Sharma SK, Schütz L (2020) Deciphering the role of trehalose in tripartite symbiosis among rhizobia, arbuscular mycorrhizal fungi, and legumes for enhancing abiotic stress tolerance in crop plants. Front Microbiol 11:509919

Shi F, Dong Y, Wang M, Qiu D (2020) Transcriptomics analyses reveal that OsMIOX improves rice drought tolerance by regulating the expression of plant hormone and sugar related genes. Plant Biotechnol Rep 14(3):339–349

Shinwari ZK, Jan SA, Nakashima K, Yamaguchi-Shinozaki K (2020) Genetic engineering approaches to understanding drought tolerance in plants. Plant Biotechnol Rep 14:151–162

Shirinbayan S, Khosravi H, Malakouti MJ (2019) Alleviation of drought stress in maize (*Zea mays*) by inoculation with Azotobacter strains isolated from semi-arid regions. Appl Soil Ecol 133:138–145

Shravani K, Triveni S, Latha P, Ramulu V, Tejashree M, Damodharachari K (2019) Influence of biofertilizers under drip fertigation system on NPK uptake and yield performance of greengram (*Vigna radiata* L.). IJCMAS 8:2409–2417

Singh M, Kumar J, Singh S, Singh VP, Prasad SM (2015) Roles of osmoprotectants in improving salinity and drought tolerance in plants: a review. Rev Environ Sci Biotechnol 14:407–426

Singh DP, Singh V, Gupta VK, Shukla R, Prabha R, Sarma BK, Patel JS (2020) Microbial inoculation in rice regulates antioxidative reactions and defense related genes to mitigate drought stress. Sci Rep 10(1):4818

Singh V, Gupta K, Singh S, Jain M, Garg R (2023) Unravelling the molecular mechanism underlying drought stress response in chickpea via integrated multi-omics analysis. Front Plant Sci 14:1156606

Song XP, Yan MX, Liang Q, Zhang XQ, Li CN, Malviya MK, Li YR (2025) Recent advances in employing plant rhizobacteria for environmental stress mitigation in plants. Plant Stress 17:100947

Sood G, Kaushal R, Sharma M (2020) Alleviation of drought stress in maize (*Zea mays* L.) by using endogenous endophyte *Bacillus subtilis* in North West Himalayas. Acta Agric Scand B Soil Plant Sci 70(5):361–370

Sorongpong S, Debnath S, Rahi P, Bera B, Pandey P (2025) Sustainable tea cultivation with a Rhizobacterial consortium: a microbiome-driven alternative to chemical fertilizers. Microorganisms 13(8):1715

Stočes M, Vaněk J, Masner J, Pavlík J (2016) Internet of things (iot) in agriculture-selected aspects. AGRIS on-line Pap Econ Inform 8:83–88

Sun X, Zhang J (2021) Bacterial exopolysaccharides: chemical structures, gene clusters and genetic engineering. Int J Biol Macromol 173:481–490

Sun W, Dou F, Li C, Ma X, Ma LQ (2021) Impacts of metallic nanoparticles and transformed products on soil health. Crit Rev Environ Sci Technol 51:973–1002

Sun H, Zhang J, Liu W, EW WX, Li H, Cui Y, Zhao D, Liu K, Du B (2022) Identification and combinatorial engineering of indole-3-acetic acid synthetic pathways in *Paenibacillus polymyxa*. iotechnol biofuels bioprod 15:8

Tariq A, Guo S, Farhat F, Shen X (2025) Engineering synthetic microbial communities: diversity and applications in soil for plant resilience. Agronomy 15:513

Terefe HA (2012) People in crises: tackling the root causes of famine in the Horn of Africa. Nor Agric Econ Res Inst, pp 1–16

Terra MDCNS, Santo RMD, Prado Júnio JAD, de Mello JM, Scolforo JRS, Fontes MAL, ter Steege H (2018) Water availability drives gradients of tree diversity, structure and functional traits in the Atlantic–Cerrado–Caatinga transition, Brazil. J Plant Ecol 11(6):803–814

Thingujam D, Gouli S, Cooray SP, Chandran KB, Givens SB, Gandhimeyyan RV, Mukhtar MS (2025) Climate-resilient crops: integrating AI, multi-omics, and advanced phenotyping to address global agricultural and societal challenges

Timmusk S, Paalme V, Pavlicek T, Bergquist J, Vangala A, Danilas T, Nevo E (2011) Bacterial distribution in the rhizosphere of wild barley under contrasting microclimates. PLoS One 6(3):e17968. https://doi.org/10.1371/journal.pone.0017968

Timmusk S, Abd El-Daim IA, Copolovici L, Tanilas T, Kännaste A, Behers L (2014a) Drought-tolerance of wheat improved by rhizosphere bacteria from harsh environments: enhanced biomass production and reduced emissions of stress volatiles

Timmusk S, Behers L, Muthoni J, Muraya A, Aronsson AC (2014b) Perspectives and challenges of microbial application for crop improvement. Front Plant Sci 5:368. https://doi.org/10.3389/fpls.2014.00368

Ullah A, Nisar M, Ali H, Hazrat A, Hayat K, Keerio AA, Ihsan M, Laiq M, Ullah S, Fahad S (2019) Drought tolerance improvement in plants: an endophytic bacterial approach. Appl Microbiol Biotechnol 103:7385–7397

Ullah F, Ali S, Siraj M, Akhtar MS, Zaman W (2025) plant microbiomes alleviate abiotic stress-associated damage in crops and enhance climate-resilient agriculture. Plants 14(12):1890

Vahdati K, Lotfi N (2013) Abiotic stress tolerance in plants with emphasizing on drought and salinity stresses in walnut. In: Vahdati K, Leslie C (eds) Abiotic stress: plant responses and applications in agriculture. IntechOpen, Rijeka, pp 307–365

Van Dijk AI, Beck HE, Crosbie RS, De Jeu RA, Liu YY, Podger GM, Viney NR (2013) The millennium drought in Southeast Australia (2001–2009): natural and human causes and implications for water resources, ecosystems, economy, and society. Water Resour Res 49(2):1040–1057

Vanderstraeten L, Depaepe T, Bertrand S, Van Der Straeten D (2019) The ethylene precursor ACC affects early vegetative development independently of ethylene signaling. Front Plant Sci 10:1591

Vardharajula S, Zulfikar Ali S, Grover M, Reddy G, Bandi V (2011) Drought-tolerant plant growth promoting *Bacillus* spp.: effect on growth, osmolytes, and antioxidant status of maize under drought stress. J Plant Interact 6(1):1–14

Vassileva M, Malusà E, Sas-Paszt L, Trzcinski P, Galvez A, Flor-Peregrin E, Vassilev N (2021) Fermentation strategies to improve soil bio-inoculant production and quality. Microorganisms 9(6):1254

Vocciante M, Grifoni M, Fusini D, Petruzzelli G, Franchi E (2022) The role of plant growth-promoting rhizobacteria (PGPR) in mitigating plant's environmental stresses. Appl Sci 12(3):1231

Vurukonda SSKP, Vardharajula S, Shrivastava M, SkZ A (2016) Enhancement of drought stress tolerance in crops by plant growth promoting rhizobacteria. Microbiol Res 184:13–24

Wang DC, Jiang CH, Zhang LN, Chen L, Zhang XY, Guo JH (2019) Biofilms positively contribute to *Bacillus amyloliquefaciens* 54-induced drought tolerance in tomato plants. Int J Mol Sci 20(24):6271

Wang S, Chen K, Zhang J, Wang J, Li H, Yang X, Shi Q (2022) Genome-wide characterization of MATE family members in *Cucumis melo* L. and their expression profiles in response to abiotic and biotic stress. Horti Plant J 8(4):474–488

Wankhade A, Wilkinson E, Britt DW, Kaundal A (2025) A review of plant–microbe interactions in the rhizosphere and the role of root exudates in microbiome engineering. Appl Sci 15:7127. https://doi.org/10.3390/app15137127

Wu S, Hu C, Tan Q, Li L, Shi K, Zheng Y, Sun X (2015) Drought stress tolerance mediated by zinc-induced antioxidative defense and osmotic adjustment in cotton (*Gossypium hirsutum*). Acta Physiol Plant 37(8):167

Wu L, Li X, Ma L, Borriss R, Wu Z, Gao X (2018) Acetoin and 2, 3-butanediol from *Bacillus amyloliquefaciens* induce stomatal closure in *Arabidopsis thaliana* and *Nicotiana benthamiana*. J Exp Bot 69(22):5625–5635

Xiong L, Lee H, Ishitani M, Zhu JK (2002) Regulation of osmotic stress-responsive gene expression by thelos6/aba1 locus in *Arabidopsis*. J Biol Chem 277(10):8588–8596

Xu J, Li XL, Luo L (2012) Effects of engineered *Sinorhizobium meliloti* on cytokinin synthesis and tolerance of alfalfa to extreme drought stress. AEM 78:8056–8061

Yang Y, Wang J, Xiu Z, Alvarez PJ (2013) Impacts of silver nanoparticles on cellular and transcriptional activity of nitrogen-cycling bacteria. Environ Toxicol Chem 32:1488–1494

Yasir M, He S, Sun G, Geng X, Pan Z, Gong W, Du X (2019) A genome-wide association study revealed key SNPs/genes associated with salinity stress tolerance in upland cotton. Genes 10(10):829

Yasmin H, Rashid U, Hassan MN, Nosheen A, Naz R, Ilyas N, Alyemeni MN (2021) Volatile organic compounds produced by *Pseudomonas pseudoalcaligenes* alleviated drought stress by modulating defense system in maize (*Zea mays* L.). Physiol Plant 172(2):896–911

Yasmin H, Bano A, Wilson NL, Nosheen A, Naz R, Hassan MN, Kennedy I (2022) Drought-tolerant *Pseudomonas* sp. showed differential expression of stress-responsive genes and induced drought tolerance in *Arabidopsis thaliana*. Physiol Plant 174:e13497. https://doi.org/10.1111/ppl.13497

Yi HS, Ahn YR, Song GC, Ghim SY, Lee S, Lee G, Ryu CM (2016) Impact of a bacterial volatile 2, 3-butanediol on *Bacillus subtilis* rhizosphere robustness. Front Microbiol 7:993

Yooyongwech S, Theerawitaya C, Samphumphuang T, Cha-um S (2013) Water-deficit tolerant identification in sweet potato genotypes (*Ipomoea batatas* (L.) Lam.) in vegetative developmental stage using multivariate physiological indices. Sci Hortic 162:242–251

You J, Zhang Y, Liu A, Li D, Wang X, Dossa K, Zhang X (2019) Transcriptomic and metabolomic profiling of drought-tolerant and susceptible sesame genotypes in response to drought stress. BMC Plant Biol 19(1):267

Zahedi SM, Karimi M, Venditti A, Zahra N, Siddique KH, Farooq M (2025) Plant adaptation to drought stress: the role of anatomical and morphological characteristics in maintaining the water status. J Soil Sci Plant Nutr 25(1):409–427

Zancarini A, Westerhuis JA, Smilde AK, Bouwmeester HJ (2021) Integration of omics data to unravel root microbiome recruitment. Curr Opin Biotechnol 70:255–261

Zandalinas SI, Balfagón D, Arbona V, Gómez-Cadenas A, Inupakutika MA, Mittler R (2016) ABA is required for the accumulation of APX1 and MBF1c during a combination of water deficit and heat stress. J Exp Bot 67(18):5381–5390

Zargar SM, Mir RA, Ebinezer LB, Masi A, Hami A, Manzoor M, Rakwal R (2022) Physiological and multi-omics approaches for explaining drought stress tolerance and supporting sustainable production of rice. Front Plant Sci 12:803603. https://doi.org/10.3389/fpls.2021.803603

Zhang W (2024) Strategic engineering of synthetic microbial communities (SynComs) for optimizing plant health and yield in agriculture. Mol Microbiol Res 14

Zhang Y, Du H, Xu F, Ding Y, Gui Y, Zhang J, Xu W (2020a) Root-bacteria associations boost rhizosheath formation in moderately dry soil through ethylene responses. Plant Physiol 183(2):780–792

Zhang Y, Li Y, Hassan MJ, Li Z, Peng Y (2020b) Indole-3-acetic acid improves drought tolerance of white clover via activating auxin, abscisic acid and jasmonic acid related genes and inhibiting senescence genes. BMC Plant Biol 20(1):150

Zhang Y, Hao Z, Feng S, Zhang X, Xu Y, Hao F (2021) Agricultural drought prediction in China based on drought propagation and large-scale drivers. Agric Water Manag 255:107028

Zhang H, Sun X, Dai M (2022) Improving crop drought resistance with plant growth regulators and rhizobacteria: mechanisms, applications, and perspectives. Plant Commun 3(1):100228

Zhang Y, Mo C, Pan Y, Yang P, Ding X, Lei Q, Kang P (2023) Responses of soil microbial survival strategies and functional changes to wet–dry cycle events. Microorganisms 11(11):2783

Zhao S, Wei H, Lin CY, Zeng Y, Tucker MP, Himmel ME, Ding SY (2016) *Burkholderia phytofirmans* inoculation-induced changes on the shoot cell anatomy and iron accumulation reveal novel components of *Arabidopsis*-endophyte interaction that can benefit downstream biomass deconstruction. Front Plant Sci 7:24

Chapter 10
Engineering the Rhizomicrobiome for Drought-Resilient Soil and Sustainable Agriculture

Izzah Shahid and Samina Mehnaz

Abstract Depleting freshwater reservoirs and worsening droughts are major threats to global agricultural productivity. Water unavailability contributes to significant yield losses by inducing physiological changes in plants and damaging root and shoot architecture. Current farming practices are unsustainable and cannot address the challenges posed by droughts. Hence, consistent and sustainable approaches are required to develop climate-resilient, high-yielding crops with improved drought tolerance. The rhizomicrobiome comprises plant-associated bacterial communities that regulate plant productivity through nutrient uptake, induction of systemic immune resistance, conferment of stress tolerance against abiotic and biotic stressors, and secretion of growth-stimulating compounds. Rhizobacteria can impart drought tolerance through the production of phytohormones, exopolysaccharides (EPS), aminocyclopropane-1-carboxylate (ACC) deaminase, volatile organic compounds (VOCs), accumulation of osmolytes, and the regulation of plant stress-responsive genes. Engineering plant-associated bacterial communities is a viable, less resource-intensive, and eco-friendly approach for mitigating the damaging effects of drought in plants. Although several plant growth-promoting bacteria have been shown to combat drought stress, the effects of these individual strains are limited, and many have not been promising in field trials. Therefore, a holistic approach is required to engineer rhizobacterial communities that can systematically compete with drought. Contemporary strategies for rhizomicrobiome modulation include "omics" techniques, enrichment of desirable microbial populations in the rhizosphere, engineering of microbial pathways to overproduce drought-

I. Shahid (✉)
Institute of Multidisciplinary Research in Applied Biology, Public University of Navarre, Pamplona, Spain
e-mail: izzah.shahid@unavarra.es

S. Mehnaz
Kauser Abdulla Malik School of Life Sciences, Forman Christian College (A Chartered University), Lahore, Pakistan

S. Mehnaz, I. Shahid (eds.), *Frontiers in Rhizoengineering*, Microorganisms for Sustainability 65, https://doi.org/10.1007/978-981-92-3665-7_10

fighting compounds, and improvisation of bacterial signaling systems. Consistent efforts to engineer rhizosphere bacterial communities for drought mitigation and the development of unique bioinocula can catalyze the achievement of global food security in the face of a changing climate.

Keywords Drought · PGPR · Rhizosphere · Rhizomicrobiome · Secondary metabolites

10.1 Introduction

Drought has become the most alarming threat to global plant productivity due to the current unprecedented climate change. Reduced precipitation rates and altered rainfall patterns cause drought-like conditions worldwide (Hussain and Sulaimon 2018). Long drought exposure can significantly impair plant growth and worsen, in combination with agronomic and edaphic factors, and deplete freshwater reservoirs (Gregersen et al. 2013). Forecasted global changes indicate that drought influences not only food and feed production but also biomass production, which impacts bioenergy crops (Quinn et al. 2015). For instance, drought stress has led to an ~21% decline in the global production of *Triticum aestivum* L. and an ~40% reduction in the yield of *Zea mays* L., which are the major food crops consumed worldwide (Zhang et al. 2018). It impairs plant growth through various physiological, anatomical, and morphological damage mechanisms. In addition to affecting agricultural productivity, drought has a significant effect on soil biotic diversity. It induces a substantial decrease in microbial biomass, limits nutrient mobility and access through the soil, and increases soil heterogeneity (Jansson and Hofmockel 2020).

Plant and soil biotic diversity resist drought stress by developing stress-tolerant strategies that enhance their survival under this grave challenge. Plants trigger certain antioxidant immune systems to eradicate toxic substances and alter the transpiration rates. Similar to plants, soil microbes defend themselves through diverse resistance mechanisms and provide additional help to their plant partners, thereby increasing their survival under drought conditions (Malik et al. 2020). Under water-starved conditions, plants increase mucilage production, alter root exudation chemistry, limit photosynthetic activity, decrease transpiration rates, activate osmoprotective modes, and increase root hydraulic conductance, facilitating antioxidant systems (Williams and de Vries 2020). These responses and effects may or may not contribute to the drought resilience of plants; nonetheless, additional help from the rhizomicrobiome can significantly affect plant performance. Rhizosphere-associated microbes can produce exopolysaccharides (EPS), antioxidants, and osmoprotective secondary metabolites; promote root elongation through phytohormone secretion; ensure holistic reorchestration of water use; and enhance nutrient acquisition via C- and N-scavenging enzymes (Vurukonda et al. 2016). These traits cannot be attributed to a particular microbial genus; thus, their role in drought resilience in crops is attributed to complex interkingdom and species-specific metabolic pathways. This highlights the need for comprehensive rhizomicrobiome engineering approaches to develop climate-smart agricultural systems. This chapter summarizes some of the fundamental and contemporary approaches used for rhizomicrobiome engineering, their scope and efficacy, and the prospects for achieving agricultural sustainability by harnessing rhizosphere microbes.

10.2 Damaging Effects of Drought on Plants

Drought causes morphological, physiological, and oxidative damage to plants. Some of the most preliminary damage can be observed in the form of abridged seedling formation and decreased rates of seed germination. Decreased germination potential, root and shoot weights, poor vegetative growth, undersized hypocotyl length, and overall reduced seedlings have been observed in several crops, including *Medicago sativa* L., *Oryza sativa* L., *Pisum sativum* L., *Zea mays* L., and *Saccharum officinarum* L. (Gregersen et al. 2013; Hatzig et al. 2018). Plants under drought stress demonstrate abnormal cell differentiation and elongation and reduced rates of mitosis, resulting in decreased plant growth. Hampered water movement through xylem tissues and adjacent cells and decreased turgor pressure under restrained water conditions reduce cell elongation and growth, compromising the number of leaves and the leaf surface area. Abnormal physiological changes in plants are indicated by morphological changes that directly affect crop yield. In particular, the stage at which a crop is exposed to drought directly influences its yield and quality. For instance, drought-hit crops exhibit a shortened pre-anthesis stage, causing reduced filling of cereal grains (Shokat et al. 2023). Meager leaf development, reduced rates of photosynthesis, and dehydrated conditions directly influence the functioning of sucrose-UDP glucosyltransferase, ADP glucose pyrophosphorylase, starch branching enzymes, and other enzymes involved in the filling of cereal grains.

Water-starved conditions in plants also directly influence the leaf water potential, stomatal conductance, transpiration rates, and temperature of the canopy and leaves. A clear decline in the transpiration rate can be observed in drought-hit plants, resulting in reduced leaf water potential, eventually intensifying the temperature of the canopy and leaves (Mahmood 2020). This phenomenon indicates higher water adeptness under drought stress to effectively use the available water to balance water uptake and accumulated dry matter. This strategy has been documented for several wheat and rice cultivars, and the expected reason for water-use efficiency is linked to stomatal closure and decreased transpiration rates under water-limited and elevated heat conditions (Melandri et al. 2020; Thakur et al. 2022). In contrast, inappropriate water adeptness was observed in *Solanum tuberosum* L. subjected to water starvation during the early stages of plantation. The plants showed poor yield and significantly reduced biomass, indicating the superiority of cereals in water adeptness over tubers (Obidiegwu et al. 2015). This observation supports the hypothesis that drought conditions substantially affect nutrient association and assimilation in plants. Plants administer several vital nutrients, such as magnesium, silicon, nitrogen, iron, and phosphorus, through roots, alongside water uptake, and channelize them to plant tissues. Limited water conditions disturb the normal nutrient uptake process, resulting in poor yield and plant biomass (Hussain et al. 2019). Owing to abnormal root architecture and atypical development under water-limited conditions, plant roots cannot retain transportable nutrients or channel them properly. Drought also influences the activity of plant-microbial associations, and thus, reduced nitrogen fixation has been observed in several leguminous plants. Drought can challenge the existing soil rhizobia and impact the survival of the inoculated rhizobia as biofertilizers (Thilakarathna et al. 2021). Reduced nodulation and nitrogenase activity in drought-affected plants are associated with a reduction in nodule O_2 permeability, hampered C supply, and inhibition of the nitrogen feedback mechanism (Nishida and Suzaki 2018). In addition, water scarcity not only influences nodulation but also impacts the survival and activities of rhizomicrobiomes, leaving plants vulnerable to disaster.

The detrimental effects of drought are not limited to leaf anatomy and structure. Physiologically, plants face the most critical fate because of the reduced rates of photosynthesis. Abridged leaf surface area, atypical leaf senescence, and compromised water levels in leaves prevent adequate photosynthesis in plants. Stomatal closure curtails the access of plants to CO_2, putting them at risk of photo-injury (Mukarram et al. 2021). Moreover, compromised water levels induce alterations in photosynthetic pigments and mechanisms and hamper the proper functioning of vital photosynthetic enzymes, causing the swelling of stromal lamellae, which in turn disrupts the normal arrangement of thylakoid membranes (Razi and Muneer 2021). A recent study also demonstrated diminished superoxide dismutase (SODs) and chlorophyll contents in durum wheat genotypes under drought exposure (Khayatnezhad and Gholamin 2021). Another study reported decreased chlorophyll fluorescence (CF), total phenolic and flavonoid contents, and antioxidant activities in lettuce plants under drought conditions (Shin et al. 2021). It also indicated that stomatal closure under drought stress upturns heat indulgence in leaves and interferes with CO_2 consumption, resulting in the generation of reactive radical species (ROS), causing oxidative damage to leaves. Water-deficient conditions concentrate and add to the viscidity of the plant cells, diminishing their original volume and resulting in the denaturation of enzymes and proteins. Consequently, increased ionic concentrations and solute levels can be toxic to photosynthetic pigments and enzymes, such as Rubisco (Hussain et al. 2018). Damaged photosynthetic activities disturb the assimilation equilibrium and affect acid invertase activity, thereby disrupting the phloem tissue mechanism. This influences the overall assimilate partitioning mechanism and the carry-over rates of sucrose. These events gradually accumulate ROS within plant tissues, which consequently causes oxidative damage to the plant. These reactive oxygen species are produced as a result of both enzymatic and non-enzymatic reactions and are primarily formed in the cellular chloroplasts (You and Chan 2015). Reactive radicals directly damage functional proteins and lipids and increase peroxidation rates. Simultaneously, ROS accumulation can trigger an enzymatic antioxidant defense system and enable cross talk between several stress-suppressive metabolic pathways. These metabolic pathways upregulate the production of glutathione reductase, catalase, peroxidases, superoxide dismutase, carotenoids, and glutathione, which scavenge free radicals and maintain cellular redox potential (Verma et al. 2019). Acquiring knowledge of the microbiome associated with drought stress and the subsequent engineering of the rhizosphere can be beneficial for creating drought-resilient crops to cope with imminent global water scarcity.

10.3 Integrated Rhizoengineering Practices for Stress Mitigation in Plants

Several methodological advances and integrated approaches have paved the way for the regulation and governance of rhizosphere-associated microbial communities. Overall, rhizo-engineering comprises several plant- and metaorganism-based approaches, in addition to engineering the rhizomicrobiome, for better plant growth and stress tolerance. Plant-based approaches include selective plant breeding and the development of high-yielding cultivars that can alter plant resistance to diseases and environmental factors. Often, cultivars with an enhanced ability to develop mutual symbiosis are selected in the process and are advantageous because they do not require infrastructural and management changes in field sites. Moreover, the development of high-yield cultivars can influence rhizosphere microbial diversity by enhancing the growth of selected microbial species present in the rhizosphere. These cultivars with improved tolerance can better resist adverse edaphic, climatic, and biological stressors and ensure improved nutrient availability to growing plants (Dubey et al. 2019). However, none of the breeding programs have evaluated the developed varieties for their interaction with rhizomicrobiomes, and their role in shaping microbial communities remains largely unknown. In addition, these cultivars have often been reported to produce undesirable results and have detrimental effects on plants under high-nutrient conditions. Another approach is to alter the levels of signaling molecules, organic root exudates, and compounds entering the soil through genetic modification. Maneuvering usually results in enhanced secretion of root exudates and quorum-sensing signaling molecules that attract plant growth-promoting microbial diversity, modify soil properties, including pH and anionic efflux of the roots, and prevent probable microbial infections (Kumar et al. 2015). Genetically engineered plants have increased microbial diversity and abundance of plant growth-promoting bacteria associated with their roots, enhanced bioremediation properties, improved plant growth parameters, endurance to unfavorable soil conditions and pathogens, and resilience against climatic stressors. However, rapid degradation and inactivation of released root exudates and significantly low levels of exudation that influence the rhizosphere microbiome have been documented as limiting factors. In addition, the targeted genes do not always lead to the accumulation of anions and enhanced efflux and require additional activation factors to perform the targeted activity. Genetic manipulation of quorum sensing networks has been shown to block or hamper the recruitment of beneficial microbes in the rhizosphere, delivering opposite results, and researchers are still perplexed about engineering plants because of the likely integration of transgenic DNA into microbial communities (Zhang et al. 2015).

The management and selection of complementary plants and microbes through crop rotation is categorized as a metaorganism-based approach. This two-way approach involves engineering plants to produce certain compounds and microbes to degrade these compounds. The benefits of these techniques include improved physicochemical properties of the soil, higher levels of nutrient cycling, elevation of organic carbon content, and the use of organic fertilizers to enhance the soil's biological activities. However, the mechanisms leading to these changes are poorly understood, and certain biosolids can be toxic to microflora, yielding contrasting results. Moreover, several agricultural inputs significantly influence microbial diversity and can be challenging to use in appropriate proportions without harming the rhizomicrobiome of the plant. These factors have led to the development of microbe-mediated strategies to increase plant resilience against edaphic and environmental stressors, especially drought, and several promising results have been reported (Fig. 10.1).

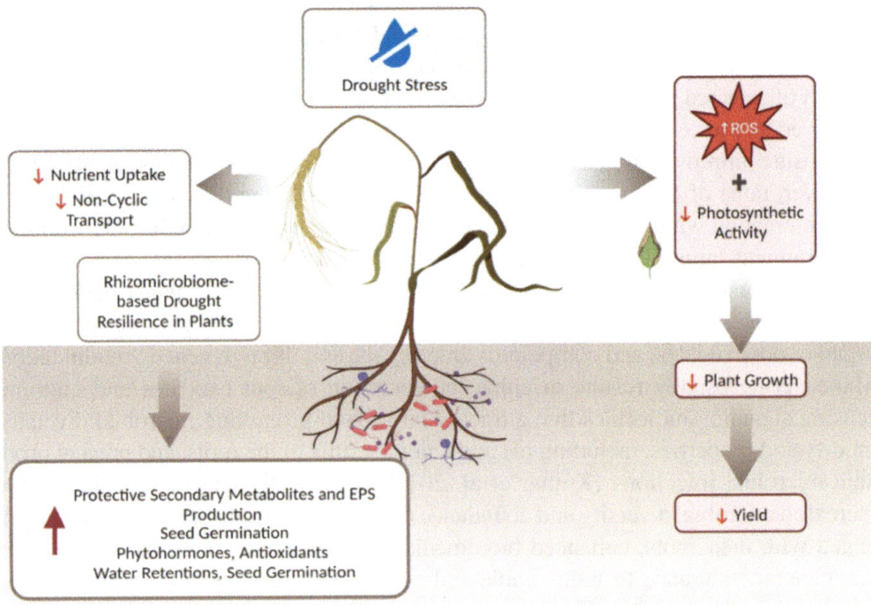

Fig. 10.1 Impact of drought on plant health and the expected role of rhizomicrobiomes in drought mitigation

10.4 Rhizomicrobiome

The rhizosphere is the most significant biological hotspot, harboring up to 10^{11} microbial cells/g root and 10^{12} functional genes/gram of soil belonging to approximately 30,000 prokaryotic inhabitants (Prosser 2015). Rhizosphere microbiota often refers to the vast diversity of microbes inhabiting plant roots and the associated soil (Uzoh and Babalola 2018). The rhizomicrobiome intervenes in and controls several biogeochemical and soil processes and is the driver of soil organic matter decomposition (Li et al. 2019). Because of its tremendous function, diversity, and magnitude, the rhizomicrobiome is referred to as the second genome of the plant, which assists in unpredictable circumstances. The abundance and diversity of specific root-associated microflora are dependent on several factors, including the constituents of root exudates, quorum sensing, inter- and intraspecies competence, survival, and previous inhabitants (Sánchez-Cañizares et al. 2017). Root-associated microflora are not limited to rhizoplane regions and can also travel to the apoplast of the cortex and the aerial parts of plants, constituting endophytic and phyllosphere microbiomes. These microbes perform divergent duties, including organic matter disintegration, rhizoremediation, bioremediation, xenobiotic degradation, toxin exclusion, pathogen elimination, and nutrient cycling. Most of these microbes are cultured and exploited for their desired functions; however, not all root-adhered microflora are culturable.

Contemporary molecular biology approaches have revealed the potential of approximately 99% unculturable microbial diversity for plant growth promotion. The remaining 1% of culturable microbial diversity comprises fungal and bacterial genera, among which the dominant genera include *Pseudomonas*, *Serratia*, *Xanthomonas*, *Enterobacter*, *Bacillus*, *Azospirillum*, *Burkholderia*, *Azotobacter*, *Chromobacterium*, and *Flavobacterium* (Dutta and Bora 2019). Moreover, the latest techniques have shown close synergism of microbial communities in exerting proper functions; hence, understanding microbial assemblages and the basis of community formation is very significant from an ecological perspective. Insights into the bacterial communities of different plants, including peas, potatoes, alfalfa, wheat, tomato, and corn, have shown the abundance and dominance of *Gammaproteobacteria*, indicating their high root-colonizing and metabolic potential (Sharma et al. 2021). These microbial communities can be engineered to increase plant resilience to several abiotic and biotic stressors and can be designed to prepare synthetic microbial communities for future applications.

10.4.1 Plant-Rhizomicrobiome Interactions

Plant roots thrive and penetrate deep into the soil, altering its physical composition, moisture levels, key nutrients, and microbial dynamics. The chemical composition of exudates encourages the recruitment of better-suited microbes that share the responsibility of increasing the plant's nutrient accessibility, increasing environmental fitness, and excluding harmful flora. This assimilated ecological alliance is acknowledged as a holobiont system, which gives rise to an inseparable and devoted setup (Vandenkoornhuyse et al. 2015). This plant-microbe alliance is mostly categorized into three distinct classes: mutualistic, which helps plants through diverse plant growth-promoting attributes; parasitic, responsible for imparting plant-damaging effects; and commensalism, which does not influence plant systems. Nonetheless, these interactions account for only the direct influence of the microbiome on plants and exclude indirect interactions occurring because of unculturable microbes and communication occurring inside plant genomes in association with close microflora. Microbes residing in the internal tissues of plants can exert numerous growth-promoting interactions, stimulate nutrient procurement, and contribute to plant immunity against certain abiotic and biotic stressors (Lebeis 2015). In contrast, rhizosphere-inhabiting microbes not only influence plants but their interactions also shape the composition and diversity of rhizomicrobiomes.

Dominant rhizosphere microbes can hunt phytopathogenic microbes and nematodes through the mechanistic production of antimicrobial metabolites and can induce systemic resistance in plants for better environmental adaptability and survival. In addition to safeguarding plants from potential predators, these microbial inhabitants are plant allies that forbear several abiotic stresses and possess remarkable biofertilization capacities. These microbes can also positively influence root progression, aid nutrient acquisition, perform rhizoremediation, and are important partners of the plant system in maintaining health (Basu et al. 2021; Vocciante et al. 2022). These plant growth-promoting rhizobacteria (PGPR) facilitate smooth nutrient supply to plants through zinc, phosphate, and potassium solubilization, nitrogen fixation, and production of plant hormones, including auxins, gibberellins, abscisic acid, cytokinins, and iron chelation. In addition, PGPB can produce diverse exopolysaccharides and active enzymes, such as ACC deaminase, which enhance plant resilience (Backer et al. 2018). Based on the type of beneficial traits, PGPB can be classified as biostimulants that promote plant growth through certain chemical hormones and regulators, biofertilizers, microbes that can solubilize and mobilize insoluble minerals to plants, and biocontrol agents that can protect plants from certain phytopathogens via the production of antibiotics and siderophores and can induce systemic resistance in plants (Santoyo et al. 2021). However, most soil microbes possess multifaceted properties and act simultaneously as biofertilizers and biocontrol agents. Likewise, many microbes can induce systemic resistance in plants against abiotic stresses through the accumulation of protective metabolites, in addition to playing their role as mineral mobilizers. Hence, researchers have focused on targeting microbes that possess manifold attributes and can enhance the resilience of plants against abiotic and biotic stressors.

10.4.2 Crafting Rhizomicrobiome for Increased Drought Resilience

As drought has become one of the gravest challenges faced by future cropping systems, researchers are working relentlessly to develop crops that require minimal irrigation or can withstand water starvation. Among these approaches, genetic manipulation of plant genomes and selective breeding have been extensively studied and have shown promising results. However, these expensive techniques have often not shown the desired outcome because of several undocumented and documented reasons, giving rise to microbiome editing for developing drought-tolerant future agronomic systems. Rhizomicrobiome crafting thus refers to the editing of root-adjoining and in-vicinity microbes to improve the endurance and yield of the plants. Several methodological advancements can be considered for crafting rhizomicrobiomes to enhance drought tolerance and active functioning under water-challenging conditions. Functional genomic approaches have enabled a deeper investigation of the microflora associated with the rhizosphere and the differentiation between active and inactive microbes under water-limited conditions. These techniques were revolutionized by the integration of omics technologies, which allow two-way verification and enhanced liberty to exploit root-associated microbes for developing climate-smart and starvation-resilient crops (Fig. 10.2). Some of these techniques for engineering and exploiting useful microbes are detailed below, and their advantages, disadvantages, potential targets, and mechanisms are outlined in Table 10.1.

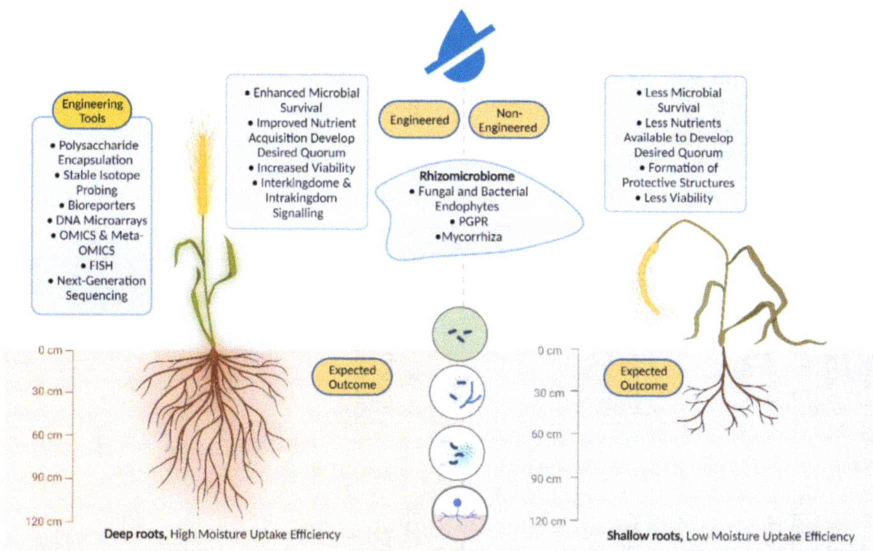

Fig. 10.2 Engineered rhizomicrobiomes promote plant health and resilience under drought conditions

Table 10.1 List of commonly used tools to understand and engineer rhizomicrobiome for enhanced drought resilience

Techniques	Strategy	Mechanism and types	Merits	Demerits	Targets	References
Stable isotope probing (SIP)	^{13}C-, ^{15}N-, $^{13}CO_2$-, ^{18}O-labeled isotopes	Generation of labeled DNA or RNA, DNA-SIP, RNA-SIP, protein-SIP, chip-SIP, FAME-SIP, SC-SIP, qSIP, HT-SIP	Insights into actively growing resilient microbes and metabolic pathways	Low yields of retrieved mRNAs, unable to provide spatial information, some of them are destructive methods	DNA, RNA, amino acids, lipids, single cells	Alcolombri et al. (2022), Nuccio et al. (2022), Papp et al. (2018)
Microarrays	Identification of gene expression	Fluorescently labeled cDNA	Analysis of differential expression of genes, high sensitivity and throughput, no special equipment needed	Time-consuming, only produces chipped genes, cross-hybridization with minimal homologous sequencing	DNA, RNA, amino acids	Babalola et al. (2020), Enespa (2022)
Bioreporters/ biosensors	In situ analysis of microbial communities	Fusion of reporter gene with promoter of interest	Indication of active colonizers and continuous monitoring of desired PGPR, PCR-independent visual inspection, high sensitivity, low cost	Limited information on reporter stability, detection limits, and toxicity. Sometimes destructive and require sample extraction from plant sources	Salicylic acid (SA), cell wall proteins, DNA, RNA, single-stranded DNA, indole-3-acetic acid, galactosidases	Razmi et al. (2019), Chu et al. (2022), Pini et al. (2017)

Genomics and metagenomics	Genome sequencing and sequencing of unculturable microbial diversity	Genome sequencing of culturable bacteria and unculturable microbial diversity using whole genome sequencing platforms including Illumina and PacBio, denaturation gel gradient electrophoresis (DGGE), terminal restriction fragment length polymorphism (t-RFLP), pyrosequencing, internal transcribed spacers (ITS) sequencing	Identification of drought-tolerant genes and indication of the role and nature of unculturable microbial diversity, tons of data to compare with the diversity of genomic sequences	Not error-prone, time-consuming, costly, diverse data interpretations	Whole genome sequencing, DNA, genes, environmental genes and genomes	Fadiji and Babalola (2020), Priya et al. (2021), López-Mondéjar et al. (2017)
Proteomics and metaproteomics	Analysis of protein expression	Analysis of accumulated protein pools for insights into active peptides mitigating drought stress	Rapid detection of active proteins/ metabolites for their potential role-play in drought amelioration	Tools for accurate quantification and sample comparison are underrepresented in metaproteomics, lack of standards	Total protein content of the cell, in situ carbon sources of community members	Herschend et al. (2017), Schiebenhoefer et al. (2019) Djemiel et al. (2022)

(Continued)

Table 10.1 (Continued)

Techniques	Strategy	Mechanism and types	Merits	Demerits	Targets	References
Metabolomics	Analysis of secondary metabolite expression and assessment of associated genes and metabolic pathways. Top-down, bottom-up, LC-MS, GC-MS, NMR, HPLC	Analysis of primary and secondary metabolites of PGPR produced in response to specific drought stress and assessment of metabolic pathways	Selection of specific and useful PGPR as bioinocula to ameliorate drought stress. Assessment of growth kinetics and rapid production of desirable metabolites	Complex data-processing, misleading identification of a single responsible compound for a particular trait	Primary and secondary metabolites	Chen et al. (2021), Gupta et al. (2022), Pang et al. (2021)
Transcriptomics and metatranscriptomics	Assessment of active differentially expressed transcripts and potential gene targets, expressed sequence tag (EST) analysis, MiSeq, RNAseq transcriptional profiles	Analysis of active genes and transcripts produced in response to drought stress. Comparative analysis of different PGPR for the production of metabolites to withstand drought	Rapid assessment of active transcripts and differentially expressed genes in response to certain drought challenges and development of microbial consortia to be used as bioinocula	Intensive computational resources and expertise, lack of fully developed analytical methodology, software tailored approach	Whole transcriptome of communicating microbial communities	Xu et al. (2021), Lucaciu et al. (2019), Shakya et al. (2019)

10.4.2.1 Encapsulation of Rhizobacteria

Encapsulation promotes the stabilization, protection, and functionalization of rhizobacteria and enhances their viability and shelf life (Arora and Mishra 2016). Encapsulated bioinocula are slowly released into the soil and can increase plant fitness under drought conditions compared to non-encapsulated bioinocula. Single-strain biofertilizers or inocula based on microbial consortia can be encapsulated using carrier materials, including carrageenan, sodium-alginate-humic acid, starch, sodium alginate, peats, clays, and polysaccharides. Nonetheless, microbial encapsulation in peat has shown a sharp decline in bacterial concentration and damaging effects on the quality of the developed formulation (Szczech and Maciorowski 2016). Similarly, liquid inocula have shown decreased microbial survival rates (Rojas-Sánchez et al. 2022). In contrast, polysaccharides have been shown to increase the survival of rhizobacteria and have promising effects in mitigating drought and other abiotic stresses in plants. Polysaccharide-encapsulated microbial fertilizers demonstrate increased resistance to acidic environments, enzymes, nucleases, electrostatic interactions, ion-induced gelation, and prebiotic properties (Saberi Riseh et al. 2021). In addition, polysaccharide-based hydrogels, such as chitosan, sodium alginate, and starch, can hold a large amount of water within the interstitial sites of their structures and show promising compatibility with plant needs exposed to drought conditions (Kabir et al. 2018). Hence, engineered bioinocula with polysaccharides can show superior survival rates, viability, and efficacy of the developed formulation and can improve plant resilience under drought conditions.

10.4.2.2 DNA Microarray

Nucleic acid arrays are widely used in gene expression studies and comprise small glass-based or silicon chips. Thousands of small oligonucleotide probes adhere to these tiny chips as spots and produce fluorescence, indicating actively transcribed genes. For this, labeled ssRNA or ssDNA from the sample of interest was obtained and subjected to hybridization on the chips under stringent conditions, and the resultant fluorescence was analyzed as a heat map (Cottier et al. 2018). Based on this technique, functional gene arrays have been developed to visualize the active rhizosphere microbiome that is amenable to engineering. Comparative quantification provides insights into the more active and abundant microbes at a particular time. The rapid development of new selective chips for targeted microarrays has expanded our understanding of complex microbial dynamics and synergism (Shi et al. 2019). Sensitive and specific microarray chips can be developed for the analysis of drought-tolerant communities and culturable microbiomes to develop bioinocula for future crops.

10.4.2.3 Fluorescent In Situ Hybridization (FISH) Method

Similar to the DNA microarray methodology, FISH quantifies and detects active nucleic acids in situ. Target cells were fixed and treated for permeation to facilitate the entry of single-stranded nucleic acid probes that hybridized with complementary sequences and emitted fluorescence. Targets and cells were visualized using flow cytometry or epifluorescence microscopy. Thus, hybridization provides an effective method for estimating and detecting active DNA under native cell conditions. This technique has been previously employed to analyze the composition of diverse plant microbiomes (Santos and Olivares 2021). However, one of the limitations of this technique is its compliance with natural habitats for investigating endophytic microbes, which hampers the laboratory isolation process of rhizobacteria. Nonetheless, recent developments have enhanced the efficiency of this method and provided insights into the active endophytic microflora. A recent investigation demonstrated the development of double-labeled oligonucleotide probes yielding DOPE-FISH single-stranded probes, which offered detailed insights into culturable bacterial taxa and their putative biological functions (Darriaut et al. 2022). Similarly, another study showed the active colonization of *Medicago truncatula* plants by an efficient endophyte, *Arthrobacter agilis*, using FISH and qPCR (Aviles-Garcia et al. 2016). In addition, recent advances in the technique have allowed the use of mRNA FISH and microautoradiography-assisted FISH for the analysis of single-cell gene expression of an active microbial community, thereby enhancing the culture possibilities of the targeted microbes. This technique can be specifically utilized to visualize active microflora under drought conditions and accurately target culturable drought-tolerant microbes.

10.4.2.4 Next-Generation Sequencing (NGS) Techniques

The first DNA sequencing method was the Sanger sequencing method, which is based on the addition of fluorescently labeled dideoxynucleotides at the terminating end of the complementary strand against the template strand. The improvement in this method with the addition of capillary electrophoresis was named the first-generation sequencing method. This method is useful for sequencing microbial genomes; however, it has a limited throughput, and the subsequent sequencing steps are labor-intensive. The introduction of next-generation sequencing tools in the early years of the twentieth century revolutionized microbial research and allowed the creation of numerous high-throughput datasets using several sequencing platforms. Famous sequencing platforms, including Illumina and Ion Torrent, are inclusive and equipped to perform thousands of individual reactions in parallel. Nonetheless, the limitation of these second-generation sequencing platforms is the production of thousands of short-read fragments that must be reassembled and subsequently characterized. Third-Generation Sequencing platforms, including Oxford Nanopore and Pacific Biosciences, overcome the limitation of short reads and can achieve read lengths of up to 10 kb. These third-generation platforms have become more accurate and specific over time and require less data processing and labor for data analysis (Hu et al. 2021). Usually, the target DNA is chopped and combined with adapter molecules. The sequences were PCR-amplified following immobilization on beads, where similar DNA molecules were clustered. Further treatment remained specific to the sequencing platform used, and the reads were characterized through the library preparation. Next-generation sequencing (NGS) technologies have significantly contributed to microbiome research. Genomic analysis of beneficial root-associated microbes allows us to decipher the potential plant growth-promoting and stress-tolerant genes and biochemical pathways that can be exploited. The inoculation of these engineered microbes with enhanced stress tolerance has shown promising results and is a widely used tool for developing biostimulants for future drought-resilient crops.

10.4.2.5 Stable Isotope Probing (SIP) Method

Stable isotope probing (SIP) refers to the labeling of certain substrates with ^{13}C, ^{15}N, $^{13}CO_2$, and ^{18}O atoms (Angel et al. 2018). The labeled substrates are assimilated by the rhizomicrobiome, yielding RNA- or DNA-based labeled biomarkers, indicating a metabolically active microbiome. This technique ensures the microbial community structure and dynamics of active metabolic functions in response to stress. SIP can provide insights into the specific metabolites, enzymes, and pathways involved in mitigating drought stress, and this information can be used to engineer the microbiome. With quantitative SIP (qSIP), actively growing microbial taxa can be detected and used for biofertilization and upregulation and editing of drought-mitigating metabolic pathways.

10.4.2.6 Omics Technologies for Rhizomicrobiome Engineering

The invention and integration of omics have completely revolutionized the study and investigation of microbes and have shown incredible achievements in engineering soil microbes. Omics technologies are categorized into different groups based on the targets and types of output datasets generated. Omics comprise genomics, transcriptomics, proteomics, and metabolomics and offer the most useful strategies for investigating the rhizomicrobiome.

Transcriptomics offers a compilation of diverse and accumulating transcripts under various circumstances in different target cells (Chauhan et al. 2019). This method indicates alterations in gene expression under fluctuating environmental conditions, suggesting the upregulation of specific genes. To analyze complex microbial communities, total RNA was extracted from plant roots and processed to obtain gene transcript information. A comparison of changes in gene expression patterns under normal and stressed conditions indicates the putative role of certain genes in the induction of systemic defense in host plants. The potential roles of certain microbial genes were determined using expressed sequence tags (ESTs). These tags comprise short base pair reads of 200–800 base pairs and are generated from cDNA clones, indicating the expressed genomic regions. Represented as a cost-effective alternative to whole-genome sequencing, basic transcriptomic analysis can be useful in identifying highly active drought-tolerant genes in rhizomicrobiomes (Moradi et al. 2021). In addition, the mechanisms of enhanced survival under abiotic and biotic stress conditions and the basis and principles of the association of microbes with roots can be investigated using transcriptomic analysis. Furthermore, transcriptomics simultaneously allows for the comparison and analysis of the gene expression of both host plants and microbes, providing a detailed RNA profile of the active genome.

High-throughput sequencing technologies, such as Illumina sequencing, offer a non-PCR-based approach for real-time examination of the transcriptional activities of complex microbial networks. Nonetheless, the annotation and curation of metadata generated from sequencing platforms remain challenging. The short sequence reads generated during the analysis can be processed either through assembly-free platforms or assembly-based methods (Martinez et al. 2016). Platforms are typically selected based on the analysis nature. In an assembly-free method, short sequence reads are directly analyzed through reference databases, whereas the assembly-based approach is more sophisticated, in which the reads are aligned to form contigs for a better understanding. The specialized alignment tools included the Basic Local Alignment Search Tool (BLAST), Burrows-Wheeler Aligner-Maximal Exact-Matches (BWA-MEM), and Bowtie 2, the results of which were annotated using Gene Ontology (GO), Clusters of Orthologous Genes (COG), SWISS-Prot, and Kyoto Encyclopedia of Genes and Genomes (KEGG). Aligned contigs can be assessed using software packages such as Cufflinks, MetaVelvet, AbySS, Trinity, and Scriptures (Dries et al. 2021). Likewise, RNA-Seq analysis is a quantitative approach for metatranscriptomic data analysis and is assisted by several workflows, such as CANEapp, an application for the comprehensive automated analysis of next-generation sequencing experiments (Velmeshev et al. 2016). Finally, the curated datasets were subjected to computational pipelines, including HUMAnN51 and MG-RAST52, which were used in combination with mapping, differential gene expression, and quality filtering. The establishment of these sophisticated bioinformatic pipelines has provided the liberty of comparative metatranscriptomics analysis, which helps assess the microbial communities associated with the same plant host under different environmental conditions and different exposures to stressors. This provides deep insights into community-wide gene expression changes, microbial assemblages, and responses under diverse circumstances. A stable, culturable, and metabolically active microbial community can be recognized and used to develop bioformulations to combat different environmental stressors. This information also reveals the secrets of microbial speciation, evolution, adaptation, and ecological fitness of rhizomicrobiomes.

In addition to transcriptomics and metatranscriptomics, genomic analysis has provided key information about root-associated microbes. Genomic analysis combined with metagenomics can provide functional insights into the active microbiome and a comprehensive view of the genetic diversity, evolution, and composition of rhizosphere microbial communities. Sequence-based metagenomic analysis provides reliable information on culturable and unculturable microbial communities through 16S rRNA analysis. Metagenomic datasets generated using sequencing platforms such as Illumina and 454-pyrosequencing were interpreted using bioinformatics pipelines, including MEGAN, CARMA, Quantitative Insights into Microbial Ecology (QIIME), Illumina reads, and PacBio reads (Escobar-Zepeda et al. 2015). Metagenomics offers a thorough understanding of the diversity and functionality of uncultured microbes at the community level, with respect to the active microbiome under specific abiotic stress. Based on metagenomic analysis of plants exposed to drought stress, rice, wheat, sugar beet, and soybean have shown differential abundance of associated microbes, indicating their protective role and complementary services for plant needs (Joshi 2018; Tsurumaru et al. 2015).

Metabolomic studies of diverse metabolites are linked to microbial community functions. Metabolomics allows the study of exudates and chemicals released by microbes in the rhizosphere rather than genes and proteins and characterizes the unique cellular pathways responsible for the release of these chemicals. Commonly used tools for metabolomic analysis include liquid chromatography or gas chromatography coupled with mass spectrometry (LC-MS/MS and GC-MS) and nuclear magnetic resonance (NMR). However, current metabolic platforms provide limited information on the temporal and spatial dynamics of microbial exudates (Lucaciu et al. 2019). Datasets produced using LC-MS, GC-MS, and NMR are usually visualized using MetaboAnalyst, GNPS, or XCMS, which offer structural information regarding the detected metabolites. Metabolomics has revealed the role of several protective secondary metabolites released in response to drought stress and the upregulation of certain metabolic pathways (Sud et al. 2016).

Proteomics is another branch of omics that provides information on active microbial proteins under specific stresses. Proteins are the most diverse and significant units of cellular function that unravel posttranslational modifications to understand complex protein structures and validate transcriptomics datasets. Among the different approaches used to study the cellular proteome, GeLC-MS, multidimensional protein identification technology (MudPIT), and 2-dimensional gel electrophoresis are some of the techniques used. Typically, proteins are first digested and subjected to separation, leading to mass spectrometric analysis (Gupta et al. 2019). Proteomic and metaproteomic approaches have provided an excellent understanding of the functional characteristics of microbial communities and the regulation of specific proteins under external stressors, such as drought. The major limitations of proteomic research are sample preparation and protein extraction because of the higher chances of degradation due to alkaloids, organic acids, polysaccharides, and polyphenols, which interfere with protein structure. Furthermore, the total proteome information is considered incomplete without complementary genomic information. Hence, more accurate proteome analysis methods are needed to verify the bulk of the data generated and to correlate them with the available genomic tools. Scientists are attempting to develop more sophisticated platforms for proteogenomics to link the genetic information of plant endophytic microbial communities with proteomic datasets. Linking genomic information with protein datasets can provide an accurate understanding of the structure and interplay of active genes under specific external stress.

10.5 Maneuvering Plant-Microbial Interactions

The rhizomicrobiome and plants continuously communicate through chemical interactions, making the holobiont a complete unit. These chemical signals ensure the harmonization of activities and responses that are beneficial to both partners. A thorough understanding of these signaling networks among all partners is indispensable for untangling the cross talk between plants and microbes for enhanced survival under stress conditions. Moreover, a complete portrayal of these metabolic activities will enable the establishment of rhizomicrobiomes with the ability to successfully colonize plant roots and harmonize the activities required at a particular time point. The bidirectional trafficking of chemical signals strengthens favored and helpful microbial flora and deters pathogens and undesirable substances (Sharma et al. 2021). Rhizosphere-inhabiting microbiomes release diffusible signaling molecules, such as N-acyl-homoserine lactones (AHLs), to regulate gene expression and quorum sensing, induce systemic resistance (ISR), and develop plant roots. Likewise, some associated microbes produce quorum-quenching signals, and ethylene-related compounds offer an ostensible opportunity to engineer microbial networks for their respective functions under abiotic and biotic stresses. In addition to engineering microbial networks, transkingdom signaling can be engineered to produce certain regulatory compounds in plants. For instance, a study showed the synthetic transkingdom signaling pathway for the production of the bacterial signal rhizopine *scyllo*-inosamine in barley and *Medicago truncatula* (Geddes et al. 2019). The synthetic signal elicited bacterial responses and fixed N_2. However, examples of these transformations are limited and require comprehensive strategies to pave the way for the success of this technology.

10.6 Prospects and Conclusions

The rhizomicrobiome is a hotspot for microbial functioning and extends its role in plant health, impacting crop yield. Nonetheless, little is known about the complex interplay between associated microbial diversity and their interactions with plant partners. Significant advances have been made in understanding these interactions through classical and contemporary investigative methods. Nonetheless, much remains hidden and needs to be deciphered using sophisticated high-throughput technologies that can clarify the recruitment of microbial communities by plants, their assemblage under specific stress circumstances, and their relationship with plant partners. This knowledge is indispensable for ensuring higher crop yields to face the challenges of global hunger due to the deteriorating quality and quantity of irrigation water, worsening droughts, and limited cultivable areas.

References

Alcolombri U, Pioli R, Stocker R, Berry D (2022) Single-cell stable isotope probing in microbial ecology. ISME Commun 2:55. https://doi.org/10.1038/s43705-022-00157-8

Angel R, Panhölzl C, Gabriel R, Herbold C, Wanek W, Richter A, Eichorst SA, Woebken D (2018) Application of stable-isotope labeling techniques for the detection of active diazotrophs. Environ Microbiol 20:44–61. https://doi.org/10.1111/1462-2920.13972

Arora NK, Mishra J (2016) Prospecting the roles of metabolites and additives in future bioformulations for sustainable agriculture. Appl Soil Ecol 107:405–407. https://doi.org/10.1016/j.apsoil.2016.07.020

Aviles-Garcia ME, Flores-Cortez I, Hernández-Soberano C, Santoyo G, Valencia-Cantero E (2016) The plant growth-promoting rhizobacterium *Arthrobacter agilis* UMCV2 endophytically colonizes Medicago truncatula. Rev Argent Microbiol 48:342–346. https://doi.org/10.1016/j.ram.2016.07.002

Babalola OO, Fadiji AE, Enagbonma BJ, Alori ET, Ayilara MS, Ayangbenro AS (2020) The nexus between plant and plant microbiome: revelation of the networking strategies. Front Microbiol 11:548037. https://doi.org/10.3389/fmicb.2020.548037

Backer R, Rokem JS, Ilangumaran G, Lamont J, Praslickova D, Ricci E, Subramanian S, Smith DL (2018) Plant growth-promoting rhizobacteria: context, mechanisms of action, and roadmap to commercialization of biostimulants for sustainable agriculture. Front Plant Sci 9:1473. https://doi.org/10.3389/fpls.2018.01473

Basu A, Prasad P, Das SN, Kalam S, Sayyed RZ, Reddy MS, El Enshasy H (2021) Plant growth promoting rhizobacteria (PGPR) as green bioinoculants: recent developments, constraints, and prospects. Sustainability 13:1140. https://doi.org/10.3390/su13031140

Chauhan PS, Lata C, Tiwari S, Chauhan AS, Mishra SK, Agrawal L, Chakrabarty D, Nautiyal CS (2019) Transcriptional alterations reveal *Bacillus amyloliquefaciens*-rice cooperation under salt stress. Sci Rep 9:11912. https://doi.org/10.1038/s41598-019-48309-8

Chen L, Schwier M, Krumbach J, Kopriva S, Jacoby RP (2021) Metabolomics in plant-microbe interactions in the roots. Adv Bot Res 98. Academic Press, London:133–161

Chu SS, Nguyen HA, Zhang J, Tabassum S, Cao H (2022) Towards multiplexed and multimodal biosensor platforms in real-time monitoring of metabolic disorders. Sensors 22:5200. https://doi.org/10.3390/s22145200

Cottier F, Srinivasan KG, Yurieva M, Liao W, Poidinger M, Zolezzi F, Pavelka N (2018) Advantages of meta-total RNA sequencing (MeTRS) over shotgun metagenomics and amplicon-based sequencing in the profiling of complex microbial communities. NPJ Biofilms Microbiomes 4:2. https://doi.org/10.1038/s41522-017-0043-x

Darriaut R, Antonielli L, Martins G, Ballestra P, Vivin P, Marguerit E, Mitter B, Masneuf-Pomarède I, Compant S, Ollat N, Lauvergeat V (2022) Soil composition and rootstock genotype drive the root associated microbial communities in young grapevines. Front Microbiol 13:1031064. https://doi.org/10.3389/fmicb.2022.1031064

Djemiel C, Dequiedt S, Karimi B, Cottin A, Horrigue W, Bailly A, Boutaleb A, Sadet-Bourgeteau S, Maron PA, Chemidlin Prevost-Boure N, Ranjard L (2022) Potential of meta-omics to provide modern microbial indicators for monitoring soil quality and securing food production. Front Microbiol 13:889788. https://doi.org/10.3389/fmicb.2022.889788

Dries R, Chen J, Del Rossi N, Khan MM, Sistig A, Yuan GC (2021) Advances in spatial transcriptomic data analysis. Genome Res 31:1706–1718. https://doi.org/10.1101/gr.275224.121

Dubey A, Malla MA, Khan F, Chowdhary K, Yadav S, Kumar A, Sharma S, Khare PK, Khan ML (2019) Soil microbiome: a key player for conservation of soil health under changing climate. Biodivers Conserv 28:2405–2429. https://doi.org/10.1007/s10531-019-01760-5

Dutta J, Bora U (2019) Rhizosphere microbiome and plant probiotics. In: Kalia VC (ed) New and future developments in microbial biotechnology and bioengineering. Elsevier, Amsterdam, pp 273–281

Enespa CP (2022) Tool and techniques study to plant microbiome current understanding and future needs: an overview. Commun Integr Biol 15:209–225. https://doi.org/10.1080/19420889.2022.2082736

Escobar-Zepeda A, Vera-Ponce de León A, Sanchez-Flores A (2015) The road to metagenomics: from microbiology to DNA sequencing technologies and bioinformatics. Front Genet 6:348. https://doi.org/10.3389/fgene.2015.00348

Fadiji AE, Babalola OO (2020) Metagenomics methods for the study of plant-associated microbial communities: a review. J Microbiol Methods 170:105860. https://doi.org/10.1016/j.mimet.2020.105860

Geddes BA, Paramasivan P, Joffrin A, Thompson AL, Christensen K, Jorrin B, Brett P, Conway SJ, Oldroyd GE, Poole PS (2019) Engineering transkingdom signaling in plants to control gene expression in rhizosphere bacteria. Nat Commun 10:3430. https://doi.org/10.1038/s41467-019-10879-y

Gregersen IB, Sørup HJ, Madsen H, Rosbjerg D, Mikkelsen PS, Arnbjerg-Nielsen K (2013) Assessing future climatic changes of rainfall extremes at small spatio-temporal scales. Clim Chang 118:783–797. https://doi.org/10.1007/s10584-012-0663-4

Gupta SK, Rai AK, Sarim KM, Sharma A, Budhlakoti N, Arora D, Verma DK, Singh DP (2019) Metaproteomic data of maize rhizosphere for deciphering functional diversity. Data Brief 27:104574. https://doi.org/10.1016/j.dib.2019.104574

Gupta S, Schillaci M, Roessner U (2022) Metabolomics as an emerging tool to study plant–microbe interactions. Emerg Top Life Sci 6:175–183. https://doi.org/10.1042/ETLS20210263

Hatzig SV, Nuppenau JN, Snowdon RJ, Schießl SV (2018) Drought stress has transgenerational effects on seeds and seedlings in winter oilseed rape (*Brassica napus* L.). BMC Plant Biol 18:1–3. https://doi.org/10.1186/s12870-018-1568-9

Herschend J, Damholt ZB, Marquard AM, Svensson B, Sørensen SJ, Hägglund P, Burmølle M (2017) A meta-proteomics approach to study the interspecies interactions affecting microbial biofilm development in a model community. Sci Rep 7:16483. https://doi.org/10.1038/s41598-017-16653-8

Hu T, Chitnis N, Monos D, Dinh A (2021) Next-generation sequencing technologies: An overview. Hum Immunol 82:801–811. https://doi.org/10.1016/j.humimm.2021.02.012

Hussain AO, Sulaimon HO (2018) Comparative study of food security in Africa amid growing population. Int J Sci Res Publ 8:6–12

Hussain HA, Hussain S, Khaliq A, Ashraf U, Anjum SA, Men S, Wang L (2018) Chilling and drought stresses in crop plants: implications, cross talk, and potential management opportunities. Front Plant Sci 9:393. https://doi.org/10.3389/fpls.2018.00393

Hussain HA, Men S, Hussain S, Chen Y, Ali S, Zhang S, Zhang K, Li Y, Xu Q, Liao C, Wang L (2019) Interactive effects of drought and heat stresses on morpho-physiological attributes, yield, nutrient uptake and oxidative status in maize hybrids. Sci Rep 9:3890. https://doi.org/10.1038/s41598-019-40362-7

Jansson JK, Hofmockel KS (2020) Soil microbiomes and climate change. Nat Rev Microbiol 18:35–46. https://doi.org/10.1038/s41579-019-0265-7

Joshi S (2018) Rhizospheric bacterial diversity in different Dalbergia sissoo Roxb. provenances. MSc thesis. GBPUAT, Pantnagar, p 102

Kabir SF, Sikdar PP, Haque B, Bhuiyan MR, Ali A, Islam MN (2018) Cellulose-based hydrogel materials: chemistry, properties and their prospective applications. Prog Biomater 7:153–174. https://doi.org/10.1007/s40204-018-0095-0

Khayatnezhad M, Gholamin R (2021) The effect of drought stress on the superoxide dismutase and chlorophyll content in durum wheat genotypes. Adv Life Sci 8:119–123

Kumar A, Dames JF, Gupta A, Sharma S, Gilbert JA, Ahmad P (2015) Current developments in arbuscular mycorrhizal fungi research and its role in salinity stress alleviation: a biotechnological perspective. Crit Rev Biotechnol 35:461–474. https://doi.org/10.3109/07388551.2014.899964

Lebeis SL (2015) Greater than the sum of their parts: characterizing plant microbiomes at the community-level. Curr Opin Plant Biol 24:82–86. https://doi.org/10.1016/j.pbi.2015.02.003

Li H, Su JQ, Yang XR, Zhu YG (2019) Distinct rhizosphere effect on active and total bacterial communities in paddy soils. Sci Total Environ 649:422–430. https://doi.org/10.1016/j.scitotenv.2018.08.373

López-Mondéjar R, Kostovčík M, Lladó S, Carro L, García-Fraile P (2017) Exploring the plant microbiome through multi-omics approaches. In: Kumar V, Kumar M, Sharma S, Prasad R (eds) Probiotics in agroecosystem. Springer, Singapore, pp 233–268

Lucaciu R, Pelikan C, Gerner SM, Zioutis C, Köstlbacher S, Marx H, Herbold CW, Schmidt H, Rattei T (2019) A bioinformatics guide to plant microbiome analysis. Front Plant Sci 10:1313. https://doi.org/10.3389/fpls.2019.01313

Mahmood YA (2020) Drought effects on leaf canopy temperature and leaf senescence in barley. Iraqi J Agric Sci 51:1684–1693

Malik AA, Martiny JB, Brodie EL, Martiny AC, Treseder KK, Allison SD (2020) Defining trait-based microbial strategies with consequences for soil carbon cycling under climate change. ISME J 14:1–9. https://doi.org/10.1038/s41396-019-0510-0

Martinez X, Pozuelo M, Pascal V, Campos D, Gut I, Gut M, Azpiroz F, Guarner F, Manichanh C (2016) MetaTrans: an open-source pipeline for metatranscriptomics. Sci Rep 6:26447. https://doi.org/10.1038/srep26447

Melandri G, Prashar A, McCouch SR, Van Der Linden G, Jones HG, Kadam N, Jagadish K, Bouwmeester H, Ruyter-Spira C (2020) Association mapping and genetic dissection of drought-induced canopy temperature differences in rice. J Exp Bot 71:1614–1627. https://doi.org/10.1093/jxb/eraa015

Moradi S, Khoshru B, Mitra D, Mahakur B, Das Mohapatra PK, Asgari Lajayer B, Ghorbanpour M (2021) Transcriptomics analyses and the relationship between plant and plant growth-promoting rhizobacteria (PGPR). In: Sharma SK, Singh UB, Sahu PK (eds) Omics science for rhizosphere biology. Springer, Singapore, pp 89–111

Mukarram M, Choudhary S, Kurjak D, Petek A, Khan MM (2021) Drought: sensing, signaling, effects and tolerance in higher plants. Physiol Plant 172:1291–1300. https://doi.org/10.1111/ppl.13423

Nishida H, Suzaki T (2018) Nitrate-mediated control of root nodule symbiosis. Curr Opin Plant Biol 44:129–136. https://doi.org/10.1016/j.pbi.2018.04.006

Nuccio EE, Blazewicz SJ, Lafler M, Campbell AN, Kakouridis A, Kimbrel JA, Wollard J, Vyshenska D, Riley R, Tomatsu A, Hestrin R (2022) HT-SIP: a semi-automated stable isotope probing pipeline identifies cross-kingdom interactions in the hyphosphere of arbuscular mycorrhizal fungi. Microbiome 10:199. https://doi.org/10.1186/s40168-022-01391-z

Obidiegwu JE, Bryan GJ, Jones HG, Prashar A (2015) Coping with drought: stress and adaptive responses in potato and perspectives for improvement. Front Plant Sci 6:542. https://doi.org/10.3389/fpls.2015.00542

Pang Z, Chen J, Wang T, Gao C, Li Z, Guo L, Xu J, Cheng Y (2021) Linking plant secondary metabolites and plant microbiomes: a review. Front Plant Sci 12:621276. https://doi.org/10.3389/fpls.2021.621276

Papp K, Hungate BA, Schwartz E (2018) Microbial rRNA synthesis and growth compared through quantitative stable isotope probing with H218O. Appl Environ Microbiol 84:e02441–e02417. https://doi.org/10.1128/AEM.02441-17

Pini F, East AK, Appia-Ayme C, Tomek J, Karunakaran R, Mendoza-Suárez M, Edwards A, Terpolilli JJ, Roworth J, Downie JA, Poole PS (2017) Bacterial biosensors for in vivo spatiotemporal mapping of root secretion. Plant Physiol 174:1289–1306. https://doi.org/10.1104/pp.16.01372

Priya P, Aneesh B, Harikrishnan K (2021) Genomics as a potential tool to unravel the rhizosphere microbiome interactions on plant health. J Microbiol Methods 185:106215. https://doi.org/10.1016/j.mimet.2021.106215

Prosser JI (2015) Dispersing misconceptions and identifying opportunities for the use of 'omics' in soil microbial ecology. Nat Rev Microbiol 13:439–446. https://doi.org/10.1038/nrmicro3468

Quinn LD, Straker KC, Guo J, Kim S, Thapa S, Kling G, Lee DK, Voigt TB (2015) Stress-tolerant feedstocks for sustainable bioenergy production on marginal land. Bioenergy Res 8(3):1081–1100. https://doi.org/10.1007/s12155-015-9580-y

Razi K, Muneer S (2021) Drought stress-induced physiological mechanisms, signaling pathways and molecular response of chloroplasts in common vegetable crops. Crit Rev Biotechnol 41:669–691. https://doi.org/10.1080/07388551.2021.1884035

Razmi A, Golestanipour A, Nikkhah M, Bagheri A, Shamsbakhsh M, Malekzadeh-Shafaroudi S (2019) Localized surface plasmon resonance biosensing of tomato yellow leaf curl virus. J Virol Methods 267:1–7. https://doi.org/10.1016/j.jviromet.2019.02.007

Rojas-Sánchez B, Guzmán-Guzmán P, Morales-Cedeño LR, Orozco-Mosqueda MD, Saucedo-Martínez BC, Sánchez-Yáñez JM, Fadiji AE, Babalola OO, Glick BR, Santoyo G (2022) Bioencapsulation of microbial inoculants: mechanisms, formulation types and application techniques. Appl Biosci 1:198–220. https://doi.org/10.3390/applbiosci1020013

Saberi Riseh R, Ebrahimi-Zarandi M, Gholizadeh Vazvani M, Skorik YA (2021) Reducing drought stress in plants by encapsulating plant growth-promoting bacteria with polysaccharides. Int J Mol Sci 22:12979. https://doi.org/10.3390/ijms222312979

Sánchez-Cañizares C, Jorrín B, Poole PS, Tkacz A (2017) Understanding the holobiont: the inter-dependence of plants and their microbiome. Curr Opin Microbiol 38:188–196. https://doi.org/10.1016/j.mib.2017.05.008

Santos LF, Olivares FL (2021) Plant microbiome structure and benefits for sustainable agriculture. Curr Plant Biol 26:100198. https://doi.org/10.1016/j.cpb.2021.100198

Santoyo G, Urtis-Flores CA, Loeza-Lara PD, Orozco-Mosqueda MD, Glick BR (2021) Rhizosphere colonization determinants by plant growth-promoting rhizobacteria (PGPR). Biology 10:475. https://doi.org/10.3390/biology10060475

Schiebenhoefer H, Van Den Bossche T, Fuchs S, Renard BY, Muth T, Martens L (2019) Challenges and promise at the interface of metaproteomics and genomics: an overview of recent progress in metaproteogenomic data analysis. Expert Rev Proteomics 16:375–390. https://doi.org/10.1080/14789450.2019.1605895

Shakya M, Lo CC, Chain PS (2019) Advances and challenges in metatranscriptomic analysis. Front Genet 10:904. https://doi.org/10.3389/fgene.2019.00904

Sharma P, Sharma MM, Malik A, Vashisth M, Singh D, Kumar R, Singh B, Patra A, Mehta S, Pandey V (2021) Rhizosphere, rhizosphere biology, and rhizospheric engineering. In: Sharma AK (ed) Plant growth-promoting microbes for sustainable biotic and abiotic stress management. Springer, Cham, pp 577–624

Shi Z, Yin H, Van Nostrand JD, Voordeckers JW, Tu Q, Deng Y, Yuan M, Zhou A, Zhang P, Xiao N, Ning D (2019) Functional gene array-based ultrasensitive and quantitative detection of microbial populations in complex communities. mSystems 4:e00296–e00219. https://doi.org/10.1128/mSystems.00296-19

Shin YK, Bhandari SR, Jo JS, Song JW, Lee JG (2021) Effect of drought stress on chlorophyll fluorescence parameters, phytochemical contents, and antioxidant activities in lettuce seedlings. Horticulturae 7:238. https://doi.org/10.3390/horticulturae7080238

Shokat S, Großkinsky DK, Singh S, Liu F (2023) The role of genetic diversity and pre-breeding traits to improve drought and heat tolerance of bread wheat at the reproductive stage. Food Energy Secur 12:e478. https://doi.org/10.1002/fes3.478

Sud M, Fahy E, Cotter D, Azam K, Vadivelu I, Burant C, Edison A, Fiehn O, Higashi R, Nair KS, Sumner S (2016) Metabolomics Workbench: An international repository for metabolomics data and metadata, metabolite standards, protocols, tutorials and training, and analysis tools. Nucleic Acids Res 44:D463–D470. https://doi.org/10.1093/nar/gkv1042

Szczech M, Maciorowski R (2016) Microencapsulation technique with organic additives for biocontrol agents. J Hortic Res 24:111–122. https://doi.org/10.1515/johr-2016-0013

Thakur V, Rane J, Nankar AN (2022) Comparative analysis of canopy cooling in wheat under high temperature and drought stress. Agronomy 12:978. https://doi.org/10.3390/agronomy12040978

Thilakarathna MS, Torkamaneh D, Bruce RW, Rajcan I, Chu G, Grainger CM, Szczyglowski K, Hill B, Raizada MN (2021) Testing whether pre-pod-fill symbiotic nitrogen fixation in soybean is subject to drift or selection over 100 years of soybean breeding. Front Agron 3:725813. https://doi.org/10.3389/fagro.2021.725813

Tsurumaru H, Okubo T, Okazaki K, Hashimoto M, Kakizaki K, Hanzawa E, Takahashi H, Asanome N, Tanaka F, Sekiyama Y, Ikeda S (2015) Metagenomic analysis of the bacterial community associated with the taproot of sugar beet. Microbes Environ 30:63–69. https://doi.org/10.1264/jsme2.ME14157

Uzoh IM, Babalola OO (2018) Rhizosphere biodiversity as a premise for application in bio-economy. Agric Ecosyst Environ 265:524–534. https://doi.org/10.1016/j.agee.2018.07.010

Vandenkoornhuyse P, Quaiser A, Duhamel M, Le Van A, Dufresne A (2015) The importance of the microbiome of the plant holobiont. New Phytol 206:1196–1206. https://doi.org/10.1111/nph.13312

Velmeshev D, Lally P, Magistri M, Faghihi MA (2016) CANEapp: a user-friendly application for automated next generation transcriptomic data analysis. BMC Genomics 17:1–3. https://doi.org/10.1186/s12864-016-2681-6

Verma G, Srivastava D, Tiwari P, Chakrabarty D (2019) ROS modulation in crop plants under drought stress. In: Hasanuzzaman M, Fotopoulos V, Nahar K, Fujita M (eds) Reactive oxygen, nitrogen and sulfur species in plants: production, metabolism, signaling and defense mechanisms. Wiley-Blackwell, Hoboken, pp 311–336

Vocciante M, Grifoni M, Fusini D, Petruzzelli G, Franchi E (2022) The role of plant growth-promoting rhizobacteria (PGPR) in mitigating plant's environmental stresses. Appl Sci 12:1231. https://doi.org/10.3390/app12031231

Vurukonda SS, Vardharajula S, Shrivastava M, SkZ A (2016) Enhancement of drought stress tolerance in crops by plant growth promoting rhizobacteria. Microbiol Res 184:13–24. https://doi.org/10.1016/j.micres.2015.12.003

Williams A, de Vries FT (2020) Plant root exudation under drought: implications for ecosystem functioning. New Phytol 225:1899–1905. https://doi.org/10.1111/nph.16223

Xu L, Pierroz G, Wipf HM, Gao C, Taylor JW, Lemaux PG, Coleman-Derr D (2021) Holo-omics for deciphering plant-microbiome interactions. Microbiome 9:1. https://doi.org/10.1186/s40168-021-01114-w

You J, Chan Z (2015) ROS regulation during abiotic stress responses in crop plants. Front Plant Sci 6:1092. https://doi.org/10.3389/fpls.2015.01092

Zhang Y, Yeo WS, Ng KH, Yap HK (2015) CRISPR/Cas9 for genome engineering: the next genomic revolution. J Biochem Mol Biol Res 1:112–117

Zhang J, Zhang S, Cheng M, Jiang H, Zhang X, Peng C, Lu X, Zhang M, Jin J (2018) Effect of drought on agronomic traits of rice and wheat: a meta-analysis. Int J Environ Res Public Health 15:839. https://doi.org/10.3390/ijerph15050839

Part V
Development of Biodegradable Inorganic Biostimulants

Chapter 11
Plant Biostimulants in Synergy: Unlocking Their Full Potential

Krystian Wolski, Julia Chmiel, and Nikodem Kuźnik

Abstract The global agricultural sector faces mounting challenges stemming from climate change, population growth, and environmental degradation, necessitating the transition to more sustainable and resilient crop production systems. Plant biostimulants, defined as substances and microorganisms that enhance physiological processes in plants, are increasingly recognized as tools for improving crop growth, stress tolerance, and nutrient use efficiency without acting as fertilizers or pesticides. This chapter presents a comprehensive synthesis of current knowledge on plant biostimulants, including their definitions, classification schemes, and modes of action. Particular emphasis is placed on the phenomenon of synergy, in which combinations of biostimulants produce effects greater than the sum of their individual contributions. Synergistic interactions are discussed involving arbuscular mycorrhizal fungi (AMF), plant growth-promoting rhizobacteria (PGPR), and various organic compounds, with reference to their capacity to enhance photosynthesis, water use efficiency, nutrient uptake, biomass accumulation, and resistance to abiotic and biotic stressors. Although promising, these interactions are still not adequately understood at the mechanistic level. Variability in results depending on crop species and environmental context presents a significant barrier to field-level predictability. The chapter highlights the need for standardized assessment protocols, field-based validation, and integrative approaches that combine molecular, physiological, and agronomic data. Advancing understanding in this area is essential for the rational design of synergistic biostimulant formulations tailored to specific agroecosystems.

Keywords Biostimulants · Synergy · Mycorrhiza · Resistance

K. Wolski · J. Chmiel · N. Kuźnik (✉)
Faculty of Chemistry, Silesian University of Technology, Gliwice, Poland
e-mail: nikodem.kuznik@polsl.pl

S. Mehnaz, I. Shahid (eds.), *Frontiers in Rhizoengineering*, Microorganisms for Sustainability 65, https://doi.org/10.1007/978-981-92-3665-7_11

11.1 Introduction

Modern agricultural systems around the world face an array of unprecedented challenges that arise from population growth, climate change, and environmental degradation. According to UN projections, the world population will reach almost 10 billion by 2050, thus significantly increasing global food demand (UNDE 2025). Meanwhile, traditional agricultural practices, which rely heavily on the intensive use of mineral fertilizers and pesticides, are increasingly regarded as unsustainable in the long term. Their excessive application leads to soil degradation, groundwater contamination, and loss of biodiversity, posing a threat to the productive capacity of future generations (Tilman et al. 2002). Given these challenges, the concept of sustainable agriculture becomes particularly relevant. It aims to ensure high productivity while preserving natural resources and minimizing adverse environmental impacts. In this context, there is growing interest in plant biostimulants, which are an innovative tool that supports the development and resistance of crops through natural modulation of plant physiological processes. Unlike fertilizers or pesticides, biostimulants do not function as direct sources of nutrients or as pesticides; instead, they enhance the endogenous defense and adaptation mechanisms of plants, thus improving metabolic processes, photosynthetic performance, and nutrient uptake (du Jardin 2015).

The most widely used biostimulants include protein hydrolysates, seaweed extracts, humic acids, phenolic compounds, and rhizosphere microorganisms such as mycorrhizal fungi and plant growth-promoting bacteria (Calvo et al. 2014). These biostimulants may influence gene expression related to immunity, water regulation, and nutrition, as well as hormonal signaling pathways. As a result, biostimulants play a key role in mitigating the effects of abiotic and biotic stresses that pose a threat to global food security (Nardi et al. 2016).

In recent years, increasing attention has been paid to studying the phenomenon of synergy in the action of biostimulants. Synergistic interactions may occur between different types of biostimulants and between biostimulants and the soil ecosystem, microbiome, and other agronomic practices. These combinations can not only amplify the effects of individual components but also reduce the need for agricultural chemical inputs, aligning with the principles of sustainable development (Rouphael and Colla 2018).

An increasing body of research indicates that the synergistic action of biostimulants has the potential to produce stronger or more consistent effects than individual applications. The mechanisms underlying this synergy include, among others, coactivation of diverse signaling pathways, enhancement of soil structure, increased abundance and activity of the rhizosphere microbiome, and improved transport and assimilation of nutrients in plant tissues. The potential result is enhanced yield and crop quality, along with improved plant health (Van Oosten et al. 2017). However, despite these promising results, biostimulants remain underutilized in conventional agricultural systems. This is due, among other factors, to the lack of sufficiently developed regulatory frameworks, the limited understanding of their mechanisms of action, and the high variability of their effectiveness depending on environmental factors and crop species.

The objective of this chapter is to critically evaluate the current scientific understanding of plant biostimulants, with an emphasis on the mechanisms underlying their synergistic interactions.

11.2 Definition of Biostimulants

Plant biostimulants are agricultural preparations that are increasingly used in crop production; however, they remain a challenging category to define and classify. Biostimulants can consist of a single substance, a group of compounds, or living microorganisms. Their mode of action involves supplying plants with bioactive compounds that regulate growth processes, improve resistance to environmental stressors, improve crop quality, and facilitate nutrient uptake (du Jardin 2015; Van Oosten et al. 2017; EBIC 2025) (Fig. 11.1). This functional definition highlights the versatility of biostimulants but also reveals the difficulty in determining whether a given substance meets the criteria for classification as a biostimulant.

In the scientific literature, they are often defined not by the nature of active ingredients but by the exclusion of well-established categories such as fertilizers and pesticides (La Torre et al. 2016). Moreover, the wide range of their constituents, including humic and fulvic acids, seaweed extracts, amino acids, and microorganisms, makes classification even more difficult. Importantly, biostimulants often exhibit synergistic effects, both in combination with other agricultural products, such as components of microelement fertilizers, and among components within their own formulations. These interactions can improve efficacy while simultaneously increasing the complexity of their classification (Yakhin et al. 2017).

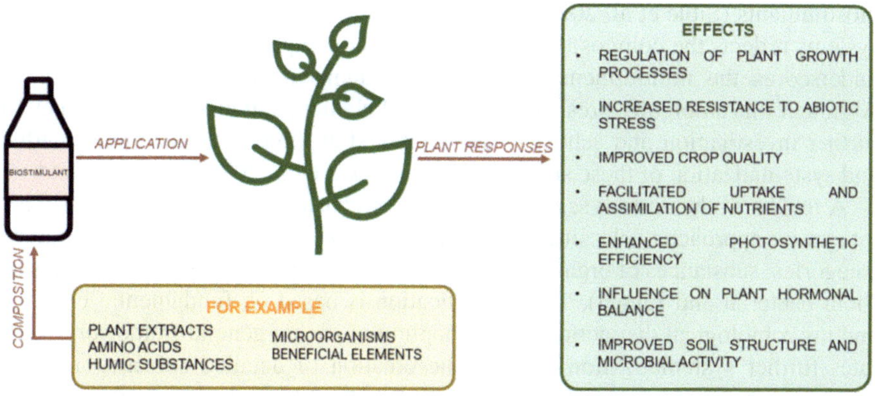

Fig. 11.1 Schematic representation of the mode of action and composition of a plant biostimulant

In the era of sustainable agriculture, which prioritizes the reduction in harmful chemical input, biostimulants represent a promising alternative. Applied at low concentrations, they do not contribute to soil, water, or air. In addition, their ability to improve tolerance to abiotic stress makes them valuable tools for helping crops withstand rapid climatic fluctuations such as drought or excessive rainfall (Van Oosten et al. 2017).

The global lack of a universally accepted definition of plant biostimulants highlights a pressing issue for researchers: the need to establish clear regulatory frameworks and standardized classification systems. This is essential not only to harmonize the production and application guidelines but also to advance the mechanistic understanding of how these substances influence plant physiology (du Jardin 2015).

11.3 Classifications of Biostimulants

11.3.1 Introduction

As the field of plant biostimulants continues to expand, there is a growing need for their systematic and standardized classification. However, this remains a significant challenge for researchers due to the continuous emergence of new biostimulant substances and their considerable diversity, which hinders the formulation of clear and universally applicable classification criteria (Franzoni et al. 2022). Furthermore, the scientific literature offers taxonomies with varying degrees of specificity, further complicating efforts to establish a uniform classification framework. Among the most commonly used criteria is the origin of the biostimulant. For example, biostimulants are often classified as bacterial, fungal, algae-derived, or derived from plant and animal waste, as well as humic substances. Some publications also include biochar and elements such as silicon and selenium within the broader definition of biostimulants (Sible et al. 2021; Asif et al. 2023). This divergence in classification systems reflects the compositional and functional complexity of biostimulants and underscores the multidimensional nature of their effects on plant physiology. Although this ambiguity presents a conceptual challenge, it also opens avenues for further investigation and scholarly debate aimed at improving our understanding and systematization of these substances (Manghwar et al. 2022).

A review of the current scientific literature suggests that the most universal and integrative approach to classification divides plant biostimulants into three principal categories: substances of organic origin, inorganic compounds, and microorganisms (both bacterial and fungal). This classification is based on fundamental chemical and microbiological distinctions. The adoption of such a general framework facilitates further systematization through the creation of detailed subcategories and allows researchers to trace the biochemical and ecological origins of individual active components (Fig. 11.2).

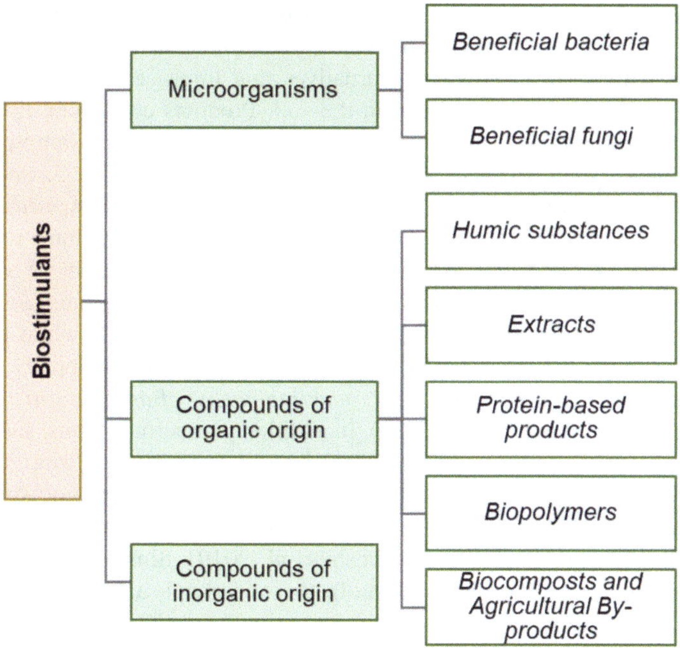

Fig. 11.2 Proposed classification of plant biostimulants

11.3.2 Microorganisms

Microorganisms constitute a fundamental category of biostimulants due to their ability to directly and indirectly enhance plant growth, nutrient acquisition, and stress tolerance. Among them, beneficial bacteria and fungi are the most studied groups, offering diverse mechanisms of action and broad applicability across agricultural systems.

11.3.2.1 Beneficial Bacteria

Bacteria comprise a broad and diverse group of microorganisms capable of promoting plant growth through the modulation of physiological and biochemical processes. These effects involve multiple metabolic pathways and positively influence plant performance. In the scientific literature, these biostimulatory bacteria are commonly referred to as plant growth-promoting rhizobacteria (PGPR). Depending on their functional roles, PGPR can be classified into groups that facilitate macro- and micronutrient availability, enhance plant tolerance to abiotic stress, and improve crop quality (Manghwar et al. 2022; Hamid et al. 2021).

The extensive use of conventional fertilizers, such as NPK formulations, may contribute significantly to environmental degradation (Asadu et al. 2024). This has driven the search for sustainable alternatives that improve the bioavailability of essential nutrients naturally present in the soil. Products containing live bacterial cultures can facilitate the uptake of key nutrients by plants. For example, nitrogen-fixing bacteria such as *Rhizobium*, *Azoarcus*, *Burkholderia*, *Mesorhizobium*, *Sinorhizobium*, *Frankia*, *Allorhizobium*, *Bradyrhizobium*, and *Azorhizobium* are widely recognized (Singh et al. 2024). Phosphorus and potassium, which often occur in poorly solube forms in soil, can be solubilized and made more bioavailable through the activity of specific bacterial taxa. Phosphate-solubilizing bacteria include *Pseudomonas*, *Rhizobium*, *Bacillus*, and *Enterobacter*, while potassium-solubilizing strains include *Bacillus edaphicus*, *Acidothiobacillus* spp., *Ferrooxidans* spp., *Pseudomonas* spp., *Bacillus mucilaginosus*, *Burkholderia* spp., and *Paenibacillus* spp. (Hamid et al. 2021). In addition to macronutrients, such as nitrogen, phosphorus, and potassium, micronutrients such as iron and zinc play crucial roles in plant development. *Pseudomonas putida* facilitates iron acquisition, which is essential for DNA synthesis, photosynthesis, and phytohormone production (Esparza-Reynoso et al. 2024; Kobayashi et al. 2019). Similarly, *Bacillus cereus* helps to mobilize zinc, which acts as a cofactor for carbonic anhydrase. This enzyme catalyzes the reactions of carbonic acid formation from CO_2 and water, which is a key step in the process of CO_2 assimilation by plants (Shakeel et al. 2015; Zhang et al. 2021).

The application of bacterial strains in biostimulants also improves plant tolerance to abiotic stresses such as drought, excessive precipitation, salinity, and heavy metal contamination. This increased tolerance is largely mediated by microbial metabolites that activate plant plant defense and stress-response mechanisms. For example, *Pseudomonas moraviensis* improves tolerance to heavy metals, *Pseudomonas putida* improves drought resistance, and *Pantoea dispersa* supports plant function at low temperatures (Hassan et al. 2017; Fadiji et al. 2022).

In addition, bacterial biostimulants can improve crop quality by enhancing nutrient uptake and stimulating the synthesis or accumulation of plant pigments. Notable examples include *Pseudomonas fluorescens*, *Lactobacillus plantarum*, and *Lactobacillus acidophilus* (Hamid et al. 2021).

Incorporating bacteria into biostimulant formulations represents a promising strategy for sustainable agriculture. However, its efficacy is highly context-dependent, as individual bacterial strains may exhibit beneficial effects only for specific plant species under defined environmental conditions. Furthermore, due to the complexity of the biochemical and metabolic mechanisms of both plants and bacteria, their mutual interactions are difficult to fully investigate. Therefore, studies at the molecular level are necessary to better understand the processes that occur between the two different domains (Hamid et al. 2021; Asadu et al. 2024).

11.3.2.2 Beneficial Fungi

Fungi play a pivotal role in soil ecosystems, influencing plant health and development through various biological interactions. Over hundreds of millions of years, they have coevolved with plants, forming intricate networks of symbiotic relationships (Willis et al. 2013). Among the microorganisms with recognized biostimulatory potential, particular attention has been paid to arbuscular mycorrhizal fungi (AMF) and species of the *Trichoderma* genus, which are widely applied in agriculture to enhance plant growth and resilience to adverse environmental conditions.

AMF, members of the Glomeromycota phylum, establish symbiotic associations with the roots of more than 80% of terrestrial plant species, including many agricultural crops, making their agricultural use particularly relevant to the need for more sustainable food production (Willis et al. 2013; Ehrlich and Harte 2015). Their hallmark feature is the ability to colonize the root cortex and form arbuscules, which constitute the principal sites of nutrient exchange between the plant and the fungus. Many AMF species also form vesicles, which primarily serve as storage structures (Farooq et al. 2015; Ganugi et al. 2019). Through this symbiosis, AMF significantly improve the uptake of phosphorus and other mineral elements that are often poorly available in soil (Plenchette et al. 2005). Furthermore, they confer increased resistance to abiotic stressors such as drought, salinity, and high temperatures by modulating ion transport and activating antioxidant enzyme systems (Estrada et al. 2013). Studies have also demonstrated their role in enhancing resistance to phytopathogens through the induction of defensive metabolite synthesis and systemic resistance signaling pathways (Bitterlich et al. 2018). In addition, AMF can improve soil structure by promoting soil particle aggregation and increasing water retention capacity (Augé 2001). Their activity is further amplified through synergistic interactions with other beneficial soil microorganisms, such as plant growth-promoting rhizobacteria, positioning them as integral components of the soil microbiome (Miransari 2011; Mpongwana et al. 2023).

Agricultural applications of AMF focus on their ability to enhance crop yield and quality. Inoculation with AMF has been shown to increase the accumulation of essential nutrients, including nitrogen, phosphorus, potassium, calcium, and magnesium, especially under stress conditions (Lee et al. 2015; Pellegrino et al. 2015). Furthermore, colonization with AMF reduces the accumulation of heavy metals in edible plant tissues, making them potentially valuable tools for the phyto stabilization of contaminated soils (Li et al. 2023; Tan et al. 2023). Their influence on crop quality also extends to the enhancement of phytochemical content, such as polyphenols and essential oils, which is particularly beneficial in the cultivation of medicinal and aromatic plants (Chen et al. 2022; de Assis et al. 2020). Another advantage lies in their potential to reduce the reliance on phosphate fertilizers, aligning with the principles of sustainable agriculture (Figueira-Galán et al. 2023).

Alongside AMF, fungi of the *Trichoderma* genus, which are widely distributed in soil and associated with plant roots, have emerged as important biostimulants. They exert multifaceted effects, including the promotion of plant growth, suppression of pathogens, and enhanced abiotic stress tolerance (Begum et al. 2019; Colla et al. 2015). *Trichoderma* spp. are particularly noted for their ability to generate induced systemic resistance, which results in increased production of antioxidant enzymes and defensive secondary metabolites (Kuwada et al. 2006). Their mycoparasitic activity also allows them to suppress phytopathogens such as *Fusarium*, *Botrytis*, and *Rhizoctonia*, offering a promising approach to biological plant protection (Poveda and Baptista 2021). In addition to biocontrol properties, *Trichoderma* enhances root system architecture, thereby improving the acquisition of nitrogen and phosphorus (Shoresh et al. 2010). This mechanism contributes to improved plant adaptation to environmental stresses such as salinity and drought by modulating water homeostasis and promoting the synthesis of osmoprotectants (Shoresh et al. 2010).

Despite their numerous benefits, the practical implementation of fungi in agriculture presents several challenges. A major limitation is the difficulty of large-scale production of AMF, as these obligate biotrophs require living host plants for propagation, complicating commercial cultivation (Dalpé and Monreal 2004). Moreover, their effectiveness is influenced by edaphic factors, nutrient availability, and the composition of the microbial community, which can vary significantly between sites (Colla et al. 2015). Although research on the mechanisms underlying the biostimulatory effects of AMF and *Trichoderma* is expanding, many aspects, particularly the molecular signaling pathways involved, remain incompletely understood (Li et al. 2023; Tan et al. 2023).

11.3.3 Compounds of Organic Origin

Organic biostimulants represent the largest and most diverse category of substances that support plant growth and development. This group encompasses both naturally occurring organic compounds and products derived from processed biological materials. A unifying characteristic of these substances is their ability to stimulate plant physiological processes by enhancing nutrient uptake, modulating water balance, increasing tolerance to environmental stress, and influencing soil microorganisms.

11.3.3.1 Humic Substances

Humic substances (HS) constitute the principal component of soil organic matter and are defined as a heterogeneous mixture of organic compounds formed through the microbial decomposition of cellular residues from living organisms (Nardi et al.

2007; Lamar 2020). They are typically classified based on their solubility at different pH levels, with three primary fractions identified:

- **Humic acids (HA):** Soluble in alkaline solutions but insoluble in acidic media; characterized by high molecular weight and chemical stability.
- **Fulvic acids (FA):** Soluble in both acidic and alkaline environments; lower in molecular weight and more mobile within the soil matrix.
- **Humins:** Insoluble in water and in a range of pH conditions; highly stable and capable of long-term binding of organic matter to soil.

Soil organic matter is formed through the decomposition and transformation of biological residues and comprises a continuum of organic compounds at different stages of decomposition. Humic substances may also be classified according to the source, extraction method, and chemical composition, all of which influence their physicochemical properties and agricultural applications (Lehmann and Kleber 2015).

11.3.3.2 Extracts

Plant and algae extracts represent a significant group of organic biostimulants used in agriculture. Based on their origin, they can be classified into two main categories: algae (or seaweed) extracts and plant extracts (Calvo et al. 2014).

11.3.3.2.1 Algae Extracts

Extracts derived from macroalgae, such as brown, red, and green algae, are rich in minerals and organic bioactive compounds, including polysaccharides (laminarin, fucoidan, and alginates), plant hormones (auxins, cytokinins, and gibberellins), polyamines, and betaines (Calvo et al. 2014; Khan et al. 2009). Modern algae extraction methods include alkaline hydrolysis, acid hydrolysis, water extraction, ultrasound-assisted extraction, enzymatic extraction, and advanced techniques such as supercritical fluid extraction. Depending on the extraction method and the algae species, the composition and potential applications can vary (Ugarte et al. 2010).

11.3.3.2.2 Plant Extracts

Plant extracts are derived from various parts of the plant, including leaves, roots, fruits, and agricultural by-products. Their chemical composition varies depending on the source and typically includes secondary metabolites such as polyphenols, flavonoids, alkaloids, and saponins (Bulgari et al. 2017; Moreno-Hernández et al. 2020). Certain plant extracts exhibit biostimulatory effects comparable to phytohormones. For example, *Withania somnifera*, traditionally used in Ayurvedic medicine, has shown potential to enhance plant growth (Gorelick et al. 2015). Studies on *Borago officinalis* leaf extracts suggest a beneficial impact on lettuce growth and quality (Bulgari et al. 2017; Moreno-Hernández et al. 2020).

Plant extract classification may consider not only botanical origin but also the extraction technique and concentration of active compounds. Depending on their chemical profile, plant extracts can be applied in various cropping systems, although their efficacy is influenced by environmental conditions and application methods (Bulgari et al. 2017; Goñi et al. 2016).

11.3.3.3 Protein-Based Products

Protein hydrolysates (PH) and amino acids represent a distinct class of organic biostimulants widely used in agriculture. They are typically classified by their origin and chemical composition.

11.3.3.3.1 Protein Hydrolysates

PH are complex mixtures of amino acids, peptides, polypeptides, and partially degraded proteins produced by the hydrolysis of plant- or animal-derived proteins (du Jardin 2015). Their classification is based on the production method, which significantly influences the composition and functional properties of the final product. The main types include:

- Enzymatic hydrolysates, produced using proteolytic enzymes, enable controlled protein breakdown while preserving the integrity of amino acids and bioactive peptides.
- Chemical hydrolysates, resulting from acid or alkaline hydrolysis, lead to partial degradation of proteins and peptides.
- Thermal hydrolysates, generated at high temperatures, often result in more extensive structural breakdown (Calvo et al. 2014; Vranova et al. 2011).

After enzymatic or chemical processing, these hydrolysates provide readily available amino acids and peptides that enhance nutrient uptake and improve plant tolerance to abiotic stressors (Schiavon et al. 2008). Protein hydrolysates are commercially available in various formulations, including powders, granules, and liquid concentrates, allowing flexible application via fertigation, foliar spraying, or seed treatment (Colla et al. 2015).

11.3.3.3.2 Amino Acids

Within this group of biostimulants, amino acids also play a critical role and can be classified according to several criteria. Based on their metabolic functions in plants, two major categories are recognized:

- Proteinogenic amino acids (e.g., glutamine, arginine, and glutamic acid), which serve as structural components of proteins and are essential for nitrogen metabolism (Colla et al. 2015).
- Non-proteinogenic amino acids, such as γ-aminobutyric acid (GABA), which are involved in the regulation of metabolic pathways and plant stress responses (Balfagón et al. 2022).

Among the specific protein amino acids and related compounds with biostimulatory properties, proline is distinguished, which participates in the mechanisms of response to osmotic stress; glycine-betaine, which stabilizes protein structures and cell membranes under abiotic stress conditions; arginine, a precursor of polyamines that regulate plant cell growth and differentiation; and glutamine, which acts as a nitrogen store and transporter and is involved in the biosynthesis of other amino acids and root development (Colla et al. 2015; Chen and Murata 2011; Rentsch et al. 2007).

Amino acids can also be classified based on their ability to chelate micronutrients, thus improving nutrient bioavailability in soil and supporting key physiological processes in plants (Colla et al. 2015; Farrell et al. 2014). The efficacy of protein hydrolysates and amino acids depends on their chemical composition and application conditions, which are crucial factors in their practical agricultural use.

11.3.3.4 Biopolymers

Biopolymers constitute a significant group of naturally derived biostimulants, increasingly applied in agriculture due to their biodegradability and their capacity to interact with plant systems. Among them, chitin and its derivative chitosan have attracted particular interest. Chitin is one of the most abundant natural polysaccharides after cellulose, composed of *N*-acetyl-D-glucosamine units linked by $\beta\,(1 \rightarrow 4)$ glycosidic bonds (Dutta et al. 2002). It occurs naturally in the exoskeletons of crustaceans and insects, as well as in the cell walls of fungi and certain algae (Ramírez et al. 2010). Partial deacetylation of chitin produces chitosan, a biopolymer with unique physicochemical properties, including a cationic charge that enables interaction with plant cell surfaces (Harish Prashanth and Tharanathan 2007). In addition to conventional chitosan, chitooligosaccharides (oligochitosans), short-chain derivatives produced by enzyme or chemical hydrolysis, are also recognized for their bioactivity (Feng et al. 2004).

Biopolymers can be classified based on their origin and extraction methods. Chitin and chitosan are extracted most commonly from animal sources, particularly crustaceans, although fungi and certain algae may also serve as alternative sources (Ramírez et al. 2010). Depending on their origin and the processing method used, chitin and chitosan can differ in molecular weight, degree of deacetylation, and solubility, which in turn affects their properties and potential applications in agriculture (Iriti et al. 2009). As a biostimulant, chitosan can enhance plant resistance to fungal and bacterial infections, improve photosynthetic performance, and strengthen interactions between plants and soil microorganisms (Shahrajabian et al. 2021).

This group of biostimulants also includes modern chitosan-based nanoformulations, which can be used as carriers for other bioactive compounds or nutrients (Malerba and Cerana 2019). Ongoing research on the properties of biopolymers continues to open new possibilities for their use in agriculture, particularly in the context of biotechnology and nanotechnology.

11.3.3.5 Biocomposts and Agricultural by-Products

Bioconversion of organic by-products with biostimulant activity represents an emerging category of plant biostimulants. These processes involve various methods to transform organic matter into compounds that support plant growth and development (Abbott et al. 2018; Poveda 2021).

11.3.3.5.1 Vermicompost

One of the most well-studied examples of bioconversion is vermicomposting, in which earthworms process organic material to produce a substance enriched with bioactive compounds. Vermicompost is a source of multiple phytohormones, including cytokinins, auxins, abscisic acid, gibberellins, and brassinosteroids, which play an important role in the regulation of plant physiological processes (Zhang et al. 2015; Wong et al. 2020). Advances in analytical techniques have enabled the identification of specific cytokinins, such as zeatin and isopentenyladenine, which are involved in the promotion of cell proliferation and the control of the plant cell cycle (Zhang et al. 2014; Wong et al. 2015).

11.3.3.5.2 Insect Larvae-Based Composts

Another form of bioconversion involves composts derived from insect larvae, such as mealworms (*Tenebrio molitor*) and black soldier flies (*Hermetia illucens*). During the degradation of organic waste, these larvae produce substantial amounts of bioactive compounds that have been shown to enhance plant resistance to environmental stresses and stimulate metabolic activity (Poveda 2021; Wang et al. 2021). In addition to phytohormones, both vermicompost and insect-based composts contain humic and fulvic acids, which promote nutrient uptake, activate plant signaling pathways, and improve overall physiological efficiency (Canellas et al. 2015). Another key group of compounds found in these materials is betaines, which contribute to resilience to stress by stabilizing enzymes and proteins, protecting lipid membranes from oxidative damage, and improving plant adaptation to adverse environmental conditions (Huang et al. 2021; Annunziata et al. 2019).

11.3.3.5.3 Agricultural by-Products

In addition to compost bioconversion, agricultural by-products are an important source of biostimulants, which can be processed into substances that support plant growth. Plant residues, including those remaining after legumes, cereals, and vegetables, contain bioactive secondary metabolites such as polyphenols, alkaloids, and flavonoids, which can influence plant immunity and metabolism (Godlewska et al. 2021). Fruit and vegetable waste can serve as raw material for the production of enzyme-modified biostimulants that support root system development and increase plant tolerance to environmental stress (Arancon and Solarte 2019).

Another group of biostimulants derived from agricultural by-products includes fermentation composts, which contain humic and fulvic acids rich in bioactive compounds and phytohormones (Huang et al. 2021). Introducing these products into the soil contributes to improving its quality, increasing the bioavailability of nutrients, and stimulating the development of beneficial microorganisms (Godlewska et al. 2021).

An important component that should also be included in the group of organic biostimulants is biochar. Biochar is a highly carbonized material formed through pyrolysis, that is, the thermochemical decomposition of biomass in the absence of oxygen (Weber and Quicker 2018). It is most commonly produced from woody biomass containing hemicellulose, cellulose, and lignin, which differ in decomposition temperature, affecting the stability of the resulting biochar (Yang et al. 2007). The use of biochar in agriculture is based on its high resistance to degradation and its ability to retain water and nutrients, which can improve plant productivity and soil properties (Biederman and Harpole 2013). As a biostimulant, biochar is characterized by long-term action focused on improving soil fertility, which distinguishes it from most biostimulants that act directly on plant growth within a single growing season (Laird 2008).

The use of bioconversion and agricultural by-products to produce biostimulants is a key element of sustainable agriculture. Vermicompost, insect-based composts, and fermented agricultural waste provide plants with valuable bioactive compounds that support their development and resistance to stress factors. The use of biostimulants derived from recycled organic materials fits within the concept of a circular economy, supporting the efficient use of natural resources and promoting environmentally friendly agricultural practices.

11.3.4 Compounds of Inorganic Origin

Inorganic compounds constitute a distinct group within the classification of biostimulants. This is the least numerous and simultaneously the most debated category, since it often includes macro- and microelemental components in the form of oxides, salts, or nanoparticles. According to the accepted definition, a compound should not be classified as a biostimulant if it performs a nutritional function. Consequently, all inorganic compounds containing macroelements, such as nitrogen, phosphorus, potassium, sulfur, calcium, or magnesium, should be excluded from the biostimulant category (Kumar et al. 2021).

The role of compounds containing microelements remains more controversial. Microelements such as boron, zinc, cobalt, manganese, copper, molybdenum, iron, chlorine, and nickel are not considered nutritional in the strict sense, but are essential for the proper physiological functioning of plants (Cakmak et al. 2023). However, as noted in the definition of biostimulants introduced at the beginning of this chapter, a key criterion is that biostimulants supply plants with additional compounds that are not strictly required for growth and development. Accordingly, compounds containing microelements should also not be classified as biostimulant elements.

Inorganic biostimulants should include only compounds based on elements referred to as *beneficial elements*. These include titanium, vanadium, silicon, aluminum, cerium, lanthanum, selenium, iodine, cobalt, and sodium. Unlike conventional macro- and micronutrients, these elements are not essential for plant survival, but enhance physiological functions and improve performance (Trejo-Téllez and Gómez-Merino 2023). Examples include compounds based on titanium dioxide, silicon, or selenium, which influence processes such as seed germination, root growth, tolerance to abiotic stress, and nutrient uptake efficiency (Bacilieri et al. 2017). In recent years, nanoparticle formulations of these compounds have received growing attention. Nanoparticles, defined by their size below 100 nm, exhibit unique physicochemical properties that enable more efficient activity in plant systems (Juárez-Maldonado et al. 2019). Silver nanoparticles, although not currently classified as macroelements, microelements, or beneficial elements, have shown the ability to regulate plant function by improving nutrient content and stimulating root development. After a comprehensive evaluation of their mechanisms of action and effects on plant growth, silver can be considered for inclusion in the group of beneficial elements (Guzmán-Báez et al. 2021).

11.4 Synergy among Biostimulants

Plant biostimulants have been known for over 50 years (Mannino 2023; Sani and Yong 2021; Kapoore et al. 2021; Zúñiga-Dávila et al. 2019). Primary research focused on the effects of biostimulants on root and green part development, as well as crop quantity and quality. The association of the activity of these substances with their stressful and beneficial effect was the basis for attributing the phenomenon of hormesis (Kuželet al. 2003). Modern research methods allow one to determine the change in the expression of plant genes (Al-Ghamdi and Elansary 2018) as a result of the action of substances, which has shown how complex the mechanisms of functioning and response to abiotic stressors (Bhupenchandra et al. 2022). In view of the

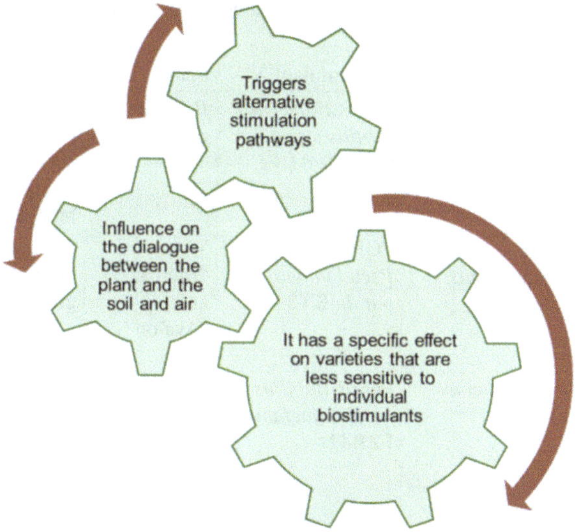

Fig. 11.3 Synergy among biostimulants

state of this knowledge, it seems natural to suspect that the introduction of two or more biostimulants may show synergy, competence, or even inhibition (Fig. 11.3). It should also be considered whether different factors or their different forms within the same class of biostimulants do not strengthen or weaken each other in their actions on the plant. In recent years, there has been a particular increase in awareness of this issue, as numerous reports have been presented, which are the basis for the development of useful knowledge and the creation of recommendations for the purposeful and cooperative application of biostimulants (Sagar et al. 2021; Anuar et al. 2023).

The studies described in the literature are generally based on the idea of synergy as a supra-additive increase in the effect of the application of two different factors related to their separate application and blinding (Table 11.1). Recently, a rapid and combinatorial method has been presented, which can contribute to an easier and more effective synergy assessment (Benito et al. 2022).

11.4.1 Synergies with AMF and PGPR

So far, AMF and PGPR have received the most attention. Since the role of AMF in plant root stimulation is well known (Sun and Shahrajabian 2023), the combination of its PGPR seems to naturally promote the action of both factors (Karthikeyan et al. 2016). But is it synergistic?

Studies on the effects of AMF (*G. aggregatum*, *G. fasciculatum*, *G. intraradices, and G. mosseae*) and *Bacteria subtilis* strain CIMAP B1 belonging to PGPR on *Pelargonium graveolens* showed a positive effect of each of the biostimulants by

Table 11.1 Examples of synergy among biostimulants

Biostimulant 1	Biostimulant 2	Plant	Synergistic effect	Stressor	References
AMF	PGPR *Bacteria subtilis* strain CIMAP B1	*Pelargonium graveolens*	Synergy in biomass yield, no effect on oil production per biomass	None	Alam et al. (2011)
AMF *Rhizophagus irregularis*	PGPR *Pseudomonas putida* KT2440	Wheat "Mercato" and "Avalon"	Synergy in biomass growth	Partially by pathogen stimulation with chitosan	Pérez-De-Luque et al. (2017)
AMF *Glomus mosseae*	PGPR *Bacillus amyloliquefaciens* FZB42)	*Nitraria tangutorum* Bobr.	Synergy in biomass yield, photosynthetic, and water and gas exchange effectiveness	Salinity	Pan et al. (2022)
AMF (native Marrocian)	PGPR *Bacillus megaterium*, *Arthrobacter globiformis*, and *Enterobacter ludwigii*	Date palm *Phoenix dactylifera* L.) "Bouf-gouss"	Synergy in root growth, chlorophyll, and carotenoids level	None	Raho et al. (2022)
AMF *Claroideoglomus etunicatum; Rhizophagus intraradices*; and *Funneliformis mosseae*	PGPR *Bacillus subtilis* BERA 71	Talh tree *Acacia gerrardii*	Synergy in nutrition components, lipids, phenols, and fiber content, and rendering the osmoregulatory system toward salinity protection	Salinity	Hashem et al. (2016)
PGRP *Pseudomonas putida* NBRIRA	PGRP *Bacillus amyloliquefaciens* NBRISN13	Chickpea *Cicer arietinum* L.	Higher growth and yields	Drought	Kumar et al. (2016)
AMF *Rhizophagus intraradices*	Saprotrophic fungi *Trichoderma aureoviride* and *Clonostachys rosea* (*C. reshni*)	Cleopatra mandarin *Citrus reshni* Hort. Ex. Tan.	Synergy in growth enhancement (shoot weight)	None	Camprubí et al. (1995)
AMF *Glomus mosseae*	*Trichoderma aureoviride*	Marigold *Tagetes erecta* L.	Synergy in biomass growth (dry matter and foliar area)	Pathogen *Pythium ultimum*	Calvet et al. (1993)

(Continued)

Table 11.1 (Continued)

Biostimulant 1	Biostimulant 2	Plant	Synergistic effect	Stressor	References
AMF *Rhizophagus irregularis BEG72*	*Trichoderma atroviride* MUCL 45632	Lettuce, tomato, and zucchini	Synergia chlorophyll SPAD index and shoot and root dry weight	None	Colla et al. (2015)
AMF *R. irregularis*	*Plant-driven protein hydrolysates*	*Eggplant "Birgah" F1*	Synergy in fruit number and weight, quality by reducing the glycoalkaloid concentration, and fruit browning potential	None	Di Miceli et al. (2023)
AMF *Funneliformis mosseae*	Dopamine	Apple trees *M. hupehensis*		Salinity	Gao et al. (2020)
Humic acid	*Microalgea Scenedesmus subspicatus* Chodat (synonym: *Desmodesmus subspicatus*)	Onion "BR-29" and "perfecta" F1	Synergy for root growth with low concentrations of both biostimulants	None	Gemin et al. (2019)
Humic acid complex® (2% humic acid/1% organic carbon, and 15% humic acid/1% fulvic acid)	Manure tea microgeo®, composed of organic compounds and active and dormant cells from various microorganisms (bacteria, yeasts, filamentous fungi, and algae), metabolites, and organomineral chelates	Hawaiian haze cannabis *Cannabis sativa* L.	Plant height, chlorophyll content, photosynthetic efficiency, aboveground biomass, and bucked biomass	Adverse weather conditions—Fewer days of vegetative growth	Da Cunha Leme Filho et al. (2024)
Chitosan	Salicylic acid	Date fruits "Khenizi"	Ripeness, polyphenol content, and shelf life of the fruit	None	Ahmed et al. (2022)
Ascophyllum nodosum seaweed extracts	5-Aminolevulinic acid	*Asparagus aethiopicus*	Branch length and fresh and dry weight	Salinity	Al-Ghamdi and Elansary (2018)

increasing the nutrient content (Alam et al. 2011). However, the merger of AMF and PGPR led to a marked increase in biomass. Although there was no effective increase in oil produced by a unit of plant mass, the amount of oil was ultimately higher due to the increase in mass biomass itself.

AMF *Rhizophagus irregularis* and PGPR Pseudomonas putida KT2440 were tested on two varieties of wheat, "Mercato" and "Avalon" (Pérez-De-Luque et al. 2017). Although the responses to root and green growth in the two cultivars varied greatly when AMF and PGPR were used separately, the combination of these biostimulants led to a significant synergistic effect and, according to the authors, provided optimal growth conditions. The authors link these effects to the mechanism of signaling soil microbes to the need to develop a beneficial rhizosphere community. In addition, additional studies have noted a synergistic effect on the increase in resistance, i.e., an increased immune reaction of plants in the form of callose deposition, to the irritating effect of chitosan as a pathogenic agent (Table 11.1).

Studies on vegetation restoration in arid regions with *Nitraria tangutorum* Bobr. revealed a limited influence of AMF or PGPR separately in high saline stress conditions (Pan et al. 2022). However, combined coinoculation with both biostimulants led to a significant improvement in soil structure, but also a clear increase in biomass, the efficiency of its photosynthesis, and the exchange of water and gases through better stomal control.

Feeding the date palm (*Phoenix dactylifera* L.) variety "Bouf-gouss" with compost based on olive pomace from AMF and PGPR made it possible to observe the effect of each biostimulant on the dry weights of the shoots and roots, the stomatal function, and the chlorophyll level in parallel with the uptake of mineral nutrients (Raho et al. 2022). Each couple contributed to at least 100% of the increase in indicators. On the other hand, the combination of both biostimulants and compost led to synergy in the growth of roots, chlorophyll, and carotenoids.

In *Acacia gerrardii*, a synergistic effect between arbuscular mycorrhizal fungi (AMF) and the PGPR Bacillussubtilis (BERA 71) was demonstrated under salinity stress conditions (Hashem et al. 2016). The observed synergy occurred regardless of the presence of stress in the form of salinity. Synergy contributed to the growth of valuable components of the plant, such as total lipid, lignin, tannins, and fiber content. The level of antioxidants and factors that affect the osmoregulatory system were also examined. Their level was also significantly higher as a result of vaccination of the AMF and PGPR pairs.

As salinity has become a serious deleterious factor and the problem is continually increasing, several studies have been reported in a review (Sagar et al. 2021). A common effect of AMF and PGPR coinoculation was observed for crop plants like sorghum, wheat, bean, potato, and maize under salinity stress. The enhancement of the osmoregulatory system was associated with overexpression of HKT and Na^+/H^+ antiporter genes and the synthesis of phytochromes.

The development and increase in AMF and PGPR activity can be achieved indirectly by creating appropriate conditions, e.g., the introduction of biochar and anaerobic digestate as waste from biogas plants in melon cultivation caused a change in the rhizosphere of the microbiome by the growth of their (e.g., *Arthrobacter*, *Mitsuaria*, or *Bacillus*) relative abundance (Ortiz-Liébana et al. 2023). The result was a significant increase in yield and fruit quality parameters (Ortiz-Liébana et al. 2022).

Studies of two strains PGPR *Pseudomonas putida* NBRIRA and *Bacillus amyloliquefaciens* NBRISN13 in chickpea cultivation (*Cicer arietinum* L.) under drought conditions have shown their synergistic interaction (Kumar et al. 2016). Coinoculation with both strains led to enhanced lateral root growth and consequently nutrient and water uptake, resulting in higher growth and yields.

11.4.2 Synergy with AMF and Other Biostimulants

Early studies of the apparently antagonistic effects of saprotrophic fungi (*Trichoderma aureoviride* and *Clonostachys rosea*) on AMF (*Rhizophagus intraradices*) revealed synergy in the growth of Cleopatra mandarin citrus seedlings (*Citrus reshni* Hort. Ex Tan.) (Camprubí et al. 1995). Coinoculation with *T. aureoviride* alone did not contribute to the increased growth of the seedlings. Inoculation with *G. intraradices*, on the other hand, was effective in promoting growth in only one of the three substrates tested. However, synergy in growth stimulation was observed in two of the three substrates tested. The authors of this work also cite other pioneering work in research on the synergy of coinoculation of saprotrophic and mycorrhizal fungi.

A very interesting interaction of organisms from the AMF groups (*Glomus mosseae*) and *Trichoderma aureoviride*, both known as plant biostimulants, was studied in the presence of the potent soil pathogen *Pythium ultimum* in marigold (*Tagetes erecta* L.) (Calvet et al. 1993). A protective effect of AMF on the plant was observed in the presence of *Pythium ultimum*, and no effect of *Trichoderma aureoviride* alone in the presence of the pathogen. However, the increase in biomass was significant, even multiplied in the presence of coinoculation with both strains, i.e., AMF and *Trichoderma aureoviride*, which the authors identify as their synergism with pathogen activity.

Studies on coinoculation *Glomus intraradices BEG72* (currently classified as *Rhizophagus irregularis BEG72) with Trichoderma atroviride MUCL 45632 demonstrated synergy of the chlorophyll SPAD index and biomass growth in tomato, lettuce, and zucchini (*Colla et al. 2015*). The results support previous findings described in the article.*

In a two-year open field study on eggplant "Birgah" F1, a joint interaction of plant-driven protein hydrolysates and AMF (R. irregularis) revealed synergy in the number and weight of marketable fruits (Di Miceli et al. 2023). *Quality of the fruits was measured as reduced glycoalkaloid concentration and their browning potential. Additionally, the efficiency of nitrogen uptake and use was significantly increased in the coinoculation of both biostimulants.*

AMF (Funneliformis mosseae) was stimulated by dopamine in the growth of apple trees under salinity conditions (Gao et al. 2020). *Since both additives are known as biostimulants, their significant promotion of root length, surface area, and diameter has been recognized as synergy and salt resistance. Furthermore, transcriptomic analyses revealed that MdTYDC overexpression improved AMF symbiosis by stimulating the production of flavonoids, including luteolin, kaempferol, and quercetin.*

An in-depth study of the interaction of chitin with AMF was carried out in *Arabidopsis thaliana* (De Tender et al. 2024). Chitin is a biostimulant only in the presence of mycorrhizal fungi. It directly affects the development of these fungi. In addition, it directly affects the germination of the seeds themselves. A possible mechanism in seeds can be based on the action of chitosan, which increases water absorption. *Linnemannia elongata* has also been shown to reduce germination efficiency. This negative effect disappeared when chitin was added to the system. The use of chitin alone reduced the mass of lettuce roots, but the addition of mushrooms together with chitin stimulated their growth. The authors postulate mechanisms involving auxin, enzymes that break down chitin to nitrogen present in fungi, which indirectly contributes to the assimilation of Ca (also from chitin, which comes from crustaceans) and Fe (e.g., through increased production of siderophores).

11.4.3 Other Examples of Synergies

Humic acid and microalgae were introduced into the onion plantation (Popa et al. 2022; Gemin et al. 2019). Synergy on root growth was observed for these biostimulants at their low concentration. Therefore, the use of small amounts of both biostimulants allows the synergistic effect to be used to minimize the number of substances used in onion cultivation.

The combined application of MicroLife Humic Acid Complex® (2% humic acid, 1% organic carbon, 15% humic acid, and 1% fulvic acid) and the commercial biofertilizer Microgeo® categorized as a manure tea composed of organic compounds, active, and dormant cells from various microorganisms (bacteria, yeasts, filamentous fungi, and algae), metabolites, and organomineral chelates on Hawaiian Haze Cannabis (*Cannabis sativa* L.) in an open field study brought new results on the synergism between such a complex combination of biostimulants and biofertilizer (Da Cunha Leme Filho et al. 2024). The use of this combination in 2020, when adverse weather conditions occurred in the form of fewer days for vegetative growth, resulted in a synergistic effect on cannabis height, chlorophyll content, photosynthetic efficiency, and biomass growth. It should be noted that the synergy was not observed in the previous year, which was much more favorable for the plants due to the weather. It is also significant that the preparations used are so complex and rich in both biostimulants and fertilizer components that it cannot be clearly stated whether the synergy results only from the interaction of biostimulants. Nevertheless, these studies show that the use of properly designed combinations of ingredients not only results in their complementary effect but also provides synergy, i.e., a more than additive effect in the development and value of the plant. The same research group also conducted an investigation to determine the relationship between phosphorus absorption and AMF inoculation on the Sun Protection Factor in the cultivation of *Anade-nanthera colubrina* (Vell.) (Falcão et al. 2024). Greenhouse studies have revealed that no synergy is observed in this case and that the two factors are independent.

Studies on the date "Khenizi" showed a synergy between the effects of salicylic acid and chitosan on ripeness, polyphenol content, and shelf life (including the number of microorganisms in the fruit) during storage (Ahmed et al. 2022).

Seaweed extract and 5-aminolevulinic acid sprays on *Asparagus aethiopicus* plants under salt stress revealed the synergy of both biostimulants (Al-Ghamdi and Elansary 2018). The beneficial effects observed as branch length and fresh and dry weight were found to be correlated with the accumulation of specific phenols, the enhancement of transpiration and photosynthetic rates, and stomatal regulation.

11.5 Summary and Outlook

The antagonistic, i.e., competency, inhibiting effect of various biostimulants should be particularly signaled to users so that their use makes sense and does not contribute to the excessive use of resources and energy. So far, the understanding of this sphere has been marginal because only in some cases the lack of synergy has been confirmed, and most often the interaction of various biostimulants is not analyzed at all. However, taking advantage of the synergy effect would not only reduce the number of agricultural treatments and energy consumption but would also have the opportunity to limit ecosystem change and contribute to "field health," which is extremely beneficial, especially in the face of progressive climate change.

In this context, future research should focus on identifying molecular markers of synergistic interactions and validating them across different plant species and soil types. Combining laboratory insights with field-based agronomic trials may pave the way for more precise and predictable biostimulant formulations. Moreover, advancing our mechanistic understanding of biostimulant synergy can inform the development of next-generation bioinoculants tailored to the complexity of the rhizosphere. These directions are essential to realize the full potential of rhizoengineering in sustainable crop production.

References

Abbott LK, Macdonald LM, Wong MTF, Webb MJ, Jenkins SN, Farrell M (2018) Potential roles of biological amendments for profitable grain production – a review. Agric Ecosyst Environ 256:34–50. https://doi.org/10.1016/j.agee.2017.12.021

Ahmed ZFR, Kaur N, Maqsood S, Schmeda-Hirschmann G (2022) Preharvest applications of chitosan, salicylic acid, and calcium chloride have a synergistic effect on quality and storability of date palm fruit (*Phoenix dactylifera* L.). Hort Sci 57:422–430. https://doi.org/10.21273/HORTSCI16416-21

Alam M, Khaliq A, Sattar A, Shukla RS, Anwar M, Dharni S (2011) Synergistic effect of arbuscular mycorrhizal fungi and *Bacillus subtilis* on the biomass and essential oil yield of rose-scented geranium (*Pelargonium graveolens*). Arch Agron Soil Sci 57:889–898. https://doi.org/10.1080/03650340.2010.498013

Al-Ghamdi AA, Elansary HO (2018) Synergetic effects of 5-aminolevulinic acid and *Ascophyllum nodosum* seaweed extracts on asparagus phenolics and stress related genes under saline irrigation. Plant Physiol Biochem 129:273–284. https://doi.org/10.1016/j.plaphy.2018.06.008

Annunziata MG, Ciarmiello LF, Woodrow P, Dell'Aversana E, Carillo P (2019) Spatial and temporal profile of glycine betaine accumulation in plants under abiotic stresses. Front Plant Sci 10:230. https://doi.org/10.3389/fpls.2019.00230

Anuar MSK, Hashim AM, Ho CL, Wong MY, Sundram S, Saidi NB, Yusof MT (2023) Synergism: biocontrol agents and biostimulants in reducing abiotic and biotic stresses in crop. World J Microbiol Biotechnol 39:123. https://doi.org/10.1007/s11274-023-03579-3

Arancon NQ, Solarte Z (2019) Vermiculture in greenhouse plants, field crop production, and hydroponics. In: Oxford research encyclopedia of environmental science. Oxford University Press, Oxford. https://doi.org/10.1093/acrefore/9780199389414.013.295

Asadu CO, Ezema CA, Ekwueme BN, Onu CE, Onoh IM, Adejoh T, Ezeorba TPC, Ogbonna CC, Otuh PI, Okoye JO, Emmanuel UO (2024) Enhanced efficiency fertilizers: overview of production methods, materials used, nutrients release mechanisms, benefits and considerations. Environ Pollut Manag 1:32–48. https://doi.org/10.1016/j.epm.2024.07.002

Asif A, Ali M, Qadir M, Karthikeyan R, Singh Z, Khangura R, Di Gioia F, Ahmed ZFR (2023) Enhancing crop resilience by harnessing the synergistic effects of biostimulants against abiotic stress. Front Plant Sci 14:1276117. https://doi.org/10.3389/fpls.2023.1276117

Augé RM (2001) Water relations, drought and vesicular-arbuscular mycorrhizal symbiosis. Mycorrhiza 11:3–42. https://doi.org/10.1007/s005720100097

Bacilieri FS, de Vasconcelos ACP, Lana RMQ, Mageste JG, Torres JLR (2017) Titanium (Ti) in plant nutrition – a review. Aust J Crop Sci 11:382–386. https://doi.org/10.21475/ajcs.17.11.04.pne202

Balfagón D, Gómez-Cadenas A, Rambla JL, Granell A, de Ollas C, Bassham DC, Mittler R, Zandalinas SI (2022) γ-Aminobutyric acid plays a key role in plant acclimation to a combination of high light and heat stress. Plant Physiol 188:2026–2038. https://doi.org/10.1093/plphys/kiac010

Begum N, Qin C, Ahanger MA, Raza S, Khan MI, Ashraf M, Ahmed N, Zhang L (2019) Role of arbuscular mycorrhizal fungi in plant growth regulation: implications in abiotic stress tolerance. Front Plant Sci 10:1068. https://doi.org/10.3389/fpls.2019.01068

Benito P, Ligorio D, Bellón J, Yenush L, Mulet JM (2022) A fast method to evaluate in a combinatorial manner the synergistic effect of different biostimulants for promoting growth or tolerance against abiotic stress. Plant Methods 18:1–17. https://doi.org/10.1186/s13007-022-00943-6

Bhupenchandra I, Chongtham SK, Devi EL, Ramesh R, Choudhary AK, Salam MD, Sahoo MR, Bhutia TL, Devi SH, Thounaojam AS, Behera C, Harish MN, Kumar A, Dasgupta M, Devi YP, Singh D, Bhagowati S, Devi CP, Singh HR, Khaba CI (2022) Role of biostimulants in mitigating the effects of climate change on crop performance. Front Plant Sci 13:967665. https://doi.org/10.3389/fpls.2022.967665

Biederman LA, Harpole WS (2013) Biochar and its effects on plant productivity and nutrient cycling: a meta-analysis. GCB Bioenergy 5:202–214. https://doi.org/10.1111/gcbb.12037
Bitterlich M, Rouphael Y, Graefe J, Franken P (2018) Arbuscular mycorrhizas: a promising component of plant production systems provided favorable conditions for their growth. Front Plant Sci 9:1329. https://doi.org/10.3389/fpls.2018.01329
Bulgari R, Morgutti S, Cocetta G, Negrini N, Farris S, Calcante A, Spinardi A, Ferrari E, Mignani I, Oberti R, Ferrante A (2017) Evaluation of borage extracts as potential biostimulant using a phenomic, agronomic, physiological, and biochemical approach. Front Plant Sci 8:935. https://doi.org/10.3389/fpls.2017.00935
Cakmak I, Brown P, Colmenero-Flores JM, Husted S, Kutman BY, Nikolic M, Rengel Z, Schmidt SB, Zhao FJ (2023) Micronutrients. In: Marschner's mineral nutrition of plants, 4th edn. Academic, London, pp 283–385. https://doi.org/10.1016/B978-0-12-819773-8.00017-4
Calvet C, Pera J, Barea JM (1993) Growth response of marigold (*Tagetes erecta* L.) to inoculation with *Glomus mosseae, Trichoderma aureoviride* and *Pythium ultimum* in a peat-perlite mixture. Plant Soil 148:1–6. https://doi.org/10.1007/BF02185378
Calvo P, Nelson L, Kloepper JW (2014) Agricultural uses of plant biostimulants. Plant Soil 383:3–41. https://doi.org/10.1007/s11104-014-2131-8
Camprubí A, Calvet C, Estaún V (1995) Growth enhancement of *Citrus reshni* after inoculation with *Glomus intraradices* and *Trichoderma aureoviride* and associated effects on microbial populations and enzyme activity in potting mixes. Plant Soil 173:233–238. https://doi.org/10.1007/BF00011460
Canellas LP, Olivares FL, Aguiar NO, Jones DL, Nebbioso A, Mazzei P, Piccolo A (2015) Humic and fulvic acids as biostimulants in horticulture. Sci Hortic 196:15–27. https://doi.org/10.1016/j.scienta.2015.09.013
Chen THH, Murata N (2011) Glycinebetaine protects plants against abiotic stress: mechanisms and biotechnological applications. Plant Cell Environ 34:1–20. https://doi.org/10.1111/j.1365-3040.2010.02232.x
Chen K, Kleijn D, Scheper J, Fijen TPM (2022) Additive and synergistic effects of arbuscular mycorrhizal fungi, insect pollination and nutrient availability in a perennial fruit crop. Agric Ecosyst Environ 325:107742. https://doi.org/10.1016/j.agee.2021.107742
Colla G, Rouphael Y, Di Mattia E, El-Nakhel C, Cardarelli M (2015) Co-inoculation of *Glomus intraradices* and *Trichoderma atroviride* acts as a biostimulant to promote growth, yield and nutrient uptake of vegetable crops. J Sci Food Agric 95:1706–1715. https://doi.org/10.1002/jsfa.6875
Da Cunha Leme Filho JF, Chim BK, Bermand C, Diatta AA, Thomason WE (2024) Effect of organic biostimulants on cannabis productivity and soil microbial activity under outdoor conditions. J Cannabis Res 6:16. https://doi.org/10.1186/s42238-024-00214-2
Dalpé Y, Monreal M (2004) Arbuscular mycorrhiza inoculum to support sustainable cropping systems. Crop Manag 3:1–11. https://doi.org/10.1094/CM-2004-0301-09-RV
de Assis RMA, Carneiro JJ, Medeiros APR, de Carvalho AA, da Cunha Honorato A, Carneiro MAC, Bertolucci SKV, Pinto JEBP (2020) Arbuscular mycorrhizal fungi and organic manure enhance growth and accumulation of citral, total phenols, and flavonoids in *Melissa officinalis* L. Ind Crop Prod 158:112981. https://doi.org/10.1016/j.indcrop.2020.112981
De Tender C, Vandecasteele M, Ommeslag S, De Zutter N, Vandenbussche E, Haegeman A, Audenaert K, Li L, Vandecasteele B, Voorthuijzen F, Maenhout K, Beirinckx S, Barneveldt R, Goormachtig S, Debode J (2024) *Linnemannia elongata*: a key species in chitin-based plant growth promotion. Phytobiomes J 8:366–377. https://doi.org/10.1094/PBIOMES-05-23-0031-R
Di Miceli G, Vultaggio L, Sabatino L, De Pasquale C, La Bella S, Consentino BB (2023) Synergistic effect of a plant-derived protein hydrolysate and arbuscular mycorrhizal fungi on eggplant grown in open fields: a two-year study. Horticulturae 9:592. https://doi.org/10.3390/horticulturae9050592

du Jardin P (2015) Plant biostimulants: definition, concept, main categories and regulation. Sci Hortic 196:3–14. https://doi.org/10.1016/j.scienta.2015.09.021
Dutta PK, Ravikumar MNV, Dutta J (2002) Chitin and chitosan for versatile applications. J Macromol Sci Part C Polym Rev 42:307–354. https://doi.org/10.1081/MC-120006451
Ehrlich PR, Harte J (2015) Opinion: to feed the world in 2050 will require a global revolution. Proc Natl Acad Sci USA 112:14743–14744. https://doi.org/10.1073/pnas.1519841112
Esparza-Reynoso S, Ayala-Rodríguez JÁ, López-Bucio J (2024) *Pseudomonas putida* configures *Arabidopsis* root architecture through modulating the sensing systems for phosphate and iron acquisition. Plant Sci 342:112028. https://doi.org/10.1016/j.plantsci.2024.112028
Estrada B, Aroca R, Barea JM, Ruiz-Lozano JM (2013) Native arbuscular mycorrhizal fungi isolated from a saline habitat improved maize antioxidant systems and plant tolerance to salinity. Plant Sci 201-202:42–51. https://doi.org/10.1016/j.plantsci.2012.11.009
European Biostimulants Industry Council (EBIC) (2025) Plant biostimulants. https://biostimulants.eu/plant-biostimulants/. Accessed 10 Jan 2025
Fadiji AE, Babalola OO, Santoyo G, Perazzolli M (2022) The potential role of microbial biostimulants in the amelioration of climate change-associated abiotic stresses on crops. Front Microbiol 12:829099. https://doi.org/10.3389/fmicb.2021.829099
Falcão EL, Barreto CB, Hijri M, Bastos Filho CJA, da Silva FSB (2024) No synergy between P and AMF inoculation to improve sun protection factor production in *Anadenanthera colubrina* (Vell.) Brenan leaves. Rhizosphere 30:100916. https://doi.org/10.1016/j.rhisph.2024.100916
Farooq M, Hussain M, Wakeel A, Siddique KHM (2015) Salt stress in maize: effects, resistance mechanisms, and management: a review. Agron Sustain Dev 35:461–481. https://doi.org/10.1007/s13593-015-0287-0
Farrell M, Prendergast-Miller M, Jones DL, Hill PW, Condron LM (2014) Soil microbial organic nitrogen uptake is regulated by carbon availability. Soil Biol Biochem 77:261–267. https://doi.org/10.1016/j.soilbio.2014.07.003
Feng J, Zhao L, Yu Q (2004) Receptor-mediated stimulatory effect of oligochitosan in macrophages. Biochem Biophys Res Commun 317:414–420. https://doi.org/10.1016/j.bbrc.2004.03.048
Figueira-Galán D, Heupel S, Duelli G, Tomasi Morgano M, Stapf D, Requena N (2023) Exploring the synergistic effects of biochar and arbuscular mycorrhizal fungi on phosphorus acquisition in tomato plants by using gene expression analyses. Sci Total Environ 884:163506. https://doi.org/10.1016/j.scitotenv.2023.163506
Franzoni G, Cocetta G, Prinsi B, Ferrante A, Espen L (2022) Biostimulants on crops: their impact under abiotic stress conditions. Horticulturae 8:189. https://doi.org/10.3390/horticulturae8030189
Ganugi P, Masoni A, Pietramellara G, Benedettelli S (2019) A review of studies from the last twenty years on plant–arbuscular mycorrhizal fungi associations and their uses for wheat crops. Agronomy 9:840. https://doi.org/10.3390/agronomy9120840
Gao T, Liu X, Shan L, Wu Q, Liu Y, Zhang Z, Ma F, Li C (2020) Dopamine and arbuscular mycorrhizal fungi act synergistically to promote apple growth under salt stress. Environ Exp Bot 178:104159. https://doi.org/10.1016/j.envexpbot.2020.104159
Gemin LG, Mógor ÁF, De Oliveira Amatussi J, Mógor G (2019) Microalgae associated to humic acid as a novel biostimulant improving onion growth and yield. Sci Hortic 256:108560. https://doi.org/10.1016/j.scienta.2019.108560
Godlewska K, Pacyga P, Michalak I, Biesiada A, Szumny A, Pachura N, Piszcz U (2021) Effect of botanical extracts on the growth and nutritional quality of field-grown white head cabbage (*Brassica oleracea* var. *capitata*). Molecules 26:1992. https://doi.org/10.3390/molecules26071992
Goñi O, Fort A, Quille P, McKeown PC, Spillane C, O'Connell S (2016) Comparative transcriptome analysis of two *Ascophyllum nodosum* extract biostimulants: same seaweed but different. J Agric Food Chem 64:2980–2989. https://doi.org/10.1021/acs.jafc.6b00621

Gorelick J, Rosenberg R, Smotrich A, Hanuš L, Bernstein N (2015) Hypoglycemic activity of withanolides and elicitated *Withania somnifera*. Phytochemistry 116:283–289. https://doi.org/10.1016/j.phytochem.2015.02.029

Guzmán-Báez GA, Trejo-Téllez LI, Ramírez-Olvera SM, Salinas-Ruíz J, Bello-Bello JJ, Alcántar-González G, Hidalgo-Contreras JV, Gómez-Merino FC (2021) Silver nanoparticles increase nitrogen, phosphorus, and potassium concentrations in leaves and stimulate root length and number of roots in tomato seedlings in a hormetic manner. Dose-Response 19:1–15. https://doi.org/10.1177/15593258211044576

Hamid B, Zaman M, Farooq S, Fatima S, Sayyed RZ, Baba ZA, Sheikh TA, Reddy MS, El Enshasy H, Gafur A, Suriani NL (2021) Bacterial plant biostimulants: a sustainable way towards improving growth, productivity, and health of crops. Sustainability 13:2856. https://doi.org/10.3390/su13052856

Harish Prashanth KV, Tharanathan RN (2007) Chitin/chitosan: modifications and their unlimited application potential – an overview. Trends Food Sci Technol 18:117–131. https://doi.org/10.1016/j.tifs.2006.10.022

Hashem A, Abd Allah EF, Alqarawi AA, Al-Huqail AA, Shah MA (2016) Induction of osmoregulation and modulation of salt stress in *Acacia gerrardii* Benth. By arbuscular mycorrhizal fungi and *Bacillus subtilis* (BERA 71). Biomed Res Int 2016:6294098. https://doi.org/10.1155/2016/6294098

Hassan TU, Bano A, Naz I (2017) Alleviation of heavy metals toxicity by the application of plant growth promoting rhizobacteria and effects on wheat grown in saline sodic field. Int J Phytoremediation 19:522–529. https://doi.org/10.1080/15226514.2016.1267696

Huang S, Zheng X, Luo L, Ni Y, Yao L, Ni W (2021) Biostimulants in bioconversion compost of organic waste: a novel booster in sustainable agriculture. J Clean Prod 319:128704. https://doi.org/10.1016/j.jclepro.2021.128704

Iriti M, Picchi V, Rossoni M, Gomarasca S, Ludwig N, Gargano M, Faoro F (2009) Chitosan antitranspirant activity is due to abscisic acid-dependent stomatal closure. Environ Exp Bot 66:493–500. https://doi.org/10.1016/j.envexpbot.2009.01.004

Juárez-Maldonado A, Ortega-Ortíz H, Morales-Díaz AB, González-Morales S, Morelos-Moreno Á, Cabrera-De la Fuente M, Sandoval-Rangel A, Cadenas-Pliego G, Benavides-Mendoza A (2019) Nanoparticles and nanomaterials as plant biostimulants. Int J Mol Sci 20:162. https://doi.org/10.3390/ijms20010162

Kapoore RV, Wood EE, Llewellyn CA (2021) Algae biostimulants: a critical look at microalgal biostimulants for sustainable agricultural practices. Biotechnol Adv 49:107754. https://doi.org/10.1016/j.biotechadv.2021.107754

Karthikeyan B, Abitha B, Henry AJ, Sa T, Joe MM (2016) Interaction of rhizobacteria with arbuscular mycorrhizal fungi (AMF) and their role in stress abetment in agriculture. In: Pagano M, Dhar PP (eds) Recent advances on mycorrhizal fungi. Springer, Cham. https://doi.org/10.1007/978-3-319-24355-9_11

Khan W, Rayirath UP, Subramanian S, Jithesh MN, Rayorath P, Hodges DM, Critchley AT, Craigie JS, Norrie J, Prithiviraj B (2009) Seaweed extracts as biostimulants of plant growth and development. J Plant Growth Regul 28:386–399. https://doi.org/10.1007/s00344-009-9103-x

Kobayashi T, Nozoye T, Nishizawa NK (2019) Iron transport and its regulation in plants. Free Radic Biol Med 133:11–20. https://doi.org/10.1016/j.freeradbiomed.2018.10.439

Kumar M, Mishra S, Dixit V, Kumar M, Agarwal L, Chauhan PS, Nautiyal CS (2016) Synergistic effect of *Pseudomonas putida* and *Bacillus amyloliquefaciens* ameliorates drought stress in chickpea (*Cicer arietinum* L.). Plant Signal Behav 11:e1071004. https://doi.org/10.1080/15592324.2015.1071004

Kumar S, Kumar S, Mohapatra T (2021) Interaction between macro- and micro-nutrients in plants. Front Plant Sci 12:665583. https://doi.org/10.3389/fpls.2021.665583

Kuwada K, Wamocho LS, Utamura M, Matsushita I, Ishii T (2006) Effect of red and green algal extracts on hyphal growth of arbuscular mycorrhizal fungi, and on mycorrhizal development and growth of papaya and passionfruit. Agron J 98:1340–1344. https://doi.org/10.2134/agronj2005.0354

Kužel S, Hruby M, Cígler P, Tlustoš P, Van PN (2003) Mechanism of physiological effects of titanium leaf sprays on plants grown on soil. Biol Trace Elem Res 91:179–189. https://doi.org/10.1385/BTER:91:2:179

La Torre A, Battaglia V, Caradonia F (2016) An overview of the current plant biostimulant legislations in different European member states. J Sci Food Agric 96:727–734. https://doi.org/10.1002/jsfa.7358

Laird DA (2008) The charcoal vision: a win–win–win scenario for simultaneously producing bioenergy, permanently sequestering carbon, while improving soil and water quality. Agron J 100:178–181. https://doi.org/10.2134/agronj2007.0161

Lamar RT (2020) Possible role for electron shuttling capacity in elicitation of PB activity of humic substances on plant growth enhancement. In: Geelen D, Xu L (eds) The chemical biology of plant biostimulants. Wiley, Hoboken, pp 97–121. https://doi.org/10.1002/9781119357254.ch4

Lee Y, Krishnamoorthy R, Selvakumar G, Kim K, Sa T (2015) Alleviation of salt stress in maize plant by co-inoculation of arbuscular mycorrhizal fungi and *Methylobacterium oryzae* CBMB20. J Korean Soc Appl Biol Chem 58:533–540. https://doi.org/10.1007/s13765-015-0072-4

Lehmann J, Kleber M (2015) The contentious nature of soil organic matter. Nature 528:60–68. https://doi.org/10.1038/nature16069

Li X, Zhou M, Shi F, Meng B, Liu J, Mi Y, Dong C, Su H, Liu X, Wang F, Wei Y (2023) Influence of arbuscular mycorrhizal fungi on mercury accumulation in rice (*Oryza sativa* L.): from enriched isotope tracing perspective. Ecotoxicol Environ Saf 255:114776. https://doi.org/10.1016/j.ecoenv.2023.114776

Malerba M, Cerana R (2019) Recent applications of chitin- and chitosan-based polymers in plants. Polymers 11:839. https://doi.org/10.3390/polym11050839

Manghwar H, Zaman W, Kumari M, Swarupa P, Kumar Kesari K, Kumar A (2022) Microbial inoculants as plant biostimulants: a review on risk status. Life 13:12. https://doi.org/10.3390/life13010012

Mannino G (2023) A new era of sustainability: plant biostimulants. Int J Mol Sci 24:16329. https://doi.org/10.3390/ijms242216329

Miransari M (2011) Interactions between arbuscular mycorrhizal fungi and soil bacteria. Appl Microbiol Biotechnol 89:917–930. https://doi.org/10.1007/s00253-010-3004-6

Moreno-Hernández JM, Benítez-García I, Mazorra-Manzano MA, Ramírez-Suárez JC, Sánchez E (2020) Strategies for production, characterization and application of protein-based biostimulants in agriculture: a review. Chil J Agric Res 80:274–289. https://doi.org/10.4067/S0718-58392020000200274

Mpongwana S, Manyevere A, Mupangwa J, Mpendulo CT, Mashamaite CV (2023) Optimizing biomass yield of three herbaceous forage legumes through dual inoculation of arbuscular mycorrhizal fungi and rhizobia. S Afr J Bot 159:61–71. https://doi.org/10.1016/j.sajb.2023.06.006

Nardi S, Muscolo A, Vaccaro S, Baiano S, Spaccini R, Piccolo A (2007) Relationship between molecular characteristics of soil humic fractions and glycolytic pathway and Krebs cycle in maize seedlings. Soil Biol Biochem 39:3138–3146. https://doi.org/10.1016/j.soilbio.2007.07.006

Nardi S, Pizzeghello D, Schiavon M, Ertani A (2016) Plant biostimulants: physiological responses induced by protein hydrolyzed-based products and humic substances in plant metabolism. Sci Agric 73:18–23. https://doi.org/10.1590/0103-9016-2015-0006

Ortiz-Liébana N, Zotti M, Barquero M, González-Andrés F (2022) An organic fertilizer 'doped' with a *Bacillus* strain improves melon and pepper yield, modifying the rhizosphere microbiome with negligible changes in the bulk soil microbiome. Agronomy 12:2620. https://doi.org/10.3390/agronomy12112620

Ortiz-Liébana N, Zotti M, Barquero M, González-Andrés F (2023) Biochar + AD exerts a biostimulant effect in the yield of horticultural crops and improves bacterial biodiversity and species richness in the rhizosphere. Sci Hortic 321:112277. https://doi.org/10.1016/j.scienta.2023.112277

Pan J, Huang CH, Peng F, Wang T, Liao J, Ma SX, You QG, Xue X (2022) Synergistic combination of arbuscular mycorrhizal fungi and plant growth-promoting rhizobacteria modulates morpho-physiological characteristics and soil structure in Nitraria tangutorum Bobr. Under saline soil conditions. Res Cold Arid Reg 14:393–402. https://doi.org/10.1016/j.rcar.2023.02.002

Pellegrino E, Öpik M, Bonari E, Ercoli L (2015) Responses of wheat to arbuscular mycorrhizal fungi: a meta-analysis of field studies from 1975 to 2013. Soil Biol Biochem 84:210–217. https://doi.org/10.1016/j.soilbio.2015.02.020

Pérez-De-Luque A, Tille S, Johnson I, Pascual-Pardo D, Ton J, Cameron DD (2017) The interactive effects of arbuscular mycorrhiza and plant growth-promoting rhizobacteria synergistically enhance host plant defences against pathogen. Sci Rep 7:16409. https://doi.org/10.1038/s41598-017-16697-4

Plenchette C, Clermont-Dauphin C, Meynard JM, Fortin JA (2005) Managing arbuscular mycorrhizal fungi in cropping systems. Can J Plant Sci 85:31–40. https://doi.org/10.4141/P03-159

Popa DG, Lupu C, Constantinescu-Aruxandei D, Oancea F (2022) Humic substances as microalgal biostimulants – implications for microalgal biotechnology. Mar Drugs 20:1–27. https://doi.org/10.3390/md20050327

Poveda J (2021) Insect frass in the development of sustainable agriculture: a review. Agron Sustain Dev 41:5. https://doi.org/10.1007/s13593-020-00656-x

Poveda J, Baptista P (2021) Filamentous fungi as biocontrol agents in olive (*Olea europaea* L.) diseases: mycorrhizal and endophytic fungi. Crop Prot 146:105672. https://doi.org/10.1016/j.cropro.2021.105672

Raho O, Boutasknit A, Anli M, Ben-Laouane R, Rahou YA, Ouhaddou R, Duponnois R, Douira A, Modafar CE, Meddich A (2022) Impact of native biostimulants/biofertilizers and their synergistic interactions on the agro-physiological and biochemical responses of date palm seedlings. Gesunde Pflanz 74:1053–1069. https://doi.org/10.1007/s10343-022-00668-5

Ramírez MA, Rodríguez AT, Alfonso L, Peniche C (2010) Chitin and its derivatives as biopolymers with potential agricultural applications. Biotecnol Apl 27:270–276

Rentsch D, Schmidt S, Tegeder M (2007) Transporters for uptake and allocation of organic nitrogen compounds in plants. FEBS Lett 581:2281–2289. https://doi.org/10.1016/j.febslet.2007.04.013

Rouphael Y, Colla G (2018) Synergistic biostimulatory action: designing the next generation of plant biostimulants for sustainable agriculture. Front Plant Sci 9:1655. https://doi.org/10.3389/fpls.2018.01655

Sagar A, Rathore P, Ramteke PW, Ramakrishna W, Reddy MS, Pecoraro L (2021) Plant growth promoting rhizobacteria, arbuscular mycorrhizal fungi and their synergistic interactions to counteract the negative effects of saline soil on agriculture: key macromolecules and mechanisms. Microorganisms 9:1491. https://doi.org/10.3390/microorganisms9071491

Sani MNH, Yong JWH (2021) Harnessing synergistic biostimulatory processes: a plausible approach for enhanced crop growth and resilience in organic farming. Biology 11:41. https://doi.org/10.3390/biology11010041

Schiavon M, Ertani A, Nardi S (2008) Effects of an alfalfa protein hydrolysate on the gene expression and activity of enzymes of the tricarboxylic acid (TCA) cycle and nitrogen metabolism in *Zea mays* L. J Agric Food Chem 56:11800–11808. https://doi.org/10.1021/jf802362g

Shahrajabian MH, Chaski C, Polyzos N, Tzortzakis N, Petropoulos SA (2021) Sustainable agriculture systems in vegetable production using chitin and chitosan as plant biostimulants. Biomolecules 11:819. https://doi.org/10.3390/biom11060819

Shakeel M, Rais A, Hassan MN, Hafeez FY (2015) Root associated *Bacillus* sp. improves growth, yield and zinc translocation for basmati rice (*Oryza sativa*) varieties. Front Microbiol 6:1286. https://doi.org/10.3389/fmicb.2015.01286

Shoresh M, Harman GE, Mastouri F (2010) Induced systemic resistance and plant responses to fungal biocontrol agents. Annu Rev Phytopathol 48:21–43. https://doi.org/10.1146/annurev-phyto-073009-114450

Sible CN, Seebauer JR, Below FE (2021) Plant biostimulants: a categorical review, their implications for row crop production, and relation to soil health indicators. Agronomy 11:1297. https://doi.org/10.3390/agronomy11071297

Singh R, Kaur S, Bhullar SS, Singh H, Sharma LK (2024) Bacterial biostimulants for climate smart agriculture practices: mode of action, effect on plant growth and roadmap for commercial products. J Sustain Agric Environ 3:e12085. https://doi.org/10.1002/sae2.12085

Sun W, Shahrajabian MH (2023) The application of arbuscular mycorrhizal fungi as microbial biostimulant, sustainable approaches in modern agriculture. Plants 12:3101. https://doi.org/10.3390/plants12173101

Tan Q, Guo Q, Wei R, Zhu G, Du C, Hu H (2023) Influence of arbuscular mycorrhizal fungi on bioaccumulation and bioavailability of as and Cd: a meta-analysis. Environ Pollut 316:120619. https://doi.org/10.1016/j.envpol.2022.120619

Tilman D, Cassman KG, Matson PA, Naylor R, Polasky S (2002) Agricultural sustainability and intensive production practices. Nature 418:671–677. https://doi.org/10.1038/nature01014

Trejo-Téllez LI, Gómez-Merino FC (2023) Editorial: beneficial elements: novel players in plant biology for innovative crop production, volume II. Front Plant Sci 14:1303462. https://doi.org/10.3389/fpls.2023.1303462

Ugarte RA, Craigie JS, Critchley AT (2010) Fucoid flora of the rocky intertidal of the Canadian Maritimes: implications for the future with rapid climate change. In: Israel A, Einav R, Seckbach J (eds) Seaweeds and their role in globally changing environments. Springer, Dordrecht, pp 69–90. https://doi.org/10.1007/978-90-481-8569-6_5

United Nations Department of Economic and Social Affairs (2025) The world population is projected to reach 9.8 billion in 2050 and 11.2 billion in 2100. https://www.un.org/en/desa/world-population-projected-reach-98-billion-2050-and-112-billion-2100. Accessed 10 Apr 202

Van Oosten MJ, Pepe O, De Pascale S, Silletti S, Maggio A (2017) The role of biostimulants and bioeffectors as alleviators of abiotic stress in crop plants. Chem Biol Technol Agric 4:5. https://doi.org/10.1186/s40538-017-0089-5

Vranova V, Rejsek K, Skene KR, Formanek P (2011) Non-protein amino acids: plant, soil and ecosystem interactions. Plant Soil 342:31–48. https://doi.org/10.1007/s11104-010-0673-y

Wang Q, Ren X, Sun Y, Zhao J, Awasthi MK, Liu T, Li R, Zhang Z (2021) Improvement of the composition and humification of different animal manures by black soldier fly bioconversion. J Clean Prod 278:123397. https://doi.org/10.1016/j.jclepro.2020.123397

Weber K, Quicker P (2018) Properties of biochar. Fuel 217:240–261. https://doi.org/10.1016/j.fuel.2017.12.054

Willis A, Rodrigues BF, Harris PJC (2013) The ecology of arbuscular mycorrhizal fungi. Crit Rev Plant Sci 32:1–20. https://doi.org/10.1080/07352689.2012.683375

Wong WS, Tan SN, Ge L, Chen X, Yong JWH (2015) The importance of phytohormones and microbes in biofertilizers. In: Arora NK (ed) Plant microbes symbiosis: applied facets. Springer, New Delhi, pp 105–158. https://doi.org/10.1007/978-3-319-24654-3_6

Wong WS, Zhong HT, Cross AT, Yong JWH (2020) Plant biostimulants in vermicomposts. In: Geelen D, Xu L (eds) The chemical biology of plant biostimulants. Wiley, Hoboken, pp 155–180. https://doi.org/10.1002/9781119357254.ch6

Yakhin OI, Lubyanov AA, Yakhin IA, Brown PH (2017) Biostimulants in plant science: a global perspective. Front Plant Sci 7:2049. https://doi.org/10.3389/fpls.2016.02049

Yang H, Yan R, Chen H, Lee DH, Zheng C (2007) Characteristics of hemicellulose, cellulose and lignin pyrolysis. Fuel 86:1781–1788. https://doi.org/10.1016/j.fuel.2006.12.013

Zhang H, Tan SN, Wong WS, Ng CYL, Teo CH, Ge L, Chen X, Yong JWH (2014) Mass spectrometric evidence for the occurrence of plant growth promoting cytokinins in vermicompost tea. Biol Fertil Soils 50:401–403. https://doi.org/10.1007/s00374-013-0846-y

Zhang H, Tan SN, Teo CH, Yew YR, Ge L, Chen X, Yong JWH (2015) Analysis of phytohormones in vermicompost using a novel combinative sample preparation strategy of ultrasound-assisted extraction and solid-phase extraction coupled with liquid chromatography–tandem mass spectrometry. Talanta 139:189–197. https://doi.org/10.1016/j.talanta.2015.02.052

Zhang Z, Li F, Nie Y, Zhang X, Zhang S, Ji X (2021) Zinc-based deep eutectic solvent – an efficient carbonic anhydrase mimic for CO_2 hydration and conversion. Sep Purif Technol 276:119446. https://doi.org/10.1016/j.seppur.2021.119446

Zúñiga-Dávila F, González-Andrés F, Ormeño-Orrillo E (eds) (2019) Microbial probiotics for agricultural systems: advances in agronomic use. Springer, Cham

Chapter 12
Engineering Biodegradable Inorganic Materials as Next-Generation Biostimulants

Salma Batool, Mohamed Jekeen Fathima Hafsa, and Malik Abdul Rahim

Abstract Biodegradable inorganic biostimulants have been developed, resulting in a transition toward sustainable agriculture to secure long-term food security and environmental health. Although traditional NPK fertilizers have been used to enhance yields in the past, their continued application has led to immense soil erosion, contamination of groundwater, and reduced productivity. This chapter examines the merging of nanotechnology and biotechnology to develop smart biostimulants that use inorganic substances such as silicates, carbonates, phosphates, and metal nanoparticles. These materials are geared toward soil ecosystem compatibility to achieve international sustainability objectives. These biostimulants are used to enhance nutrient uptake using microbial synthesis and green formulation approaches. They also control rhizospheric microbes and stimulate plant antioxidant defense mechanisms against abiotic stresses, such as drought, salinity, and heavy metal toxicity. Nanoencapsulation and controlled release systems are technological advances that are necessary for better delivery and less environmental runoff. These systems can release active chemicals at a designated time or according to environmental conditions. Irrespective of technical obstacles in standardization and regulatory uncertainties, the shift to intelligent and data-driven biohybrid systems provides a disruptive avenue for precise agriculture. This would make these formulations eco-friendly, as biodegradable polymers would be used as nanovehicles. Future trends show the development of smart sensors and Internet of Things (IoT) technologies that ensure stable fields and reduce waste to a minimum when using biostimulants. This transition from using traditional organics to new inorganic technologies is a major milestone toward creating a circular bioeconomy. Modern agriculture can reduce nutrient deficiency by addressing biodegradability and accurate delivery, which safeguards soil biodiversity.

S. Batool (✉) · M. J. F. Hafsa · M. A. Rahim
Department of Basic and Applied Chemistry, Faculty of Science & Technology, University of Central Punjab, Lahore, Pakistan
e-mail: salma.batool@ucp.edu.pk

S. Mehnaz, I. Shahid (eds.), *Frontiers in Rhizoengineering*, Microorganisms for Sustainability 65, https://doi.org/10.1007/978-981-92-3665-7_12

Keywords Biodegradable biostimulants · Nanobiofertilizers · Rhizoengineering · Sustainable agriculture · Inorganic nanoparticles · Microbial synthesis · Abiotic stress mitigation · Precision agriculture

12.1 Introduction

12.1.1 The Development of Biostimulants: Organics to Inorganics

The development of biostimulants in agriculture has advanced significantly, from traditional organic materials to new inorganic technologies. Historically, synthetic NPK chemical fertilizers have played a pivotal role in agriculture by resulting in better yields, but they have also contributed to soil contamination, degradation, and fertility decline. This has led to the development of biofertilizers derived from natural microorganisms that nourish plants and kill plant pathogens. To meet the demand for sustainability, biotechnology has introduced plant growth-promoting rhizobacteria (PGPR) that can enhance soil fertility and efficiency in terms of crop yield. Another significant step was the utilization of rhizobacteria as natural nanofactories for the microbial synthesis of nanoparticles as potential biostimulants and fertilizers. Combining biotechnology and nanotechnology, nanobiofertilizers improved solving the limitations in microbial stability and nutrient bioavailability (Patel et al. 2023). The synergism of inorganic nanoformulations with microbes results in better absorption of nutrients and eco-friendliness (Rautela et al. 2021). The switch from organic to inorganic substances, mainly nanosized carriers and loaded nutrients, has led to the development of smart and eco-friendly biostimulants.

12.1.2 The Importance of Biodegradability in Modern Agriculture

In light of current commercial agricultural products, biodegradability is a crucial aspect, particularly for nanotechnology-based fertilizers. The remnants of the prevailing chemical fertilizers in water and soil cause prolonged toxicity and environmental destruction. (Ghormade et al. 2011).

The use of biodegradable polymers in nanoformulations as nanovehicles carrying inorganic biostimulants ensures soil ecosystem compatibility, decreases environmental risks, and promotes global sustainability goals. Moreover, biologically produced nanofertilizers made by microbial means are comparatively more compatible with the ecosystem and reduce the toxicity of production(Ijaz et al. 2020; Ying et al. 2022). Inorganic biostimulants (IBS) designed by incorporating biodegradable nanoparticles are a sustainable and eco-friendly approach to mitigate nutrient deficiency and improve yield in the agricultural sector.

12.1.3 The Inorganic Biostimulants in Rhizoengineering and Sustainable Farming

Biostimulants of inorganic origin, particularly nanofertilizers, hold great potential in rhizoengineering and sustainable agriculture. These substances enhance the interaction between plant roots and beneficial microorganisms by releasing nutrients in the rhizosphere (Iqbal et al. 2019). This precise usage increases the efficiency of nutrient use, reduces leaching loss, and promotes the colonization of roots by the population of PGPR and other beneficial microbes (Rautela et al. 2021).

Inorganic nanoparticles added to bionanofertilizers are nutrient delivery systems, microbial activity, and secondary metabolite stimulators that improve plant development and resilience (Panichikkal et al. 2019; Arora et al. 2024). Such developments contribute to firmer and controlled rhizospheric conditions, which are crucial for soil health and sustainable agricultural production (Table 12.1). Furthermore, the application of nano-based approaches is in accordance with the broader goals in the field of precision agriculture, where biosensors and tailor-made nutrient solutions are applied to improve resource efficiency (Kah et al. 2018). The introduction of IBS in rhizoengineering is a step toward attaining agroecological balance and food security at the global level (Arora et al. 2024).

Table 12.1 Efficacy of inorganic biostimulants (IBS) in rhizoengineering

Inorganic biostimulant	Efficacy	References
Silicon (monosilicic acid)	Growth, enhanced nutrient access, stress resistance in rhizosphere; increased root microbe interactions	Verma et al. (2022)
Zinc oxide nanoparticles	Increased germination, root development, and antioxidative and alleviated biotic/abiotic stress	Verma et al. (2024)
Titanium dioxide nanoparticles	Enhanced chlorophyll, antioxidant enzyme activity, enhanced growth, and alleviation of salt/drought stress	Khalid et al. (2024)
Selenium nanoparticles	Increased germination, antioxidant status, and growth; candidate multifunctional biostimulant	Bărbieru et al. (2019)
Copper-containing formulations/Cu NPs	In association with vegetal biostimulants, higher yield and photosynthetic performance	Rouphael et al. (2023)
Biogenic ZnO-NPs	Stimulated in vitro growth, elevated chlorophyll/carotenoids, and improved micropropagation parameters	Regni et al. (2024)

12.2 Basic Principles of Biodegradable Inorganic Biostimulants

12.2.1 Definition and Essential Characteristics

Biodegradable IBS are new agricultural inputs developed through the combination of inorganic nanosubstances (e.g., hydroxyapatite or silica nanoparticles) and biodegradable organic substances (e.g., humic or fulvic acids). These formulations do not act as conventional fertilizers by directly supplying nutrients. Instead, they trigger physiological processes in plants to enhance nutrient uptake, improve resistance to abiotic stress, and eventually improve the quality and quantity of crops (Du Jardin 2015). According to the European Biostimulants Industry Council (EBIC), biostimulants refer to substances and/or microorganisms that can stimulate natural processes when applied to plants or the rhizosphere to enhance nutrient uptake, nutrient efficiency, abiotic stress tolerance, and crop quality (Xu and Geelen 2018).

Giordana et al. (2023) stated that the possibility of developing multifunctional and nontoxic nanocomposites by selecting suitable solids can be suitable for use in agronomy, with multiple possible beneficial effects, including nourishing crops, improving the plant, and improving the soil.

12.2.2 Inorganic Matrices

The inorganic matrices used in their formulation are of great significance for the effectiveness and biodegradability of IBS. Hydroxyapatite ($Ca_{10}(PO_4)_6(OH)_2$) (HP) is characterized by its high phosphate content and compatibility with biological systems. HP nanoparticles are known to be efficient slow-release phosphorus fertilizers, which are more efficient when surface-engineered with plant-friendly additives, such as citric acid or humic acids (Giordana et al. 2023).

Nano-calcium carbonate (NCC) is known to have exceptional biocompatibility and low toxicity, with insignificant adverse effects on the photosynthesis of plants (Boyjoo et al. 2014). Plants absorb NCC better than standard calcium carbonate particles because of its smaller particle size and high adhesion (Hua et al. 2015). The main constituent of NCC, calcium, plays a crucial role in the growth and development of plants, cell wall formation, environmental stress resistance, and photosynthetic processes to a large degree (Tantawy et al. 2014).

Other commonly used substrates include silica nanoparticles (SiO_2), which are commonly produced from agricultural waste, such as rice husks. Amorphous silica particles provide mechanical stability to organic biostimulants and increase plant resistance to pathogens and abiotic stress (Giordana et al. 2023).

The thermoprocessing of biological waste products, such as salmon bones (Salmo salar), can be used to produce calcium phosphate (CaP) particles in a circular economy approach to the production of biostimulants. CaP materials can be modified with the addition of humic substances (HS) to form a biostimulant, which

Table 12.2 Important biodegradable inorganic biostimulants (IBS)

Material type	Examples	Functions in biostimulants	Reference
Silicates	Diatomaceous earth, biosilica	Enhance photosynthesis, structural support	Samani et al. (2024)
Carbonates	Magnesium carbonate, calcium carbonate	Increased biomass, source of nutrient, antifungal activity	Nelwamondo et al. (2023)
Phosphates	KH_2PO_4, nanophosphate from agro-waste	Slow-release P delivery, improved nutrient efficiency	Noruzi et al. (2023)
Metal nanoparticles	Zinc, iron (Fe), copper (Cu)	Micronutrient supply, promoted growth	Anuta et al. (2025)

can be used to improve their performance in agricultural applications. The synergistic effect of CaP and HS is aimed at improving nutrient absorption, particularly increasing phosphorus (P) absorption, because of the limited bioavailability of mineral CaP in neutral and alkaline conditions (Adamiano et al. 2021).

Zinc oxide nanoparticles (ZnO NPs) are among the most widely used nanoparticles in the agricultural industry (Timoshenko et al. 2021). ZnO NPs have specific antimicrobial effects that can alter key microbial activities in the soil, including mineralization, nitrogen fixation, and plant-stimulating activities. In contrast to bulk ZnO particles (which have a correspondingly large spherical diameter of above 100 nm and are largely insoluble in water), ZnO NPs are more readily and to a larger extent dissolved in soils because of thermodynamic and kinetic considerations (Strekalovskaya et al. 2024), as summarized in Table 12.2.

As Xu and Geelen (2018) emphasized, the concept of a circular economy is aimed at turning waste products and objects that have reached their end of use into new resources, which will support inorganic biostimulant usage in accordance with environmental and regulatory objectives.

12.2.3 Mechanisms of Biodegradation in Soil and Rhizosphere

The degradability and behavior of these biostimulants in soil environments largely depend on the solubility of the inorganic matrix, pH of the rhizosphere, and the interaction between the biostimulants and the organic coating (Fig. 12.1). Silica nanoparticles, especially those derived through the digestion of rice husk, are in amorphous forms and highly porous in nature, which prolong their interaction with soil microbiota and retain water (Giordana et al. 2023).

The biostimulatory effect of the organic component also promotes biodegradation by accelerating the colonization of microbes and activating rhizospheric enzymes. The interaction between the mineral degradation process and the organic process ensures that the material breaks down without the formation of residues and enhances its appropriateness for sustainable and regenerative farming (Giordana et al. 2023).

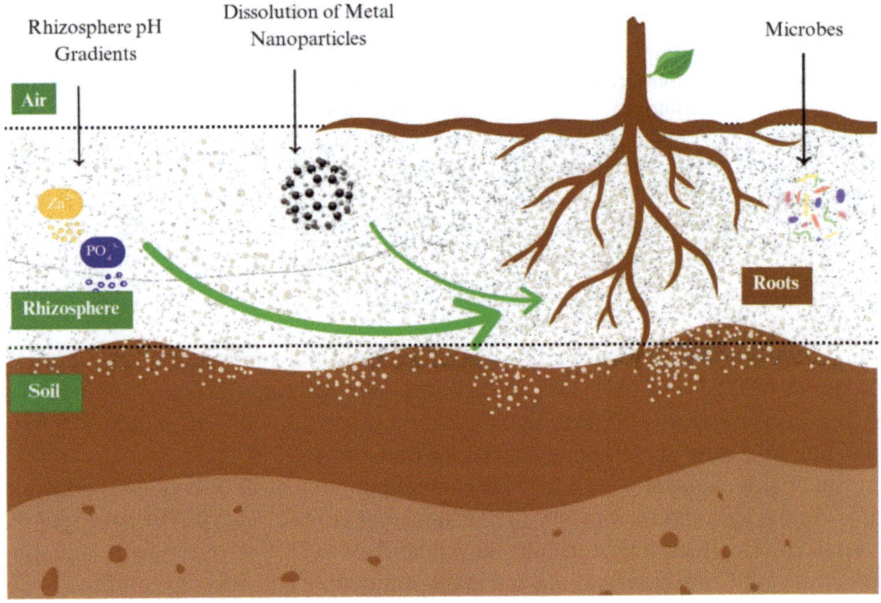

Fig. 12.1 Action of engineered biostimulant particles with soil and rhizosphere

12.3 Technological Advancements in Biodegradable Inorganic Formulations

12.3.1 Nanoformulations: Improve Efficiency and Degradability

Nanoformulations enhance the efficiency and degradability of biostimulants by enhancing their delivery, persistence, and uptake by plants (Table 12.3, Fig. 12.2). Nanoencapsulation enhances both the activity and duration of active substances and minimizes environmental losses (Magnabosco et al. 2023). Biodegradable IBS are nanoformulations that involve the use of nanomaterials to hold significant nutrients and ensure that they are not degraded by environmental stressors (Eliaspour et al. 2020).

Nanoparticles are small, and their surface characteristics allow them to be absorbed and internalized more efficiently by plants, leading to high bioavailability (Toksha et al. 2021; Mahapatra et al. 2022). Nanofertilizers have a high surface-to-volume ratio, which offers the possibility of absorbing high nutrient levels with low application quantities, thereby reducing environmental runoff and toxicity (Mahapatra et al. 2022). Moreover, the application of green synthesis strategies, such as using plant extracts or microorganisms, enhances the production of biodegradable nanoparticles, which is in line with the goal of eco-sustainability (Zohora et al. 2017).

Table 12.3 Advantages of nanoformulations of biostimulants

Feature	Benefit	Reference
Small particle size	Enhanced plant absorption and internalization	Mahapatra et al. (2022)
Surface-to-volume ratio	Higher nutrient efficiency with lower doses	Toksha et al. (2021)
Nanoencapsulation	Reduced environmental loss, prolonged activity	Eliaspour et al. (2020)
Green synthesis (e.g., plant based)	Biodegradability and eco-compatibility	Zohora et al. (2017)

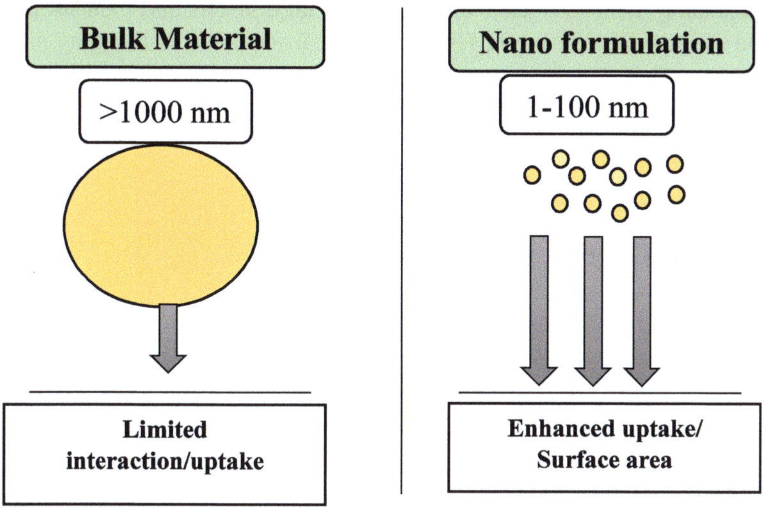

Fig. 12.2 Comparison of bulk material and nanoformulation behavior in agrochemical applications

12.3.2 Biohybrid Systems: The Combination of Biopolymers and Microbial Elements

The combination of inorganic nanoparticles, organic biopolymers, and microbial components develops biohybrid systems to enhance their compatibility and functionality. Such hybrid materials leverage the structural and biochemical properties of biopolymers, such as proteins or polysaccharides, which leads to better dispersion, adhesion, and environmental safety (Hao et al. 2020a, b). By including microbial metabolites or bio-derived carriers, plant resistance and nutrient-use efficiency can be increased, thereby enhancing synergistic interactions (Adamu et al. 2021). These technologies not only allow the distribution of biostimulants in an environmentally friendly manner but also limit the reliance on synthetic carriers, which expands the idea of a circular bioeconomy (Magnabosco et al. 2023).

12.3.3 *Controlled Release Systems and Effective Delivery Systems*

In biostimulant formulations, controlled-release and smart delivery systems are a major breakthrough, enabling the release of active chemicals at a certain time, location, or in response to environmental conditions (Hao et al. 2020a, b). Such systems employ bio-based encapsulants to release nutrients over time (Manjunatha et al. 2016). The coatings also ensure that nutrients are provided over a period, reducing losses due to leaching and volatilization (Mukherjee et al. 2016). These techniques also reduce the degradation of payloads caused by light, pH, or microbial activity, which enhances their resistance to environmental disasters and their usefulness in the field (Yang et al. 2020). pH-sensitive nanocarriers and UV-blocking nanostructures are materials that facilitate the tight regulation of the release kinetics of the maximum availability of nutrients or bioactives at the most critical growth stages, as shown in Table 12.4 (Magnabosco et al. 2023).

12.4 Plant Growth and Stress Reducing Mechanisms

12.4.1 *Improved Uptake of Nutrients and Mobilization*

Biodegradable *IBS* enhances nutrient absorption and mobilization in plants in numerous ways (Table 12.5). These chemicals enhance root system engineering, which creates a greater effective soil exploration region and access to macro- and micronutrients (Colla et al. 2014; Colla et al. 2017). Certain biostimulants stimulate

Table 12.4 Controlled release of biostimulants

Delivery system type	Trigger/property	Function	References
Nanocarriers which are pH-sensitive	pH changes in the root system	Target nutrient release	Magnabosco et al. (2023)
UV-blocking nanostructures	Sunlight exposure	Protection from degradation due to UV	Yang et al. (2020)
Biobased encapsulants	Moisture/soil interaction	Slow, steady nutrient availability	Manjunatha et al. (2016)

Table 12.5 Biostimulant effects on nutrient uptake and mobilization in plants

Mechanism	Effect on plant	Supporting studies
Root architecture enhancement	Greater soil exploration, improved nutrient access	Colla et al. (2014); Colla et al. (2017)
Enzyme activation (e.g., nitrogen metabolism)	Better N use, micronutrient mobilization	(Rouphael and Colla (2018)
Chelating/complexation properties	Enhanced micronutrient uptake (Fe, Zn, etc.)	(Du Jardin (2015)

the action of important enzymes that enhance the solubility and movement of nutrients, such as phosphorus, iron, and zinc (Colla et al. 2015; Rouphael and Colla 2018).

Moreover, the chelating effect of several biostimulants and micronutrient complexation enhances root uptake, especially under nutrient-deprived conditions such as phosphorous limitation and iron deficiency (Colla and Rouphael 2015; Du Jardin 2015). Microbial agents induce metabolic interactions in the rhizosphere, promoting nutrient bioavailability by facilitating nutrient mobilization (De Pascale et al. 2017).

12.4.2 Rhizospheric Microbiota Regulation

Biostimulants are significant in controlling rhizosphere microbiota by enhancing the establishment of beneficial microbial communities in the plant root zone (Colla et al. 2017). This type of modulation boosts nutrient cycling, root colonization, and stress resistance (Rouphael and Colla 2018). Microbial inoculants such as plant growth-promoting rhizobacteria (PGPR) include *Bacillus, Pseudomonas, Azospirillum, Rhizobium, and arbuscular mycorrhizal fungi*, which are widely applied to boost nutrient cycling, root colonization, and plant growth by various mechanisms such as nutrient solubility, phytohormone synthesis, and resistance to stress (Van Oosten et al. 2017; Rouphael and Colla 2018).

Recent studies have proposed that several bacteria related to the rhizosphere can survive and remain metabolically active, even in environments treated with nanoparticle-based formulations. Evidence from in vitro studies shows that certain nanoparticles do not completely stop the growth of rhizosphere bacteria, including *Bacillus* species and *Pseudomonas,* suggesting microbial tolerance to some extent under controlled conditions (Shoala 2025). Bio-based treatments also enhance the relationship between plants, microbes, and soil. These interactions aid in the formation of an effective rhizosphere, making plants more efficient in nutrient intake and contributing to plant vigor (Nephali et al. 2020).

12.4.3 Plant Activation of the Antioxidant Defense Systems

One of the main ways through which biodegradable *IBS* enhance plant resistance to abiotic and oxidative stresses is by stimulating antioxidant defense systems (Rouphael et al. 2017). These biostimulants enhance both enzymatic and nonenzymatic antioxidant pathways, allowing plants to cope with oxidative stress (Tuteja et al. 2011; Choudhury et al. 2017). Enzymatic processes, catalase, superoxide dismutase, and peroxidases are stimulated, which contributes to the elimination of harmful reactive oxygen species (Szymańska et al. 2017; He et al. 2018). Simultaneously, nonenzymatic antioxidants, such as ascorbic acid, glutathione, and carotenoids, act as scavengers, maintaining redox balance. This twofold activation prevents oxidative damage of biological elements, especially during abiotic stress conditions (Arbona et al. 2003; de Pinto and De Gara 2004).

12.4.4 Reducing Abiotic Stresses

Biostimulants ensure that plants are immune to abiotic stresses by improving their biochemical and physiological resilience (Table 12.6). They also increase the osmotic adjustment process in drought and salinity and preserve osmoprotectants, such as glycine betaine and proline, which have the ability to preserve cell turgor and membrane integrity (Du Jardin 2015). Some of these formulations stimulate hormonal homeostasis through the activation of stress- and metabolic reconfiguration-responsive gene signaling pathways (Van Oosten et al. 2017). They may also assist in the detoxification of heavy metals by making them less bioavailable or by activating antioxidant enzymes of stress defense (Shehu et al. 2019).

12.5 Biosafety and Environmental Impact

12.5.1 Degrading Products and Their Fate

The environmental behavior of biodegradable inorganic biostimulants is inherently related to their origin and composition. Because these inputs are often obtained from a broad spectrum of biological and inorganic sources, such as agro-industrial residues, composts, and marine algae, their decomposition in the environment differs considerably (Calvo et al. 2014). Most of these sources are in line with sustainable agricultural practices and the concepts of the circular economy, and most of them break down into nontoxic substances that enhance soil organic matter and microbial well-being (Rouphael and Colla 2020). Nevertheless, it is not easy to predict the degradation rates and pathways because of the lack of compositional consistency owing to variations in raw materials and processing procedures. This heterogeneity requires context-dependent studies to comprehend the long-term and environmental fates of biostimulant leftovers following their administration (Brown and Saa 2015).

Table 12.6 Abiotic stress mitigation using biostimulants

Stress type	Biostimulant mechanism	Outcome	Reference
Drought and salinity	Osmoprotectant concentration (e.g., proline)	Maintains cell turgor, stability of membranes	Du Jardin (2015)
Heavy metal toxicity	Activation of detox enzymes	Reduced metal bioavailability, increased tolerance	Shehu et al. (2019)
General stress	Hormonal modulation, gene regulation	Enhanced stress-responsive pathways	Van Oosten et al. (2017)

12.5.2 Evaluation of Ecotoxicity and Soil Health Impact

Although biostimulants are promoted because of their sustainability, their effects on soil ecosystems should be considered carefully. Several classes, such as humic acids and protein hydrolysates, have been shown to enhance positive microbial communities, in addition to enhancing soil structure and fertility (Luziatelli et al. 2019). However, the environmental effects of biostimulants cannot be considered benign. In addition, improperly defined formulations or overuse can pose ecotoxicological risks (Bhupenchandra et al. 2022). The United States Environmental Protection Agency (US EPA) implemented an ecological risk assessment framework that supports safe application thresholds using soil and ecosystem safety endpoints, and the Organization for Economic Co-operation and Development (OECD) guidelines that have been used across the European regulatory systems are examples of standardized regulatory frameworks that are commonly used in the ecotoxicological assessment of biostimulants (Mendes da Silva and Andrade–Vieira 2025; Shaffer et al. 2025). Biosafety requires that biostimulant formulations be evaluated based not only on their agronomic efficacy but also on their effects on biodiversity in the soil, long-term ecological resilience, and nutrients (Brown and Saa 2015).

12.5.3 Field Applications Risk Mitigation Strategies

Dosage is one of the factors to be considered because certain biostimulants, especially those derived from animal protein hydrolysates obtained from agro-industrial by-products that are rich in collagen, such as residual materials from meat processing and leather scraps, have been found to be phytotoxic above the optimal range of concentration of use (Nardi et al. 2016). Other measures, such as labeling, using standardized application methods, and compatibility tests with other agrochemicals should also be performed to reduce risks. In addition, the efficacy and safety testing should be defined legally and have standard procedures to prevent product integrity and regulatory loopholes (Bhupenchandra et al. 2022). To integrate biostimulants into precision farming and sustainable field management, it is essential to develop a formulation that enhances environmental stability and reduces off-target effects (Rouphael and Colla 2020).

12.6 Case Studies and Applications

12.6.1 Silica-Based Biostimulants Model System

Siliceous natural nanomaterials (SNNMs) (diatomaceous earth, zeolites, etc.) are plant protective substances, soil improvers, and biostimulants owing to their porous nanostructure and high absorbency (Constantinescu-Aruxandei et al. 2020). SNNMs release soluble silicon (such as silicic acid), enhancing the efficiency of photosynthesis and nutrient use (Riotte et al. 2018; Sandhya et al. 2018), thereby benefiting plant health (Balakhnina et al. 2015). The application of SNNMs via foliar means was found to reduce the population of pests, such as tomato leaf miners, offering a biocompatible alternative to synthetic pesticides (De Smedt et al. 2016).

Biosilica nanoparticles (SiNPs) prepared using rice husks were incorporated into an alginate glycerol sorbitol polymer to form a seed coating film that significantly enhanced germination, root growth, and metabolism under salt stress. Biogenic SiNPs (approximately 50 nm) had a high specific surface area, as well as controlled silicon release, which is a slow-releasing physiological biostimulant (Tritean et al. 2024; Naikoo et al. 2025).

12.6.2 *Biodegradable Systems of Nanozinc and Nanophosphate*

The combination of humic compounds and microbial consortia application increased nutrient usage efficiency and encouraged eco-friendly rhizoengineering (Table 12.7). A study of common beans for 2 years in the Brazilian savannah showed that the introduction of nano-Zn at 1.5 kg ha^{-1} using the foliage together with the coinoculation of *Rhizobium tropici* and *Bacillus subtilis* led to significant increases in grain yield, plant height, shoot dry mass, and protein content. This combination also enhanced the leaf chlorophyll index and Zn utilization efficiencies, such as uptake, recovery, and allocation to grains, which eventually increased the predicted daily Zn consumption using bio-fortified beans. Synergistic effects are shown in nanofertilizers and microbial inoculants by increasing crop nutrition and yield, which contribute to biofortification and decrease the reliance on synthetic inputs, particularly in micronutrient-deficient areas (Jalal et al. 2023).

Similarly, biodegradable nanophosphate fertilizers produced using agricultural waste, including date palm pits, have shown promising results in enhancing nutrient productivity and sustainability. The two nano-derived phosphate fertilizer formulations, nDPF1 and nDPF2, formed once the nanostructured date pits were treated

Table 12.7 Case study comparison of two nanobiostimulant systems

Feature	Nanozinc system	Nanophosphate system
Main ingredients	Nano Zn + *Rhizobium* + *Bacillus*	Date pit-derived phosphate + KH_2PO_4 + MgO
Application method	Foliar spray	Soil amendment
Major benefits	Higher protein, Zn uptake, biofortification	Moisture retention, gradual P release
Crop improvement	Common bean: yield and nutrient density	Maize: biomass and nutrient uptake
Reference	Jalal et al. (2023)	Jia et al. (2022)

with KH_2PO_4 and MgO, significantly improved the water retention and nutrient release properties in sandy soils. Remarkably, the use of nDPF2 in soil increased its moisture retention by 5.6 times over soil receiving traditional fertilizers and contributed to prolonged water supply under dry conditions. Other systems exhibited slow, extended excretion of phosphorus, K, and Mg, reducing the loss of nutrients through leaching and increasing nutrient uptake. In pot trials, nDPF1 and nDPF2 were far better than conventional fertilizers in enhancing maize biomass and nutrient uptake, confirming their use in sustainable agriculture. Waste agricultural nanocarriers are a circular economy strategy that can address the issue of nutrient management and increase environmental safety (Elsabagh et al. 2024).

12.6.3 Use of Combined Microbial-Inorganic Biostimulant Formulations

Several studies have demonstrated that a composite biostimulant produced by a combination of humic compounds of vermicompost and plant growth-promoting bacteria (PGPB) increases plant resilience and fruit yield (Calvo et al. 2014). This combined expression could enhance food security and the autonomy of farmers in agricultural systems with low input. Humic compounds and microbial consortia coexist to exhibit synergistic effects in utilization, increasing the efficiency of nutrient usage and promoting environmentally friendly rhizoengineering (Canellas et al. 2022).

12.7 Challenges and Opportunities

12.7.1 Technical Problems in the Development of Biostimulants

Various technical difficulties have been reported due to the advancement of biodegradable IBS, which impede their optimization and wider acceptance in agriculture. This is a major disadvantage because the procedure for developing nanoparticles does not have a standardized protocol that regulates their uniformity and performance in different applications, particularly because of their size, shape, and surface properties (Yadav et al. 2023). This situation worsens because of the incomplete knowledge regarding the exact molecular and biochemical processes through which biostimulants provide a positive impact on plants (Li et al. 2022).

An additional challenge is introduced due to the complexity of plant reactions, as biostimulants may have reliable results in a specific crop or situation, but they may have no or weak results in another (Rodrigues et al. 2017; Chaudhari et al. 2023; Hernández-Herrera et al. 2023). In addition, the creation and production methods of precision biostimulants, such as nanoformulations, require specialized facilities and machinery, which makes their production resource-intensive and limits their scalability (Yadav et al. 2023).

12.7.2 Compliance Requirements and Regulations

Regulatory uncertainty remains a key hurdle to the development of biostimulants and nano-based fertilizers. One of the key issues is the lack of a standardized definition of biostimulants or nanomaterials at the global level; therefore, regulations vary greatly across countries and regions (Yadav et al. 2023). Such regulatory imbalance makes it harder to approve and certify manufacturers, but it is also very expensive to comply with the corresponding markets, which encompass multiple jurisdictions (Khoulati et al. 2025).

Furthermore, current regulatory systems often lack clear pathways for assessing, categorizing, and evaluating the risks of these new products, which may impair innovation and reduce the attractiveness of new products to the private sector (Gonçalves et al. 2023). In spite of the fact that attempts to develop science-based regulatory practices are underway such as those of the European Union and international advocacy groups, the pace of its development is slow, which indicates the need to unite the activities aimed at making certification procedures easier and ensuring safety while fostering innovation (Khoulati et al. 2025).

12.7.3 Commercialization Challenges

The practical and financial challenges of biodegradable IBS, especially those based on nanotechnology, are varied and challenging. The steep prices of production caused by the need to use specific equipment and processes make nanoformulations more expensive than traditional fertilizers (Jakhar et al. 2022). Moreover, the lack of awareness and knowledge of nano-enabled products among people further increases skepticism and unwillingness toward the adoption of new technologies among farmers (Fellet et al. 2021). There are also challenges associated with the transfer of laboratory-level research to the field level, which must guarantee the stability, effectiveness, and performance of products in the challenging conditions of various agricultural conditions (Behera et al. 2021). Also, commercialization is not only about production but also investment in training, distribution, and education so that end-users can use it appropriately (Khoulati et al. 2025). The combination of these factors increases the economic risks associated with the implementation of biostimulant programs and stresses the need to conduct powerful demonstration studies, effective communication, and policy support to create a reputation and need in the market (Table 12.8).

Table 12.8 Biostimulant commercialization: major challenges and opportunities

Challenge	Description	Possible solutions	References
High production cost	Specialized equipment for nanoformulations	Scale-up, use of waste materials	Jakhar et al. (2022)
Regulatory ambiguity	Inconsistent definitions and safety standards	Global harmonization, science-driven policies	Khoulati et al. (2025)
Farmer skepticism	Limited awareness and adoption	Education, demonstration trials	Fellet et al. (2021)
Different field performance	Context-dependent results	Specific site testing and customization	Behera et al. (2021)

12.7.4 Future Trends: Environmentally Friendly Biostimulants

Future trends in biostimulants are gradually aligned with the ideas of precision agriculture, sustainability, and digital integration. One of the most promising technologies is nanoencapsulation, which improves the stability, solubility, and controlled release of active compounds, thereby increasing the efficiency of delivery and reducing environmental losses (Orozco-Mosqueda et al. 2021; Hung et al. 2023; Nathaniel et al. 2023). Concurrently, the advancement of gene-editing systems, such as CRISPR/Cas systems, has enabled the development of crop varieties that respond more responsively to biostimulant use, allowing products to be customized to crops (Meerkerk 2020; Dobrei et al. 2022). The combination of intelligent sensors with Internet of Things (IoT) technologies offers new possibilities for live monitoring of field and crop conditions and remote use of biostimulants with minimal input waste and maximum productivity (Hawes et al. 2021; Baghdadi et al. 2022; Alsudays et al. 2024). Additionally, the combination of soil, plant, and climate data with omics-centered technologies, such as genomics, metabolomics, and proteomics, with artificial intelligence and machine learning technologies is expected to enable the creation of biostimulant compositions characterized by a high level of customization (Tahoun et al. 2022; Alsudays et al. 2024; Martins et al. 2024). These trends are a move toward smart, eco-friendly, and data-driven solutions for sustainable agriculture.

References

Adamiano A, Fellet G, Vuerich M, Scarpin D, Carella F, Piccirillo C, Jeon JR, Pizzutti A, Marchiol L, Iafisco M (2021) Calcium phosphate particles coated with humic substances: a potential plant biostimulant from circular economy. Molecules 26:2810. https://doi.org/10.3390/molecules26092810

Adamu A, Ahmad K, Siddiqui Y, Ismail IS, Asib N, Bashir Kutawa A, Adzmi F, Ismail MR, Berahim Z (2021) Ginger essential oils-loaded nanoemulsions: potential strategy to manage bacterial leaf blight disease and enhanced rice yield. Molecules 26:3902. https://doi.org/10.3390/molecules26133902

Alsudays IM, Alshammary FH, Alabdallah NM, Alatawi A, Alotaibi MM, Alwutayd KM, Alharbi MM, Alghanem SM, Alzuaibr FM, Gharib HS (2024) Applications of humic and fulvic acid under saline soil conditions to improve growth and yield in barley. BMC Plant Biol 24:191. https://doi.org/10.1186/s12870-024-04880-1

Anuta V, Blidaru A, Dinu-Pîrvu CE, Fierascu RC, Fierascu I, Toma DI, Popa L, Ghica MV, Prisada RM (2025) Metal-based nanoparticles with biostimulatory effects: harnessing nanotechnology for enhanced agricultural sustainability. Materials 18:3142. https://doi.org/10.3390/ma18133142

Arbona V, Flors V, Jacas J, García-Agustín P, Gómez-Cadenas A (2003) Enzymatic and non-enzymatic antioxidant responses of *Carrizo citrange*, a salt-sensitive citrus rootstock, to different levels of salinity. Plant Cell Physiol 44:388–394. https://doi.org/10.1093/pcp/pcg047

Arora PK, Tripathi S, Omar RA, Chauhan P, Sinhal VK, Singh A, Srivastava A, Garg SK, Singh VP (2024) Next-generation fertilizers: the impact of bionanofertilizers on sustainable agriculture. Microb Cell Factories 23:254. https://doi.org/10.1186/s12934-024-02514-5

Baghdadi A, Della Lucia MC, Borella M, Bertoldo G, Ravi S, Zegada-Lizarazu W, Chiodi C, Pagani E, Hermans C, Stevanato P (2022) A dual-omics approach for profiling plant responses to biostimulant applications under controlled and field conditions. Front Plant Sci 13:983772. https://doi.org/10.3389/fpls.2022.983772

Balakhnina TI, Bulak P, Matichenkov VV, Kosobryukhov AA, Włodarczyk TM (2015) The influence of Si-rich mineral zeolite on the growth processes and adaptive potential of barley plants under cadmium stress. Plant Growth Regul 75:557–565. https://doi.org/10.1007/s10725-014-0021-y

Bărbieru OG, Dimitriu L, Călin M, Răut I, Constantinescu-Aruxandei D, Oancea F (2019) Plant biostimulants based on selenium nanoparticles biosynthesized by *Trichoderma* strains. In: Proceedings of MDPI. MDPI, Basel

Behera B, Supraja KV, Paramasivan B (2021) Integrated microalgal biorefinery for the production and application of biostimulants in circular bioeconomy. Bioresour Technol 339:125588. https://doi.org/10.1016/j.biortech.2021.125588

Bhupenchandra I, Chongtham SK, Devi EL, Choudhary AK, Salam MD, Sahoo MR, Bhutia TL, Devi SH, Thounaojam AS, Behera C (2022) Role of biostimulants in mitigating the effects of climate change on crop performance. Front Plant Sci 13:967665. https://doi.org/10.3389/fpls.2022.967665

Boyjoo Y, Pareek VK, Liu J (2014) Synthesis of micro and nano-sized calcium carbonate particles and their applications. J Mater Chem A 2(14270–14):288. https://doi.org/10.1039/C4TA02070G

Brown P, Saa S (2015) Biostimulants in agriculture. Front Plant Sci 6:671. https://doi.org/10.3389/fpls.2015.00671

Calvo P, Nelson L, Kloepper JW (2014) Agricultural uses of plant biostimulants. Plant Soil 383:3–41. https://doi.org/10.1007/s11104-014-2131-8

Canellas LP, Canellas NO, da Silva RM, Spaccini R, Mota GP, Olivares FL (2022) Biostimulants using humic substances and plant-growth-promoting bacteria: effects on cassava (*Manihot esculentus*) and okra (*Abelmoschus esculentus*) yield. Agronomy 13:80. https://doi.org/10.3390/agronomy13010080

Chaudhari V, Patel N, Barot D, Nadoda N (2023) Role of bio-stimulants in solanaceous vegetable crops: a review. Int J Adv Biochem Res 7:178–183

Choudhury FK, Rivero RM, Blumwald E, Mittler R (2017) Reactive oxygen species, abiotic stress and stress combination. Plant J 90:856–867. https://doi.org/10.1111/tpj.13529

Colla G, Rouphael Y (2015) Biostimulants in horticulture Sci Hortic 196:1–2. https://doi.org/10.1016/j.scienta.2015.10.044

Colla G, Rouphael Y, Canaguier R, Svecova E, Cardarelli M (2014) Biostimulant action of a plant-derived protein hydrolysate produced through enzymatic hydrolysis. Front Plant Sci 5:448. https://doi.org/10.3389/fpls.2014.00448

Colla G, Rouphael Y, Di Mattia E, El-Nakhel C, Cardarelli M (2015) Co-inoculation of *Glomus intraradices* and *Trichoderma atroviride* acts as a biostimulant to promote growth, yield and nutrient uptake of vegetable crops. J Sci Food Agric 95:1706–1715. https://doi.org/10.1002/jsfa.6875

Colla G, Hoagland L, Ruzzi M, Cardarelli M, Bonini P, Canaguier R, Rouphael Y (2017) Biostimulant action of protein hydrolysates: unraveling their effects on plant physiology and microbiome. Front Plant Sci 8:2202. https://doi.org/10.3389/fpls.2017.02202

Constantinescu-Aruxandei D, Lupu C, Oancea F (2020) Siliceous natural nanomaterials as biorationals—plant protectants and plant health strengtheners. Agronomy 10:1791. https://doi.org/10.3390/agronomy10111791

De Pascale S, Rouphael Y, Colla G (2017) Plant biostimulants: innovative tool for enhancing plant nutrition in organic farming. Eur J Hortic Sci 82:277–285. https://doi.org/10.17660/eJHS.2017.82.6.2

de Pinto MC, De Gara L (2004) Changes in the ascorbate metabolism of apoplastic and symplastic spaces are associated with cell differentiation. J Exp Bot 55:2559–2569. https://doi.org/10.1093/jxb/erh253

De Smedt C, Van Damme V, De Clercq P, Spanoghe P (2016) Insecticide effect of zeolites on the tomato leafminer *Tuta absoluta* (*Lepidoptera: Gelechiidae*). Insects 7:72. https://doi.org/10.3390/insects7040072

Dobrei A, Andras Sauca V, Nistor E, Borca F, Dobrei A (2022) Research concerning the economic efficiency of stimulants and foliar fertilizers in table grape varieties. Sci Pap Ser B Hortic 66

Du Jardin P (2015) Plant biostimulants: definition, concept, main categories and regulation. Sci Hortic 196:3–14. https://doi.org/10.1016/j.scienta.2015.09.021

Eliaspour S, Seyed Sharifi R, Shirkhani A, Farzaneh S (2020) Effects of biofertilizers and iron nano-oxide on maize yield and physiological properties under optimal irrigation and drought stress conditions. Food Sci Nutr 8:5985–5998. https://doi.org/10.1002/fsn3.1885

Elsabagh SS, Elkhatib EA, Rashad M (2024) Eco-friendly nano-enabled fertilizers derived from date industry waste for sustainable and controlled-release of P, K and Mg nutrients: sorption mechanisms, controlled-release performance and kinetics. Bioresour Bioprocess 11:3. https://doi.org/10.1186/s40643-024-00721-3

Fellet G, Pilotto L, Marchiol L, Braidot E (2021) Tools for nano-enabled agriculture: fertilizers based on calcium phosphate, silicon, and chitosan nanostructures. Agronomy 11:1239. https://doi.org/10.3390/agronomy11061239

Ghormade V, Deshpande MV, Paknikar KM (2011) Perspectives for nano-biotechnology enabled protection and nutrition of plants. Biotechnol Adv 29:792–803. https://doi.org/10.1016/j.biotechadv.2011.06.005

Giordana A, Malandrino M, Zambon A, Lusvardi G, Operti L, Cerrato G (2023) Biostimulants derived from organic urban wastes and biomasses: an innovative approach. Front Chem 11:969865. https://doi.org/10.3389/fchem.2023.969865

Gonçalves J, Freitas J, Fernandes I, Silva P (2023) Microalgae as biofertilizers: a sustainable way to improve soil fertility and plant growth. Sustainability 15:12413. https://doi.org/10.3390/su151612413

Hao L, Gong L, Chen L, Guan M, Zhou H, Qiu S, Wen H, Chen H, Zhou X, Akbulut M (2020a) Composite pesticide nanocarriers involving functionalized boron nitride nanoplatelets for pH-responsive release and enhanced UV stability. Chem Eng J 396:125233. https://doi.org/10.1016/j.cej.2020.125233

Hao L, Lin G, Wang H, Wei C, Chen L, Zhou H, Chen H, Xu H, Zhou X (2020b) Preparation and characterization of zein-based nanoparticles via ring-opening reaction and self-assembly as aqueous nanocarriers for pesticides. J Agric Food Chem 68:9624–9635. https://doi.org/10.1021/acs.jafc.0c02547

Hawes C, Iannetta PP, Squire GR (2021) Agroecological practices for whole-system sustainability. CABI Rev 16:1–12. https://doi.org/10.1079/PAVSNNR202116034

He M, He CQ, Ding NZ (2018) Abiotic stresses: general defenses of land plants and chances for engineering multistress tolerance. Front Plant Sci 9:1771. https://doi.org/10.3389/fpls.2018.01771

Hernández-Herrera RM, González-González MF, Velasco-Ramírez AP, Velasco-Ramírez SF, Santacruz-Ruvalcaba F, Zamora-Natera JF (2023) Seaweed extract components are correlated with the seeds germination and growth of tomato seedlings. Seeds 2:436–448. https://doi.org/10.3390/seeds2040032

Hua KH, Wang HC, Chung RS, Hsu JC (2015) Calcium carbonate nanoparticles can enhance plant nutrition and insect pest tolerance. J Pestic Sci 40:208–213. https://doi.org/10.1584/jpestics.D15-011

Hung SHW, Huang TC, Lai YC, Wu IC, Liu CH, Huarng YF, Hwang HH, Chiang EPI, Kuo CH, Huang CC (2023) Endophytic biostimulants for smart agriculture: *Burkholderia seminalis* 869 T2 benefits heading leafy vegetables in-field management in Taiwan. Agronomy 13:967. https://doi.org/10.3390/agronomy13040967

Ijaz I, Gilani E, Nazir A, Bukhari A (2020) Detail review on chemical, physical and green synthesis, classification, characterizations and applications of nanoparticles. Green Chem Lett Rev 13:223–245. https://doi.org/10.1080/17518253.2020.1802517

Iqbal M, Umar S, Mahmooduzzafar F (2019) Nano-fertilization to enhance nutrient use efficiency and productivity of crop plants. In: Husen A, Iqbal M (eds) Nanomaterials and plant potential. Springer, Cham, pp 473–505

Jakhar AM, Aziz I, Kaleri AR, Hasnain M, Haider G, Ma J, Abideen Z (2022) Nano-fertilizers: a sustainable technology for improving crop nutrition and food security. NanoImpact 27:100411. https://doi.org/10.1016/j.impact.2022.100411

Jalal A, Mortinho ES, da Silva Oliveira CE, Fernandes GC, Junior EF, de Lima BH, Moreira A, Nogueira TAR, Galindo FS, Filho MCMT (2023) Nano-zinc and plant growth-promoting bacteria is a sustainable alternative for improving productivity and agronomic biofortification of common bean. Chem Biol Technol Agric 10:77. https://doi.org/10.1186/s40538-023-00451-2

Jia X, Shi N, Tang W, Su Z, Chen H, Tang Y, Sun B, Zhao L (2022) Nano-hydroxyapatite increased soil quality and boosted beneficial soil microbes. Plant Nano Biol 1:100002. https://doi.org/10.1016/j.plana.2022.100002

Kah M, Kookana RS, Gogos A, Bucheli TD (2018) A critical evaluation of nanopesticides and nanofertilizers against their conventional analogues. Nat Nanotechnol 13:677–684. https://doi.org/10.1038/s41565-018-0131-1

Khalid MF, Jawaid MZ, Nawaz M, Shakoor RA, Ahmed T (2024) Employing titanium dioxide nanoparticles as biostimulant against salinity: improving antioxidative defense and reactive oxygen species balancing in eggplant seedlings. Antioxidants 13:1209. https://doi.org/10.3390/antiox13101209

Khoulati A, Ouahhoud S, Taibi M, Ezrari S, Mamri S, Merah O, Hakkou A, Addi M, Maleb A, Saalaoui E (2025) Harnessing biostimulants for sustainable agriculture: innovations, challenges, and future prospects. Discov Agric 3:56. https://doi.org/10.1007/s44279-025-00156-8

Li J, Van Gerrewey T, Geelen D (2022) A meta-analysis of biostimulant yield effectiveness in field trials. Front Plant Sci 13:836702. https://doi.org/10.3389/fpls.2022.836702

Luziatelli F, Ficca AG, Colla G, Baldassarre Švecová E, Ruzzi M (2019) Foliar application of vegetal-derived bioactive compounds stimulates the growth of beneficial bacteria and enhances microbiome biodiversity in lettuce. Front Plant Sci 10:60. https://doi.org/10.3389/fpls.2019.00060

Magnabosco P, Masi A, Shukla R, Bansal V, Carletti P (2023) Advancing the impact of plant biostimulants to sustainable agriculture through nanotechnologies. Chem Biol Technol Agric 10:117. https://doi.org/10.1186/s40538-023-00496-3

Mahapatra DM, Satapathy KC, Panda B (2022) Biofertilizers and nanofertilizers for sustainable agriculture: phycoprospects and challenges. Sci Total Environ 803:149990. https://doi.org/10.1016/j.scitotenv.2021.149990

Manjunatha S, Biradar D, Aladakatti YR (2016) Nanotechnology and its applications in agriculture: a review. J Farm Sci 29:1–13

Martins EM, Pillajo JQ, Jones ML (2024) Humic and fulvic acids promote growth and flowering in petunias at low and optimal fertility. HortScience 59:235–244. https://doi.org/10.21273/HORTSCI17603-23

Meerkerk L (2020) Economic feasibility & future prospects of microbials. MSc thesis. Wageningen University & Research, Wageningen

Mendes da Silva L, Andrade–Vieira LF (2025) Ecotoxicological bioassays with terrestrial plants: a holistic view of standards, guidelines, and protocols. J Toxic Environ Health, Part B 28:183–221. https://doi.org/10.1080/10937404.2025.2445100

Mukherjee R, Kumar R, Sinha A, Lama Y, Saha AK (2016) A review on synthesis, characterization, and applications of nano zero valent iron (nZVI) for environmental remediation. Crit Rev Environ Sci Technol 46:443–466. https://doi.org/10.1080/10643389.2015.1103832

Naikoo NB, Chesti MH, Bhat MA, Mir AH, Bashir O, Bhat TA, Mir MS, Amin Z, Gull A, Ayoub L (2025) Biostimulants towards soil health improvement: a review. Agric Rev 46:1–8. https://doi.org/10.18805/ag-IE-2526

Nardi S, Pizzeghello D, Schiavon M, Ertani A (2016) Plant biostimulants: physiological responses induced by protein hydrolyzed-based products and humic substances in plant metabolism. Sci Agric 73:18–23. https://doi.org/10.1590/0103-9016-2015-0006

Nathaniel H, Franzén D, Lingegård S, Franzen F, Söderqvist T, Gröndahl F (2023) Using stakeholder perceptions to deepen the understanding of beachcast governance and management practices on Gotland, Sweden. Ocean Coast Manag 239:106583. https://doi.org/10.1016/j.ocecoaman.2023.106583

Nelwamondo AM, Kaningini AG, Ngmenzuma TY, Maseko ST, Maaza M, Mohale KC (2023) Biosynthesis of magnesium oxide and calcium carbonate nanoparticles using *Moringa oleifera* extract and their effectiveness on the growth, yield and photosynthetic performance of groundnut (*Arachis hypogaea* L.) genotypes. Heliyon 9:e20000. https://doi.org/10.1016/j.heliyon.2023.e20000

Nephali L, Piater LA, Dubery IA, Patterson V, Huyser J, Burgess K, Tugizimana F (2020) Biostimulants for plant growth and mitigation of abiotic stresses: a metabolomics perspective. Meta 10:505. https://doi.org/10.3390/metabo10120505

Noruzi M, Hadian P, Soleimanpour L, Ma'mani L, Shahbazi K (2023) Hydroxyapatite nanoparticles: an alternative to conventional phosphorus fertilizers in acidic culture media. Chem Biol Technol Agric 10:71. https://doi.org/10.1186/s40538-023-00444-1

Orozco-Mosqueda MDC, Flores A, Rojas-Sánchez B, Urtis-Flores CA, Morales-Cedeño LR, Valencia-Marin MF, Chávez-Avila S, Rojas-Solis D, Santoyo G (2021) Plant growth-promoting bacteria as bioinoculants: attributes and challenges for sustainable crop improvement. Agronomy 11:1167. https://doi.org/10.3390/agronomy11061167

Panichikkal J, Thomas R, John JC, Radhakrishnan E (2019) Biogenic gold nanoparticle supplementation to plant beneficial *Pseudomonas monteilii* was found to enhance its plant probiotic effect. Curr Microbiol 76:503–509. https://doi.org/10.1007/s00284-019-01660-x

Patel C, Singh J, Karunakaran A, Ramakrishna W (2023) Evolution of nano-biofertilizer as a green technology for agriculture. Agriculture 13:1865. https://doi.org/10.3390/agriculture13101865

Rautela I, Dheer P, Thapliyal P, Shah D, Joshi M, Upadhyay S, Gururani P, Sinha VB, Gaurav N, Sharma MD (2021) Current scenario and future perspectives of nanotechnology in sustainable agriculture and food production. Plant Cell Biotechnol Mol Biol 22:99–121

Regni L, Del Buono D, Micheli M, Facchin SL, Cesarini A, Priolo D, Proietti P (2024) Biogenic zinc oxide nanoparticles improve in vitro growth of blueberries. Horticulturae 10:1234. https://doi.org/10.3390/horticulturae10121234

Riotte J, Sandhya K, Prakash NB, Audry S, Zambardi T, Chmeleff J, Buvaneshwari S, Meunier JD (2018) Origin of silica in rice plants and contribution of diatom earth fertilization: insights from isotopic Si mass balance in a paddy field. Plant Soil 423:481–501. https://doi.org/10.1007/s11104-017-3547-8

Rodrigues LA, Alves CZ, Rego CHQ, Silva TRBD, Silva JBD (2017) Humic acid on germination and vigor of corn seeds. Rev Caatinga 30:149–154. https://doi.org/10.1590/1983-21252017v30n117rc

Rouphael Y, Colla G (2018) Synergistic biostimulatory action: designing the next generation of plant biostimulants for sustainable agriculture. Front Plant Sci 9:1655. https://doi.org/10.3389/fpls.2018.01655

Rouphael Y, Colla G (2020) Biostimulants in agriculture. Front Plant Sci 11:40. https://doi.org/10.3389/fpls.2020.00040

Rouphael Y, Cardarelli M, Bonini P, Colla G (2017) Synergistic action of a microbial-based biostimulant and a plant derived-protein hydrolysate enhances lettuce tolerance to alkalinity and salinity. Front Plant Sci 8:131. https://doi.org/10.3389/fpls.2017.00131

Rouphael Y, Carillo P, Ciriello M, Formisano L, El-Nakhel C, Ganugi P, Fiorini A, Miras Moreno B, Zhang L, Cardarelli M (2023) Copper boosts the biostimulant activity of a vegetal-derived protein hydrolysate in basil: morpho-physiological and metabolomics insights. Front Plant Sci 14:1235686. https://doi.org/10.3389/fpls.2023.1235686

Samani M, Ahlawat YK, Golchin A, Alikhani HA, Baybordi A, Mishra S, Şimşek Ö (2024) Nano silica's role in regulating heavy metal uptake in *Calendula officinalis*. BMC Plant Biol 24:598. https://doi.org/10.1186/s12870-024-05254-3

Sandhya K, Prakash NB, Meunier JD (2018) Diatomaceous earth as source of silicon on the growth and yield of rice in contrasted soils of Southern India. J Soil Sci Plant Nutr 18:344–360. https://doi.org/10.4067/S0718-95162018005001201

Shaffer RM, Lee AL, Nachman R, Christensen K, Bateson TF (2025) A perspective from US Environmental Protection Agency (EPA) scientists: how your epidemiologic analyses can inform the human health risk assessment process. Environ Health Perspect 133:045001. https://doi.org/10.1289/EHP13745

Shehu AA, Alsamadany H, Alzahrani Y (2019) β-Aminobutyric acid (BABA) priming and abiotic stresses: a review. Int J Biosci 14:450–459

Shoala T (2025) Exploring the role of nanotechnology in the management of abiotic stress. In: Abd-Elsalam KA, Mohamed HI (eds) Nanobiotechnology in agri-food sector. Springer, Cham, pp 45–61

Strekalovskaya EI, Perfileva AI, Krutovsky KV (2024) Zinc oxide nanoparticles in the "soil–bacterial community–plant" system: impact on the stability of soil ecosystems. Agronomy 14:1588. https://doi.org/10.3390/agronomy14071588

Szymańska R, Ślesak I, Orzechowska A, Kruk J (2017) Physiological and biochemical responses to high light and temperature stress in plants. Environ Exp Bot 139:165–177. https://doi.org/10.1016/j.envexpbot.2017.05.002

Tahoun AMA, El-Enin MMA, Mancy AG, Sheta MH, Shaaban A (2022) Integrative soil application of humic acid and foliar plant growth stimulants improves soil properties and wheat yield and quality in nutrient-poor sandy soil of a semiarid region. J Soil Sci Plant Nutr 22:2857–2871. https://doi.org/10.1007/s42729-022-00859-9

Tantawy A, Salama Y, Abdel-Mawgoud M, Ghoname A (2014) Comparison of chelated calcium with nano calcium on alleviation of salinity negative effects on tomato plants. Middle East J Agric Res 3:912–916

Timoshenko A, Kolesnikov S, Rajput VD, Minkina T (2021) Effects of zinc-oxide nanoparticles on soil microbial community and their functionality. In: Rajput VD, Minkina T, Sushkova S, Mandzhieva S (eds) Zinc-based nanostructures for environmental and agricultural applications. Elsevier, Amsterdam, pp 267–284

Toksha B, Sonawale VAM, Vanarase A, Bornare D, Tonde S, Hazra C, Kundu D, Satdive A, Tayde S, Chatterjee A (2021) Nanofertilizers: a review on synthesis and impact of their use on crop yield and environment. Environ Technol Innov 24:101986. https://doi.org/10.1016/j.eti.2021.101986

Tritean N, Trică B, Dima ŞO, Capră L, Gabor RA, Cimpean A, Oancea F, Constantinescu-Aruxandei D (2024) Mechanistic insights into the plant biostimulant activity of a novel

formulation based on rice husk nanobiosilica embedded in a seed coating alginate film. Front Plant Sci 15:1349573. https://doi.org/10.3389/fpls.2024.1349573
Tuteja N, Gill SS, Tuteja R (2011) Plant responses to abiotic stresses: shedding light on salt, drought, cold and heavy metal stress. In: Tuteja N, Gill SS (eds) Omics and plant abiotic stress tolerance. Bentham Science, Sharjah, pp 39–64
Van Oosten MJ, Pepe O, De Pascale S, Silletti S, Maggio A (2017) The role of biostimulants and bioeffectors as alleviators of abiotic stress in crop plants. Chem Biol Technol Agric 4:5. https://doi.org/10.1186/s40538-017-0089-5
Verma KK, Song XP, Li DM, Singh M, Wu JM, Singh RK, Sharma A, Zhang BQ, Li YR (2022) Silicon and soil microorganisms improve rhizospheric soil health with bacterial community, plant growth, performance and yield. Plant Signal Behav 17:2104004. https://doi.org/10.1080/15592324.2022.2104004
Verma N, Kaushal P, Sidhu AK (2024) Harnessing biological synthesis: zinc oxide nanoparticles for plant biotic stress management. Front Chem 12:1432469. https://doi.org/10.3389/fchem.2024.1432469
Xu L, Geelen D (2018) Developing biostimulants from agro-food and industrial by-products. Front Plant Sci 9:1567. https://doi.org/10.3389/fpls.2018.01567
Yadav A, Yadav K, Abd-Elsalam KA (2023) Exploring the potential of nanofertilizers for a sustainable agriculture. Plant Nano Biol 5:100044. https://doi.org/10.1016/j.plana.2023.100044
Yang J, Zhou Z, Liang Y, Tang J, Gao Y, Niu J, Dong H, Tang R, Tang G, Cao Y (2020) Sustainable preparation of microcapsules with desirable stability and bioactivity using phosphonium ionic liquid as a functional additive. ACS Sustain Chem Eng 8:13440–13448. https://doi.org/10.1021/acssuschemeng.0c04297
Ying S, Guan Z, Ofoegbu PC, Clubb P, Rico C, He F, Hong J (2022) Green synthesis of nanoparticles: current developments and limitations. Environ Technol Innov 26:102336. https://doi.org/10.1016/j.eti.2022.102336
Zohora N, Kumar D, Yazdani M, Rotello VM, Ramanathan R, Bansal V (2017) Rapid colorimetric detection of mercury using biosynthesized gold nanoparticles. Colloids Surf A Physicochem Eng Asp 532:451–457. https://doi.org/10.1016/j.colsurfa.2017.04.036

MIX
Papier aus verantwortungsvollen Quellen
Paper from responsible sources
FSC® C105338

If you have any concerns about our products,
you can contact us on
ProductSafety@springernature.com

In case Publisher is established outside the EU,
the EU authorized representative is:
Springer Nature Customer Service Center GmbH
Europaplatz 3, 69115 Heidelberg, Germany

Printed by Libri Plureos GmbH
in Hamburg, Germany